T0297045

CENTRAL SIMPLE ALGEBRAS AND GALOIS COHOMOLOGY

The first comprehensive, modern introduction to the theory of central simple algebras over arbitrary fields, this book starts from the basics and reaches such advanced results as the Merkurjev–Suslin theorem, a culmination of work initiated by Brauer, Noether, Hasse and Albert and the starting point of current research in motivic cohomology theory by Voevodsky, Suslin, Rost and others.

Assuming only a solid background in algebra, the text covers the basic theory of central simple algebras, methods of Galois descent and Galois cohomology, Severi–Brauer varieties, and techniques in Milnor K-theory and K-cohomology, leading to a full proof of the Merkurjev–Suslin theorem and its application to the characterization of reduced norms. The final chapter rounds off the theory by presenting the results in positive characteristic, including the theorems of Bloch–Gabber–Kato and Izhboldin.

This second edition has been carefully revised and updated, and contains important additional topics.

Philippe Gille is a Research Director (CNRS) at Camille Jordan Institute, Lyon, France. He has written numerous research papers on linear algebraic groups and related structures.

Tamás Szamuely is a Research Adviser at the Alfréd Rényi Institute of Mathematics of the Hungarian Academy of Sciences and a Professor at Central European University, Budapest, Hungary. He is the author of Galois Groups and Fundamental Groups, also published in this series, as well as numerous research papers.

Central Simple Algebras and Galois Cohomology

Second Edition

PHILIPPE GILLE

Centre National de la Recherche Scientifique (CNRS),
Institut Camille Jordan, Lyon

TAMÁS SZAMUELY

Alfréd Rényi Institute of Mathematics,
Hungarian Academy of Sciences, Budapest

CAMBRIDGE
UNIVERSITY PRESS

CAMBRIDGE
UNIVERSITY PRESS

University Printing House, Cambridge CB2 8BS, United Kingdom

One Liberty Plaza, 20th Floor, New York, NY 10006, USA

477 Williamstown Road, Port Melbourne, VIC 3207, Australia

314-321, 3rd Floor, Plot 3, Splendor Forum, Jasola District Centre, New Delhi - 110025, India

79 Anson Road, #06-04/06, Singapore 079906

Cambridge University Press is part of the University of Cambridge.

It furthers the University's mission by disseminating knowledge in the pursuit of education, learning and research at the highest international levels of excellence.

www.cambridge.org
Information on this title: www.cambridge.org/9781316609880
DOI: 10.1017/9781316661277

First published 2006
Second edition 2017

A catalogue record for this publication is available from the British Library

ISBN 978-1-107-15637-1 Hardback
ISBN 978-1-316-60988-0 Paperback

Contents

v

Contents

Preface

This book provides a comprehensive and up-to-date introduction to the theory of central simple algebras over arbitrary fields, emphasizing methods of Galois cohomology and (mostly elementary) algebraic geometry. The central result is the Merkurjev–Suslin theorem. As we see it today, this fundamental theorem is at the same time the culmination of the theory of Brauer groups of fields initiated by Brauer, Noether, Hasse and Albert in the 1930s, and a starting point of motivic cohomology theory, a domain which is at the forefront of current research in algebraic geometry and K-theory – suffice it here to mention the recent spectacular results of Voevodsky, Suslin, Rost and others. As a gentle ascent towards the Merkurjev–Suslin theorem, we cover the basic theory of central simple algebras, methods of Galois descent and Galois cohomology, Severi–Brauer varieties, residue maps and, finally, Milnor K-theory and K-cohomology. These chapters also contain a number of noteworthy additional topics. The last chapter of the book rounds off the theory by presenting the results in positive characteristic. For an overview of the contents of each chapter we refer to their introductory sections.

Prerequisites. The book should be accessible to a graduate student or a non-specialist reader with a solid training in algebra including Galois theory and basic commutative algebra, but no homological algebra. Some familiarity with algebraic geometry is also helpful. Most of the text can be read with a basic knowledge corresponding to, say, the first volume of Shafarevich's text. To help the novice, we summarize in an appendix the results from algebraic geometry we need. The first three sections of Chapter 8 require some familiarity with schemes, and in the proof of one technical statement we are forced to use techniques from Quillen K-theory. However, these may be skipped in a first reading by those willing to accept some 'black boxes'.

Acknowledgments

Our first words of thanks go to Jean-Louis Colliot-Thélène and Jean-Pierre Serre, from whom we learned much of what we know about the subject and who, to our great joy, have also been the most assiduous readers of the manuscript, and suggested many improvements. Numerous other colleagues helped us with their advice during the preparation of the text, or spotted inaccuracies in previous versions. Thanks are due to Spencer Bloch, Jean-Benoît Bost, Irene Bouw, Gábor Braun, Ferenc Bródy, Jérôme Burési, Baptiste Calmès, Mathieu Florence, Ofer Gabber, Skip Garibaldi, Luc Illusie, Bruno Kahn, Max-Albert Knus, David Leep, David Madore, Alexander Merkurjev, Ján Mináč, Arturo Pianzola, Peter Roquette, Joël Riou, Christophe Soulé, Jean-Pierre Tignol, Burt Totaro and Stefan Wewers.

Parts of the book formed the basis of a graduate course by the first author at Université de Paris-Sud and of a lecture series by the two authors at the Alfréd Rényi Institute. We thank both audiences for their pertinent questions and comments, and in particular Endre Szabó who shared his geometric insight with us. Most of the book was written while the first author visited the Rényi Institute in Budapest with a Marie Curie Intra-European Fellowship. The support of the Commission and the hospitality of the Institute are gratefully acknowledged. Last but not least, we are indebted to Diana Gillooly for assuring us a smooth and competent publishing procedure.

Note on the second edition

The first edition of our book has been well received by the mathematical community, and we have received a lot of feedback from experts and graduate students alike. Based partly on their comments, we have tried to correct in this new edition all misprints and inaccuracies known to us. We have updated the text in order to take into account the most important developments during the last ten years, and have also included new material. The most substantial changes to the text of the first edition are as follows.

- We have considerably expanded the material in Section 2.6 on reduced norms.
- There is a new Section 2.8 on a different approach to period-index questions from that of the first edition (which remains in Chapter 4), mainly based on recent work of Antieau and Williams.
- There is a new Section 2.9 on central simple algebras over complete discretely valued fields. Compared to the previous section, this one is much more traditional.

- Section 6.3 has been thoroughly revised: it now includes statements over arbitrary complete discretely valued fields.
- Section 8.6 now includes a theorem of Merkurjev on the generators of the p-torsion in the Brauer group of fields of characteristic different from p.
- Arguably the most important addition to the text is contained in three new sections (8.7–8.9) devoted to the cohomological characterization of reduced norms. This is also a major result in the groundbreaking paper of Merkurjev and Suslin. More recently, it has played a key role in the study of cohomological invariants of algebraic groups. Our approach partly differs from the original one.
- Section 9.1 now includes a recent result of M. Florence on the symbol length in positive characteristic.
- The discussion of the differential symbol in Section 9.5 has been extended to cover mod p^i differential symbols with values in logarithmic de Rham–Witt groups as well.
- There is a new Section 9.8 devoted to Izhboldin's theorem on p-torsion in Milnor K-theory of fields of characteristic p, a fundamental result that was only briefly mentioned in the first edition.

Thanks are due to Menny Aka, Alexis Bouthier, Gábor Braun, Eric Brussel, Jean-Louis Colliot-Thélène, Anamaria Costache, J. P. Ding, Christian Hirsch, Ofer Gabber, Qing Liu, Ján Mináč, Joël Riou, Damian Rössler, Jean-Pierre Serre, Endre Szabó, Charles Vial and Tim Wouters for their comments on the first edition. We are also indebted to Ben Antieau, Cyril Demarche, David Harari, János Kollár, Joël Riou, Jean-Pierre Tignol, Burt Totaro and Gergely Zábrádi for their help with the present one.

1

Quaternion algebras

As a prelude to the book, we present here our main objects of study in the simplest case, that of quaternion algebras. Many concepts that will be ubiquitous in what follows, such as division algebras, splitting fields or norms appear here in a concrete and elementary context. Another important notion we shall introduce is that of the conic associated with a quaternion algebra; these are the simplest examples of Severi–Brauer varieties, objects to which a whole chapter will be devoted later. In the second part of the chapter two classic theorems from the 1930s are proven: a theorem of Witt asserting that the associated conic determines a quaternion algebra up to isomorphism, and a theorem of Albert that gives a criterion for the tensor product of two quaternion algebras to be a division algebra. The discussion following Albert's theorem will lead us to the statement of one of the main theorems proven later in the book, that of Merkurjev concerning division algebras of period 2.

The basic theory of quaternion algebras goes back to the nineteenth century. The original references for the main theorems of the last two sections are Witt [1] and Albert [1], [5], respectively.

1.1 Basic properties

In this book we shall study finite-dimensional algebras over a field. Here by an algebra over a field k we mean a k-vector space equipped with a not necessarily commutative but associative k-linear multiplication. All k-algebras will be tacitly assumed to have a unit element.

Historically the first example of a finite-dimensional noncommutative algebra over a field was discovered by W. R. Hamilton during a walk with his wife (presumably doomed to silence) on 16 October 1843. It is the *algebra of quaternions*, a 4-dimensional algebra with basis $1, i, j, k$ over the field $\mathbf{R}$ of real numbers, the multiplication being determined by the rules

$$i^2 = -1, \quad j^2 = -1, \quad ij = -ji = k.$$

This is in fact a *division algebra* over $\mathbf{R}$, which means that each nonzero element x has a two-sided multiplicative inverse, i.e. an element y with $xy = yx = 1$. Hamilton proved this as follows.

For a quaternion $q = x + yi + zj + wk$, introduce its *conjugate*

$$\bar{q} = x - yi - zj - wk$$

and its *norm* $N(q) = q\bar{q}$. A computation gives $N(q) = x^2 + y^2 + z^2 + w^2$, so if $q \neq 0$, the quaternion $\bar{q}/N(q)$ is an inverse for q.

We now come to an easy generalization of the above construction. *Henceforth in this chapter, unless otherwise stated, k will denote a field of characteristic not 2.*

Definition 1.1.1 For any two elements $a, b \in k^\times$ define the *(generalized) quaternion algebra* (a, b) as the 4-dimensional k-algebra with basis $1, i, j, ij$, multiplication being determined by

$$i^2 = a, \quad j^2 = b, \quad ij = -ji.$$

One calls the set $\{1, i, j, ij\}$ a *quaternion basis* of (a, b).

Remark 1.1.2 The isomorphism class of the algebra (a, b) depends only on the classes of a and b in $k^\times/k^{\times 2}$, because the substitution $i \mapsto ui, j \mapsto vj$ induces an isomorphism

$$(a, b) \xrightarrow{\sim} (u^2 a, v^2 b)$$

for all $u, v \in k^\times$. This implies in particular that the algebra (a, b) is isomorphic to (b, a); indeed, mapping $i \mapsto abj, j \mapsto abi$ we get

$$(a, b) \cong (a^2 b^3, a^3 b^2) \cong (b, a).$$

Given an element $q = x + yi + zj + wij$ of the quaternion algebra (a, b), we define its *conjugate* by

$$\bar{q} = x - yi - zj - wij.$$

The map $(a, b) \to (a, b)$ given by $q \mapsto \bar{q}$ is an anti-automorphism of the k-algebra (a, b), i.e. it is a k-vector space automorphism of (a, b) satisfying $\overline{(q_1 q_2)} = \bar{q}_2 \bar{q}_1$. Moreover, we have $\bar{\bar{q}} = q$; an anti-automorphism with this property is called an *involution* in ring theory.

We define the *norm* of $q = x + yi + zj + wij$ by $N(q) = q\bar{q}$. A calculation yields

$$N(q) = x^2 - ay^2 - bz^2 + abw^2 \in k, \qquad (1.1)$$

so $N : (a, b) \to k$ is a *nondegenerate quadratic form*. The computation

$$N(q_1 q_2) = q_1 q_2 \bar{q}_2 \bar{q}_1 = q_1 N(q_2) \bar{q}_1 = N(q_1) N(q_2)$$

shows that the norm is a multiplicative function, and the same argument as for Hamilton's quaternions yields:

Lemma 1.1.3 *An element q of the quaternion algebra (a, b) is invertible if and only if it has nonzero norm. Hence (a, b) is a division algebra if and only if the norm $N : (a, b) \to k$ does not vanish outside 0.*

Remark 1.1.4 In fact, one can give an intrinsic definition of the conjugation involution (and hence of the norm) on a quaternion algebra (a, b) which does not depend on the choice of the basis $(1, i, j, ij)$. Indeed, call an element q of (a, b) a *pure quaternion* if $q^2 \in k$ but $q \notin k$. A straightforward computation shows that a nonzero $q = x + yi + zj + wij$ is a pure quaternion if and only if $x = 0$. Hence a general q can be written uniquely as $q = q_1 + q_2$ with $q_1 \in k$ and q_2 pure, and conjugation is given by $\bar{q} = q_1 - q_2$. Moreover, a pure quaternion q satisfies $N(q) = -q^2$.

Example 1.1.5 (The matrix algebra $M_2(k)$) Besides the classical Hamilton quaternions, the other basic example of a quaternion algebra is the k-algebra $M_2(k)$ of 2×2 matrices. Indeed, the assignment

$$i \mapsto I := \begin{bmatrix} 1 & 0 \\ 0 & -1 \end{bmatrix}, \quad j \mapsto J := \begin{bmatrix} 0 & b \\ 1 & 0 \end{bmatrix}$$

defines an isomorphism $(1, b) \cong M_2(k)$, because the matrices

$$\mathrm{Id} = \begin{bmatrix} 1 & 0 \\ 0 & 1 \end{bmatrix}, \quad I = \begin{bmatrix} 1 & 0 \\ 0 & -1 \end{bmatrix}, \quad J = \begin{bmatrix} 0 & b \\ 1 & 0 \end{bmatrix} \quad \text{and} \quad IJ = \begin{bmatrix} 0 & b \\ -1 & 0 \end{bmatrix}$$
$$(1.2)$$

generate $M_2(k)$ as a k-vector space, and they satisfy the relations

$$I^2 = \mathrm{Id}, \quad J^2 = b\,\mathrm{Id}, \quad IJ = -JI.$$

Definition 1.1.6 A quaternion algebra over k is called *split* if it is isomorphic to $M_2(k)$ as a k-algebra.

Proposition 1.1.7 *For a quaternion algebra (a, b) the following statements are equivalent.*

1. *The algebra (a, b) is split.*
2. *The algebra (a, b) is not a division algebra.*
3. *The norm map $N : (a, b) \to k$ has a nontrivial zero.*
4. *The element b is a norm from the field extension $k(\sqrt{a})|k$.*

Of course, instead of (4) another equivalent condition is that a is a norm from the field extension $k(\sqrt{b})|k$.

Proof The implication (1) $\Rightarrow$ (2) is obvious and (2) $\Rightarrow$ (3) was proven in Lemma 1.1.3. For (3) $\Rightarrow$ (4) we may assume a is not a square in k, for otherwise the claim is obvious. Take a nonzero quaternion $q = x + yi + zj + wij$ with norm 0. Then equation (1.1) implies $(z^2 - aw^2)b = x^2 - ay^2$, and so in particular $z^2 - aw^2 = (z + \sqrt{a}w)(z - \sqrt{a}w) \neq 0$, for otherwise a would be a square in k. Denoting by $N_{K|k}$ the field norm from $K = k(\sqrt{a})$ we get

$$b = N_{K|k}(x + \sqrt{a}y)N_{K|k}(z + \sqrt{a}w)^{-1},$$

whence (4) by multiplicativity of $N_{K|k}$. Finally, we shall show assuming (4) that $(a, b) \cong (1, 4a^2)$, whence (1) by the isomorphism in Example 1.1.5. To see this, we may again assume that a is not a square in k. If b is a norm from K, then so is b^{-1}, so by (4) and our assumption on a we find $r, s \in k$ satisfying $b^{-1} = r^2 - as^2$. Putting $u = rj + sij$ thus yields $u^2 = br^2 - abs^2 = 1$. Moreover, one verifies that $ui = -iu$, which implies that the element $v = (1 + a)i + (1 - a)ui$ satisfies $uv = (1 + a)ui + (1 - a)i = -vu$ and $v^2 = (1 + a)^2a - (1 - a)^2a = 4a^2$. Passing to the basis $(1, u, v, uv)$ thus gives the required isomorphism $(a, b) \cong (1, 4a^2)$. □

Remark 1.1.8 Over a field of characteristic 2 one defines the generalized quaternion algebra $[a, b)$ by the presentation

$$[a, b) = \langle i, j \mid i^2 + i = a, \; j^2 = b, \; ij = ji + j \rangle$$

where $a \in k$ and $b \in k^\times$. This algebra has properties analogous to those in the above proposition (see Exercise 4).

1.2 Splitting over a quadratic extension

We now prove a structure theorem for division algebras of dimension 4. Recall first that the centre $Z(A)$ of a k-algebra A is the k-subalgebra consisting of elements $x \in A$ satisfying $xy = yx$ for all $y \in A$. By assumption we have $k \subset Z(A)$; if this inclusion is an equality, one says that A is *central* over k. If A is a division algebra, then $Z(A)$ is a field. We then have:

Proposition 1.2.1 *A 4-dimensional central division algebra D over k is isomorphic to a quaternion algebra.*

We first prove:

Lemma 1.2.2 *If D contains a commutative k-subalgebra isomorphic to a nontrivial quadratic field extension $k(\sqrt{a})$ of k, then D is isomorphic to a quaternion algebra (a, b) for suitable $b \in k^{\times}$.*

Proof A k-subalgebra as in the lemma contains an element q with $q^2 = a \in k$. By assumption, q is not in the centre k of D and hence the inner automorphism of D given by $x \mapsto q^{-1}xq$ has exact order 2. As a k-linear automorphism of D, it thus has -1 as an eigenvalue, which means that there exists $r \in D$ such that $qr + rq = 0$. This relation shows that $r \notin k(q)$ (for otherwise r and q would commute), and therefore 1 and r form a basis of D as a 2-dimensional $k(q)$-vector space. It follows that the elements $1, q, r, qr$ form a k-basis of D and moreover they are fixed by the k-linear automorphism $x \mapsto r^{-2}xr^2$. Thus r^2 belongs to the centre of D, which is k by assumption. The lemma follows by setting $r^2 = b \in k^{\times}$. □

Proof of Proposition 1.2.1 Let d be an element of $D \setminus k$. As D is finite dimensional over k, the powers $\{1, d, d^2, \dots\}$ are linearly dependent, so there is a polynomial $f \in k[x]$ with $f(d) = 0$. As D is a division algebra, it has no zero divisors and we may assume f irreducible. This means there is a k-algebra homomorphism $k[x]/(f) \to D$ which realizes the field $k(d)$ as a k-subalgebra of D. Now the degree $[k(d) : k]$ cannot be 1 as $d \notin k$, and it cannot be 4 as D is not commutative. Hence $[k(d) : k] = 2$, and the lemma applies. □

The crucial ingredient in the above proof was the existence of a quadratic extension $k(\sqrt{a})$ contained in D. Observe that the algebra $D \otimes_k k(\sqrt{a})$ then splits over $k(\sqrt{a})$. In fact, it follows from basic structural results to be proven in the next chapter (Lemma 2.2.2 and Wedderburn's theorem) that any 4-dimensional central k-algebra for which there exists a quadratic extension of k with this splitting property is a division algebra or a matrix algebra.

It is therefore interesting to characterize those quadratic extensions of k over which a quaternion algebra splits.

Proposition 1.2.3 *Consider a quaternion algebra A over k, and fix an element $a \in k^\times \setminus k^{\times 2}$. The following statements are equivalent:*

1. *A is isomorphic to the quaternion algebra (a, b) for some $b \in k^\times$.*
2. *The $k(\sqrt{a})$-algebra $A \otimes_k k(\sqrt{a})$ is split.*
3. *A contains a commutative k-subalgebra isomorphic to $k(\sqrt{a})$.*

Proof To show $(1) \Rightarrow (2)$, note that $(a, b) \otimes_k k(\sqrt{a})$ is none but the quaternion algebra (a, b) defined over the field $k(\sqrt{a})$. But a is a square in $k(\sqrt{a})$, so $(a, b) \cong (1, b)$, and the latter algebra is isomorphic to $M_2(k(\sqrt{a}))$ by Example 1.1.5. Next, if A is split, the same argument shows that (1) always holds, so to prove $(3) \Rightarrow (1)$ one may assume A is nonsplit, in which case Lemma 1.2.2 applies.

The implication $(2) \Rightarrow (3)$ is easy in the case when $A \cong M_2(k)$: one chooses an isomorphism $M_2(k) \cong (1, a)$ as in Example 1.1.5 and takes the subfield $k(J)$, where J is the basis element with $J^2 = a$. We now assume A is nonsplit, and extend the quaternion norm N on A to $A \otimes_k k(\sqrt{a})$ by base change. Applying part (3) of Proposition 1.1.7 to $A \otimes_k k(\sqrt{a})$ one gets that there exist elements $q_0, q_1 \in A$, not both 0, with $N(q_0 + \sqrt{a}q_1) = 0$. Denote by $B : A \otimes_k k(\sqrt{a}) \times A \otimes_k k(\sqrt{a}) \to k(\sqrt{a})$ the symmetric bilinear form associated with N (recall that $B(x, y) = (N(x + y) - N(x) - N(y))/2$ by definition, hence $B(x, x) = N(x)$). We get

$$0 = B(q_0 + \sqrt{a}q_1, q_0 + \sqrt{a}q_1) = N(q_0) + aN(q_1) + 2\sqrt{a}B(q_0, q_1).$$

Now note that since $q_0, q_1 \in A$, the elements $B(q_0, q_1)$ and $N(q_0) + aN(q_1)$ both lie in k. So it follows from the above equality that

$$N(q_0) = -aN(q_1) \quad \text{and} \quad 2B(q_0, q_1) = q_0\bar{q}_1 + q_1\bar{q}_0 = 0.$$

Here $N(q_0), N(q_1) \neq 0$ as A is nonsplit. The element $q_2 := q_0\bar{q}_1 \in A$ satisfies

$$q_2^2 = q_0\bar{q}_1q_0\bar{q}_1 = -q_0\bar{q}_0q_1\bar{q}_1 = -N(q_0)N(q_1) = aN(q_1)^2.$$

The square of the element $q := q_2N(q_1)^{-1}$ is then precisely a, so mapping $\sqrt{a}$ to q embeds $k(\sqrt{a})$ into A. $\square$

We conclude this section by another characterization of the quaternion norm.

Proposition 1.2.4 *Let (a, b) be a quaternion algebra over a field k, and let $K = k(\sqrt{a})$ be a quadratic splitting field for (a, b). Then for all $q \in (a, b)$ and all K-isomorphisms $\phi : (a, b) \otimes_k K \xrightarrow{\sim} M_2(K)$ we have $N(q) = \det(\phi(q))$.*

Proof First note that $\det(\phi(q))$ does not depend on the choice of ϕ. Indeed, if $\psi : (a,b) \otimes_k K \overset{\sim}{\to} M_2(K)$ is a second isomorphism, then $\phi \circ \psi^{-1}$ is an automorphism of $M_2(K)$. But it is well known that all K-automorphisms of $M_2(K)$ are of the form $M \to CMC^{-1}$ for some invertible matrix C (check this by hand or see Lemma 2.4.1 for a proof in any dimension), and that the determinant map is invariant under such automorphisms.

Now observe that by definition the quaternion norm on $(a,b) \otimes_k K$ restricts to that on (a,b). Therefore to prove the proposition it is enough to embed (a,b) into $M_2(K)$ via ϕ and check that on $M_2(K)$ the quaternion norm (which is intrinsic by Remark 1.1.4) is given by the determinant. For this, consider a basis of $M_2(K)$ as in (1.2) with $b = 1$ and write

$$\begin{bmatrix} a_1 & a_2 \\ a_3 & a_4 \end{bmatrix} = \left(\frac{a_1 + a_4}{2}\right) \begin{bmatrix} 1 & 0 \\ 0 & 1 \end{bmatrix} + \left(\frac{a_1 - a_4}{2}\right) \begin{bmatrix} 1 & 0 \\ 0 & -1 \end{bmatrix} +$$
$$+ \left(\frac{a_2 + a_3}{2}\right) \begin{bmatrix} 0 & 1 \\ 1 & 0 \end{bmatrix} + \left(\frac{a_2 - a_3}{2}\right) \begin{bmatrix} 0 & 1 \\ -1 & 0 \end{bmatrix}.$$

Then equation (1.1) yields

$$N\left(\begin{bmatrix} a_1 & a_2 \\ a_3 & a_4 \end{bmatrix}\right) = \left(\frac{a_1 + a_4}{2}\right)^2 - \left(\frac{a_1 - a_4}{2}\right)^2 - \left(\frac{a_2 + a_3}{2}\right)^2 + \left(\frac{a_2 - a_3}{2}\right)^2$$

$$= a_1 a_4 - a_2 a_3 = \det\left(\begin{bmatrix} a_1 & a_2 \\ a_3 & a_4 \end{bmatrix}\right).$$

$\square$

1.3 The associated conic

We now introduce another important invariant of a quaternion algebra (a,b), the *associated conic* $C(a,b)$. By definition, this is the projective plane curve defined by the homogeneous equation

$$ax^2 + by^2 = z^2 \tag{1.3}$$

where x, y, z are the homogeneous coordinates in the projective plane $\mathbf{P}^2$. In the case of $(1,1) \overset{\sim}{\to} M_2(k)$ we get the usual circle

$$x^2 + y^2 = z^2.$$

Remark 1.3.1   In fact, the conic $C(a, b)$ is canonically attached to the algebra (a, b) and does not depend on the choice of a basis. To see why, note first that the conic $C(a, b)$ is isomorphic to the conic $ax^2 + by^2 = abz^2$ via the substitution $x \mapsto by, y \mapsto ax, z \mapsto abz$ (after substituting, divide the equation by ab). But $ax^2 + by^2 - abz^2$ is exactly the square of the pure quaternion $xi + yj + zij$ and hence is intrinsically defined by Remark 1.1.4.

This observation also shows that if two quaternion algebras (a, b) and (c, d) are isomorphic as k-algebras, then the conics $C(a, b)$ and $C(c, d)$ are also isomorphic over k. Indeed, constructing an isomorphism $(a, b) \cong (c, d)$ is equivalent to finding a k-basis in (a, b) that satisfies the multiplicative rule in (c, d).

Recall from algebraic geometry that the conic $C(a, b)$ is said to have a k-*rational point* if there exist $x_0, y_0, z_0 \in k$, not all zero, that satisfy equation (1.3) above.

We can now give a complement to Proposition 1.1.7.

Proposition 1.3.2   *The quaternion algebra (a, b) is split if and only if the conic $C(a, b)$ has a k-rational point.*

Proof   If (x_0, y_0, z_0) is a k-rational point on $C(a, b)$ with $y_0 \neq 0$, then $b = (z_0/y_0)^2 - a(x_0/y_0)^2$ and part (4) of Proposition 1.1.7 is satisfied. If y_0 happens to be 0, then x_0 must be nonzero and we get similarly that a is a norm from the extension $k(\sqrt{b})|k$. Conversely, if $b = r^2 - as^2$ for some $r, s \in k$, then $(s, 1, r)$ is a k-rational point on $C(a, b)$.   □

Remark 1.3.3   Again, the proposition has a counterpart in characteristic 2; see Exercise 4.

Example 1.3.4   For $a \neq 1$, the projective conic $ax^2 + (1 - a)y^2 = z^2$ has the k-rational point $(1, 1, 1)$, hence the quaternion algebra $(a, 1 - a)$ splits by the proposition. This innocent-looking fact is a special case of the so-called *Steinberg relation* for symbols that we shall encounter later.

Remark 1.3.5   It is a well-known fact from algebraic geometry that a smooth projective conic defined over a field k is isomorphic to the projective line $\mathbf{P}^1$ over k if and only if it has a k-rational point. The isomorphism is given by taking the line joining a point P of the conic to some fixed k-rational point O and then taking the intersection of this line with $\mathbf{P}^1$ embedded as, say, some coordinate axis in $\mathbf{P}^2$. In such a way we get another equivalent condition for the splitting of a quaternion algebra, which will be substantially generalized later.

In the remainder of this section we give examples of how Proposition 1.3.2 can be used to give easy proofs of splitting properties of quaternion algebras over special fields.

Example 1.3.6 Let k be the finite field with q elements (q odd). Then any quaternion algebra (a, b) over k is split.

To see this, it suffices by Proposition 1.3.2 to show that the conic $C(a, b)$ has a k-rational point. We shall find a point (x_0, y_0, z_0) with $z_0 = 1$. As the multiplicative group of k is cyclic of order $q - 1$, there are exactly $1 + (q-1)/2$ squares in k, including 0. Thus the sets $\{ax^2 : x \in k\}$ and $\{1 - by^2 \mid y \in k\}$ both have cardinality $1 + (q - 1)/2$, hence must have an element in common.

The next two examples concern the field $k(t)$ of rational functions over a field k, which is by definition the fraction field of the polynomial ring $k[t]$. Note that sending t to 0 induces a k-homomorphism $k[t] \rightarrow k$; we call it the *specialization* map attached to t.

Example 1.3.7 Let (a, b) be a quaternion algebra over k. Then (a, b) is split over k if and only if $(a, b) \otimes_k k(t)$ is split over $k(t)$.

Here necessity is obvious. For sufficiency, we assume given a point (x_t, y_t, z_t) of $C(a, b)$ defined over $k(t)$. As the equation (1.3) defining $C(a, b)$ is homogeneous, we may assume after multiplication by a suitable element of $k(t)$ that x_t, y_t, z_t all lie in $k[t]$ and one of them has a nonzero constant term. Then specialization gives a k-point $(x_t(0), y_t(0), z_t(0))$ of $C(a, b)$.

Finally we give an example of a splitting criterion for a quaternion algebra over $k(t)$ that does not come from k.

Example 1.3.8 For $a \in k^\times$ the $k(t)$-algebra (a, t) is split if and only if a is a square in k.

Here sufficiency is contained in Example 1.1.5. For necessity, assume given a $k(t)$-point (x_t, y_t, z_t) of $C(a, b)$ as above. Again we may assume x_t, y_t, z_t are all in $k[t]$. If x_t and z_t were both divisible by t, then equation (1.3) would imply the same for y_t, so after an eventual division we may assume they are not. Then setting $t = 0$ gives $a x_t^2(0) = z_t(0)^2$ and so $a = x_t^2(0)^{-1} z_t(0)^2$ is a square.

1.4 A theorem of Witt

In this section we prove an elegant theorem which characterizes isomorphisms of quaternion algebras by means of the function fields of the associated conics. Recall that the function field of an algebraic curve C is the field $k(C)$ of rational functions defined over some Zariski open subset of C. In the concrete case of a conic $C(a, b)$ as in the previous section, the simplest way to define it is to take the fraction field of the integral domain $k[x, y]/(ax^2 + by^2 - 1)$ (this is also the function field of the affine curve of equation $ax^2 + by^2 = 1$).

A crucial observation for the sequel is the following.

Remark 1.4.1 The quaternion algebra $(a, b) \otimes_k k(C(a, b))$ is always split over $k(C(a, b))$. Indeed, the conic $C(a, b)$ always has a point over this field, namely $(x, y, 1)$ (where we also denote by x, y their images in $k(C(a, b))$). This point is called the *generic point* of the conic.

Now we can state the theorem.

Theorem 1.4.2 (Witt) *Let $Q_1 = (a_1, b_1)$, $Q_2 = (a_2, b_2)$ be quaternion algebras, and let $C_i = C(a_i, b_i)$ be the associated conics. The algebras Q_1 and Q_2 are isomorphic over k if and only if the function fields $k(C_1)$ and $k(C_2)$ are isomorphic over k.*

Remark 1.4.3 It is known from algebraic geometry that two smooth projective curves are isomorphic if and only if their function fields are. Thus the theorem states that *two quaternion algebras are isomorphic if and only if the associated conics are isomorphic as algebraic curves.*

In Chapter 5 we shall prove a broad generalization of the theorem, due to Amitsur. We now begin the proof by the following easy lemma.

Lemma 1.4.4 *If (a, b) is a quaternion algebra and $c \in k^\times$ is a norm from the field extension $k(\sqrt{a})|k)$, then $(a, b) \cong (a, bc)$.*

Proof By hypothesis, we may write $c = x^2 - ay^2$ with $x, y \in k$. Hence we may consider c as the norm of the quaternion $q = x + yi + 0j + 0ij$ and set $J = qj = xj + yij$. Then J is a pure quaternion satisfying

$$iJ + Ji = 0, \quad J^2 = -N(J) = -N(q)N(j) = bc,$$

and $1, i, J, iJ$ is a basis of (a, b) over k (by a similar argument as in the proof of Lemma 1.2.1). The lemma follows. $\square$

Proof of Theorem 1.4.2 Necessity follows from Remark 1.3.1, so it is enough to prove sufficiency. If both Q_1 and Q_2 are split, the theorem is obvious. So we may assume one of them, say Q_1, is nonsplit. By Remark 1.4.1 the algebra $Q_1 \otimes_k k(C_1)$ is split, hence so is the algebra $Q_1 \otimes_k k(C_2)$ by assumption. If Q_2 is split, then $k(C_2)$ is a rational function field, and therefore Q_1 is also split by Example 1.3.7.

So we may assume that both algebras are nonsplit. In particular a_1 is not a square in k, and the algebra $Q_1 \otimes_k L$ becomes split over the quadratic extension $L := k(\sqrt{a_1})$. For brevity's sake, we write C instead of C_1 in what follows. The field $L(C) = L \otimes_k k(C)$ is the function field of the curve C_L obtained by extension of scalars from C; this curve is isomorphic to the projective line over L, and hence $L(C)$ is isomorphic to the rational function field $L(t)$. As $Q_2 \otimes_k L(C)$ is split over $L(C)$ by assumption, Example 1.3.7 again yields that $Q_2 \otimes_k L$ must be split over L. Proposition 1.2.3 then implies that $Q_2 \xrightarrow{\sim} (a_1, c)$ for some $c \in k^\times$. As $Q_2 \otimes_k k(C)$ is split over $k(C)$ (again by assumption and Remark 1.4.1), it follows from Proposition 1.1.7 that $c = N_{L(C)/k(C)}(f)$ for some $f \in L(C)^\times$.

Our goal is to identify the function f in order to compute c. Recall (e.g. from Section A.4 of the Appendix) that the group $\mathrm{Div}(C_L)$ of divisors on C_L is defined as the free abelian group generated by the closed points of C_L (in this case they correspond to irreducible polynomials in $L(t)$, plus a point at infinity). There is a *divisor map* $\mathrm{div} : L(C)^\times \to \mathrm{Div}(C_L)$ associating to a function the divisor given by its zeroes and poles, and a *degree map* given by $\sum m_i P_i \mapsto \sum m_i [\kappa(P_i) : L]$, where $\kappa(P_i)$ is the residue field of the closed point P_i. The two maps fit into an exact sequence

$$0 \to L(C)^\times / L^\times \xrightarrow{\mathrm{div}} \mathrm{Div}(C_L) \xrightarrow{\deg} \mathbf{Z} \to 0, \qquad (1.4)$$

corresponding in our case to the decomposition of rational functions into products of irreducible polynomials and their inverses.

The Galois group $\mathrm{Gal}\,(L|K) = \{1, \sigma\}$ acts on this exact sequence as follows (see Remark A.4.5 of the Appendix). On $L(C)$ it acts via its action on L (but note that under the isomorphism $L(C) \cong L(t)$ this action does *not* induce the similar action on the right-hand side!). On $\mathrm{Div}(C_L)$ it acts by sending a closed point P to its conjugate $\sigma(P)$. Finally, it acts trivially on $\mathbf{Z}$, making the maps of the sequence $\mathrm{Gal}\,(L|K)$-equivariant.

Now consider the map $(1 + \sigma) : \mathrm{Div}(C_L) \to \mathrm{Div}(C_L)$. By additivity of the divisor map, we have

$$(1 + \sigma)\mathrm{div}(f) = \mathrm{div}(f\sigma(f)) = \mathrm{div}(c) = 0,$$

as c is a constant. On the other hand, as σ has order 2, we have a natural direct sum decomposition

$$\mathrm{Div}(C_L) = \left(\bigoplus_{P=\sigma(P)} \mathbf{Z}P \right) \oplus \left(\bigoplus_{P \neq \sigma(P)} \mathbf{Z}P \right),$$

where σ acts trivially on the first summand, and exchanges P and $\sigma(P)$ in the second. Writing $\mathrm{div}(f) = E_1 + E_2$ according to this decomposition, we get

$$0 = (1+\sigma)\mathrm{div}(f) = 2E_1 + (1+\sigma)E_2.$$

This implies that $E_1 = 0$ and E_2 is of the form $\sum(m_i P_i - m_i \sigma(P_i))$ for some closed points $P_1, \ldots, P_r$ and $m_i \neq 0$. Setting $D = \sum m_i P_i$, we may therefore write

$$\mathrm{div}(f) = (1-\sigma)D.$$

Let d be the degree of D. The point $P_0 := (1 : 0 : \sqrt{a_1})$ is an L-rational point of our conic C_L, whose equation is $a_1 x^2 + b_1 y^2 = z^2$. Exact sequence (1.4) therefore shows that there exists $g \in L(C)^\times$ such that

$$D - dP_0 = \mathrm{div}(g).$$

It follows that

$$\mathrm{div}(f) = \mathrm{div}(g\sigma(g)^{-1}) + d(1-\sigma)P_0.$$

Replacing f by $f\sigma(g)g^{-1}$ gives a function still satisfying $c = N_{L(C)|k(C)}(f)$, but with

$$\mathrm{div}(f) = d(1-\sigma)P_0. \tag{1.5}$$

We are now able to identify the function f up to a constant. We first claim that the rational function $h := (z - \sqrt{a_1}x)y^{-1} \in L(C)^\times$ satisfies

$$\mathrm{div}(h) = (1 : 0 : \sqrt{a_1}) - (1 : 0 : -\sqrt{a_1}) = (1-\sigma)P_0. \tag{1.6}$$

Indeed, let $P = (x_0 : y_0 : z_0)$ be a pole of h (over an algebraic closure $\bar{k}$). Then we must have $y_0 = 0$ and hence $P = (1 : 0 : \pm\sqrt{a_1})$; in particular, P is an L-rational point. But by the equation of C we have $h = b_1 y(z + \sqrt{a_1}z)^{-1}$, so $(1 : 0 : \sqrt{a_1})$ is a zero of h and not a pole. Therefore $(1 : 0 : -\sqrt{a_1})$ is the only pole of h and similarly $(1 : 0 : \sqrt{a_1})$ is its only zero. Comparing formulae (1.5) and (1.6), we get from the left exactness of sequence (1.4) that $f = c_0 h^d$ for some constant $c_0 \in L^\times$. We compute

$$c = N_{L(C)|k(C)}(f) = N_{L|k}(c_0)N_{L(C)|k(C)}(h)^d$$

$$= N_{L|k}(c_0)\left(\frac{z^2 - a_1 x^2}{y^2}\right)^d = N_{L|k}(c_0)b_1^d.$$

So Lemma 1.4.4 implies

$$Q_2 \cong (a_1, c) \cong (a_1, b_1^d).$$

By our assumption Q_2 is nonsplit, so d is odd and $Q_2 \cong (a_1, b_1)$, as desired. $\square$

1.5 Tensor products of quaternion algebras

Now we step forward and consider higher dimensional k-algebras, where k is still assumed to be a field of characteristic not 2. The simplest of these are *biquaternion algebras*, which are by definition those k-algebras that are isomorphic to a tensor product of two quaternion algebras over k.

We begin with two lemmas that are very helpful in calculations. The first is well known:

Lemma 1.5.1 *The tensor product of two matrix algebras $M_n(k)$ and $M_m(k)$ over k is isomorphic to the matrix algebra $M_{nm}(k)$.*

Proof Perhaps the simplest proof is to note that given k-endomorphisms $\phi \in \mathrm{End}_k(k^n)$ and $\psi \in \mathrm{End}_k(k^m)$, the pair (ϕ, ψ) induces an element $\phi \otimes \psi$ of $\mathrm{End}_k(k^n \otimes_k k^m)$. The resulting map $\mathrm{End}_k(k^n) \otimes \mathrm{End}_k(k^m) \to \mathrm{End}_k(k^n \otimes_k k^m)$ is obviously injective, and it is surjective e.g. by dimension reasons. $\square$

Lemma 1.5.2 *Given elements $a, b, b' \in k^\times$ we have an isomorphism*

$$(a, b) \otimes_k (a, b') \xrightarrow{\sim} (a, bb') \otimes_k M_2(k).$$

Proof Denote by $(1, i, j, ij)$ and $(1, i', j', i'j')$ quaternion bases of (a, b) and (a, b'), respectively, and consider the k-subspaces

$$A_1 = k(1 \otimes 1) \oplus k(i \otimes 1) \oplus k(j \otimes j') \oplus k(ij \otimes j'),$$

$$A_2 = k(1 \otimes 1) \oplus k(1 \otimes j') \oplus k(i \otimes i'j') \oplus k\big((-b'i) \otimes i'\big)$$

of $(a, b) \otimes_k (a, b')$. One checks that A_1 and A_2 are both closed under multiplication and hence are subalgebras of $(a, b) \otimes_k (a, b')$. By squaring the basis elements $i \otimes 1, j \otimes j'$ and $1 \otimes j', i \otimes i'j'$ we see that A_1 and A_2 are isomorphic to the quaternion algebras (a, bb') and $(b', -a^2 b')$, respectively. But this latter algebra is isomorphic to $(b', -b')$, which is split because the conic $C(b', -b')$ has the k-rational point $(1, 1, 0)$.

Now consider the map $\rho : A_1 \otimes_k A_2 \to (a, b) \otimes_k (a, b')$ induced by the k-bilinear map $(x, y) \to xy$. Inspection reveals that all standard basis elements of $(a, b) \otimes_k (a, b')$ lie in the image of ρ, so it is surjective and hence induces the required isomorphism for dimension reasons. $\square$

Corollary 1.5.3 *For a quaternion algebra (a, b) the tensor product algebra $(a, b) \otimes_k (a, b)$ is isomorphic to the matrix algebra $M_4(k)$.*

Proof The case $b = b'$ of the previous lemma and Example 1.1.5 give

$$(a, b) \otimes_k (a, b) \cong (a, b^2) \otimes_k M_2(k) \cong (a, 1) \otimes_k M_2(k) \cong M_2(k) \otimes_k M_2(k),$$

and we conclude by Lemma 1.5.1. $\square$

A biquaternion algebra $A = Q_1 \otimes_k Q_2$ is equipped with an involution σ defined as the product of the conjugation involutions on Q_1 and Q_2, i.e. by setting $\sigma(q_1 \otimes q_2) = \bar{q}_1 \otimes \bar{q}_2$ and extending by linearity. We remark that the involution σ is not canonical but depends on the decomposition $A \cong Q_1 \otimes_k Q_2$. For $i = 1, 2$ denote by Q_i^- the subspace of pure quaternions in Q_i (cf. Remark 1.1.4).

Lemma 1.5.4 *Let V be the k-subspace of A consisting of elements satisfying $\sigma(a) = -a$, and W the subspace of those with $\sigma(a) = a$. One has a direct sum decomposition $A = V \oplus W$, and moreover one may write*

$$V = (Q_1^- \otimes_k k) \oplus (k \otimes_k Q_2^-) \quad and \quad W = k \oplus (Q_1^- \otimes_k Q_2^-).$$

Proof One has $V \cap W = 0$. Moreover, there are natural inclusions

$$(Q_1^- \otimes_k k) \oplus (k \otimes_k Q_2^-) \subset V \quad and \quad k \oplus (Q_1^- \otimes_k Q_2^-) \subset W.$$

For dimension reasons these must be isomorphisms and $V \oplus W$ must be the whole of A. $\square$

Denote by N_1 and N_2 the quaternion norms on Q_1 and Q_2, respectively, and consider the quadratic form

$$\phi(x, y) = N_1(x) - N_2(y) \tag{1.7}$$

on V, called an *Albert form* of A. Again it depends on the decomposition $A \cong Q_1 \otimes_k Q_2$.

Theorem 1.5.5 (Albert) *For a biquaternion algebra $A \cong Q_1 \otimes_k Q_2$ over k, the following statements are equivalent:*

1. *The algebra A is not a division algebra.*
2. *There exist $a, b, b' \in k^\times$ such that $Q_1 \xrightarrow{\sim} (a, b)$ and $Q_2 \xrightarrow{\sim} (a, b')$.*
3. *The Albert form (1.7) has a nontrivial zero on A.*

Proof For the implication (2) $\Rightarrow$ (3), note that the assumption in (2) implies that there exist pure quaternions $q_i \in Q_i^-$ with $q_i^2 = -N_i(q_i) = a$ for $i = 1, 2$, and hence $\phi(q_1, q_2) = 0$. For (3) $\Rightarrow$ (1), assume there is a nontrivial relation $\phi(q_1, q_2) = 0$ in pure quaternions. Note that q_1 and q_2 commute, because the components Q_1 and Q_2 centralize each other in the tensor product $Q_1 \otimes_k Q_2$. Hence we have $0 = \phi(q_1, q_2) = q_1^2 - q_2^2 = (q_1 + q_2)(q_1 - q_2)$, which implies that A cannot be a division algebra.

For the hardest implication (1) $\Rightarrow$ (2) assume (2) is false, and let us prove that $A \cong Q_1 \otimes_k Q_2$ is a division algebra. If (2) is false, then both Q_1 and Q_2 are division algebras (otherwise say $b' = 1$ and suitable a, b will do). Denote by K_i a quadratic extension of K contained in Q_i for $i = 1, 2$. By our assumption that (2) is false, Proposition 1.2.3 implies that K_1 splits Q_1 but not Q_2, and similarly for K_2. Therefore both $K_1 \otimes_k Q_2$ and $K_2 \otimes_k Q_1$ are division algebras. It will suffice to show that each nonzero $\alpha \in A$ has a left inverse α_l, for then the conjugate $\alpha_r := \sigma(\sigma(\alpha)_l)$ is a right inverse for α, and $\alpha_l = \alpha_l \alpha \alpha_r = \alpha_r$. Moreover, it is enough to find $\alpha^* \in A$ such that $\alpha^* \alpha$ is a nonzero element lying in either $K_1 \otimes_k Q_2$ or $Q_1 \otimes_k K_2$, for then $\alpha^* \alpha$ has a left inverse, and so does α. Fix a quaternion basis $\{1, i, j, ij\}$ for Q_2 such that $K_2 = k(j)$. We can then write

$$\alpha = (\beta_1 + \beta_2 j) + (\beta_3 + \beta_4 j)ij$$

with suitable $\beta_i \in Q_1$. We may assume that $\gamma := \beta_3 + \beta_4 j \neq 0$, for otherwise α lies in $Q_1 \otimes_k K_2$ already. Then γ^{-1} exists in $Q_1 \otimes_k K_2$, and after replacing α by $\gamma^{-1}\alpha$ we are reduced to the case when $\alpha = \beta_1 + \beta_2 j + ij$. If β_1 and β_2 commute, then $k(\beta_1, \beta_2)$ is either k or a quadratic extension $K|k$ contained in Q_1. Thus $\alpha \in Q_2$ or $\alpha \in K \otimes_k Q_2$, and we are done in this case. So we may assume $\beta_1 \beta_2 - \beta_2 \beta_1 \neq 0$. We then contend that $\alpha^* := \beta_1 - \beta_2 j - ij$ is a good choice. Indeed, we compute

$$\alpha^* \alpha = (\beta_1 - \beta_2 j - ij)(\beta_1 + \beta_2 j + ij) = (\beta_1 - \beta_2 j)(\beta_1 + \beta_2 j) - (ij)^2 =$$
$$= \beta_1^2 - \beta_2^2 j^2 - (ij)^2 + (\beta_1 \beta_2 - \beta_2 \beta_1)j,$$

where the second equality holds since ij commutes with β_1, β_2 (for the same reason as above), and anticommutes with j. Since j^2 and $(ij)^2$ lie in k and $\beta_1\beta_2 - \beta_2\beta_1 \neq 0$, this shows $\alpha^*\alpha \in (Q_1 \otimes_k K_2) \setminus \{0\}$, as required. □

Remark 1.5.6   The above proof, taken from Lam [1], is a variant of Albert's original argument. For other proofs of the theorem, valid in all characteristics, see Knus [1] as well as Tits [1] (for the equivalence (1) ⇔ (2)).

The theorem makes it possible to give concrete examples of biquaternion division algebras, such as the following one.

Example 1.5.7   Let k be a field of characteristic $\neq 2$ as usual, and let F be the purely transcendental extension $k(t_1, t_2, t_3, t_4)$. Then the biquaternion algebra

$$(t_1, t_2) \otimes_F (t_3, t_4)$$

is a division algebra over F.

To see this, we have to check that the Albert form has no nontrivial zero. Assume it does. Then by formula (1.1) for the quaternion norm we have a nontrivial solution of the equation

$$- t_1 x_1^2 - t_2 x_2^2 + t_1 t_2 x_{1,2}^2 + t_3 x_3^2 + t_4 x_4^2 - t_3 t_4 x_{3,4}^2 = 0 \qquad (1.8)$$

in the variables $x_1, x_2, x_{1,2}, x_3, x_4, x_{3,4}$. By multiplying with a rational function we may assume $x_1, x_2, x_{1,2}, x_3, x_4, x_{3,4}$ are all in $k(t_1, t_2, t_3)[t_4]$ and one of them is not divisible by t_4.

Assume that $x_1, x_2, x_{1,2}, x_3$ are all divisible by t_4. Then t_4^2 must divide $t_4 x_4^2 - t_3 t_4 x_{3,4}^2$, so t_4 divides $x_4^2 - t_3 x_{3,4}^2$. Setting $t_4 = 0$ produces a solution of the equation $x^2 - t_3 y^2 = 0$ with $x, y \in k(t_1, t_2, t_3)$ not both 0, which implies that t_3 is a square in $k(t_1, t_2, t_3)$; this is a contradiction. So one of the $x_1, x_2, x_{1,2}, x_3$ is not divisible by t_4, and by setting $t_4 = 0$ in equation (1.8) we get a nontrivial solution of the equation

$$-t_1 y_1^2 - t_2 y_2^2 + t_1 t_2 y_{1,2}^2 + t_3 y_3^2 = 0$$

with entries in $k(t_1, t_2, t_3)$. A similar argument as before then shows that there is a nontrivial solution of the equation

$$-t_1 z_1^2 - t_2 z_2^2 + t_1 t_2 z_{1,2}^2 = 0$$

over the field $k(t_1, t_2)$. Applying the same trick one last time, we see that the equation

$$t_1 w_1^2 = 0$$

has a nontrivial solution in $k(t_1)$, which finally yields a contradiction.

In general, we say that a finite-dimensional division algebra D over a field k has *period 2* if $D \otimes_k D$ is isomorphic to a matrix algebra over k. Quaternion algebras have this property by Corollary 1.5.3. Also, applying Lemma 1.5.1 we see that tensor products of division algebras of period 2 are again of period 2.

According to Proposition 1.2.1, a 4-dimensional central division algebra over k is in fact a quaternion algebra. Moreover, in 1932 Albert proved that a 16-dimensional central division algebra of period 2 is isomorphic to a biquaternion algebra. Thus it was plausible to conjecture that a central division algebra of period 2 and dimension 4^m is always a tensor product of m quaternion algebras. However, in 1979 Amitsur, Rowen and Tignol [1] produced a 64-dimensional central division algebra of period 2 which is not a tensor product of quaternion algebras.

Therefore the following theorem of Merkurjev [1], which is one of the highlights of this book, is all the more remarkable.

Theorem 1.5.8 (Merkurjev) *Let D be a central division algebra of period 2 over a field k. There exist positive integers m_1, m_2, n and quaternion algebras $Q_1, ..., Q_n$ over k such that there is an isomorphism*

$$D \otimes_k M_{m_1}(k) \cong Q_1 \otimes_k Q_2 \otimes_k \cdots \otimes_k Q_n \otimes M_{m_2}(k).$$

EXERCISES

1. Let Q be a quaternion algebra over k. Show that the conjugation involution is the only linear map $\sigma : Q \to Q$ such that $\sigma(1) = 1$ and $\sigma(q)q \in k$ for all $q \in Q$.

2. Show that a quaternion algebra is split if and only if it has a basis (e, f, g, h) in which the norm is given by $(xe + yf + zg + wh) \mapsto xy - zw$. (In the language of quadratic forms, this latter property means that the norm form is *hyperbolic*.)

3. Let Q be a quaternion algebra over k, and let $K|k$ be a quadratic extension embedded as a k-subalgebra in Q. Verify that one has $N(q) = N_{K|k}(q)$ for all $q \in K$, where N is the quaternion norm and $N_{K|k}$ is the field norm. [*Hint:* Extend a suitable k-basis of K to a quaternion basis of Q.]

4. Let k be a field of characteristic 2, and let $[a, b)$ be the quaternion algebra of Remark 1.1.8. Show that the following are equivalent:
   - $[a, b) \cong M_2(k)$.
   - $[a, b)$ is not a division algebra.
   - The element b is a norm from the extension $k(\alpha)|k$, where α is a root of the equation $x^2 + x = a$.
   - The projective conic $ax^2 + by^2 = z^2 + zx$ has a k-rational point.

5. Let again k be a field of characteristic 2, and consider the $k(t)$-algebra $[a, t)$.
   (a) By analogy with Example 1.3.8, show that $[a, t)$ is split if and only if $a = u^2 + u$ for some $u \in k$.

(b) Using Exercise 4, deduce that the conic $x^2 + ty^2 = z^2 + zx$ over the field $\mathbf{F}_2(t)$ has no $\mathbf{F}_2(t)$-rational point, whereas it has a rational point over the purely inseparable extension $\mathbf{F}_2(\sqrt{t})$.

6. Determine those prime numbers p for which the quaternion algebra $(-1, p)$ is split over the field $\mathbf{Q}$ of rational numbers.

7. (Chain lemma) Assume that the quaternion algebras (a, b) and (c, d) are isomorphic. Show that there exists an $e \in k^\times$ such that

$$(a, b) \cong (e, b) \cong (e, d) \cong (c, d).$$

[*Hint:* Consider the symmetric bilinear form $B(q_1, q_2) := \frac{1}{2}(q_1 \bar{q}_2 + q_2 \bar{q}_1)$ on the subspace $B_0 \subset (a, b)$ of elements $q \in (a, b)$ satisfying $q + \bar{q} = 0$. Note that $i, j, I, J \in B_0$, where $1, i, j, ij$ and $1, I, J, IJ$ are the standard bases of $(a, b) \cong (c, d)$ with $i^2 = a$, $j^2 = b$, $ij = -ji$ and $I^2 = c$, $J^2 = d$, $IJ = -JI$. Take an element $\varepsilon \in B_0 \setminus \{0\}$ with $B(\varepsilon, j) = B(\varepsilon, J) = 0$ and set $e = \varepsilon^2$.]

2

Central simple algebras and Galois descent

In this chapter we treat the basic theory of central simple algebras from a modern viewpoint. The main point we would like to emphasize is that, as a consequence of Wedderburn's theorem, we may characterize central simple algebras as those finite-dimensional algebras which become isomorphic to some full matrix ring over a finite extension of the base field. We then show that this extension can in fact be chosen to be a Galois extension, which enables us to exploit a powerful theory in our further investigations, that of Galois descent. Using descent we can give elegant treatments of such classical topics as the construction of reduced norms or the Skolem–Noether theorem. The main invariant concerning central simple algebras is the Brauer group, which classifies all finite-dimensional central division algebras over a field. Using Galois descent, we shall identify it with a certain first cohomology set equipped with an abelian group structure.

The foundations of the theory of central simple algebras go back to the great algebraists of the dawn of the twentieth century; we merely mention here the names of Wedderburn, Dickson and Emmy Noether. The Brauer group appears in the pioneering paper of the young Richard Brauer [1]. Though Galois descent had been implicitly used by algebraists in the early years of the twentieth century and Châtelet had considered special cases in connection with Diophantine equations, it was André Weil who first gave a systematic treatment with applications to algebraic geometry in mind (Weil [2]). The theory in the form presented below was developed by Jean-Pierre Serre, and finally found a tantalizing generalization in the general descent theory of Grothendieck ([1], [2]).

2.1 Wedderburn's theorem

Let k be a field. We assume throughout that all k-algebras under consideration are finite dimensional over k. A k-algebra A is called *simple* if it has

19

no (two-sided) ideal other than 0 and A. Recall moreover from the previous chapter that A is *central* if its centre equals k.

Here are the basic examples of central simple algebras.

Example 2.1.1 A division algebra over k is obviously simple. Its centre is a field (indeed, inverting the relation $xy = yx$ gives $y^{-1}x^{-1} = x^{-1}y^{-1}$ for all $y \in D, x \in Z(D)$). Hence D is a central simple algebra over $Z(D)$.

As concrete examples (besides fields), we may cite nonsplit quaternion algebras: these are central over k by definition and division algebras by Proposition 1.1.7.

The next example shows that split quaternion algebras are also simple.

Example 2.1.2 If D is a division algebra over k, the ring $M_n(D)$ of $n \times n$ matrices over D is simple for all $n \geq 1$. Checking this is an exercise in matrix theory. Indeed, we have to show that the two-sided ideal $\langle M \rangle$ in $M_n(D)$ generated by a nonzero matrix M is $M_n(D)$ itself. Consider the matrices E_{ij} having 1 as the j-th element of the i-th row and zero elsewhere. Since each element of $M_n(D)$ is a D-linear combination of the E_{ij}, it suffices to show that $E_{ij} \in \langle M \rangle$ for all i, j. But in view of the relation $E_{ki}E_{ij}E_{jl} = E_{kl}$ we see that it is enough to show $E_{ij} \in \langle M \rangle$ for *some* i, j. Now choose i, j so that the j-th element in the i-th row of M is a nonzero element m. Then $m^{-1}E_{ii}ME_{jj} = E_{ij}$, and we are done.

Noting the easy fact that in a matrix ring the centre can only contain scalar multiples of the identity matrix, we get that $M_n(D)$ is a central simple algebra over $Z(D)$.

The main theorem on simple algebras over a field provides a converse to the above example.

Theorem 2.1.3 (Wedderburn) *Let A be a finite-dimensional simple algebra over a field k. Then there exist an integer $n \geq 1$ and a division algebra $D \supset k$ so that A is isomorphic to the matrix ring $M_n(D)$. Moreover, the division algebra D is uniquely determined up to isomorphism.*

The proof will follow from the next two lemmas. Before stating them, let us recall some basic facts from module theory. First, a nonzero A-module M is *simple* if it has no A-submodules other than 0 and M.

Example 2.1.4 Let us describe the simple left modules over $M_n(D)$, where D is a division algebra. For all $1 \leq r \leq n$, consider the left ideal $I_r \subset M_n(D)$

formed by matrices $M = [m_{ij}]$ with $m_{ij} = 0$ for $j \neq r$. A simple argument with the matrices E_{ij} of Example 2.1.2 shows that the I_r are *minimal* left ideals with respect to inclusion, i.e. simple $M_n(D)$-modules. Moreover, we have $M_n(D) = \bigoplus I_r$ and the I_r are all isomorphic as $M_n(D)$-modules. Finally, if M is a simple $M_n(D)$-module, it must be a quotient of $M_n(D)$, but then the induced map $\bigoplus I_r \to M$ must induce an isomorphism with some I_r. Thus all simple left $M_n(D)$-modules are isomorphic to (say) I_1.

Next, an *endomorphism* of a left A-module M over a ring A is an A-homomorphism $M \to M$; these form a ring $\mathrm{End}_A(M)$ where addition is given by the rule $(\phi+\psi)(x) = \phi(x)+\psi(x)$ and multiplication by composition of maps. If A is a k-algebra, then so is $\mathrm{End}_A(M)$, for multiplication by an element of k defines an element in the centre of $\mathrm{End}_A(M)$. In the case when A is a division algebra, M is a left vector space over A, so the usual argument from linear algebra shows that choosing a basis of M induces an isomorphism of $\mathrm{End}_A(M)$ with a matrix algebra. More precisely, define the *opposite algebra* A° of A as the k-algebra with the same underlying k-vector space as A, but in which the product of two elements x, y is given by the element yx with respect to the product in A. Then $\mathrm{End}_A(M) \cong M_n(A^\circ)$, where n is the dimension of M over A.

The module M is equipped with a left module structure over $\mathrm{End}_A(M)$, multiplication being given by the rule $\phi \cdot x = \phi(x)$ for $x \in M, \phi \in \mathrm{End}_A(M)$.

Lemma 2.1.5 (Schur) *Let M be a simple module over a k-algebra A. Then* $\mathrm{End}_A(M)$ *is a division algebra.*

Proof The kernel of a nonzero endomorphism $M \to M$ is an A-submodule different from M, hence it is 0. Similarly, its image must be the whole of M. Thus it is an isomorphism, which means it has an inverse in $\mathrm{End}_A(M)$. □

Now let M be a left A-module with endomorphism ring $E = \mathrm{End}_A(M)$. As remarked above, M is naturally a left E-module, hence one may also consider the endomorphism ring $\mathrm{End}_E(M)$. One defines a ring homomorphism $\lambda_M :$ $A \to \mathrm{End}_E(M)$ by sending $a \in A$ to the endomorphism $x \mapsto ax$ of M. This is indeed an E-endomorphism, for if $\phi : M \to M$ is an element of E, one has $\phi \cdot ax = \phi(ax) = a\phi(x) = a\phi \cdot x$ for all $x \in M$.

Lemma 2.1.6 (Rieffel) *Let L be a nonzero left ideal in a simple k-algebra A, and put $E = \mathrm{End}_A(L)$. Then the map $\lambda_L : A \to \mathrm{End}_E(L)$ defined above is an isomorphism.*

Note that in a ring A a left ideal is none but a submodule of the left A-module A.

Proof   Since $\lambda_L \neq 0$, its kernel is a proper two-sided ideal of A. But A is simple, so λ_L is injective. For surjectivity, we show first that $\lambda_L(L)$ is a left ideal in $\mathrm{End}_E(L)$. Indeed, take $\phi \in \mathrm{End}_E(L)$ and $l \in L$. Then $\phi \cdot \lambda_L(l)$ is the map $x \mapsto \phi(lx)$. But for all $x \in L$, the map $y \mapsto yx$ is an A-endomorphism of L, i.e. an element of E. As ϕ is an E-endomorphism, we have $\phi(lx) = \phi(l)x$, and so $\phi \cdot \lambda_L(l) = \lambda_L(\phi(l))$.

Now observe that the right ideal LA generated by L is a two-sided ideal, hence $LA = A$. In particular, we have $1 = \sum l_i a_i$ with $l_i \in L$, $a_i \in A$. Hence for $\phi \in \mathrm{End}_E(L)$ we have $\phi = \phi \cdot 1 = \phi\lambda_L(1) = \sum \phi\lambda_L(l_i)\lambda_L(a_i)$. But since $\lambda_L(L)$ is a left ideal, we have here $\phi\lambda_L(l_i) \in \lambda_L(L)$ for all i, and thus $\phi \in \lambda_L(A)$. $\square$

Proof of Theorem 2.1.3   As A is finite dimensional, a descending chain of left ideals must stabilize. So let L be a minimal left ideal; it is then a simple A-module. By Schur's lemma, $E = \mathrm{End}_A(L)$ is a division algebra, and by Rieffel's lemma we have an isomorphism $A \cong \mathrm{End}_E(L)$. The discussion before Lemma 2.1.5 then yields an isomorphism $\mathrm{End}_E(L) \cong M_n(E^\circ)$, where n is the dimension of L over E (it is finite as L is already finite dimensional over k). Setting $D = E^\circ$ we thus get an isomorphism $A \cong M_n(D)$.

For the unicity statement, assume that D and D' are division algebras for which $A \cong M_n(D) \cong M_m(D')$ with suitable integers n, m. By Example 2.1.4, the minimal left ideal L then satisfies $D^n \cong L \cong D'^m$, whence a chain of isomorphisms $D \cong \mathrm{End}_A(D^n) \cong \mathrm{End}_A(L) \cong \mathrm{End}_A(D'^m) \cong D'$. $\square$

Corollary 2.1.7   *Let k be an algebraically closed field. Then every central simple k-algebra is isomorphic to $M_n(k)$ for some $n \geq 1$.*

Proof   By the theorem it is enough to see that there is no finite-dimensional division algebra $D \supset k$ other than k. For this, let d be an element of $D\backslash k$. As in the proof of Corollary 1.2.1 we see that there is an irreducible polynomial $f \in k[x]$ and a k-algebra homomorphism $k[x]/(f) \to D$ whose image contains d. But k being algebraically closed, we have $k[x]/(f) \cong k$. $\square$

2.2 Splitting fields

The last corollary enables one to give an alternative characterization of central simple algebras.

Theorem 2.2.1 *Let k be a field and A a finite-dimensional k-algebra. Then A is a central simple algebra if and only if there exist an integer $n > 0$ and a finite field extension $K|k$ so that $A \otimes_k K$ is isomorphic to the matrix ring $M_n(K)$.*

We first prove:

Lemma 2.2.2 *Let A be a finite-dimensional k-algebra, and $K|k$ an algebraic field extension. The algebra A is central simple over k if and only if $A \otimes_k K$ is central simple over K.*

Proof If I is a nontrivial (two-sided) ideal of A, then $I \otimes_k K$ is a nontrivial ideal of $A \otimes_k K$ (e.g. for dimension reasons); similarly, if A is not central, then neither is $A \otimes_k K$. Thus if $A \otimes_k K$ is central simple, then so is A.

To prove the converse, we first reduce to the case of a finite extension $K|k$. For this, we write $K|k$ as a union of its finite subextensions $K'|K$ and observe that every ideal $I \subset A \otimes_k K$ is the union of the ideals $I \cap K' \subset A \otimes_k K'$; moreover, we have $Z(A \otimes_k K) \subset Z(A \otimes_k K')$ for any K'. Next, using Wedderburn's theorem we may assume that $A = D$ is a division algebra. Under this assumption, if $w_1, \dots, w_n$ is a k-basis of K, then $1 \otimes w_1, \dots, 1 \otimes w_n$ yields a D-basis of $D \otimes_k K$ as a left D-vector space. Given an element $x = \sum \alpha_i (1 \otimes w_i)$ in the centre of $D \otimes_k K$, for all $d \in D$ the relation $x = (d^{-1} \otimes 1) x (d \otimes 1) = \sum (d^{-1}\alpha_i d)(1 \otimes w_i)$ implies $d^{-1}\alpha_i d = \alpha_i$ by the linear independence of the $1 \otimes w_i$. As D is central over k, the α_i must lie in k, so $D \otimes_k K$ is central over K. Now if J is a nonzero ideal in $D \otimes_k K$ generated by elements $z_1, \dots, z_r$, we may assume the z_i to be D-linearly independent and extend them to a D-basis of $D \otimes_k K$ by adjoining some of the $1 \otimes w_i$, say $1 \otimes w_{r+1}, \dots, 1 \otimes w_n$. Thus for $1 \leq i \leq r$ we may write

$$1 \otimes w_i = \sum_{j=r+1}^{n} \alpha_{ij}(1 \otimes w_j) + y_i,$$

where y_i is some D-linear combination of the z_i and hence an element of J. Here $y_1, \dots, y_r$ are D-linearly independent (because so are $1 \otimes w_1, \dots, 1 \otimes w_r$), so they form a D-basis of J. As J is a two-sided ideal, for all $d \in D$ we must have $d^{-1}y_i d \in J$ for $1 \leq i \leq r$, so there exist $\beta_{il} \in D$ with $d^{-1}y_i d = \sum \beta_{il} y_l$. We may rewrite this relation as

$$(1 \otimes w_i) - \sum_{j=r+1}^{n} (d^{-1}\alpha_{ij}d)(1 \otimes w_j) = \sum_{l=1}^{r} \beta_{il}(1 \otimes w_l) - \sum_{l=1}^{r} \beta_{il} \sum_{j=r+1}^{n} \alpha_{lj}(1 \otimes w_j),$$

from which we get as above, using the independence of the $1 \otimes w_j$, that $\beta_{ii} = 1$, $\beta_{il} = 0$ for $l \neq i$ and $d^{-1}\alpha_{ij}d = \alpha_{ij}$, i.e. $\alpha_{ij} \in k$ as D is central. This means that J can be generated by elements of K (viewed as a k-subalgebra of $D \otimes_k K$ via the embedding $w \mapsto 1 \otimes w$). As K is a field, we must have $J \cap K = K$, so $J = D \otimes_k K$. This shows that $D \otimes_k K$ is simple. $\qquad \square$

Proof of Theorem 2.2.1 Sufficiency follows from the above lemma and Example 2.1.2. For necessity, note first that denoting by $\bar{k}$ an algebraic closure of k, the lemma together with Corollary 2.1.7 imply that $A \otimes_k \bar{k} \cong M_n(\bar{k})$ for some n. Now observe that for every finite field extension K of k contained in $\bar{k}$, the inclusion $K \subset \bar{k}$ induces an injective map $A \otimes_k K \to A \otimes_k \bar{k}$ and $A \otimes_k \bar{k}$ arises as the union of the $A \otimes_k K$ in this way. Hence for a sufficiently large finite extension $K|k$ contained in $\bar{k}$ the algebra $A \otimes_k K$ contains the elements $e_1, \ldots, e_{n^2} \in A \otimes_k \bar{k}$ corresponding to the standard basis elements of $M_n(\bar{k})$ via the isomorphism $A \otimes_k \bar{k} \cong M_n(\bar{k})$, and moreover the elements a_{ij} occurring in the relations $e_i e_j = \sum a_{ijl} e_l$ defining the product operation are also contained in K. Mapping the e_i to the standard basis elements of $M_n(K)$ then induces a K-isomorphism $A \otimes_k K \cong M_n(K)$. $\qquad \square$

Corollary 2.2.3 *If A is a central simple k-algebra, its dimension over k is a square.*

Definition 2.2.4 A field extension $K|k$ over which $A \otimes_k K$ is isomorphic to $M_n(K)$ for suitable n is called a *splitting field* for A. We shall also employ the terminology A *splits over* K or K *splits* A.

The integer $\sqrt{\dim_k A}$ is called the *degree* of A.

Given Theorem 2.2.1, we can easily prove:

Lemma 2.2.5 *If A and B are central simple k-algebras split by K, then so is $A \otimes_k B$.*

Proof In view of the isomorphism $(A \otimes_k K) \otimes_K (B \otimes_k K) \cong (A \otimes_k B) \otimes_k K$ and Theorem 2.2.1, it is enough to verify the isomorphism of matrix algebras $M_n(K) \otimes_K M_m(K) \cong M_{nm}(K)$. This was done in Lemma 1.5.1. $\qquad \square$

An important example is given by the opposite algebra A° of a k-algebra A, introduced in the previous section. If A is central simple over k, then so is A°. By the lemma, their tensor product is again central simple, but more is true:

Proposition 2.2.6 *There is a canonical isomorphism $A \otimes_k A^\circ \xrightarrow{\sim} \mathrm{End}_k(A)$ of k-algebras. Consequently, $A \otimes_k A^\circ$ is isomorphic to the matrix algebra $M_{n^2}(k)$, where n is the degree of A.*

Proof Define a k-linear map $A \otimes_k A^\circ \to \mathrm{End}_k(A)$ by sending $\sum a_i \otimes b_i$ to the k-linear map $x \mapsto \sum a_i x b_i$. This map is manifestly nonzero, and hence injective, because $A \otimes_k A^\circ$ is simple by Lemma 2.2.5. Thus it is an isomorphism for dimension reasons. $\qquad\square$

We now come to a theorem that will be crucial for our considerations to come.

Theorem 2.2.7 *Every central division algebra D of degree n over an infinite field k is split by a separable extension $K|k$ of degree n. Moreover, such a K may be found among the k-subalgebras of D.*

Remark 2.2.8 Over a finite field every central simple algebra is split. This statement is equivalent to another famous theorem of Wedderburn according to which every finite division algebra is commutative. We shall give a proof of this theorem in Remark 6.2.7; see also Weil [3], Chapter I, Theorem 1 for a short elementary proof due to Witt. Without using the theorem, we can still make a straightforward remark: since a finite field is perfect, every central simple algebra over it is split by a finite separable extension.

Theorem 2.2.7 is an immediate consequence of Propositions 2.2.9 and 2.2.10 below that are interesting in their own right.

Proposition 2.2.9 *If a central simple k-algebra A of degree n contains a k-subalgebra K which is a degree n field extension of k, then A splits over K.*

Proof Let A° be the opposite algebra to A. By Proposition 2.2.6 we have an isomorphism $A \otimes_k A^\circ \cong \mathrm{End}_k(A)$. If K is as above, the inclusion $K \subset A$ induces an inclusion $K \subset A^\circ$ by commutativity of K, whence also an injection $\iota : A \otimes_k K \to \mathrm{End}_k(A)$. Viewing A as a K-vector space with K acting via right multiplication, the construction of the map $A \otimes_k A^\circ \to \mathrm{End}_k(A)$ in the proof of Proposition 2.2.6 shows that the image of ι lies in $\mathrm{End}_K(A)$. By definition, we have $\mathrm{End}_K(A) \cong M_n(K)$; in particular, it has dimension n^2 over K. On the other hand, we have $\dim_K(A \otimes_k K) = \dim_k(A) = n^2$, so the map $\iota : A \otimes_k K \to \mathrm{End}_K(A)$ is an isomorphism. $\qquad\square$

Proposition 2.2.10 *If D is as in the theorem, then D contains an element $a \in D$ such that the field extension $k(a)|k$ is separable of degree n.*

Proof By Corollary 2.1.7 there is an isomorphism $\phi : D \otimes_k \bar{k} \cong M_d(\bar{k})$. Identify $M_d(\bar{k})$ with points of the affine space $\mathbf{A}_{\bar{k}}^{d^2}$ via the standard basis, and consider the subset U of matrices with separable characteristic polynomial. These form a Zariski open set because if we identify the set of degree d monic polynomials in $\bar{k}[x]$ with points of $\mathbf{A}_{\bar{k}}^d$ via $x^d + a_{d-1}x^{d-1} + \cdots + a_0 \mapsto (a_{d-1}, \ldots, a_0)$ the separable polynomials correspond to the Zariski open set given by the nonvanishing of the discriminant; the set U is the preimage of this open set by the morphism $\mathbf{A}_{\bar{k}}^{d^2} \to \mathbf{A}_{\bar{k}}^d$ sending a matrix to its characteristic polynomial, modulo the above identifications. Now identify the elements of $D \otimes_k \bar{k}$ with the points of $\mathbf{A}_{\bar{k}}^{d^2}$ in such a way that the elements of D correspond to the k-points of the affine space $\mathbf{A}_k^{d^2}$. As k is infinite, the open subset $\phi^{-1}(U) \subset \mathbf{A}_{\bar{k}}^{d^2}$ contains a k-rational point of $\mathbf{A}^{d^2}$ (this fact holds for every nonempty Zariski open subset and is promptly verified by reducing to the case $d = 1$). This point in turn corresponds to an element $a \in D$. By construction, over $\bar{k}$ it yields a matrix $M = \phi(a \otimes 1)$ whose characteristic polynomial P has distinct roots, hence P is also the minimal polynomial of M. This minimal polynomial is the same as that of the $\bar{k}$-linear extension L_M of the left multiplication map $L_a : D \to D$, $x \mapsto ax$ to $M_d(\bar{k})$ via ϕ, as L_M is given via left multiplication by the block diagonal matrix $\mathrm{diag}(M, \ldots, M)$. But the minimal polynomial does not change by base extension (Lang [3], Chapter XIV, Corollary 2.2), so the k-linear map L_a also has the separable polynomial P as its minimal polynomial; in particular, P has coefficients in k. Finally, the minimal polynomial of the map L_a is the same as the minimal polynomial of $a \in D$ over k, which is irreducible as D is a division algebra. We conclude as in the proof of Proposition 1.2.1. $\quad\square$

Corollary 2.2.11 (Noether, Köthe) *A central simple k-algebra has a splitting field that is finite and separable over k.*

Proof Combine Theorem 2.2.7 (and Remark 2.2.8) with Wedderburn's theorem. $\quad\square$

Corollary 2.2.12 *A finite-dimensional k-algebra A is a central simple algebra if and only if there exist an integer $n > 0$ and a finite Galois field extension $K|k$ so that $A \otimes_k K$ is isomorphic to the matrix ring $M_n(K)$.*

Proof The 'if' part is contained in Theorem 2.2.1. The 'only if' part follows from the previous corollary together with the well-known fact from Galois theory according to which every finite separable field extension embeds in a finite Galois extension. □

Remarks 2.2.13

1. It is important to bear in mind that if A is a central simple k-algebra of degree n which does not split over k but splits over a finite Galois extension $K|k$ with group G, then the isomorphism $A \otimes_k K \cong M_n(K)$ is *not* G-equivariant if we equip $M_n(K)$ with the usual action of G coming from its action on K. Indeed, were it so, we would get an isomorphism $A \cong M_n(k)$ by taking G-invariants.
2. It is not always possible to realize a Galois splitting field as a k-subalgebra in a central simple algebra, as shown by a famous counterexample by Amitsur (see Amitsur [2] or Pierce [1]; see also Brussel [1] for counterexamples over $\mathbf{Q}(t)$ and $\mathbf{Q}((t))$). Central simple algebras containing a Galois splitting field are called *crossed products* in the literature.

2.3 Galois descent

Corollary 2.2.12 makes it possible to classify central simple algebras using methods of Galois theory. Here we present such a method, known as *Galois descent*.

We shall work in a more general context, that of *vector spaces V equipped with a tensor Φ of type (p, q)*. By definition, Φ is an element of the tensor product $V^{\otimes p} \otimes_k (V^*)^{\otimes q}$, where $p, q \geq 0$ are integers and V^* is the dual space $\mathrm{Hom}_k(V, k)$. Note the natural isomorphism

$$V^{\otimes p} \otimes_k (V^*)^{\otimes q} \cong \mathrm{Hom}_k(V^{\otimes q}, V^{\otimes p})$$

coming from the general formula $\mathrm{Hom}_k(V, k) \otimes_k W \cong \mathrm{Hom}_k(V, W)$.

Examples 2.3.1 The following special cases will be the most important for us:

• The trivial case $\Phi = 0$ (with any p, q). This is just V with no additional structure.
• $p = 1, q = 1$. In this case Φ is given by a k-linear endomorphism of V.
• $p = 0, q = 2$. Then Φ is a sum of tensor products of k-linear functions, i.e. a k-bilinear form $V \otimes_k V \to k$.
• $p = 1, q = 2$. This case corresponds to a k-bilinear map $V \otimes_k V \to V$.

Note that the theory of associative algebras is contained in the last example, for the multiplication in such an algebra A is given by a k-bilinear map $A \otimes_k A \to A$ satisfying the associativity condition.

So consider pairs (V, Φ) of k-vector spaces equipped with a tensor of fixed type (p, q) as above. A k-isomorphism between two such objects (V, Φ) and (W, Ψ) is given by a k-isomorphism $f : V \xrightarrow{\sim} W$ of k-vector spaces such that $f^{\otimes q} \otimes (f^{*-1})^{\otimes q} : V^{\otimes p} \otimes_k (V^*)^{\otimes q} \to W^{\otimes p} \otimes_k (W^*)^{\otimes q}$ maps Φ to Ψ. Here $f^* : W^* \xrightarrow{\sim} V^*$ is the k-isomorphism induced by f.

Now fix a finite Galois extension $K|k$ with Galois group $G = \mathrm{Gal}\,(K|k)$. Denote by V_K the K-vector space $V \otimes_k K$ and by Φ_K the tensor induced on V_K by Φ. In this way we associate with (V, Φ) a K-object (V_K, Φ_K). We say that (V, Φ) and (W, Ψ) *become isomorphic over* K if there exists a K-isomorphism between (V_K, Φ_K) and (W_K, Ψ_K). In this situation, (W, Ψ) is also called a $(K|k)$-*twisted form* of (V, Φ) or a *twisted form* for short.

Now Galois theory enables one to classify k-isomorphism classes of twisted forms as follows. Given a k-automorphism $\sigma : K \to K$, tensoring by V gives a k-automorphism $V_K \to V_K$, which we again denote by σ. Each K-linear map $f : V_K \to W_K$ induces a map $\sigma(f) : V_K \to W_K$ defined by $\sigma(f) = \sigma \circ f \circ \sigma^{-1}$. If f is a K-isomorphism from (V_K, Φ_K) to (W_K, Ψ_K), then so is $\sigma(f)$. The map $f \to \sigma(f)$ preserves composition of automorphisms, hence we get a *left action* of $G = \mathrm{Gal}\,(K|k)$ on the group $\mathrm{Aut}_K(\Phi)$ of K-automorphisms of (V_K, Φ_K). Moreover, given two k-objects (V, Φ) and (W, Ψ) as well as a K-isomorphism $g : (V_K, \Phi_K) \xrightarrow{\sim} (W_K, \Psi_K)$, one gets a map $G \to \mathrm{Aut}_K(\Phi)$ associating $a_\sigma = g^{-1} \circ \sigma(g)$ to $\sigma \in G$. The map a_σ satisfies the fundamental relation

$$a_{\sigma\tau} = a_\sigma \cdot \sigma(a_\tau) \quad \text{for all} \quad \sigma, \tau \in G. \tag{2.1}$$

Indeed, we compute

$$a_{\sigma\tau} = g^{-1} \circ \sigma(\tau(g)) = g^{-1} \circ \sigma(g) \circ \sigma(g^{-1}) \circ \sigma(\tau(g)) = a_\sigma \cdot \sigma(a_\tau).$$

Next, let $h : (V_K, \Phi_K) \xrightarrow{\sim} (W_K, \Psi_K)$ be another K-isomorphism, defining $b_\sigma := h^{-1} \circ \sigma(h)$ for $\sigma \in G$. Then a_σ and b_σ are related by

$$a_\sigma = c^{-1} b_\sigma \sigma(c), \tag{2.2}$$

where c is the K-automorphism $h^{-1} \circ g$. We abstract this in a general definition:

Definition 2.3.2 Let G be a group and A another (not necessarily commutative) group on which G acts on the left, i.e. there is a map $G \times A \to A$ sending a pair $(\sigma, a) \in G \times A$ to $\sigma(a) \in A$ so that the equalities $\sigma(ab) = \sigma(a)\sigma(b)$

and $\sigma\tau(a) = \sigma(\tau(a))$ hold for all $\sigma, \tau \in G$ and $a, b \in A$. Then a *1-cocycle* of G with values in A is a map $\sigma \mapsto a_\sigma$ from G to A satisfying the relation (2.1) above. Two 1-cocycles a_σ and b_σ are called *equivalent* or *cohomologous* if there exists $c \in A$ such that the relation (2.2) holds.

One defines the *first cohomology set* $H^1(G, A)$ of G with values in A as the quotient of the set of 1-cocycles by the equivalence relation (2.2). It is a *pointed set*, i.e. a set equipped with a distinguished element coming from the trivial cocycle $\sigma \mapsto 1$, where 1 is the identity element of A. We call this element the *base point*.

In our concrete situation, we see that the class $[a_\sigma]$ in $H^1(G, \operatorname{Aut}_K(\Phi))$ of the 1-cocycle a_σ associated with the K-isomorphism $g : (V_K, \Phi_K) \xrightarrow{\sim} (W_K, \Psi_K)$ depends only on (W, Ψ) but not on the map g. This enables us to state the main theorem of this section.

Theorem 2.3.3 *For a k-object (V, Φ) consider the pointed set $TF_K(V, \Phi)$ of twisted $(K|k)$-forms of (V, Φ), the base point being given by (V, Φ). Then the map $(W, \Psi) \to [a_\sigma]$ defined above yields a base point preserving bijection*

$$TF_K(V, \Phi) \leftrightarrow H^1(G, \operatorname{Aut}_K(\Phi)).$$

Before proving the theorem, we give some immediate examples, leaving the main application (that to central simple algebras) to the next section.

Example 2.3.4 (Hilbert's Theorem 90) Consider first the case when V has dimension n over k and Φ is the trivial tensor. Then $\operatorname{Aut}_K(\Phi)$ is just the group $\operatorname{GL}_n(K)$ of invertible $n \times n$ matrices. On the other hand, two n-dimensional vector k-spaces that are isomorphic over K are isomorphic already over k, so we get:

$$H^1(G, \operatorname{GL}_n(K)) = \{1\}. \tag{2.3}$$

This statement is due to Speiser. The case $n = 1$ is usually called Hilbert's Theorem 90 in the literature, though Hilbert only considered the case when $K|k$ is a cyclic extension of degree n. In this case, denoting by σ a generator of $G = \operatorname{Gal}(K|k)$, every 1-cocycle is determined by its value a_σ on σ. Applying the cocycle relation (2.1) inductively we get $a_{\sigma^i} = a_\sigma \sigma(a_\sigma) \ldots \sigma^{i-1}(a_\sigma)$ for all $1 \le i \le n$. In particular, for $i = n$ we get $a_\sigma \sigma(a_\sigma) \ldots \sigma^{n-1}(a_\sigma) = a_1 = 1$ (here the second equality again follows from the cocycle relation applied with $\sigma = \tau = 1$). But $a_\sigma \sigma(a_\sigma) \ldots \sigma^{n-1}(a_\sigma)$ is by definition the norm of a_σ for the extension $K|k$. Now formula (2.3) together with the coboundary relation (2.2) imply the original form of Hilbert's Theorem 90:

In a cyclic field extension $K|k$ with $\mathrm{Gal}\,(K|k) = <\sigma>$ each element of norm 1 is of the form $\sigma(c)c^{-1}$ with some $c \in K$.

Example 2.3.5 (Quadratic forms)   As another example, assume k is of characteristic different from 2, and take V to be n-dimensional and Φ a tensor of type $(0,2)$ coming from a nondegenerate symmetric bilinear form $<,>$ on V. Then $\mathrm{Aut}_K(\Phi)$ is the group $O_n(K)$ of orthogonal matrices with respect to $<,>$ and we get from the theorem that there is a base point preserving bijection

$$TF_K(V, <, >) \leftrightarrow H^1(G, O_n(K)).$$

This bijection is important for the classification of quadratic forms.

To prove the theorem, we construct an inverse to the map $(W, \Psi) \mapsto [a_\sigma]$. This is based on the following general construction.

Construction 2.3.6   Let A be a group equipped with a left action by another group G. Suppose further that X is a set on which both G and A act in a compatible way, i.e. we have $\sigma(a(x)) = (\sigma(a))(\sigma(x))$ for all $x \in X$, $a \in A$ and $\sigma \in G$. Assume finally given a 1-cocycle $\sigma \mapsto a_\sigma$ of G with values in A. Then we define the *twisted action of G on X by the cocycle a_σ* via the rule

$$(\sigma, x) \mapsto a_\sigma(\sigma(x)).$$

This is indeed a G-action, for the cocycle relation yields

$$a_{\sigma\tau}(\sigma\tau(x)) = a_\sigma\sigma(a_\tau)(\sigma\tau(x)) = a_\sigma\sigma(a_\tau\tau(x)).$$

If X is equipped with some algebraic structure (e.g. it is a group or a vector space), and G and A act on it by automorphisms, then the twisted action is also by automorphisms. The notation ${}_aX$ will mean X equipped with the twisted G-action by the cocycle a_σ.

Remark 2.3.7   Readers should be warned that the above construction can only be carried out on the level of cocycles and *not* on that of cohomology classes: equivalent cocycles give rise to different twisted actions in general. For instance, take $G = \mathrm{Gal}\,(K|k)$, $A = X = \mathrm{GL}_n(K)$, acting on itself by inner automorphisms. Then twisting the usual G-action on $\mathrm{GL}_n(K)$ by the trivial cocycle $\sigma \mapsto 1$ does not change anything, whereas if $\sigma \mapsto a_\sigma$ is a 1-cocycle with a_σ a noncentral element for some σ, then $a_\sigma^{-1}\sigma(x)a_\sigma \neq \sigma(x)$ for a noncentral x, so the twisted action is different. But a 1-cocycle $G \to \mathrm{GL}_n(K)$ is equivalent to the trivial cocycle by Example 2.3.4.

Now the idea is to take a cocycle a_σ representing some cohomology class in $H^1(G, \mathrm{Aut}_K(\Phi))$ and to apply the above construction with $G = \mathrm{Gal}\,(K|k)$, $A = \mathrm{Aut}_K(\Phi)$ and $X = V_K$. The main point is then to prove that taking the invariant subspace $(_a V_K)^G$ under the twisted action of G yields a twisted form of (V, Φ).

We show this first when Φ is trivial (i.e. we in fact prove Hilbert's Theorem 90). The statement to be checked then boils down to:

Lemma 2.3.8 (Speiser) *Let $K|k$ be a finite Galois extension with group G, and V a K-vector space equipped with a* semi-linear *G-action, i.e. a G-action satisfying*

$$\sigma(\lambda v) = \sigma(\lambda)\sigma(v) \quad \text{for all } \sigma \in G, v \in V \text{ and } \lambda \in K.$$

Then the natural map

$$\lambda : V^G \otimes_k K \to V$$

is an isomorphism, where the superscript G denotes invariants under G.

Before proving the lemma, let us recall a consequence of Galois theory. Let $K|k$ be a Galois extension as in the lemma, and consider two copies of K, the first one equipped with trivial G-action, and the second one with the action of G as the Galois group. Then the tensor product $K \otimes_k K$ (endowed with the G-action given by $\sigma(a \otimes b) \cong a \otimes \sigma(b)$) decomposes as a direct sum of copies of K:

$$K \otimes_k K \cong \bigoplus_{\sigma \in G} K e_\sigma, \tag{2.4}$$

where G acts on the right-hand side by permuting the basis elements e_σ. To see this, write $K = k[x]/(f)$ with f some monic irreducible polynomial $f \in k[x]$, and choose a root α of f in K. As $K|k$ is Galois, f splits in $K[x]$ as a product of linear terms of the form $(x - \sigma(\alpha))$ for $\sigma \in G$. Thus using a special case of the Chinese Remainder Theorem for rings (which is easy to prove directly) we get

$$K \otimes_k K \cong K[x]/(f) \cong K[x]/(\prod_{\sigma \in G}(x - \sigma(\alpha)) \cong \bigoplus_{\sigma \in G} K[x]/(x - \sigma(\alpha)),$$

whence a decomposition of the required form.

Proof Consider the tensor product $V \otimes_k K$, where the second factor K carries trivial G-action and V the G-action of the lemma. It will be enough to prove that the map $\lambda_K : (V \otimes_k K)^G \otimes_k K \to V \otimes_k K$ is an isomorphism. Indeed, by our assumption about the G-actions we have $(V \otimes_k K)^G \cong$

$V^G \otimes_k K$, and hence we may identify λ_K with the map $(V^G \otimes_k K) \otimes_k K \to V \otimes_k K$ obtained by tensoring with K. Therefore if λ had a nontrivial kernel A (resp. a nontrivial cokernel B), then λ_K would have a nontrivial kernel $A \otimes_k K$ (resp. a nontrivial cokernel $B \otimes_k K$).

Now observe that the decomposition (2.4) of $K \otimes_k K$ induces a decomposition of the $K \otimes_k K$-module $V \otimes_k K$ as a direct sum of K-vector spaces $e_\sigma(V \otimes_k K)$. Identifying $e_1(V \otimes_k K)$ with a K-vector space W with trivial G-action, we obtain a $K[G]$-module isomorphism $V \otimes_k K \cong W^{\oplus |G|}$, with G acting on the right-hand side by permutation of the factors. Under this identification the elements in $(V \otimes_K K)^G$ correspond to the diagonal elements $(w, \dots, w)$ with some $w \in W$. On the other hand, multiplication by e_σ on $V \otimes_k K$ corresponds to setting the components of a vector in $W^{\oplus |G|}$ to 0 except the one indexed by σ. It follows that the $K \otimes_k K$-submodule of $V \otimes_k K$ generated by $(V \otimes_k K)^G$ contains all elements corresponding to vectors in $W^{\oplus |G|}$ of the form $(0, \dots, 0, w, 0, \dots, 0)$. This shows that $\lambda_K : (V \otimes_k K)^G \otimes_k K \to V \otimes_k K$ is surjective, and injectivity follows from the injectivity of the diagonal map $W \to W^{\oplus |G|}$.   □

Proof of Theorem 2.3.3   As indicated above, we take a 1-cocycle a_σ representing some cohomology class in $H^1(G, \mathrm{Aut}_K(\Phi))$ and consider the invariant subspace $W := ({}_a V_K)^G$. Next observe that $\sigma(\Phi_K) = \Phi_K$ for all $\sigma \in G$ (as Φ_K comes from the k-tensor Φ) and also $a_\sigma(\Phi_K) = \Phi_K$ for all $\sigma \in G$ (as $a_\sigma \in \mathrm{Aut}_K(\Phi)$). Hence $a_\sigma \sigma(\Phi_K) = \Phi_K$ for all $\sigma \in G$, which means that Φ_K comes from a k-tensor on W. Denoting this tensor by Ψ, we have defined a k-object (W, Ψ). Speiser's lemma yields an isomorphism $W \otimes_k K \cong V_K$, and by construction this isomorphism identifies Ψ_K with Φ_K. Thus (W, Ψ) is indeed a twisted form of (V, Φ). If $a_\sigma = c^{-1} b_\sigma \sigma(c)$ with some 1-cocycle $\sigma \mapsto b_\sigma$ and $c \in \mathrm{Aut}_K(\Phi)$, we get from the definitions $({}_b V_K)^G = c(W)$, which is a k-vector space isomorphic to W. To sum up, we have a well-defined map $H^1(G, \mathrm{Aut}_K(\Phi)) \to TF_K(V, \Phi)$. The kind reader will check that this map is the inverse of the map $(W, \Psi) \mapsto [a_\sigma]$ of the theorem.   □

Remark 2.3.9   There is an obvious variant of the above theory, where instead of a single tensor Φ one considers a whole family of tensors on V. The K-automorphisms to be considered are then those preserving all tensors in the family, and twisted forms are vector spaces W isomorphic to V over K such that the family of tensors on W_K goes over to that on V_K via the K-isomorphism. The descent theorem in this context is stated and proven in the same way as Theorem 2.3.3.

2.4 The Brauer group

Now we come to the classification of central simple algebras. First we recall a well-known fact about matrix rings:

Lemma 2.4.1 *Over a field K all automorphisms of the matrix ring $M_n(K)$ are inner, i.e. given by $M \mapsto CMC^{-1}$ for some invertible matrix C.*

Proof Consider the minimal left ideal I_1 of $M_n(K)$ described in Example 2.1.4, and take an automorphism $\lambda \in \mathrm{Aut}(M_n(K))$. Replacing λ by a conjugate with a suitable matrix, we may actually assume $\lambda(I_1) = I_1$. Let $e_1, \ldots, e_n$ be the standard basis of K^n. Mapping a matrix $M \in I_1$ to Me_1 induces an isomorphism $I_1 \cong K^n$ of K-vector spaces, and thus λ induces an automorphism of K^n. As such, it is given by an invertible matrix C. We get that for all $M \in M_n(K)$, the endomorphism of K^n defined in the standard basis by $\lambda(M)$ has matrix CMC^{-1}, whence the lemma. $\square$

Corollary 2.4.2 *The automorphism group of $M_n(K)$ is the projective general linear group $\mathrm{PGL}_n(K)$.*

Proof There is a natural homomorphism $\mathrm{GL}_n(K) \to \mathrm{Aut}(M_n(K))$ mapping $C \in \mathrm{GL}_n(K)$ to the automorphism $M \mapsto CMC^{-1}$. It is surjective by the lemma, and its kernel consists of the centre of $\mathrm{GL}_n(K)$, i.e. the subgroup of scalar matrices. $\square$

Now take a finite Galois extension $K|k$ as before, and let $CSA_K(n)$ denote the set of k-isomorphism classes of central simple k-algebras of degree n split by K. We regard it as a pointed set, the base point being the class of the matrix algebra $M_n(k)$.

Theorem 2.4.3 *There is a base point preserving bijection*

$$CSA_K(n) \leftrightarrow H^1(G, \mathrm{PGL}_n(K)).$$

Proof By Corollary 2.2.12 the central simple k-algebras of degree n are precisely the twisted forms of the matrix algebra $M_n(k)$. To see this, note that as explained in Example 2.3.1, an n^2-dimensional k-algebra can be considered as an n^2-dimensional k-vector space equipped with a tensor of type (1,2) satisfying the associativity condition. But on a twisted form of $M_n(k)$ the tensor defining the multiplication automatically satisfies the associativity condition. Hence Theorem 2.3.3 applies and yields a bijection of pointed sets $CSA_K(n) \leftrightarrow H^1(G, \text{Aut}(M_n(K)))$. The theorem now follows by Corollary 2.4.2. $\qquad\qquad\qquad\qquad\qquad\qquad\qquad\qquad\qquad\qquad\qquad\square$

Our next goal is to classify all central simple k-algebras split by K by means of a single cohomology set. It should carry a product operation, since by virtue of Lemma 2.2.5 the tensor product induces a natural commutative and associative product operation

$$CSA_K(n) \times CSA_K(m) \to CSA_K(mn). \qquad (2.5)$$

Via the bijection of Theorem 2.4.3 we obtain a corresponding product operation

$$H^1(G, \text{PGL}_n(K)) \times H^1(G, \text{PGL}_m(K)) \to H^1(G, \text{PGL}_{nm}(K)) \qquad (2.6)$$

on cohomology sets. To define this product directly, note that the map

$$\text{End}_K(K^n) \otimes \text{End}_K(K^m) \to \text{End}_K(K^n \otimes K^m)$$

given by $(\phi, \psi) \mapsto \phi \otimes \psi$ restricts to a product operation

$$\text{GL}_n(K) \times \text{GL}_m(K) \to \text{GL}_{nm}(K)$$

on invertible matrices which preserves scalar matrices, whence a product

$$\text{PGL}_n(K) \times \text{PGL}_m(K) \to \text{PGL}_{nm}(K).$$

This induces a natural product on cocycles, whence the required product operation (2.6).

Next observe that for all $n, m > 0$ there are natural injective maps $\text{GL}_n(K) \to \text{GL}_{nm}(K)$ mapping a matrix $M \in \text{GL}_n(K)$ to the block matrix given by m copies of M placed along the diagonal and zeroes elsewhere. As usual, these pass to the quotient modulo scalar matrices and finally induce maps

$$\lambda_{mn} : H^1(G, \text{PGL}_m(K)) \to H^1(G, \text{PGL}_{mn}(K))$$

on cohomology. Via the bijection of Theorem 2.4.3, the class of a central simple algebra A in $H^1(G, \text{PGL}_m(K))$ is mapped to the class of $A \otimes_k M_n(k)$ by λ_{mn}.

Lemma 2.4.4 *The maps λ_{mn} are injective for all $m, n > 0$.*

Proof Assume A and A' are central simple k-algebras with $A \otimes_k M_n(k) \cong A' \otimes_k M_n(k)$. By Wedderburn's theorem they are matrix algebras over division algebras D and D', respectively, hence so are $A \otimes_k M_n(k)$ and $A' \otimes_k M_n(k)$.

But then $D \cong D'$ by the unicity statement in Wedderburn's theorem, so finally $A \cong A'$ by dimension reasons. □

The lemma prompts the following construction.

Construction 2.4.5 Two central simple k-algebras A and A' are called *Brauer equivalent* or *similar* if $A \otimes_k M_m(k) \cong A' \otimes_k M_{m'}(k)$ for some $m, m' > 0$. This defines an equivalence relation on the union of the sets $CSA_K(n)$. We denote the set of equivalence classes by Br $(K|k)$ and the union of the sets Br $(K|k)$ for all finite Galois extensions by Br (k).

Remarks 2.4.6 Brauer equivalence enjoys the following basic properties.

1. One sees from the definition that each Brauer equivalence class contains (up to isomorphism) a unique division algebra. Thus we can also say that Br $(K|k)$ classifies division algebras split by K.
2. It follows from Wedderburn's theorem and the previous remark that if A and B are two Brauer equivalent k-algebras of the same dimension, then $A \cong B$.

The set Br $(K|k)$ (and hence also Br (k)) is equipped with a product operation induced by the product (2.5) on central simple k-algebras; indeed, the tensor product of k-algebras manifestly preserves Brauer equivalence.

Proposition 2.4.7 *The sets* Br $(K|k)$ *and* Br (k) *equipped with the above product operation are abelian groups.*

Proof Basic properties of the tensor product imply that the product operation is commutative and associative. If A represents a class in Br $(K|k)$, the class of the opposite algebra A° yields an inverse in view of Proposition 2.2.6. □

Definition 2.4.8 We call Br $(K|k)$ equipped with the above product operation the *Brauer group of k relative to K* and Br (k) the *Brauer group* of k.

Now define the set $H^1(G, \mathrm{PGL}_\infty)$ as the union for all n of the sets $H^1(G, \mathrm{PGL}_n(K))$ via the inclusion maps λ_{mn}, equipped with the product operation coming from (2.6) (which is manifestly compatible with the maps λ_{mn}). Also, observe that for a Galois extension $L|k$ containing K, the natural surjection Gal $(L|k) \to$ Gal $(K|k)$ induces injective maps

$$H^1(\mathrm{Gal}\,(K|k), \mathrm{PGL}_n(K)) \rightarrow H^1(\mathrm{Gal}\,(L|k), \mathrm{PGL}_n(K))$$

for all n, and hence also injections

$$\iota_{LK} : H^1(\mathrm{Gal}\,(K|k), \mathrm{PGL}_\infty) \rightarrow H^1(\mathrm{Gal}\,(L|k), \mathrm{PGL}_\infty).$$

Fixing a separable closure k_s of k, we define $H^1(k, \mathrm{PGL}_\infty)$ as the union over all Galois extensions $K|k$ contained in k_s of the groups $H^1(\mathrm{Gal}\,(K|k), \mathrm{PGL}_\infty)$ via the inclusion maps ι_{LK}. The arguments above then yield:

Proposition 2.4.9   *The sets $H^1(G, \mathrm{PGL}_\infty)$ and $H^1(k, \mathrm{PGL}_\infty)$ equipped with the product operation coming from (2.6) are abelian groups, and there are natural group isomorphisms*

$$\mathrm{Br}\,(K|k) \cong H^1(G, \mathrm{PGL}_\infty) \quad and \quad \mathrm{Br}\,(k) \cong H^1(k, \mathrm{PGL}_\infty).$$

Remark 2.4.10   The sets $H^1(G, \mathrm{PGL}_\infty)$ are not cohomology sets of G in the sense defined so far, but may be viewed as cohomology sets of G with values in the *direct limit* of the groups $\mathrm{PGL}_n(K)$ via the maps λ_{mn}. Still, this coefficient group is fairly complicated. Later we shall identify $\mathrm{Br}\,(K|k)$ with the second cohomology *group* of G with values in the multiplicative group $K^\times$, a group that is much easier to handle.

2.5 Cyclic algebras

We are now in the position to introduce a class of algebras that will play a central role in this book, that of cyclic algebras. These are generalizations of quaternion algebras to arbitrary degree. As we shall see below, in the case when the base field contains a primitive m-th root of unity ω, a degree m cyclic algebra may be described by the presentation

$$\langle x, y \mid x^m = a, \ y^m = b, \ xy = \omega y x \rangle$$

generalizing that of quaternion algebras in the case $m = 2$. However, it is not straightforward to show that such a presentation defines a central simple algebra. We therefore start with a general construction yielding certain twisted forms of matrix algebras, and then derive the above presentation as a special case.

Construction 2.5.1 (Cyclic algebras)   Let $K|k$ be a cyclic Galois extension with Galois group $G \cong \mathbf{Z}/m\mathbf{Z}$. In the sequel we fix one such isomorphism

$\chi : G \xrightarrow{\sim} \mathbf{Z}/m\mathbf{Z}$; it is a character of G. Furthermore, let $b \in k^\times$ be given. We associate with these data a central simple algebra over k which is a $K|k$-twisted form of the matrix algebra $M_m(k)$. To do so, consider the matrix

$$\widetilde{F}(b) = \begin{bmatrix} 0 & 0 & \cdots & 0 & b \\ 1 & 0 & \cdots & 0 & 0 \\ 0 & 1 & \cdots & 0 & 0 \\ \vdots & & \ddots & & \vdots \\ 0 & 0 & \cdots & 1 & 0 \end{bmatrix} \in \mathrm{GL}_m(k).$$

We denote by $F(b)$ its image in the group $\mathrm{PGL}_m(k)$. A computation shows that $\widetilde{F}(b)^m = b \cdot I_m$, and hence $F(b)^m = 1$; in fact, the element $F(b)$ has exact order m in $\mathrm{PGL}_m(k)$.

Now consider the homomorphism $\mathbf{Z}/m\mathbf{Z} \to \mathrm{PGL}_m(k)$ defined by sending 1 to $F(b)$. Embedding $\mathrm{PGL}_m(k)$ into $\mathrm{PGL}_m(K)$ and composing by χ we thus get a 1-cocycle

$$z(b) : G \to \mathrm{PGL}_m(K).$$

We now equip the matrix algebra $M_m(K)$ with the twisted G-action $_{z(b)}M_m(K)$ coming from $z(b)$ (see Construction 2.3.6) and take G-invariants. By Theorem 2.4.3 (and its proof), the resulting k-algebra is a central simple algebra split by K. We denote it by (χ, b), and call it the *cyclic algebra* associated with χ and b.

We now come to the definition of cyclic algebras originally proposed by Dickson.

Proposition 2.5.2   *The algebra (χ, b) can be described by the following presentation. There is an element $y \in (\chi, b)$ such that (χ, b) is generated as a k-algebra by K and y, subject to the relations*

$$y^m = b, \quad \lambda y = y\sigma(\lambda) \tag{2.7}$$

for all $\lambda \in K$, where σ is the generator of G mapped to 1 by χ.

In particular, we see that K is a commutative k-subalgebra in (χ, b) which is *not* contained in the centre.

Proof   Denote by A the k-algebra given by the presentation of the proposition and define a k-algebra homomorphism $j : A \to M_m(K)$ by setting

$$j(y) = \widetilde{F}(b) \quad \text{and} \quad j(\lambda) = \mathrm{diag}(\lambda, \sigma(\lambda), \cdots, \sigma^{m-1}(\lambda)) \quad \text{for } \lambda \in K$$

(where diag($\dots$) means the diagonal matrix with the indicated entries), and extending k-linearly. To see that this is indeed a homomorphism, one checks by direct computation that the relation

$$j(\lambda)\widetilde{F}(b) = \widetilde{F}(b)j(\sigma(\lambda)) \tag{2.8}$$

holds for all $\lambda \in K$; the relation $\widetilde{F}(b)^m = b$ has already been noted above. Next we check that the image of j lands in (χ, b). For this, recall that by definition the elements of $_{z(b)}M_m(K)^G$ are those matrices M which satisfy $\widetilde{F}(b)\sigma(M)\widetilde{F}(b)^{-1} = M$. This relation is obviously satisfied by $j(y) = \widetilde{F}(b)$ as it is in $M_m(k)$, and for the $j(\lambda)$ it follows from relation (2.8) above, which proves the claim. Finally, we have to check that j is an isomorphism. For dimension reasons it is enough to check surjectivity, which in turn can be done after tensoring by K. The image of $j \otimes \mathrm{id}_K$ in $(\chi, b) \otimes_k K \cong M_m(K)$ is the K-subalgebra generated by $\widetilde{F}(b)$ and the diagonal subalgebra $K \oplus \cdots \oplus K$. If $E_{i,j}$ is the usual basis of $M_m(K)$, it therefore remains to check that the $E_{i,j}$'s belong to this subalgebra for $i \neq j$. This is achieved by computing $E_{i,j} = \widetilde{F}(b)^{i-j}E_{j,j}$ for $i \neq j$. $\qquad\square$

The following proposition provides a kind of a converse to the previous one.

Proposition 2.5.3 *Assume that A is a central simple k-algebra of degree m containing a k-subalgebra K which is a cyclic Galois field extension of degree m. Then A is isomorphic to a cyclic algebra given by a presentation of the form (2.7).*

A more general statement with a conceptual proof will be given in Corollary 4.7.7. We include a computational proof here. The crucial point is the following statement.

Lemma 2.5.4 *Under the assumptions of the proposition there exists $y \in A^\times$ such that*

$$y^{-1}xy = \sigma(x)$$

for all $x \in K$, where σ is a generator of $G = \mathrm{Gal}\,(K|k)$.

Proof In order to avoid confusing notation, we take another extension $\widetilde{K}$ of k isomorphic to K and put $\widetilde{G} := \mathrm{Gal}\,(\widetilde{K}|k)$. The algebra $A \otimes_k \widetilde{K}$ is split by Proposition 2.2.9. The embedding $K \otimes_k \widetilde{K} \to A \otimes_k \widetilde{K}$ is $\widetilde{G}$-equivariant, where $\widetilde{G}$ acts on $K \otimes_k \widetilde{K}$ via the second factor. On the other hand, the group G acts on $K \otimes_k \widetilde{K}$ via the first factor, and the two actions commute. As seen before the proof of Lemma 2.3.8, under the isomorphism $K \otimes_k \widetilde{K} \cong \widetilde{K}^m$ the action of G corresponds to permuting the components on the right-hand side. Under the diagonal embedding $K \otimes_k \widetilde{K} \to A \otimes_k \widetilde{K} \cong M_m(\widetilde{K})$ we may identify permutation of the components of the diagonal with conjugation by a permutation matrix, so we find an element $y \in \mathrm{GL}_m(\widetilde{K}) \cong (A \otimes_k \widetilde{K})^\times$ satisfying

$$\sigma(x) = y^{-1} x y \text{ for all } x \in K \otimes_k \widetilde{K}. \tag{2.9}$$

We now show that we may choose y in the subgroup $A^\times \subset (A \otimes_k \widetilde{K})^\times$, which will conclude the proof of the lemma.

For all $\widetilde{\tau} \in \widetilde{G}$ and $x \in K$ (where we view K embedded into $K \otimes_k \widetilde{K}$ via the first factor), we have

$$\sigma(x) = \sigma(\widetilde{\tau}(x)) = \widetilde{\tau}(\sigma(x)) = \widetilde{\tau}(y^{-1})\widetilde{\tau}(x)\widetilde{\tau}(y) = \widetilde{\tau}(y)^{-1} x \widetilde{\tau}(y),$$

using that the two actions commute and that $\widetilde{G}$ acts trivially on K. Thus $z_{\widetilde{\tau}} := y \widetilde{\tau}(y)^{-1}$ satisfies $z_{\widetilde{\tau}}^{-1} x z_{\widetilde{\tau}} = x$ for all $x \in K$. It follows that $z_{\widetilde{\tau}}$ lies in $Z_A(K) \otimes_k \widetilde{K}$, where $Z_A(K)$ stands for the centralizer of K in A. The natural embedding $K \to Z_A(K)$ is an isomorphism, as one sees by passing to the split case and counting dimensions. Thus the function $\widetilde{\tau} \mapsto z_{\widetilde{\tau}}$ has values in $(K \otimes_k \widetilde{K})^\times$, and moreover it is a 1-cocycle for $\widetilde{G}$ by construction.

Now observe that the group $H^1(\widetilde{G}, (K \otimes_k \widetilde{K})^\times)$ is trivial. Indeed, the group $(K \otimes_k \widetilde{K})^\times$ is the automorphism group of the $K \otimes_k \widetilde{K}$-algebra $K \otimes_k \widetilde{K}$, so by Theorem 2.3.3 the group $H^1(\widetilde{G}, (K \otimes_k \widetilde{K})^\times)$ classifies those K-algebras B for which $B \otimes_k \widetilde{K} \cong K \otimes_k \widetilde{K}$. But these K-algebras must be isomorphic to K by dimension reasons, whence the claim. In view of this claim we find $y_0 \in (K \otimes_k \widetilde{K})^\times$ such that $y\widetilde{\tau}(y)^{-1} = y_0\widetilde{\tau}(y_0)^{-1}$ for all $\widetilde{\tau} \in \widetilde{G}$. Up to replacing y by $y_0^{-1}y$ in the equation (2.9), we may thus assume that $\widetilde{\tau}(y) = y$ for all $\widetilde{\tau}$, i.e. $y \in A^\times$, as required. $\square$

Proof of Proposition 2.5.3 We first prove that the element y of the previous lemma satisfies $y^m \in k$. To see this, apply formula (2.9) to $\sigma(x)$ in place of x, with $x \in K$. It yields $\sigma^2(x) = y^{-2}xy^2$, so iterating $m - 1$ times we obtain $x = \sigma^m(x) = y^{-m}xy^m$. Thus y^m commutes with all $x \in K$ and hence lies in K by the equality $Z_A(K) = K$ noted above. Now apply (2.9) with $x = y^m$ to obtain $\sigma(y^m) = y^m$, i.e. $y^m \in k$.

Setting $b := y^m$, to conclude the proof it remains to show that the elements of K and the powers of y generate A. For this it suffices to check that the elements $1, y, \ldots, y^{m-1}$ are K-linearly independent in A, where K acts by right multiplication. If not, take a nontrivial K-linear relation $\Sigma y^i \lambda_i = 0$ with a minimal number of nonzero coefficients. After multiplying by a power of

y we may assume λ_0 and some other λ_j are not 0. Choose $c \in K^\times$ with $c \neq \sigma(c)$. Using equation (2.9) and its iterates we may write $\Sigma y^i \sigma^i(c) \lambda_i = c \left(\Sigma y^i \lambda_i \right) = 0$. It follows that $\Sigma y^i (c\lambda_i - \sigma^i(c)\lambda_i) = 0$ is a shorter nontrivial relation, a contradiction. $\square$

In special cases one gets even nicer presentations for cyclic algebras. One of these is when m is invertible in k, and k contains a primitive m-th root of unity ω. In this case, for $a, b \in k^\times$ define the k-algebra $(a, b)_\omega$ by the presentation

$$(a, b)_\omega = \langle x, y \, | \, x^m = a, \; y^m = b, \; xy = \omega yx \rangle.$$

In the case $m = 2, \omega = -1$ one gets back the generalized quaternion algebras of the previous chapter.

Another case is when k is of characteristic $p > 0$ and $m = p$. In this case for $a \in k$ and $b \in k^\times$ consider the presentation

$$[a, b) = \langle x, y \, | \, x^p - x = a, \; y^p = b, \; xy = y(x + 1) \rangle.$$

Note that the equation $x^p - x = a$ defines a cyclic Galois extension of degree p whose Galois group is given by the substitutions $\alpha \mapsto \alpha + i \; (0 \leq i \leq p-1)$ for some root α. In the case $p = 2$ this definition is coherent with that of Remark 1.1.8.

Corollary 2.5.5

1. *Assume that k contains a primitive m-th root of unity and that we may write K in the form $K = k(\sqrt[m]{a})$ with some m-th root of an element $a \in k$. Let $\chi : Gal(K|k) \cong \mathbf{Z}/m\mathbf{Z}$ be the isomorphism sending the automorphism $\sigma : \sqrt[m]{a} \mapsto \omega \sqrt[m]{a}$ to 1. Then for all $b \in k^\times$ there is an isomorphism of k-algebras*

$$(a, b)_\omega \cong (\chi, b).$$

2. *Similarly, assume that k has characteristic $p > 0$, $m = p$ and $K|k$ is a cyclic Galois extension defined by a polynomial $x^p - x + a$ for some $a \in k$. Fix a root α of $x^p - x - a$ and let $\chi : Gal(K|k) \cong \mathbf{Z}/p\mathbf{Z}$ be the isomorphism sending the automorphism $\sigma : \alpha \mapsto \alpha + 1$ to 1. Then for all $b \in k^\times$ there is an isomorphism of k-algebras*

$$[a, b) \cong (\chi, b).$$

In particular, $(a, b)_\omega$ and $[a, b)$ are central simple algebras split by K.

Proof In (1), one gets the required isomorphism by choosing as generators of (χ, b) the element $x = \sqrt[m]{a}$ and the y given by the proposition above. In (2), one chooses $x = \alpha$ and y as in the proposition. $\square$

Remark 2.5.6 In fact, we shall see later that according to Kummer theory (Corollary 4.3.9) in the presence of a primitive m-th root of unity one may write an arbitrary degree m cyclic Galois extension $K|k$ in the form $K = k(\sqrt[m]{a})$, as in the corollary above. Similarly, Artin–Schreier theory (Remark 4.3.13 (1)) shows that a cyclic Galois extension of degree p in characteristic $p > 0$ is generated by a root of some polynomial $x^p - x - a$.

In the previous chapter we have seen that the class of a nonsplit quaternion algebra has order 2 in the Brauer group. More generally, the class of a cyclic division algebra $(a, b)_\omega$ as above has order m; we leave the verification of this fact as an exercise to the reader. Thus the class of a tensor product of degree m cyclic algebras has order dividing m in the Brauer group. The remarkable fact is the converse:

Theorem 2.5.7 (Merkurjev–Suslin) *Assume that k contains a primitive m-th root of unity ω. Then a central simple k-algebra whose class has order dividing m in $\mathrm{Br}\,(k)$ is Brauer equivalent to a tensor product*

$$(a_1, b_1)_\omega \otimes_k \cdots \otimes_k (a_i, b_i)_\omega$$

of cyclic algebras.

This generalizes Merkurjev's theorem from the end of Chapter 1. In fact, Merkurjev and Suslin found this generalization soon after the first result of Merkurjev. It is this more general statement whose proof will occupy a major part of this book.

Remark 2.5.8 One cannot replace 'Brauer equivalence' with 'isomorphism' in the theorem. We have quoted a counterexample with $m = 2$ at the end of Chapter 1; for examples with m an odd prime and $i = 2$, see Jacob [1] and Tignol [1].

Here is an interesting corollary of the Merkurjev–Suslin theorem of which no elementary proof is known presently.

Corollary 2.5.9 *For k and A as in the theorem above, there exist elements $a_1, ..., a_i \in k^\times$ such that the extension $k(\sqrt[m]{a_1}, ..., \sqrt[m]{a_i})|k$ splits A. In particular, A is split by a Galois extension with solvable Galois group.*

2.6 Reduced norms and traces

We now discuss a construction which generalizes the quaternion norm encountered in the previous chapter.

Construction 2.6.1 (Reduced norms and traces)   Let A be a central simple k-algebra of degree n. Take a finite Galois splitting field $K|k$ with group G, and choose a K-isomorphism $\phi : M_n(K) \xrightarrow{\sim} A \otimes_k K$. Recall that the isomorphism ϕ is not compatible with the action of G. However, if we twist the usual action of G on $M_n(K)$ by the 1-cocycle $\sigma \mapsto a_\sigma$ with $a_\sigma = \phi^{-1} \circ \sigma(\phi)$ associated with A by the descent construction, then we get an isomorphism $_aM_n(K) \xrightarrow{\sim} A \otimes_k K$ that is already G-equivariant, whence an isomorphism $(_aM_n(K))^G \cong A$.

Now consider the determinant map $\det : M_n(K) \to K$. For all $\sigma \in G$, lifting a_σ to an invertible matrix $C_\sigma \in \mathrm{GL}_n(K)$ we get

$$\det(C_\sigma \sigma(M) C_\sigma^{-1}) = \det(\sigma(M)) = \sigma(\det(M)) \qquad (2.10)$$

by multiplicativity of the determinant and its compatibility with the usual G-action. Bearing in mind that the twisted G-action on $_aM_n(K)$ is given by $(\sigma, M) \to a_\sigma \sigma(M) a_\sigma^{-1}$, this implies that the map $\det : {}_aM_n(K) \to K$ is compatible with the action of G. So by taking G-invariants and using the isomorphism above we get a map $\mathrm{Nrd} : A \to k$, called the *reduced norm map*. On the subgroup $A^\times$ of invertible elements of A it restricts to a group homomorphism $\mathrm{Nrd} : A^\times \to k^\times$.

The above construction does not depend on the choice of ϕ, for changing ϕ amounts to replacing a_σ by an equivalent cocycle, i.e. replacing the matrix C_σ above by some $D^{-1}C_\sigma \sigma(D)$, which does not affect the expression in (2.10). The construction does not depend on the choice of K either, as one sees by embedding two Galois splitting fields K, L into a bigger Galois extension $M|k$.

By performing the above construction using the trace of matrices instead of the determinant, one gets a homomorphism $\mathrm{Trd} : A \to k$ of additive groups called the *reduced trace map*.

The reduced norm map is a generalization of the norm map for quaternion algebras, as one sees from Proposition 1.2.4. Just like the quaternion norm, it enjoys the following property:

Proposition 2.6.2   *In a central simple k-algebra A an element $a \in A$ is invertible if and only if $\mathrm{Nrd}(a) \neq 0$. Hence A is a division algebra if and only if Nrd restricts to a nowhere vanishing map on $A \setminus 0$.*

Proof If a is invertible, it corresponds to an invertible matrix via any iso-morphism $\phi : A \otimes_K K \cong M_n(K)$, which thus has nonzero determinant. For the converse, consider ϕ as above and assume an element $a \in A$ maps to a matrix with nonzero determinant. It thus has an inverse $b \in M_n(K)$. Now in any ring the multiplicative inverse of an element is unique (indeed, if b' is another inverse, one has $b = bab' = b'$), so for an automorphism $\sigma_A \in \mathrm{Aut}_k(A \otimes_k K)$ coming from the action of an element $\sigma \in \mathrm{Gal}\,(K|k)$ on K we have $\sigma_A(b) = b$. As A is the set of fixed elements of all the σ_A, this implies $b \in A$. □

We now elucidate the relation with other norm and trace maps. Recall that given a finite-dimensional k-algebra A, the *norm* and the *trace* of an element $a \in A$ are defined as follows: one considers the k-linear mapping $L_a : A \to A$ given by $L_a(x) = ax$ and puts

$$N_{A|k}(a) := \det(L_a), \qquad \mathrm{tr}_{A|k}(a) := \mathrm{tr}\,(L_a).$$

By definition, these norm and trace maps are insensitive to change of the base field.

Proposition 2.6.3 *Let A be a central simple k-algebra of degree n.*

1. *One has $N_{A|k} = (\mathrm{Nrd}_A)^n$ and $\mathrm{tr}_{A|k} = n\,\mathrm{Trd}_A$.*
2. *Assume that K is a commutative k-subalgebra of A which is a degree n field extension of k. For any $x \in K$ one has*

$$\mathrm{Nrd}_A(x) = N_{K|k}(x) \quad and \quad \mathrm{Trd}_A(x) = \mathrm{tr}_{K|k}(x).$$

Proof To prove (1) we may assume, up to passing to a splitting field of A, that $A = M_n(k)$. The required formulae then follow from the fact that for $M \in M_n(k)$, the matrix of the multiplication-by-M map L_M with respect to the standard basis of $M_n(k)$ is the block diagonal matrix $\mathrm{diag}(M, \ldots, M)$.

To check (2), note first that as a K-vector space the algebra A is isomorphic to the direct power K^n. For $x \in K$ we thus have $N_{A|k}(x) = (N_{K|k}(x))^n$ and $\mathrm{tr}_{A|k}(x) = n\,\mathrm{tr}_{K|k}(x)$. By part (1) there exists an n-th root of unity $\omega(x)$ such that $\mathrm{Nrd}_A(x) = \omega(x)N_{K|k}(x)$. To show that $\omega(x) = 1$ we use the following trick. Performing base change from k to $k(t)$ and applying the previous formula to $t + x \in K(t)^\times$ yields the equality

$$\mathrm{Nrd}_A(t + x) = \omega(t + x)N_{K|k}(t + x).$$

Since $\mathrm{Nrd}_A(t + x)$ and $N_{K|k}(t + x)$ are monic polynomials in t, we obtain $\omega(t + x) = 1$, and therefore $\mathrm{Nrd}_A(t + x) = N_{K|k}(t + x)$. We then get the desired formula $\mathrm{Nrd}_A(x) = N_{K|k}(x)$ by specializing this polynomial identity to $t = 0$. To handle the trace formula $\mathrm{Trd}_A(x) = \mathrm{tr}_{K|k}(x)$, it then suffices to look at the coefficients of t in the polynomial identity $\mathrm{Nrd}_A(1 + tx) = N_{K|k}(1 + tx)$.   □

In the remainder of this section we investigate the image of the reduced norm map $\mathrm{Nrd} : A^\times \to k^\times$ for a central simple k-algebra A. The first observation is that the image depends only on the class of A in $\mathrm{Br}\,(k)$.

Lemma 2.6.4   *If A and B are Brauer equivalent central simple k-algebras, then*

$$\mathrm{Nrd}(A^\times) = \mathrm{Nrd}(B^\times).$$

Before embarking on the proof, let us recall some facts from linear algebra.

Facts 2.6.5   Given a not necessarily commutative ring R, a matrix in $\mathrm{GL}_n(R)$ is called *elementary* if all of its diagonal entries are equal to 1 and moreover it has at most one nonzero off-diagonal entry. We denote by $E_{ij}(r)$ the elementary matrix with r in the i-th row and j-th column. Left and right multiplication of a given matrix $M \in M_n(R)$ by elementary matrices correspond to the usual row/column operations on M.

When M is upper (resp. lower) triangular with 1's in the diagonal, it can be reduced to the identity matrix by row operations annihilating above-diagonal (resp. below-diagonal) entries one by one. Hence such an M is a product of elementary matrices over an arbitrary ring R.

If moreover R is a division ring and $M \in M_n(R)$, then the same proof as over fields shows that M can be reduced to upper triangular form by row/column operations. Together with the remark of the previous paragraph this implies that for a division ring R the group $\mathrm{GL}_n(R)$ is generated by diagonal matrices and elementary matrices.

Proof of Lemma 2.6.4   In view of Wedderburn's theorem it is enough to prove that for a central division algebra D the equality $\mathrm{Nrd}(D^\times) = \mathrm{Nrd}(\mathrm{GL}_n(D))$ holds for all $n \geq 1$. By the facts recalled above, for this it will suffice to verify the following equalities:

1. $\mathrm{Nrd}(\mathrm{diag}(d_1, \ldots, d_n)) = \mathrm{Nrd}(d_1) \ldots \mathrm{Nrd}(d_n)$ for $d_1, \ldots, d_n \in D^\times$;
2. $\mathrm{Nrd}(E_{ij}(d)) = 1$ for all elementary matrices in $\mathrm{GL}_n(D)$.

To do so, we may pass to a splitting field $K|k$ of D. The choice of a K-isomorphism $D \otimes_k K \cong M_d(K)$ induces isomorphisms

$$\mathrm{GL}_n(D \otimes_k K) \cong \mathrm{GL}_n(M_d(K)) \cong \mathrm{GL}_{nd}(K)$$

that are compatible with taking reduced norms. The first equality then follows from the formula for the determinant of a block diagonal matrix, and the second from the fact that an elementary matrix in $\mathrm{GL}_n(D)$ maps in $\mathrm{GL}_{nd}(K)$ to an (upper or lower) triangular matrix with 1's in the diagonal. □

Next we prove a compatibility result for field extensions.

Proposition 2.6.6 *Let A be a central simple k-algebra over a field k. Given a finite field extension $K|k$, we have an inclusion*

$$N_{K|k}\big(\mathrm{Nrd}((A \otimes_k K)^\times)\big) \subset \mathrm{Nrd}(A^\times)$$

of subgroups in $k^\times$.

Proof Set $d := \deg(A)$ and $n = [K : k]$. The k-linear action of K on itself by multiplication defines an embedding $K \hookrightarrow \mathrm{End}_k(K)$, whence an embedding $\rho : K \hookrightarrow M_n(k)$ after fixing an isomorphism $\mathrm{End}_k(K) \cong M_n(k)$. Base changing by A yields an embedding $\rho_A : A \otimes_k K \hookrightarrow A \otimes_k M_n(k) \cong M_n(A)$. We shall prove the equality

$$N_{K|k} \circ \mathrm{Nrd}_{A \otimes_k K} = \mathrm{Nrd}_{M_n(A)} \circ \rho_A \qquad (2.11)$$

from which the proposition will follow in view of Lemma 2.6.4.

First we verify a compatibility for the algebra norms:

$$N_{K|k} \circ (N_{A \otimes_k K | K})^n = N_{M_n(A)|k} \circ \rho_A. \qquad (2.12)$$

To do so, it suffices to establish the equalities

$$(N_{A \otimes_k K | K})^n = N_{M_n(A)|K} \circ \rho_A \qquad (2.13)$$

and

$$N_{K|k} \circ N_{M_n(A)|K} = N_{M_n(A)|k}, \qquad (2.14)$$

with $M_n(A)$ viewed as a K-algebra via the composite map $K \hookrightarrow M_n(k) \to M_n(k) \otimes_k A$. The equality (2.13) results from the fact that $M_n(A)$ is a free left module of rank n over $A \otimes_k K$ via ρ_A (as so is $M_n(k)$ over K via ρ). Equality (2.14) is more difficult to prove; it is a general transitivity property of algebra norms for which we refer to Jacobson [2], vol. I, §7.4 or Cassels–Fröhlich [1], Chapter II, Appendix A.

Now formula (2.12) together with Proposition 2.6.3 (1) give

$$N_{K|k} \circ (\mathrm{Nrd}_{A \otimes_k K | K})^{dn} = (\mathrm{Nrd}_{M_n(A)})^{dn} \circ \rho_A,$$

so by multiplicativity of $N_{K|k}$ we have for all $x \in A \otimes_k K$ an equality

$$N_{K|k}(\mathrm{Nrd}_{A \otimes_k K}(x)) = \omega(x)\mathrm{Nrd}_{M_n(A)}(\rho_A(x))$$

with some dn-th root of unity $\omega(x)$. One shows $\omega(x) = 1$ by performing base change to $k(t)$ and using the same specialization argument as in the proof of Proposition 2.6.3 (2).   □

Corollary 2.6.7 *If K is a splitting field for A, then $N_{K|k}(K^\times) \subset \mathrm{Nrd}(A^\times)$.*

Proof For a split algebra over K the reduced norm map equals the determinant, and therefore its image is $K^\times$. Now apply the proposition.   □

We now have the following characterization of the image of the reduced norm map.

Proposition 2.6.8 *Let A be a central simple k-algebra. An element $c \in k^\times$ is a reduced norm from $A^\times$ if and only if $c \in N_{K|k}(K^\times)$ for a finite separable splitting field K of A that can be embedded in A as a k-subalgebra.*

For the proof we need a lemma.

Lemma 2.6.9 *Let A be a central division algebra of degree d defined over an infinite field k. The subset $S \subset A^\times$ of elements of the form λc with $\lambda \in k^\times$ and c an element of the commutator subgroup $[A^\times, A^\times]$ is Zariski dense in the affine space $\mathbf{A}_{\bar{k}}^{d^2}$ defined by the $\bar{k}$-vector space $A \otimes_k \bar{k}$.*

Proof As in the proof of Proposition 2.2.10, we first identify the elements of the matrix algebra $M_d(\bar{k})$ with $\mathbf{A}_{\bar{k}}^{d^2}$ by means of the standard basis. The elements of $\mathrm{GL}_d(\bar{k})$ correspond to the dense Zariski open subset $V \subset \mathbf{A}_{\bar{k}}^{d^2}$ given by the nonvanishing of the determinant. Since $\mathrm{SL}_d(\bar{k})$ is the commutator subgroup of $\mathrm{GL}_d(\bar{k})$ (see e.g. Lang [3], Chapter XIII, Theorems 8.3 and 9.2), we have $\mathrm{GL}_d(\bar{k}) = \bar{k}^\times \mathrm{SL}_d(\bar{k}) = \bar{k}^\times [\mathrm{GL}_d(\bar{k}), \mathrm{GL}_d(\bar{k})]$. Hence we may identify V with the image by the surjective map $\mu : \bar{k}^\times \times \mathrm{GL}_d(\bar{k}) \times \mathrm{GL}_d(\bar{k}) \to \mathrm{GL}_d(\bar{k})$ given by $(\lambda, x, y) \mapsto \lambda[x, y]$. Now choose an isomorphism $\phi : A \otimes_k \bar{k} \xrightarrow{\sim} M_d(\bar{k})$ and consider the affine space $\mathbf{A}_{\bar{k}}^{d^2}$ defined by $A \otimes_k \bar{k}$. As k is infinite, the subset $k^\times \times A^\times \times A^\times$ is Zariski dense in $\bar{k}^\times \times \phi^{-1}(V) \times \phi^{-1}(V)$ by the same argument as in the proof of Proposition 2.2.10. It follows that the image of $k^\times \times A^\times \times A^\times$ by the morphism $\phi^{-1} \circ \mu \circ \phi$ is dense in $\phi^{-1}(V)$. As ϕ is a $\bar{k}$-algebra isomorphism, this image is none but S.   □

Proof of Proposition 2.6.8 The 'if' statement follows from Corollary 2.6.7. To prove the 'only if' statement, we may assume using Lemma 2.6.4 that A is a division algebra. The case where $\deg(A) = 1$ is obvious, so we may also assume $d := \deg(A) > 1$; this case may only occur if k is infinite by Remark 2.2.8.

Now fix $a \in A^{\times}$. By the lemma, the set $S := k^{\times}[A^{\times}, A^{\times}]$ is Zariski dense in the affine d^2-space $\mathbf{A}_{\bar{k}}^{d^2}$ defined by $A \otimes_k \bar{k}$, hence so is $aS := \{as : s \in S\}$. As in the proof of Proposition 2.2.10, consider the open subset $\phi^{-1}(U) \subset \mathbf{A}_{\bar{k}}^{d^2}$ corresponding to matrices with separable characteristic polynomial via the isomorphism $\phi : A \xrightarrow{\sim} M_d(\bar{k})$. The intersection $aS \cap \phi^{-1}(U)$ is nonempty, and by the proof of Proposition 2.2.10 a point $as \in aS \cap \phi^{-1}(U)$ corresponds to an element of $A^{\times}$ such that the field extension $k(as)|k$ is separable of degree d. The field $K := k(as)$ is preserved if we multiply as by an element of $k^{\times}$, so we may assume $s \in [A^{\times}, A^{\times}]$. By Lemma 2.2.9 the subfield K splits A, and by Proposition 2.6.3 (2) we have $\mathrm{Nrd}(as) = N_{K|k}(as)$. But since $s \in [A^{\times}, A^{\times}]$, we have $\mathrm{Nrd}(s) = 1$ by multiplicativity of the reduced norm, whence $\mathrm{Nrd}(a) = \mathrm{Nrd}(as) = N_{K|k}(as)$ as required. □

The proposition together with Corollary 2.6.7 yields:

Corollary 2.6.10 *The subgroup* $\mathrm{Nrd}(A^{\times})$ *of* $k^{\times}$ *is generated by the subgroups* $N_{K|k}(K^{\times})$, *with* $K|k$ *running over the finite separable field extensions which split* A.

2.7 A basic exact sequence and applications

In this section we establish a formal proposition which, combined with the descent method, is a main tool in computations.

Proposition 2.7.1 *Let* G *be a group and*

$$1 \to A \to B \to C \to 1$$

an exact sequence of groups equipped with a G-action, *the maps being* G-homomorphisms. *Then there is an exact sequence of pointed sets*

$$1 \to A^G \to B^G \to C^G \to H^1(G, A) \to H^1(G, B) \to H^1(G, C).$$

By definition, an exact sequence of pointed sets is a sequence in which the kernel of each map equals the image of the previous one, the kernel being the subset of elements mapping to the base point. Note that, in contrast to the case of groups, a map with trivial kernel is *not* necessarily injective.

Proof The only nonobvious points are the definition of the map $\delta : C^G \to H^1(G, A)$ and the exactness of the sequence at the third and fourth terms. To define δ, take an element $c \in C^G$ and lift it to an element $b \in B$ via the surjection $B \to C$. For all $\sigma \in G$ the element $b^{-1}\sigma(b)$ maps to 1 in C because $c = \sigma(c)$ by assumption, so it lies in A. Immediate calculations then show that the map $\sigma \mapsto b^{-1}\sigma(b)$ is a 1-cocycle and that modifying b by an element of A yields an equivalent cocycle, whence a well-defined map δ as required, sending elements coming from B^G to 1. The relation $\delta(c) = 1$ means by definition that $b^{-1}\sigma(b) = a^{-1}\sigma(a)$ for some $a \in A$, so c lifts to the G-invariant element ba^{-1} in B. This shows the exactness of the sequence at the third term, and the composition $C^G \to H^1(G, A) \to H^1(G, B)$ is trivial by construction. Finally, that a cocycle $\sigma \mapsto a_\sigma$ with values in A becomes trivial in $H^1(G, B)$ means that $a_\sigma = b^{-1}\sigma(b)$ for some $b \in B$, and modifying $\sigma \mapsto a_\sigma$ by an A-coboundary we may choose b so that its image c in C is fixed by G; moreover, the cohomology class of $\sigma \mapsto a_\sigma$ depends only on c. $\square$

As a first application, we derive a basic theorem on central simple algebras.

Theorem 2.7.2 (Skolem–Noether) *All automorphisms of a central simple algebra are inner, i.e. given by conjugation by an invertible element.*

Proof Let A be a central simple k-algebra of degree n and K a finite Galois splitting field of A. Denoting by $A^\times$ the subgroup of invertible elements of A and using Lemma 2.4.1 we get an exact sequence

$$1 \to K^\times \to (A \otimes_k K)^\times \to \mathrm{Aut}_K(A \otimes_k K) \to 1$$

of groups equipped with a $G = \mathrm{Gal}\,(K|k)$-action, where the second map maps an invertible element to the inner automorphism it defines. Proposition 2.7.1 then yields an exact sequence

$$1 \to k^\times \to A^\times \to \mathrm{Aut}_k(A) \to H^1(G, K^\times),$$

where the last term is trivial by Hilbert's Theorem 90. The theorem follows. $\square$

As another application, we derive from Proposition 2.7.1 a cohomological interpretation of reduced norms. First a piece of notation: for a central simple algebra A, we denote by $\mathrm{SL}_1(A)$ the multiplicative subgroup of elements of reduced norm 1.

Proposition 2.7.3 *Let A be a central simple k-algebra split by a finite Galois extension $K|k$ of group G. There is a canonical bijection of pointed sets*

$$H^1(G, \mathrm{SL}_1(A \otimes_k K)) \leftrightarrow k^\times / \mathrm{Nrd}(A^\times).$$

For the proof we need a generalization of Example 2.3.4.

Lemma 2.7.4 *For A, K and G as above, we have $H^1(G, (A \otimes_k K)^\times) = 1$.*

Proof Let M be a left A-module with $\dim_k M = \dim_k A$. Then M is isomorphic to the left A-module A. Indeed, since $A \cong M_n(D)$ by Wedderburn's theorem, it is isomorphic to a direct sum of the minimal left ideals I_r introduced in Example 2.1.4; these are all isomorphic simple A-modules. As M is finitely generated over A, there is a surjection $A^N \to M$ for some $N > 0$, so M must be isomorphic to a direct sum of copies of I_r as well and hence isomorphic to A for dimension reasons.

Now observe that multiplication by an element of A is an endomorphism of M as a k-vector space. By the second example in Example 2.3.1 combined with Remark 2.3.9, the module M can thus be considered as a k-object (M, Φ) to which the theory of Section 2.3 applies. The $A \otimes_k K$-module $M \otimes_k K$ has the same dimension over K as $A \otimes_k K$, and is therefore isomorphic to $A \otimes_k K$ by the above. In other words, M is a twisted form of A, and all twisted forms of the A-module A as a k-vector space with additional structure are obtained in this way. But by the first paragraph they are all isomorphic to A.

Finally, an automorphism of $A \otimes_k K$ as a left module over itself is given by right multiplication by an invertible element. Hence via Theorem 2.3.3 (more precisely, its variant in Remark 2.3.9) the absence of twisted forms translates as the triviality of $H^1(G, (A \otimes_k K)^\times)$. □

Proof of Proposition 2.7.3 Applying Proposition 2.7.1 to the exact sequence

$$1 \to \mathrm{SL}_1(A \otimes_k K) \to (A \otimes_k K)^\times \xrightarrow{\mathrm{Nrd}} K^\times \to 1$$

we get an exact sequence

$$A^\times \xrightarrow{\mathrm{Nrd}} k^\times \to H^1(G, \mathrm{SL}_1(A \otimes_k K)) \to H^1(G, (A \otimes_k K)^\times),$$

where the last term is trivial by the lemma above, so we obtain a surjective map $k^\times \twoheadrightarrow H^1(G, \mathrm{SL}_1(A \otimes_k K))$. Now assume $c_1, c_2 \in k^\times$ have the same image in $H^1(G, \mathrm{SL}_1(A \otimes_k K))$. By construction of the exact sequence this means that the c_i lift to elements $a_i \in (A \otimes_k K)^\times$ satisfying $a_1^{-1}\sigma(a_1) = a_2^{-1}\sigma(a_2)$ for all $\sigma \in G$. In other words, we have $a_2 = (a_2 a_1^{-1})a_1$ with $a_2 a_1^{-1}$ fixed by the G-action and hence lying in $A^\times$, which implies $c_2 = \mathrm{Nrd}(a_2 a_1^{-1}) c_1$. $\qquad\square$

We now come back to the situation of Proposition 2.7.1. In the case when A is contained in the centre of B, the exact sequence of the proposition can be extended on the right by the second cohomology group of A. In the next chapter we shall define cohomology groups $H^i(G, A)$ of arbitrary degree for an abelian group A equipped with a G-action. Here we content ourselves with giving a direct definition in the case $i = 2$, in the spirit of Definition 2.3.2.

Definition 2.7.5 Let G be a group, and A an abelian group equipped with a G-action (written multiplicatively). A *2-cocycle* of G with values in A is a map $(\sigma, \tau) \mapsto a_{\sigma,\tau}$ from $G \times G$ to A satisfying the relation

$$\sigma(a_{\tau,\upsilon}) a_{\sigma\tau,\upsilon}^{-1} a_{\sigma,\tau\upsilon} a_{\sigma,\tau}^{-1} = 1 \qquad (2.15)$$

for all $\sigma, \tau, \upsilon \in G$. These form an abelian group $Z^2(G, A)$ for the multiplication induced from that of A. Two 2-cocycles $a_{\sigma,\tau}$ and $a'_{\sigma,\tau}$ are *cohomologous* if $a_{\sigma,\tau} a'^{-1}_{\sigma,\tau}$ is a *2-coboundary*, i.e. it is of the form $(\sigma, \tau) \mapsto a_\sigma \sigma(a_\tau) a_{\sigma\tau}^{-1}$ with some map $\sigma \mapsto a_\sigma$ from G to A. One checks that 2-coboundaries are 2-cocycles and form a subgroup in $Z^2(G, A)$. Thus the set $H^2(G, A)$ of cohomology classes of 2-cocycles is an abelian group, the *second cohomology group* of G with values in A.

Proposition 2.7.6 *Let G be a group, and*

$$1 \to A \to B \to C \to 1$$

an exact sequence of groups equipped with a G-action, such that B and C are not necessarily commutative, but A is commutative and contained in the centre of B. Then there is an exact sequence of pointed sets

$$1 \to A^G \to B^G \to C^G \to H^1(G, A) \to H^1(G, B) \to H^1(G, C) \to H^2(G, A).$$

Proof The sequence was constructed until the penultimate term in Proposition 2.7.1. To define the map $\partial : H^1(G, C) \to H^2(G, A)$, take a 1-cocycle $\sigma \mapsto c_\sigma$ representing a class in $H^1(G, C)$, and lift each c_σ to an element $b_\sigma \in B$. The cocycle relation for $\sigma \mapsto c_\sigma$ implies that for all $\sigma, \tau \in G$ the element $b_\sigma \sigma(b_\tau) b_{\sigma\tau}^{-1}$ maps to 1 in C, hence comes from an element $a_{\sigma,\tau} \in A$. The function $(\sigma, \tau) \mapsto a_{\sigma,\tau}$ depends only on the class of $\sigma \mapsto c_\sigma$ in $H^1(G, C)$. Indeed, if we replace it by an equivalent cocycle $\sigma \mapsto c^{-1} c_\sigma \sigma(c)$, lifting c to $b \in B$ replaces $a_{\sigma\tau}$ by $(b^{-1} b_\sigma \sigma(b))(\sigma(b^{-1}) \sigma(b_\tau) \sigma\tau(b))(\sigma\tau(b)^{-1} b_{\sigma,\tau}^{-1} b) = b^{-1} a_{\sigma,\tau} b$, which equals $a_{\sigma,\tau}$ because A is central in B. A straightforward calculation, which we leave to the readers, shows that $(\sigma, \tau) \mapsto a_{\sigma\tau}$ satisfies the 2-cocycle relation (2.15). Finally, replacing b_σ by another lifting $a_\sigma b_\sigma$ replaces $a_{\sigma,\tau}$ by $a_\sigma b_\sigma \sigma(a_\tau b_\tau) b_{\sigma\tau}^{-1} a_{\sigma\tau}^{-1} = a_\sigma \sigma(a_\tau) a_{\sigma\tau}^{-1} a_{\sigma,\tau}$, which has the same class in $H^2(G, A)$ (notice that we have used again that A is central in B). This defines the map ∂, and at the same time shows that it is trivial on the image of $H^1(G, B)$.

Finally, in the above notation, a class in $H^1(G, C)$ represented by $\sigma \mapsto c_\sigma$ is in the kernel of ∂ if the 2-cocycle $(\sigma, \tau) \mapsto b_\sigma \sigma(b_\tau) b_{\sigma\tau}^{-1}$ equals a 2-coboundary $(\sigma, \tau) \mapsto a_\sigma \sigma(a_\tau) a_{\sigma\tau}^{-1}$. Replacing b_σ by the equivalent lifting $a_\sigma^{-1} b_\sigma$ we may assume $b_\sigma \sigma(b_\tau) b_{\sigma\tau}^{-1} = 1$, which means that $\sigma \mapsto b_\sigma$ is a 1-cocycle representing a cohomology class in $H^1(G, B)$. $\qquad \square$

Remarks 2.7.7

1. Readers should be warned that the proposition does not hold in the above form when A is not contained in the centre of B. Instead, one has to work with twists of A as in Serre [4], §I.5.6.
2. When B and C are commutative, the exact sequence of the proposition is part of the long exact sequence for group cohomology that we shall establish in Proposition 3.1.9.

We conclude this section with an application of Proposition 2.7.6 to the Brauer group. Let $K|k$ be a finite Galois extension of fields with group G, and m a positive integer. Applying the proposition to the exact sequence of G-groups

$$1 \to K^\times \to \mathrm{GL}_m(K) \to \mathrm{PGL}_m(K) \to 1$$

we get an exact sequence of pointed sets

$$H^1(G, \mathrm{GL}_m(K)) \longrightarrow H^1(G, \mathrm{PGL}_m(K)) \xrightarrow{\delta_m} H^2(G, K^\times). \qquad (2.16)$$

Now recall the maps $\lambda_{mn} : H^1(G, \mathrm{PGL}_m(K)) \to H^1(G, \mathrm{PGL}_{mn}(K))$ introduced before Lemma 2.4.4.

Lemma 2.7.8 *The diagram*

$$
\begin{array}{ccc}
H^1(G, \mathrm{PGL}_m(K)) & \xrightarrow{\delta_m} & H^2(G, K^\times) \\
\downarrow {\lambda_{mn}} & & \downarrow {\mathrm{id}} \\
H^1(G, \mathrm{PGL}_{mn}(K)) & \xrightarrow{\delta_{mn}} & H^2(G, K^\times)
\end{array}
$$

commutes for all $m, n > 0$.

Proof   A 1-cocycle $\sigma \mapsto c_\sigma$ representing a class in $H^1(G, \mathrm{PGL}_m(K))$ is mapped by δ_m to a 2-cocycle $a_{\sigma,\tau} = b_\sigma \sigma(b_\tau) b_{\sigma\tau}^{-1}$ by the construction of the previous proof, where b_σ is given by some invertible matrix M_σ and $a_{\sigma,\tau}$ is the identity matrix I_m multiplied by some scalar $\mu_{\sigma,\tau} \in K^\times$. Performing the same construction for the image of $\sigma \mapsto c_\sigma$ by λ_{mn} means replacing M_σ by the block matrix with n copies of M_σ along the diagonal, which implies that the scalar matrix we obtain by taking the associated 2-cocycle is $\mu_{\sigma,\tau} I_{mn}$.   □

By the lemma, taking the union of the pointed sets $H^1(G, \mathrm{PGL}_m(K))$ with respect to the maps λ_{mn} yields a map

$$\delta_\infty : H^1(G, \mathrm{PGL}_\infty) \to H^2(G, K^\times).$$

Equip the set $H^1(G, \mathrm{PGL}_\infty)$ with the group structure defined in Proposition 2.4.9. In Theorem 4.4.1 we shall prove that δ_∞ is an isomorphism. Here we establish a weaker statement which is already sufficient for a number of interesting applications.

Proposition 2.7.9   *The map δ_∞ is an injective group homomorphism.*

Proof   To show that δ_∞ preserves multiplication, take cohomology classes $c_m \in H^1(G, \mathrm{PGL}_m(K))$ and $c_n \in H^1(G, \mathrm{PGL}_n(K))$. With notations as in the previous proof, the classes $\delta_m(c_m)$ and $\delta_n(c_n)$ are represented by 2-cocycles of the form $(\sigma, \tau) \to \mu_{\sigma,\tau} I_m$ and $(\sigma, \tau) \to \nu_{\sigma,\tau} I_n$, respectively. From the fact that the product c_{mn} of $\lambda_{nm}(c_n)$ and $\lambda_{mn}(c_m)$ in $H^1(G, \mathrm{PGL}_{mn}(K))$ is induced by tensor product of linear maps we infer that $\delta_{mn}(c_{mn})$ is represented by a 2-cocycle mapping (σ, τ) to the tensor product of the linear maps given by multiplication by $\mu_{\sigma,\tau}$ and $\nu_{\sigma,\tau}$, respectively. But this tensor product is none but multiplication by $\mu_{\sigma,\tau} \nu_{\sigma,\tau}$, which was to be seen.

Once we know that δ_∞ is a group homomorphism, for injectivity it is enough to show that the map δ_m in exact sequence (2.16) has trivial kernel for all m. This follows from the exact sequence in view of the triviality of $H^1(G, \mathrm{GL}_m(K))$ (Example 2.3.4).   □

2.8 Index and period

We now make use of the techniques of the previous section to establish basic results of Brauer concerning two important invariants for central simple algebras. We shall assume throughout that the base field k is infinite; indeed, we have mentioned in Remark 2.2.8 that the Brauer group of a finite field is trivial, so the discussion to follow is vacuous in that case.

The first of the announced invariants is the following.

Definition 2.8.1 Let A be a central simple algebra over a field k. The *index* $\mathrm{ind}_k(A)$ of A over k is defined to be the degree of D over k, where D is the division algebra for which $A \cong M_n(D)$ according to Wedderburn's theorem. We shall drop the subscript k from the notation when clear from the context.

Remarks 2.8.2

1. For a division algebra index and degree are one and the same thing. In general the index divides the degree of the algebra.
2. The index of a central simple k-algebra A depends only on the class of A in the Brauer group $\mathrm{Br}\,(k)$. Indeed, this class depends only on the division algebra D associated with A by Wedderburn's theorem, and the index is by definition an invariant of D.

    Wedderburn's theorem also shows that the degrees of the central simple k-algebras Brauer equivalent to A are exactly the multiples of $\mathrm{ind}(A)$.
3. We have $\mathrm{ind}(A) = 1$ if and only if A is split.

Proposition 2.8.3 *Given a central simple algebra A of index d, we have the divisibility relation*

$$\mathrm{ind}(A^{\otimes r}) \mid \binom{d}{r}$$

for all $r \leq d$.

Proof Let $K|k$ be a finite Galois extension with group G splitting A. Consider the r-th exterior power $\wedge^r K^d$ of the K-vector space K^d; it is a K-vector space of dimension $\binom{d}{r}$. The standard $\mathrm{GL}_d(K)$-action on K^d extends to the wedge product by setting $\phi(v_1 \wedge \cdots \wedge v_r) = \phi(v_1) \wedge \cdots \wedge \phi(v_r)$ for $\phi \in \mathrm{GL}_d(K)$, $v_i \in K^d$, whence a homomorphism $\mathrm{GL}_d(K) \to \mathrm{GL}(\wedge^r K^d)$. In the case when ϕ is multiplication by a scalar $\lambda \in K^\times$, its image in $\mathrm{GL}(\wedge^r K^d)$ becomes multiplication by λ^r. We thus obtain a commutative diagram of groups with G-action

$$
\begin{array}{ccccccccc}
1 & \longrightarrow & K^\times & \longrightarrow & \mathrm{GL}_d(K) & \longrightarrow & \mathrm{PGL}_d(K) & \longrightarrow & 1 \\
& & \downarrow{\scriptstyle r} & & \downarrow & & \downarrow & & \\
1 & \longrightarrow & K^\times & \longrightarrow & \mathrm{GL}(\wedge^r K^d) & \longrightarrow & \mathrm{PGL}(\wedge^r K^d) & \longrightarrow & 1
\end{array}
$$

inducing a commutative diagram

$$
\begin{array}{ccc}
H^1(G, \mathrm{PGL}_d(K)) & \longrightarrow & H^2(G, K^\times) \\
\downarrow & & \downarrow{\scriptstyle r} \\
H^1(G, \mathrm{PGL}(\wedge^r K^d)) & \longrightarrow & H^2(G, K^\times)
\end{array}
$$

where the horizontal maps are the boundary maps introduced before Lemma 2.7.8. By Propositions 2.4.9 and 2.7.9 we have an injection $\mathrm{Br}\,(K|k) \hookrightarrow H^2(G, K^\times)$. Since A has index d, its class $[A]$ in $H^2(G, K^\times)$ is in the image of the boundary map $\delta_d : H^1(G, \mathrm{PGL}_d(K)) \to H^2(G, K^\times)$. By commutativity of the diagram above, the class $r[A] \in H^2(G, K^\times)$ comes from $H^1(G, \mathrm{PGL}(\wedge^r K^d))$, and is therefore representable by an algebra of degree $\binom{d}{r}$. On the other hand, we have $r[A] = [A^{\otimes r}]$ by Propositions 2.4.9 and 2.7.9, which concludes the proof. $\qquad\square$

Remark 2.8.4 In §10.A of Knus–Merkurjev–Rost–Tignol [1] there is a ring-theoretic construction of a central simple algebra of degree $\binom{d}{r}$ representing the class in $H^1(G, \mathrm{PGL}(\wedge^r K^d))$ used in the above proof.

Applying the proposition with $d = r$, we obtain

Corollary 2.8.5 *If A has index d, then $A^{\otimes d}$ is split. Consequently, the Brauer group is a torsion abelian group.*

The above corollary enables us to define the second main invariant of Brauer classes.

Definition 2.8.6 The *period* (or *exponent*) of a central simple k-algebra A is the order of its class in $\mathrm{Br}\,(k)$. We denote it by $\mathrm{per}(A)$.

The basic relations between the period and the index are the following.

Theorem 2.8.7 (Brauer) *Let A be a central simple k-algebra.*

1. *The period $\mathrm{per}(A)$ divides the index $\mathrm{ind}(A)$.*
2. *The period $\mathrm{per}(A)$ and the index $\mathrm{ind}(A)$ have the same prime factors.*

Part (1) of the theorem is just a reformulation of Corollary 2.8.5. We now give a proof of part (2) based on ideas of Antieau and Williams [2]. It uses a lemma from elementary number theory.

Lemma 2.8.8 *Given a pair (m, n) of positive integers with $m|n$, there exist integers $q_0, \ldots, q_l$ with $1 \leq q_i < n$ and $(q_i, m) = 1$ for all i such that moreover the greatest common divisor d of the binomial coefficients*

$$\binom{n}{q_0}, \ldots, \binom{n}{q_l}$$

has exactly the same prime divisors as m.

Proof Let $p_1, \ldots, p_l$ be those prime divisors of n that do not divide m, and for each i let $p_i^{s_i}$ be the largest power of p_i dividing n. We show that the set $\{1, p_1^{s_1}, \ldots, p_l^{s_l}\}$ will do.

The binomial coefficient $\binom{n}{p_i^{s_i}}$ is the coefficient of $x^{p_i^{s_i}}$ in the expansion of $(1 + x)^n$. If q is a divisor of n which is a power of a prime p, then

$$(1 + x)^n = ((1 + x)^q)^{n/q} \equiv (1 + x^q)^{n/q} \bmod p.$$

Therefore if $q = p$ and is not among the p_i, then it divides $\binom{n}{p_i^{s_i}}$. On the other hand, if $q = p_i^{s_i}$, then the coefficient of $x^{p_i^{s_i}}$ is not divisible by p_i. $\qquad\square$

Proof of Theorem 2.8.7(2) Fix a Galois splitting field $K|k$ for A with group G. Write m for the period and n for the index of A. Consider integers $q_0, \ldots, q_l$ as in the lemma, and pick positive integers r_i with $q_i r_i \equiv 1 \bmod m$. For each i define the K-vector space

$$W_i := (\wedge^{q_i}(K^n))^{\otimes r_i}.$$

As in the proof of Proposition 2.8.3, for each i we have homomorphisms $\mathrm{GL}_n(K) \to \mathrm{GL}(W_i)$ sitting in commutative diagrams

$$
\begin{array}{ccccccccc}
1 & \longrightarrow & K^\times & \longrightarrow & \mathrm{GL}_n(K) & \longrightarrow & \mathrm{PGL}_n(K) & \longrightarrow & 1 \\
& & {\scriptstyle q_i r_i}\downarrow & & \downarrow & & \downarrow & & \\
1 & \longrightarrow & K^\times & \longrightarrow & \mathrm{GL}(W_i) & \longrightarrow & \mathrm{PGL}(W_i) & \longrightarrow & 1.
\end{array}
$$

By assumption, the class $[A]$ of A in $\mathrm{Br}\,(K|k)$ comes from $H^1(G, \mathrm{PGL}_n(K))$, so we obtain, as in the proof of Proposition 2.8.3, that $q_i r_i[A]$ comes from $\alpha_i \in H^1(G, \mathrm{PGL}(W_i))$. But A has period m, so $q_i r_i[A] = [A]$ by our assumption on r_i.

The dimension of W_i is $d_i := \binom{n}{q_i}^{r_i}$. By Remark 2.8.2 (2) this integer is divisible by the index n of A. It follows that n divides the greatest common divisor d of the d_i. By the lemma d has the same prime divisors as m, hence the prime divisors of n must be among those of m. Since $m|n$ by part (1) of the theorem, the statement follows. $\qquad\square$

Remark 2.8.9 In Remark 4.5.7 we shall give another proof of the theorem based on methods of Galois cohomology. The above proof uses only the techniques developed so far, and at the same time can be made to work (with some modifications) in a much more general setting, that of Azumaya algebras in a connected locally ringed topos. It should be noted that in the general setting there is no analogue of Wedderburn's theorem, and therefore the index is defined as the greatest common divisor of the degrees of algebras representing a given Brauer class. It is not necessarily true that the lowest of these degrees equals the index (Antieau and Williams [1]).

Corollary 2.8.10 *We have the divisibility relations*

$$\mathrm{per}(A) \mid \mathrm{ind}(A) \mid \mathrm{per}(A)^{\mathrm{ind}(A)-1}.$$

Proof Let p be a prime, and p^r (resp. p^s) the largest powers of p dividing $\mathrm{ind}(A)$ (resp. $\mathrm{per}(A)$). By the theorem it suffices to show $r \leq s(p^r - 1)$, an inequality that holds for all $r, s \geq 1$. □

Remarks 2.8.11

1. The bound $\mathrm{ind}(A) \mid \mathrm{per}(A)^{\mathrm{ind}(A)-1}$ in the corollary is very far from being optimal. For instance, for a division algebra of prime degree the period equals the index.
2. For a beautiful geometric proof of Corollary 2.8.10 (and hence of Theorem 2.8.7) see Kollár [1], §6.

Remarks 2.8.12 The relation between the period and index of all central simple algebras over a given field k has been much studied in recent years.

1. It is conjectured by Michael Artin [1] that over C_2-fields (see Remark 6.2.2 for this notion) the period should always equal the index. The conjecture is now known to hold for $\mathbf{F}_q((t))$, $\mathbf{F}_q(t)$ and their finite extensions (see Corollary 6.3.8 as well as Remarks 6.5.4 and 6.5.5), function fields of surfaces over an algebraically closed field (de Jong [1], Lieblich [1]), and completions of the latter at smooth points (Colliot-Thélène/Ojanguren/Parimala [1]).
2. Colliot-Thélène and Lieblich have proposed an extension of Artin's conjecture, namely that for a central simple algebra A over a C_i-field $\mathrm{ind}(A)$ should divide $\mathrm{per}(A)^{i-1}$. Evidence for this more general conjecture is provided by work of Lieblich: he proved $\mathrm{ind}(A)|\mathrm{per}(A)^2$ when A is defined over the function field of a surface over a finite field (Lieblich [3]), and

ind$(A)|$per$(A)^d$ for function fields of curves defined over fields of the form $k((t_1))\ldots((t_d))$ with k algebraically closed of characteristic not dividing per(A) (Lieblich [2]).

3. Another important result on this topic is that of Saltman [4], who proved that for an algebra A over the function field of a curve over a p-adic field $\mathbf{Q}_p$ one has per$(A)|$ind$(A)^2$, provided that per(A) is prime to p. Lieblich [2] has extended this result to function fields of curves over higher-dimensional local fields or maximal unramified extensions of local fields.

As an application of the above, we finally prove the following primary decomposition result.

Proposition 2.8.13 (Brauer) *Let D be a central division algebra over k. Consider the primary decomposition*

$$\mathrm{ind}(D) = p_1^{m_1} p_2^{m_2} \cdots p_r^{m_r}.$$

Then we may find central division algebras D_i ($i = 1, .., r$) such that

$$D \cong D_1 \otimes_k D_2 \otimes_k \cdots \otimes_k D_r$$

and $\mathrm{ind}(D_i) = p_i^{m_i}$ *for $i = 1, .., r$. Moreover, the D_i are uniquely determined up to isomorphism.*

The proof below is taken from Antieau and Williams [3]. It again uses a lemma from elementary number theory.

Lemma 2.8.14 *Let d be a positive integer, and p a prime divisor of d. Denote by v_p the p-adic valuation. There exists a positive integer r satisfying*

1. *$r \equiv 0 \bmod dp^{-v_p(d)}$ and $r \equiv 1 \bmod p^{v_p(d)}$;*
2. *$v_p(\binom{r+d-1}{r}) = v_p(d)$.*

Proof Choose an integer s with $s > \log_p(d)$, and then r satisfying $r \equiv 0 \bmod dp^{-v_p(d)}$ and $r \equiv 1 \bmod p^s$. Since by assumption $s > v_p(d)$, the first condition holds. For the second condition it will be enough to check

$$v_p((r+d-1)!) - v_p(r!) = v_p(d!) \tag{2.17}$$

as $v_p(d!) - v_p((d-1)!) = v_p(d)$. Recall Legendre's formula for the p-adic valuation of factorials:

$$v_p(m!) = \frac{m - S(m)}{p-1}$$

where $S(m)$ is the sum of digits in the p-adic expansion of an integer m. Equality (2.17) follows by applying the formula to $r + d - 1$, r and d and noting that the assumptions $r \equiv 1 \bmod p^s$ and $s > \log_p(d)$ imply $S(r - 1 + d) = S(r - 1) + S(d) = S(r) - 1 + S(d)$. $\square$

Proof of Proposition 2.8.13   The Brauer group is torsion (Corollary 2.8.5), so it splits into p-primary components:

$$\mathrm{Br}\,(k) = \bigoplus_p \mathrm{Br}\,(k)\{p\}.$$

In this decomposition the class of D decomposes as a sum

$$[D] = [D_1] + [D_2] + \cdots + [D_r]$$

where the D_i are division algebras with $[D_i] \in \mathrm{Br}\,(k)\{p_i\}$ for some primes p_i. By Theorem 2.8.7 (2) the index of each D_i is a power of p_i. The tensor product $A = D_1 \otimes_k D_2 \otimes_k \cdots \otimes_k D_r$ has degree $\prod_i \mathrm{ind}(D_i)$ over k and its index equals that of D by Remark 2.8.2 (2), so $\mathrm{ind}(D)$ divides $\prod_i \mathrm{ind}(D_i)$.

We now show that actually $\mathrm{ind}(D) = \prod_i \mathrm{ind}(D_i)$ holds. This will imply that the k-algebras D and $D_1 \otimes_k D_2 \otimes_k \cdots \otimes_k D_r$ have the same Brauer class and the same dimension, hence they are isomorphic as claimed. The uniqueness of the D_i holds for the same reason.

We thus have to show that we have $v_{p_i}(\mathrm{ind}(D)) \geq v_{p_i}(\mathrm{ind}(D_i))$ for all i; we do it for $i = 1$. Apply Lemma 2.8.14 with $d := \mathrm{ind}(D)$ and $p := p_1$ to find an integer $r > 0$. For a Galois splitting field K of D consider the r-th symmetric power $S^r(K^d)$, i.e. the quotient of the tensor power $(K^d)^{\otimes r}$ by the natural action of the symmetric group S_r permuting the terms. The termwise action of $\mathrm{GL}_d(K)$ on $(K^d)^{\otimes r}$ induces an action on $S^r(K^d)$, whence a homomorphism $\mathrm{GL}_d(K) \to \mathrm{GL}(S^r(K^d))$. Exactly as in the proof of Proposition 2.8.3, this homomorphism induces a commutative diagram

$$
\begin{array}{ccc}
H^1(G, \mathrm{PGL}_d(K)) & \longrightarrow & H^2(G, K^\times) \\
\downarrow & & \downarrow r \\
H^1(G, \mathrm{PGL}(S^r(K^d))) & \longrightarrow & H^2(G, K^\times).
\end{array}
$$

As D is represented by a class in $H^1(G, \mathrm{PGL}_d(K))$, the diagram shows that $r[D] \in \mathrm{Br}\,(K|k) \subset H^2(G, K^\times)$ comes from a class in $H^1(G, \mathrm{PGL}(S^r(K^d)))$. Since $\mathrm{per}(D_i)|\mathrm{per}(D)|\mathrm{ind}(D)$ for all i, property (1) of the integer r in the lemma implies $r[D_i] = 0$ for $i > 1$ and $r[D_1] = [D_1]$, whence $r[D] = [D_1]$ and $[D_1]$ comes from $H^1(G, \mathrm{PGL}(S^r(K^d)))$. On the other hand, $\dim_K(S^r(K^d)) = \binom{r+d-1}{r}$, and therefore $v_{p_1}(\mathrm{ind}(D_1)) \leq v_{p_1}(\mathrm{ind}(D))$ by property (2) of r in the lemma, as required. $\square$

Remark 2.8.15 Again, we shall give another proof of the proposition in Remark 4.5.8, but the formula $\mathrm{ind}(D) = \prod_i \mathrm{ind}(D_i)$ obtained above holds in the more general setting of Azumaya algebras. However, a general Azumaya algebra may not have a primary decomposition as in the proposition (Antieau and Williams [1]).

2.9 Central simple algebras over complete discretely valued fields

In this section K denotes a field complete with respect to a discrete valuation, with perfect residue field κ. Our goal is to prove:

Theorem 2.9.1 *Every central simple algebra A over K is split by a finite unramified extension of K.*

The bulk of the proof of the theorem is contained in the following proposition.

Proposition 2.9.2 *Every central division algebra D of degree $d > 1$ over K contains a K-subalgebra L that is an unramified field extension of K of degree > 1.*

We prove the proposition following the method of Serre [2], §XII.2. The key tool is the extension of the valuation on K to D. By definition, a discrete valuation on a division algebra D is a map $w : D \to \mathbf{Z} \cup \{\infty\}$ satisfying the same properties as in the commutative case (see Section A.6 of the Appendix). The elements satisfying $w(x) \geq 0$ form a subring $A_w \subset D$ in which the set M_w of elements with $w(x) > 0$ is a two-sided ideal. Fixing an element $\pi \in D$ such that $w(\pi)$ is the positive generator of the subgroup $w(D \setminus \{0\}) \subset \mathbf{Z}$, we may write each $m \in M_w$ in the form $m = b\pi$ with $b \in A_w$.

Lemma 2.9.3 *If D and K are as in the proposition, the discrete valuation v of K extends to a unique discrete valuation on D, given by the formula*

$$w = \frac{1}{d} v \circ \mathrm{Nrd}_D.$$

Moreover, D is complete with respect to w.

Proof We first show that if $L|K$ is a field extension contained in D, then for $x \in L$ we have

$$\frac{1}{d}\, v(\mathrm{Nrd}_D(x)) = \frac{1}{[L:K]}\, v(N_{L|K}(x)). \qquad (2.18)$$

Indeed, Proposition 2.6.3 (1) gives $N_{D|K} = (\mathrm{Nrd}_D)^d$. On the other hand, viewing D as an L-vector space of dimension $d^2/[L:K]$, we have

$$N_{D|K}(x) = N_{L|K}(x)^{d^2/[L:K]}$$

for $x \in L$. It follows that we have an equality

$$\mathrm{Nrd}_D(x) = \omega\, N_{L|K}(x)^{d/[L:K]}$$

in K with some d-th root of unity ω, whence formula (2.18) follows after applying v.

Applying formula (2.18) with $L = K$ gives $w|_K = v$. Applying it to the subfields $K(x) \subset D$ generated by each $x \in D$ and comparing with Proposition A.6.8 (2) describing the unique extension of v to finite field extensions implies the uniqueness of the extension to D. Next, we check that w as defined above is a discrete valuation. The implication $w(x) = \infty \Rightarrow x = 0$ follows from Proposition 2.6.2, and the formula $w(xy) = w(x) + w(y)$ from the multiplicativity of the reduced norm. The property $w(x + y) \ge \min(w(x), w(y))$ reduces to $w(1 + x^{-1}y) \ge \min(1, w(x^{-1}y))$ after subtracting $w(x)$. The latter can be checked in the field $L = K(x^{-1}y)$ where it holds because, as remarked above, formula (2.18) implies that $w|_L$ is a multiple of the unique extension of v to L. Finally, the completeness of D with respect to the w-adic topology is proven as in the commutative case. $\qquad\square$

Proof of Proposition 2.9.2　Extend the valuation v of K to a discrete valuation w of D as in the lemma above. If the statement of the proposition does not hold, then for each finite field extension $L|K$ contained in D the valuation $w|_L$ has residue field equal to that of v. In particular, this holds for the subfield $L = K(b)$ generated by any $b \in A_w$. Fixing b, we thus find a_0 in the ring of integers A_v of K with $b - a_0 \in M_w$. Fixing moreover a generator π as before the statement of the lemma we may write

$$b = a_0 + b_1\pi$$

with some $b_1 \in A_w$. Repeating the procedure with b_1 in place of b and continuing in the same way, we construct inductively for each $N > 0$ elements $a_N \in A_v$ and $b_N \in A_w$ satisfying

$$b = \sum_{i=0}^{N-1} a_i\pi^i + b_N\pi^N.$$

We infer that b is in the closure of the subfield $K(\pi) \subset D$ for the w-adic topology on D. But $K(\pi)$ is closed in D (this holds for any linear subspace in a finite-dimensional normed vector space over a complete valued field and is easily checked by taking coordinates), whence $b \in K(\pi)$. Since b was arbitrary here and for every $x \in D$ we have $x\pi^m \in A_w$ for m large enough, we conclude $D \subset K(\pi)$, contradicting the assumption that the centre of D is K. $\qquad\square$

Proof of Theorem 2.9.1 We use induction on the index d of A, the case $d=1$ being obvious. Using Wedderburn's theorem we may assume that A is a division algebra of degree d. Applying Proposition 2.9.2, we find a nontrivial unramified field extension $L|K$ that embeds in A over K. The L-algebra $A \otimes_K L$ is not a division algebra because it contains $L \otimes_K L$ which is a product of copies of L. Thus $\mathrm{ind}(A \otimes_K L) < \mathrm{ind}(A) = d$, and therefore $A \otimes_K L$ splits over an unramified extension $M|L$ by the inductive assumption. But $M|K$ is again an unramified extension (see Remark A.6.9), which concludes the proof. $\qquad\square$

Corollary 2.9.4 *We have* $\mathrm{Br}\,(K_{nr}) = 0$, *where* K_{nr} *denotes the maximal unramified extension of* K.

Proof A similar reasoning as in the proof of Theorem 2.2.1 shows that every central simple algebra over K_{nr} is defined over some finite unramified extension $L|K$ contained in K_{nr}. By the theorem, it is split by a finite unramified extension $M|L$, still contained in K_{nr}. $\qquad\square$

Remarks 2.9.5

1. In Serre [2], §XII.2, Proposition 2, a stronger form of the theorem is proven, namely that A is split by an unramified extension of degree $\mathrm{ind}(A)$. We shall prove this in Proposition 6.3.8, but only in the special case when the residue field κ is finite.
2. A comprehensive treatment of valuations on division algebras (not necessarily discrete) is given in the book of Tignol and Wadsworth [1].

2.10 K_1 of central simple algebras

The main result of this section is a classical theorem of Wang on commutator subgroups of division algebras. Following the present-day viewpoint, we discuss it within the framework of the K-theory of rings. Therefore we first define the group K_1 for a ring.

Construction 2.10.1   Given a not necessarily commutative ring R with unit and a positive integer n, consider the group $\mathrm{GL}_n(R)$ of $n \times n$ invertible matrices over R. For each n there are injective maps $i_{n,n+1} : \mathrm{GL}_n(R) \to \mathrm{GL}_{n+1}(R)$ given by

$$i_{n,n+1}(A) := \begin{bmatrix} A & 0 \\ 0 & 1 \end{bmatrix}. \tag{2.19}$$

Let $\mathrm{GL}_\infty(R)$ be the union of the tower of embeddings

$$\mathrm{GL}_1(R) \subset \mathrm{GL}_2(R) \subset \mathrm{GL}_3(R) \subset \cdots ,$$

given by the maps $i_{n,n+1}$. (Note that this definition of GL_∞ is *not* compatible with the definition of PGL_∞ introduced in Section 2.4.)

We define the group $K_1(R)$ as the quotient of $\mathrm{GL}_\infty(R)$ by its commutator subgroup $[GL_\infty(R), \mathrm{GL}_\infty(R)]$. This group is sometimes called the *Whitehead group* of R. It is functorial with respect to ring homomorphisms, i.e. a map $R \to R'$ of rings induces a map $K_1(R) \to K_1(R')$.

For calculations the following description of the commutator subgroup $[GL_\infty(R), \mathrm{GL}_\infty(R)]$ is useful. Recall the elementary matrices $E_{ij}(r)$ introduced before the proof of Lemma 2.6.4, and denote by $E_n(R)$ the subgroup of $\mathrm{GL}_n(R)$ generated by elementary matrices. The maps $i_{n,n+1}$ preserve these subgroups, whence a subgroup $E_\infty(R) \subset \mathrm{GL}_\infty(R)$.

Proposition 2.10.2 (Whitehead's lemma)   *The subgroup $E_\infty(R)$ is precisely the commutator subgroup $[\mathrm{GL}_\infty(R), \mathrm{GL}_\infty(R)]$ of $\mathrm{GL}_\infty(R)$.*

Proof   The relation $E_{ij}(r) = [E_{ik}(r), E_{kj}(1)]$ for distinct i, j and k is easily checked by matrix multiplication and shows that $E_\infty(R)$ is contained in $[E_\infty(R), E_\infty(R)] \subset [GL_\infty(R), \mathrm{GL}_\infty(R)]$. To show $[GL_\infty(R), \mathrm{GL}_\infty(R)] \subset E_\infty(R)$, we embed $\mathrm{GL}_n(R)$ into $\mathrm{GL}_{2n}(R)$ and for $A, B \in \mathrm{GL}_n(R)$ compute

$$\begin{bmatrix} ABA^{-1}B^{-1} & 0 \\ 0 & 1 \end{bmatrix} = \begin{bmatrix} AB & 0 \\ 0 & B^{-1}A^{-1} \end{bmatrix} \begin{bmatrix} A^{-1} & 0 \\ 0 & A \end{bmatrix} \begin{bmatrix} B^{-1} & 0 \\ 0 & B \end{bmatrix}.$$

All terms on the right are of similar shape. Denoting by I_n the identity matrix, another computation shows that

$$\begin{bmatrix} A & 0 \\ 0 & A^{-1} \end{bmatrix} = \begin{bmatrix} I_n & A \\ 0 & I_n \end{bmatrix} \begin{bmatrix} I_n & 0 \\ -A^{-1} & I_n \end{bmatrix} \begin{bmatrix} I_n & A \\ 0 & I_n \end{bmatrix} \begin{bmatrix} 0 & -I_n \\ I_n & 0 \end{bmatrix}$$
$$(2.20)$$

and similarly for the other terms. The first three terms on the right-hand side are upper or lower triangular matrices with 1's in the diagonal, so they are products of elementary matrices by Facts 2.6.5. For the fourth, notice that

$$\begin{bmatrix} 0 & -I_n \\ I_n & 0 \end{bmatrix} = \begin{bmatrix} I_n & -I_n \\ 0 & I_n \end{bmatrix} \begin{bmatrix} I_n & 0 \\ I_n & I_n \end{bmatrix} \begin{bmatrix} I_n & -I_n \\ 0 & I_n \end{bmatrix},$$

so we again conclude by applying Facts 2.6.5. $\qquad\square$

Using Whitehead's lemma we may easily calculate K_1-groups of fields.

Proposition 2.10.3 *For a field k the natural map $k^\times = \mathrm{GL}_1(k) \to \mathrm{GL}_\infty(k)$ induces an isomorphism $k^\times \xrightarrow{\sim} K_1(k)$.*

Proof We first show surjectivity. It is well known from linear algebra that a matrix in $M_n(k)$ may be put in diagonal form by means of elementary row and column operations, i.e. by multiplication with suitable elementary matrices. Thus by Whitehead's lemma each element of $K_1(k)$ may be represented by some diagonal matrix. But any diagonal matrix may be expressed as a product of a diagonal matrix of the form $\mathrm{diag}(b, 1, \ldots, 1)$ and diagonal matrices of the form $\mathrm{diag}(1, \ldots, 1, a, a^{-1}, 1, \ldots, 1)$. The same matrix calculation that establishes formula (2.20) shows that the latter are products of elementary matrices, so the class in $K_1(k)$ is represented by $\mathrm{diag}(b, 1, \ldots, 1)$, whence the required surjectivity.

To show injectivity one considers the determinant maps $\mathrm{GL}_n(k) \to k^\times$. They are compatible with the transition maps $i_{n,n+1}$, and therefore they define a homomorphism $\det_\infty : \mathrm{GL}_\infty(k) \to k^\times$, which is a splitting of the surjection $k^\times \to K_1(k)$ studied above. The proposition follows. $\qquad\square$

Remarks 2.10.4

1. More generally, for a division ring D one can consider the map

$$D^\times / [D^\times, D^\times] \to K_1(D)$$

induced by the map $D^\times = \mathrm{GL}_1(D) \to \mathrm{GL}_\infty(D)$ and show that it is an isomorphism. The proof of surjectivity goes by the same argument as above (since diagonalization of matrices by elementary row and column transformations also works over a division ring). The proof of injectivity is also the same, except that one has to work with the *Dieudonné determinant*, a non-commutative generalization of the usual determinant map (see e.g. Pierce [1], §16.5).

2. Another, much easier, generalization is the following: the isomorphism of the proposition also holds for finite direct products $k_1 \times \cdots \times k_r$ of fields. This follows from the proposition and the general formula

$$K_1(R \times R') \cong K_1(R) \times K_1(R'),$$

valid for arbitrary rings R and R', which is a consequence of the definition of K_1.

Consider now for $n, m \geq 1$ the maps $i_{n,nm} : M_n(R) \to M_{nm}(R)$ given by

$$i_{n,m}(A) := \begin{bmatrix} A & 0 \\ 0 & I_{nm-n} \end{bmatrix}.$$

By functoriality, the map $i_{1,m}$ induces a map $K_1(R) \to K_1(M_m(R))$.

Lemma 2.10.5 *The above map $K_1(R) \to K_1(M_m(R))$ is an isomorphism.*

This map is sometimes called the Morita isomorphism because of its relation with Morita equivalence in ring theory. In the case of a central simple algebra A it shows that the isomorphism class of $K_1(A)$ only depends on the Brauer class of A.

Proof For all $n \geq 1$, the diagram

$$
\begin{array}{ccc}
\mathrm{GL}_n(R) & \xrightarrow{\ i_{1,m*}\ } & \mathrm{GL}_n(M_m(R)) \cong \mathrm{GL}_{nm}(R) \\
{\scriptstyle i_{n,nm}}\downarrow & & {\scriptstyle i_{n,nm}}\downarrow \qquad {\scriptstyle i_{nm,nm^2}}\downarrow \\
\mathrm{GL}_{nm}(R) & \xrightarrow{\ i_{1,m*}\ } & \mathrm{GL}_{nm}(M_m(R)) \cong \mathrm{GL}_{nm^2}(R)
\end{array}
$$

commutes, so the map $\mathrm{GL}_\infty(R) \to \mathrm{GL}_\infty(M_m(R))$ is an isomorphism. This isomorphism preserves the commutator subgroups, whence the lemma. $\square$

The lemma enables us to construct a norm map for K_1 of k-algebras.

Construction 2.10.6 Let A be a k-algebra and $K|k$ a field extension of degree n. Denote by A_K the base change $A \otimes_k K$. We construct a norm map $N_{K|k} : K_1(A_K) \to K_1(A)$ as follows. Fixing an isomorphism $\phi : \mathrm{End}_k(K) \cong M_n(k)$ gives rise to a composite map

$$\phi_* : A_K = A \otimes_k K \to A \otimes_k \mathrm{End}_k(K) \xrightarrow{\ id \otimes \phi\ } A \otimes_k M_n(k) \cong M_n(A).$$

We then define the norm map $N_{K|k}$ as the composite

$$K_1(A_K) \xrightarrow{\phi_*} K_1(M_n(A)) \xrightarrow{\sim} K_1(A),$$

where the second map is the inverse of the isomorphism of the previous lemma. Since the conjugation action of the group $\mathrm{GL}_n(k)$ (and even of $\mathrm{GL}_n(A)$) on $M_n(A)$ induces a trivial action on $K_1(M_n(A))$, we conclude that the map above is independent of the choice of ϕ.

Proposition 2.10.7 *In the situation above the composite map*

$$K_1(A) \to K_1(A_K) \xrightarrow{N_{K|k}} K_1(A)$$

is multiplication by $n = [K : k]$.

Proof The composite $K_1(A) \to K_1(A_K) \to K_1(M_n(A))$ is induced by the map $A \to A \otimes_k M_n(k)$ sending a matrix M to the block diagonal matrix $\mathrm{diag}(M, \ldots, M)$. The same argument with formula (2.20) as in the proof of Proposition 2.10.3 shows that the class of $\mathrm{diag}(M, \ldots, M)$ in $K_1(A)$ equals that of $\mathrm{diag}(M^n, 1, \ldots, 1)$, whence the claim. □

We now focus on the case of a central simple k-algebra A and construct reduced norm maps on K_1-groups. Given an integer $n \geq 1$, we denote by $\mathrm{Nrd}_n : \mathrm{GL}_n(A) \to k^\times$ the composite

$$\mathrm{GL}_n(A) \cong \mathrm{GL}_1(M_n(A)) \xrightarrow{\mathrm{Nrd}_{M_n(A)}} k^\times.$$

Lemma 2.10.8 *For all integers $n \geq 1$, the diagram*

$$
\begin{array}{ccc}
\mathrm{GL}_n(A) & \xrightarrow{i_{n,n+1}} & \mathrm{GL}_{n+1}(A) \\
\mathrm{Nrd}_n \downarrow & & \downarrow \mathrm{Nrd}_{n+1} \\
k^\times & \xrightarrow{\mathrm{id}} & k^\times
\end{array}
$$

commutes.

Proof By the construction of reduced norm maps, it is enough to check commutativity after base change to a Galois splitting field of A. There the diagram becomes

$$
\begin{array}{ccc}
\mathrm{GL}_{nm}(k) & \xrightarrow{i_{nm,(n+1)m}} & \mathrm{GL}_{(n+1)m}(k) \\
\det \downarrow & & \downarrow \det \\
k^\times & \xrightarrow{\mathrm{id}} & k^\times,
\end{array}
$$

and commutativity is straightforward. □

By the lemma, the collection of reduced norm homomorphisms $\mathrm{Nrd}_n :$ $\mathrm{GL}_n(A) \to k^\times$ gives rise to a map $\mathrm{Nrd}_\infty : \mathrm{GL}_\infty(A) \to k^\times$, which induces a map

$$\mathrm{Nrd} : K_1(A) \to k^\times$$

called the *reduced norm map* for K_1. By construction, its composite with the natural map $A^\times \to K_1(A)$ induced by $A^\times = \mathrm{GL}_1(A) \to \mathrm{GL}_\infty(A)$ is the usual reduced norm $\mathrm{Nrd} : A^\times \to k^\times$.

Remark 2.10.9   In the split case $A = M_m(k)$ we call the reduced norm $K_1(M_m(k)) \to k^\times$ the determinant. Note that since $k^\times \cong K_1(k)$, we also have a map $i_{1,m} : k^\times \to K_1(M_m(k))$ in the other direction, which is moreover an isomorphism by Lemma 2.10.5. The end of the proof of Proposition 2.10.3 shows that the map $i_{1,m}$ and the determinant are inverse to each other.

To proceed further, we need the following compatibility property.

Proposition 2.10.10   *For a central simple k-algebra A and a finite field extension $K|k$ the diagram*

$$
\begin{array}{ccc}
K_1(A_K) & \xrightarrow{\;N_{K|k}\;} & K_1(A) \\
{\scriptstyle \mathrm{Nrd}_{A_K}}\big\downarrow & & \big\downarrow{\scriptstyle \mathrm{Nrd}_A} \\
K^\times & \xrightarrow{\;N_{K|k}\;} & k^\times
\end{array}
$$

commutes.

Proof   Again this can be checked after base change to a Galois splitting field of A. After such a base change the field K may not remain a field any more, but may become a finite product of fields. Still, the definition of the norm map $N_{K|k} : K_1(A_K) \to K_1(A)$ immediately generalizes to this setting, so we are reduced to checking the commutativity of the diagram

$$
\begin{array}{ccc}
K_1(M_m(K)) & \xrightarrow{\;N_{K|k}\;} & K_1(M_m(k)) \\
{\scriptstyle \det}\big\downarrow & & \big\downarrow{\scriptstyle \det} \\
K^\times & \xrightarrow{\;N_{K|k}\;} & k^\times
\end{array}
$$

where m is the degree of A. By Lemma 2.10.5 and Remark 2.10.4 (2) both vertical maps are isomorphisms. Furthermore, by Remark 2.10.9 the inverse maps are given by the maps $i_{1,m}$ of Lemma 2.10.5. Thus it remains to note that the diagram

$$K_1(K) \xrightarrow{N_{K|k}} K_1(k)$$

$$i_{1,m} \downarrow \qquad\qquad i_{1,m} \downarrow$$

$$K_1(M_m(K)) \xrightarrow{N_{K|k}} K_1(M_m(k)),$$

commutes by construction of the norm maps, and that its upper horizontal map becomes via the isomorphism of Proposition 2.10.3 the composite map

$$K^\times \to \operatorname{End}_k(K) \xrightarrow{\det} k,$$

which is indeed the usual field (or algebra) norm $N_{K|k}$. □

Remark 2.10.11 Proposition 2.10.10 and the surjectivity part of Remark 2.10.4 (1) yield another proof of Proposition 2.6.6.

Denote by $SK_1(A)$ the kernel of the reduced norm map $\operatorname{Nrd} : K_1(A) \to k^\times$. Proposition 2.10.10 shows that for each finite extension $K|k$ there is a norm map

$$N_{K|k} : SK_1(A_K) \to SK_1(A).$$

We now come to the main theorem of this section.

Theorem 2.10.12 (Wang) *If A is a central simple k-algebra of prime degree p, then $SK_1(A) = 0$.*

Proof The case when A is split is immediate from Lemma 2.10.5 and Proposition 2.10.3, so we may assume that A is a division algebra. By Remark 2.10.4 (1) the natural map $A^\times / [A^\times, A^\times] \to K_1(A)$ is surjective, so each element of $SK_1(A)$ may be represented by some element $a \in A^\times$ of trivial reduced norm. If $a \notin k$, let $L \subset A$ be the k-subalgebra generated by a; it is a degree p field extension of k. Otherwise take L to be any degree p extension of k contained in A. By Proposition 2.6.3 (2) we have $N_{L|k}(a) = 1$. The algebra $A_L := A \otimes_k L$ contains the subalgebra $L \otimes_k L$, which is not a division algebra, hence neither is A_L. Since $\deg_L(A_L) = p$, Wedderburn's theorem shows that $A \otimes_k L$ must be split. By the split case we have $SK_1(A_L) = 0$, hence the composite map

$$SK_1(A) \to SK_1(A_L) \xrightarrow{N_{L|k}} SK_1(A)$$

is trivial. Proposition 2.10.7 then implies that $p\,SK_1(A) = 0$. We now distinguish two cases.

Case 1: The extension $L|k$ is separable. Take a Galois closure $\widetilde{L}|k$ of L and denote by $K|k$ the fixed field of a p-Sylow subgroup in $\operatorname{Gal}(\widetilde{L}|k)$. Since

$\mathrm{Gal}\,(\widetilde{L}|k)$ is a subgroup of the symmetric group S_p, the extension $\widetilde{L}|K$ is a cyclic Galois extension of degree p. By Proposition 2.10.7 the composite

$$SK_1(A) \to SK_1(A_K) \xrightarrow{N_{K|k}} SK_1(A)$$

is multiplication by $[K : k]$, which is prime to p. On the other hand, we know that $pSK_1(A) = 0$, so the map $SK_1(A) \to SK_1(A_K)$ is injective. Up to replacing k by K and L by $\widetilde{L}$, we may thus assume that $L|k$ is cyclic of degree p. Let σ be a generator of $\mathrm{Gal}\,(L|k)$. According to the classical form of Hilbert's Theorem 90 (Example 2.3.4), there exists $c \in L^\times$ satisfying $a = c^{-1}\sigma(c)$. On the other hand, L is a subfield of A, which has degree p over k, so by Lemma 2.5.4 we find $b \in A^\times$ with $b^{-1}cb = \sigma(c)$. Hence $a = c^{-1}\sigma(c) = c^{-1}b^{-1}cb$ is a commutator in $A^\times$, and as such yields a trivial element in $SK_1(A)$.

Case 2: The extension $L|k$ is purely inseparable. In this case $N_{L|k}(a) = x^p = 1$ and thus $(a-1)^p = 0$. Since A is a division algebra, we must have $a = 1$, and the result follows. □

Remarks 2.10.13

1. With a little more knowledge of the theory of central simple algebras the theorem can be generalized to division algebras of arbitrary squarefree degree. See Chapter 4, Exercise 7.

2. In the same paper (Wang [1]) that contains the above theorem, Wang showed that over a number field the group $SK_1(A)$ is trivial for an arbitrary central simple algebra A. However, this is not so over an arbitrary field. Platonov [1] constructed examples of algebras A of degree p^2 for all primes p such that $SK_1(A) \neq 0$. For further work on $SK_1(A)$, see Merkurjev [4] and Suslin [3].

EXERCISES

1. (a) Prove that the tensor product $D_1 \otimes_k D_2$ of two division algebras of coprime degrees is a division algebra. [*Hint:* Apply Rieffel's lemma to a minimal left ideal L in $D_1 \otimes_k D_2$. Then show that $\dim_k(D_1 \otimes_k D_2) = \dim_k L$.]
   (b) Give another proof of Proposition 2.8.13 using Theorem 2.8.7(2) and question (a).

2. Show that a central simple algebra A of degree n over a field k is split if and only if it contains a k-subalgebra isomorphic to the direct product $k^n = k \times \cdots \times k$. [*Hint:* for the nontrivial implication, let $e_1, ..., e_n \in A$ be the images of the standard basis elements of k^n under the embedding $k^n \hookrightarrow A$. By passing to the algebraic closure, verify that Ae_1 is a simple left A-module and that the natural map $k \to \mathrm{End}_A(Ae_1)$ is an isomorphism, then apply Rieffel's lemma.]

3. Determine the cohomology set $H^1(G, \mathrm{SL}_n(K))$ for a finite Galois extension $K|k$ with group G.

4. Let $K|k$ be a finite Galois extension with group G, and let $B(K) \subset \mathrm{GL}_2(K)$ be the subgroup of upper triangular matrices.

   (a) Identify the quotient $\mathrm{GL}_2(K)/B(K)$ as a G-set with $\mathbf{P}^1(K)$, the set of K-points of the projective line.
   (b) Show that $H^1(G, B(K)) = 1$. [*Hint:* Exploit Proposition 2.7.1.]
   (c) Denote by K^+ the additive group of K. Show that $H^1(G, K^+) = 1$. [*Hint:* Observe that sending an element $a \in K^+$ to the 2×2 matrix (a_{ij}) with $a_{11} = a_{22} = 1, a_{21} = 0$ and $a_{12} = a$ defines a G-equivariant embedding $K^+ \to B(K)$.]

5. Let k be a field containing a primitive m-th root of unity ω. Take $a, b \in k^\times$ satisfying the condition in Proposition 2.5.5 (1). Prove that the class of the cyclic algebra $(a, b)_\omega$ has order dividing m in the Brauer group of k.

6. Show that the class of the cyclic algebra $(a, 1 - a)_\omega$ is trivial in the Brauer group for all $a \in k^\times$.

7. Show that the following are equivalent for a central simple k-algebra A:

   - A is split.
   - The reduced norm map $\mathrm{Nrd} : (A \otimes_k F)^\times \to F^\times$ is surjective for all field extensions $F|k$.
   - t is a reduced norm from the algebra $A \otimes_k k((t))$.

8. Let A be a central simple k-algebra of degree n. Assume that there exists a finite extension $K|k$ of degree prime to n that is a splitting field of A. Show that A is split. [*Hint:* Use the previous exercise.]

3

Techniques from group cohomology

In order to pursue our study of Brauer groups, we need some basic notions from the cohomology theory of groups with abelian coefficient modules. This is a theory which is well documented in the literature; we only establish here the facts we shall need in what follows, for the ease of the reader. In particular, we establish the basic exact sequences, construct cup-products and study the maps relating the cohomology of a group to that of a subgroup or a quotient. In accordance with the current viewpoint in homological algebra, we emphasize the use of complexes and projective resolutions, rather than that of explicit cocycles and the technique of dimension-shifting (though the latter are also very useful).

As already said, the subject matter of this chapter is fairly standard and almost all facts may already be found in the first monograph written on homological algebra, that of Cartan and Eilenberg [1]. Some of the constructions were first developed with applications to class field theory in view. For instance, Shapiro's lemma first appears in a footnote to Weil [1], then with a (two-page) proof in Hochschild–Nakayama [1].

3.1 Definition of cohomology groups

Let G be a group. By a *(left) G-module* we shall mean an abelian group A equipped with a left action by G. Notice that this is the same as giving a left module over the integral group ring $\mathbf{Z}[G]$: indeed, for elements $\sum n_\sigma \sigma \in \mathbf{Z}[G]$ and $a \in A$ we may define $(\sum n_\sigma \sigma)a := \sum n_\sigma \sigma(a)$ and conversely, a $\mathbf{Z}[G]$-module structure implies in particular the existence of 'multiplication-by-σ' maps on A for all $\sigma \in G$. We say that A is a *trivial* G-module if G acts trivially on A, i.e. $\sigma a = a$ for all $\sigma \in G$ and $a \in A$. By a G-homomorphism we mean a homomorphism $A \to B$ of abelian groups compatible with the

G-action. Denote by $\mathrm{Hom}_G(A, B)$ the set of G-homomorphisms $A \to B$; it is an abelian group under the natural addition of homomorphisms. Recall also that we denote by A^G the subgroup of G-invariant elements in a G-module A.

We would like to define for all G-modules A and all integers $i \geq 0$ abelian groups $H^i(G, A)$ subject to the following three properties.

1. $H^0(G, A) = A^G$ for all G-modules A.
2. For all G-homomorphisms $A \to B$ there exist canonical maps

$$H^i(G, A) \to H^i(G, B)$$

   for all $i \geq 0$.
3. Given a short exact sequence

$$0 \to A \to B \to C \to 0$$

   of G-modules, there exists an infinite long exact sequence

$$\cdots \to H^i(G, A) \to H^i(G, B) \to H^i(G, C) \to H^{i+1}(G, A) \to \cdots$$

   of abelian groups, starting from $i = 0$.

In other words, we would like to generalize the $H^1(G, A)$ introduced in the previous chapter to higher dimensions, and in particular we would like to continue the long exact sequence of Proposition 2.7.1 to an infinite sequence. This is known to be possible only when A is commutative; for noncommutative A reasonable definitions have been proposed only for $i = 2$ and 3, but we shall not consider them here.

To construct the groups $H^i(G, A)$ we begin by some reminders concerning left modules over a ring R which is not necessarily commutative but has a unit element 1. Recall that a *(cohomological) complex* $A^\bullet$ of R-modules is a sequence of R-module homomorphisms

$$\dots \cdots \xrightarrow{d^{i-1}} A^i \xrightarrow{d^i} A^{i+1} \xrightarrow{d^{i+1}} A^{i+2} \xrightarrow{d^{i+2}} \cdots$$

for all $i \in \mathbf{Z}$, satisfying $d^{i+1} \circ d^i = 0$ for all i. For $i < 0$ we shall also use the convention $A_{-i} := A^i$. We introduce the notation

$$Z^i(A^\bullet) := \ker(d^i), \quad B^i(A^\bullet) := \mathrm{Im}\,(d^{i-1}) \quad \text{and}$$
$$H^i(A^\bullet) := Z^i(A^\bullet)/B^i(A^\bullet).$$

The complex $A^\bullet$ is said to be *acyclic* or *exact* if $H^i(A^\bullet) = 0$ for all i.

A *morphism of complexes* $\phi : A^\bullet \to B^\bullet$ is a collection of homomorphisms $\phi^i : A^i \to B^i$ for all i such that the diagrams

$$
\begin{array}{ccc}
A^i & \longrightarrow & A^{i+1} \\
\phi^i \downarrow & & \downarrow \phi^{i+1} \\
B^i & \longrightarrow & B^{i+1}
\end{array}
$$

commute for all i. By this defining property, a morphism of complexes $A^\bullet \to B^\bullet$ induces maps $H^i(A^\bullet) \to H^i(B^\bullet)$ for all i. A *short exact sequence of complexes* is a sequence of morphisms of complexes

$$0 \to A^\bullet \to B^\bullet \to C^\bullet \to 0$$

such that the sequences

$$0 \to A^i \to B^i \to C^i \to 0$$

are exact for all i. Now we have the following basic fact which gives the key to the construction of cohomology groups satisfying property 3 above.

Proposition 3.1.1 *Let*

$$0 \to A^\bullet \to B^\bullet \to C^\bullet \to 0$$

be a short exact sequence of complexes of R-modules. Then there is a long exact sequence

$$\cdots \to H^i(A^\bullet) \to H^i(B^\bullet) \to H^i(C^\bullet) \xrightarrow{\partial} H^{i+1}(A^\bullet) \to H^{i+1}(B^\bullet) \to \cdots$$

The map ∂ is usually called the *connecting homomorphism* or the *(co)-boundary map*.

For the proof of the proposition we need the following equally basic lemma.

Lemma 3.1.2 (The Snake Lemma) *Given a commutative diagram of R-modules*

$$
\begin{array}{ccccccc}
A & \longrightarrow & B & \longrightarrow & C & \longrightarrow & 0 \\
\downarrow \alpha & & \downarrow \beta & & \downarrow \gamma & & \\
0 & \longrightarrow & A' & \longrightarrow & B' & \longrightarrow & C'
\end{array}
$$

with exact rows, there is an exact sequence

$$\ker(\alpha) \to \ker(\beta) \to \ker(\gamma) \to \operatorname{coker}(\alpha) \to \operatorname{coker}(\beta) \to \operatorname{coker}(\gamma).$$

Proof The construction of all maps in the sequence is immediate, except for the map $\partial : \ker(\gamma) \to \mathrm{coker}\,(\alpha)$. For this, lift $c \in \ker(\gamma)$ to $b \in B$. By commutativity of the right square, the element $\beta(b)$ maps to 0 in C', hence it comes from a unique $a' \in A'$. Define $\partial(c)$ as the image of a' in $\mathrm{coker}\,(\alpha)$. Two choices of b differ by an element $a \in A$ which maps to 0 in $\mathrm{coker}\,(\alpha)$, so ∂ is well defined. Checking exactness is left as an exercise to the readers. □

Proof of Proposition 3.1.1 Applying the Snake Lemma to the diagram

$$
\begin{array}{ccccccc}
A^i/B^i(A^\bullet) & \longrightarrow & B^i/B^i(B^\bullet) & \longrightarrow & C^i/B^i(C^\bullet) & \longrightarrow & 0 \\
\downarrow{\scriptstyle\alpha} & & \downarrow{\scriptstyle\beta} & & \downarrow{\scriptstyle\gamma} & & \\
0 \longrightarrow Z^{i+1}(A^\bullet) & \longrightarrow & Z^{i+1}(B^\bullet) & \longrightarrow & Z^{i+1}(C^\bullet) & &
\end{array}
$$

yields a long exact sequence

$$
H^i(A^\bullet) \to H^i(B^\bullet) \to H^i(C^\bullet) \to H^{i+1}(A^\bullet) \to H^{i+1}(B^\bullet) \to H^{i+1}(C^\bullet),
$$

and the proposition is obtained by splicing these sequences together. □

We also have to recall the notion of *projective* R-modules. By definition, these are R-modules P for which the natural map $\mathrm{Hom}(P, A) \to \mathrm{Hom}(P, B)$ given by $\lambda \to \alpha \circ \lambda$ is surjective for every *surjection* $\alpha : A \to B$.

Lemma 3.1.3

1. *The R-module R is projective.*
2. *Arbitrary direct sums of projective modules are projective.*

Proof For the first statement, given an R-homomorphism $\lambda : R \to B$ and a surjection $A \to B$, lift λ to an element of $\mathrm{Hom}(R, A)$ by lifting $\lambda(1)$ to an element of A. The second statement is immediate from the compatibility of Hom-groups with direct sums in the first variable. □

Recall also that a *free R-module* is by definition an R-module isomorphic to a (possibly infinite) direct sum of copies of the R-module R. The above lemma then yields:

Corollary 3.1.4 *A free R-module is projective.*

Example 3.1.5 Given an R-module A, define a free R-module $F(A)$ by taking an infinite direct sum of copies of R indexed by the elements of A. One has a surjection $\pi_A : F(A) \to A$ induced by mapping 1_a to a, where 1_a is the element of $F(A)$ with 1 in the component corresponding to $a \in A$ and 0 elsewhere.

As a first application of this example, we prove the following lemma:

Lemma 3.1.6 *An R-module P is projective if and only if there exist an R-module M and a free R-module F with $P \oplus M \cong F$.*

Proof For sufficiency, extend a map $\lambda : P \to B$ to F by defining it to be 0 on M and use projectivity of F. For necessity, take F to be the free R-module $F(P)$ associated with P in the above example. We claim that we have an isomorphism as required, with $M = \ker(\pi_P)$. Indeed, as P is projective, we may lift the identity map of P to a map $\pi : P \to F(P)$ with $\pi_P \circ \pi = \mathrm{id}_P$. □

For each R-module A there exist *projective resolutions*, i.e. infinite exact sequences

$$\cdots \to P_2 \to P_1 \to P_0 \to A \to 0$$

with P_i projective. One may take, for instance, P_0 to be the free R-module $F(A)$ defined in the example above; in particular, we get a surjection $p_0 : P_0 \to A$. Once P_i and $p_i : P_i \to P_{i-1}$ are defined (with the convention $P_{-1} = A$), one defines P_{i+1} and p_{i+1} by applying the same construction to $\ker(p_i)$ in place of A.

Now the basic fact concerning projective resolutions is:

Lemma 3.1.7 *Assume given a diagram*

$$\cdots \longrightarrow P_2 \xrightarrow{p_2} P_1 \xrightarrow{p_1} P_0 \xrightarrow{p_0} A \longrightarrow 0$$
$$\downarrow{\alpha}$$
$$\cdots \longrightarrow B_2 \xrightarrow{b_2} B_1 \xrightarrow{b_1} B_0 \xrightarrow{b_0} B \longrightarrow 0$$

where the upper row is a projective resolution of the R-module A and the lower row is an exact sequence of R-modules. Then there exist maps $\alpha_i : P_i \to B_i$ for all $i \geq 0$ making the diagram

$$\cdots \longrightarrow P_2 \xrightarrow{p_2} P_1 \xrightarrow{p_1} P_0 \xrightarrow{p_0} A \longrightarrow 0$$
$$\downarrow{\alpha_2} \qquad \downarrow{\alpha_1} \qquad \downarrow{\alpha_0} \qquad \downarrow{\alpha}$$
$$\cdots \longrightarrow B_2 \xrightarrow{b_2} B_1 \xrightarrow{b_1} B_0 \xrightarrow{b_0} B \longrightarrow 0$$

commute. Moreover, if (α_i) and (β_i) are two collections with this property, there exist maps $\gamma_i : P_i \to B_{i+1}$ for all $i \geq -1$ (with the conventions $P_{-1} = A$, $\alpha_{-1} = \beta_{-1} = \alpha$) satisfying

$$\alpha_i - \beta_i = \gamma_{i-1} \circ p_i + b_{i+1} \circ \gamma_i. \tag{3.1}$$

Proof To construct α_i, assume that the α_j are already defined for $j < i$, with the convention $\alpha_{-1} = \alpha$. Observe that $\mathrm{Im}\,(\alpha_{i-1} \circ p_i) \subset \mathrm{Im}\,(b_i)$; this is immediate for $i = 0$ and follows from $b_{i-1} \circ \alpha_{i-1} \circ p_i = \alpha_{i-2} \circ p_{i-1} \circ p_i = 0$ for $i > 0$ by exactness of the lower row. Hence by the projectivity of P_i we may define α_i as a preimage in $\mathrm{Hom}(P_i, B_i)$ of the map $\alpha_{i-1} \circ p_i : P_i \to \mathrm{Im}\,(b_i)$. For the second statement, define $\gamma_{-1} = 0$ and assume γ_j defined for $j < i$ satisfying (3.1) above. This implies $\mathrm{Im}\,(\alpha_i - \beta_i - (\gamma_{i-1} \circ p_i)) \subset \mathrm{Im}\,(b_{i+1})$ because

$$b_i \circ (\alpha_i - \beta_i - (\gamma_{i-1} \circ p_i)) = (\alpha_{i-1} - \beta_{i-1}) \circ p_i - b_i \circ \gamma_{i-1} \circ p_i = \gamma_{i-2} \circ p_{i-1} \circ p_i = 0,$$

so, again using the projectivity of P_i, we may define γ_i as a preimage of $\alpha_i - \beta_i - (\gamma_{i-1} \circ p_i) \in \mathrm{Hom}(P_i, \mathrm{Im}\,(b_{i+1}))$ in $\mathrm{Hom}(P_i, B_{i+1})$, again using the projectivity of P_i. □

Now we can construct the cohomology groups $H^i(G, A)$.

Construction 3.1.8 Let G be a group and A a G-module. Take a projective resolution $P_\bullet = (\cdots \to P_2 \xrightarrow{p_2} P_1 \xrightarrow{p_1} P_0)$ of the trivial G-module $\mathbf{Z}$. Consider the sequence $\mathrm{Hom}_G(P_\bullet, A)$ defined by

$$\mathrm{Hom}_G(P_0, A) \to \mathrm{Hom}_G(P_1, A) \to \mathrm{Hom}_G(P_2, A) \to \cdots$$

where the maps $\mathrm{Hom}_G(P_i, A) \to \mathrm{Hom}_G(P_{i+1}, A)$ are defined by $\lambda \mapsto \lambda \circ p_{i+1}$. The fact that $P_\bullet$ is a complex of G-modules implies that $\mathrm{Hom}_G(P_\bullet, A)$ is a complex of abelian groups; we index it by defining $\mathrm{Hom}_G(P_i, A)$ to be the term in degree i. We may now put

$$H^i(G, A) := H^i(\mathrm{Hom}_G(P_i, A))$$

for $i \geq 0$.

Proposition 3.1.9 *The groups $H^i(G, A)$ satisfy properties 1–3 postulated at the beginning of this section, and they do not depend on the choice of the resolution $P_\bullet$.*

Proof Notice first that $\mathrm{Hom}_G(\mathbf{Z}, A) \cong A^G$, the isomorphism arising from sending a G-homomorphism $\phi : \mathbf{Z} \to A$ to $\phi(1)$. On the other hand, every G-homomorphism $\mathbf{Z} \to A$ lifts to $\lambda_0 : P_0 \to A$ inducing the trivial homomorphism by composition with p_1. Conversely, each such λ_0 defines an element of $\mathrm{Hom}_G(\mathbf{Z}, A)$, whence property 1. Property 2 is immediate from the construction and property 3 follows from Proposition 3.1.1 applied to the sequence of complexes

$$0 \to \mathrm{Hom}_G(P_\bullet, A) \to \mathrm{Hom}_G(P_\bullet, B) \to \mathrm{Hom}_G(P_\bullet, C) \to 0,$$

which is exact because the P_i are projective. For the second statement, let $Q_\bullet$ be another projective resolution of $\mathbf{Z}$ and apply Lemma 3.1.7 with $A = \mathbf{Z}$, $B^\bullet = Q_\bullet$ and $\alpha = \mathrm{id}$. We get maps $\alpha_i : P_i \to Q_i$ inducing $\alpha_i^* : H^i(\mathrm{Hom}_G(Q_i, A)) \to H^i(\mathrm{Hom}_G(P_i, A))$ on cohomology. Exchanging the roles of the resolutions $P_\bullet$ and $Q_\bullet$ we also get maps $\beta_i : Q_i \to P_i$ inducing $\beta_i^* : H^i(\mathrm{Hom}_G(P_i, A)) \to H^i(\mathrm{Hom}_G(Q_i, A))$. We show that the compositions $\alpha_i^* \circ \beta_i^*$ and $\beta_i^* \circ \alpha_i^*$ are identity maps. By symmetry it is enough to do this for the first one. Apply the second statement of Lemma 3.1.7 with P_i in place of B^i and the maps $\beta_i \circ \alpha_i$ and id_{P_i} in place of the α_i and β_i of the lemma. We get $\gamma_i : P_i \to P_{i+1}$ satisfying $\beta_i \circ \alpha_i - \mathrm{id}_{P_i} = \gamma_{i-1} \circ p_i + p_{i+1} \circ \gamma_i$, whence $\lambda \circ \beta_i \circ \alpha_i - \lambda = \lambda \circ \gamma_{i-1} \circ p_i$ for a map $\lambda \in \mathrm{Hom}_G(P_i, A)$ satisfying $\lambda \circ p_{i+1} = 0$. This means precisely that $\lambda \circ \beta_i \circ \alpha_i - \lambda$ is in the image of the map $\mathrm{Hom}_G(P_{i-1}, A) \to \mathrm{Hom}_G(P_i, A)$, i.e. $(\beta_i \circ \alpha_i)^* = \alpha_i^* \circ \beta_i^*$ equals the identity map of $H^i(\mathrm{Hom}_G(P_i, A))$. $\square$

Remarks 3.1.10

1. The above construction is a special case of that of Ext-groups in homological algebra: for two R-modules M and N these are defined by $\mathrm{Ext}^i(M, N) := H^i(\mathrm{Hom}_R(P_\bullet, N))$ with a projective resolution $P_\bullet$ of M. The same argument as above shows independence of the choice of $P_\bullet$. In this parlance we therefore get $H^i(G, A) = \mathrm{Ext}^i_{\mathbf{Z}[G]}(\mathbf{Z}, A)$.

2. Similarly, given two R-modules M and N, one defines the R-modules $\mathrm{Tor}_i^R(M, N) := H_i(P_\bullet \otimes_R N)$ with a projective resolution $P_\bullet$ of M. There is a long exact sequence

$$\cdots \to \mathrm{Tor}_i^R(M, N') \to \mathrm{Tor}_i^R(M, N) \to \mathrm{Tor}_i^R(M, N'') \to \mathrm{Tor}_{i-1}^R(M, N') \to \cdots$$

   associated with every short exact sequence $0 \to N' \to N \to N'' \to 0$ of R-modules. Applying it with $R = \mathbf{Z}$ and the exact sequence $0 \to \mathbf{Z} \to \mathbf{Z} \to \mathbf{Z}/m\mathbf{Z} \to 0$ for an integer $m > 0$ shows that $\mathrm{Tor}_1^{\mathbf{Z}}(\mathbf{Z}/m\mathbf{Z}, N)$ is isomorphic to ${}_m N = \{x \in N : mx = 0\}$, noting that $\mathrm{Tor}_i^{\mathbf{Z}}(\mathbf{Z}, N) = 0$ for $i > 0$. For more details, see the book of Weibel [1].

In the case $R = \mathbf{Z}[G]$ and $M = \mathbf{Z}$ one may use the groups $\mathrm{Tor}_i^{\mathbf{Z}[G]}(\mathbf{Z}, N)$ to define a homology theory for G-modules, but we shall not use it in this book.

3. It follows from the definition that cohomology groups satisfy certain natural functorial properties. Namely, if

$$
\begin{array}{ccc}
A & \longrightarrow & B \\
\downarrow & & \downarrow \\
A' & \longrightarrow & B'
\end{array}
$$

is a commutative diagram of G-modules, then the associated diagrams

$$
\begin{array}{ccc}
H^i(G, A) & \longrightarrow & H^i(G, B) \\
\downarrow & & \downarrow \\
H^i(G, A') & \longrightarrow & H^i(G, B')
\end{array}
$$

commute for all $i \geq 0$. Moreover, given a commutative diagram

$$
\begin{array}{ccccccccc}
0 & \longrightarrow & A & \longrightarrow & B & \longrightarrow & C & \longrightarrow & 0 \\
& & \downarrow & & \downarrow & & \downarrow & & \\
0 & \longrightarrow & A' & \longrightarrow & B' & \longrightarrow & C' & \longrightarrow & 0
\end{array}
$$

of short exact sequences, the diagrams

$$
\begin{array}{ccc}
H^i(G, C) & \longrightarrow & H^{i+1}(G, A) \\
\downarrow & & \downarrow \\
H^i(G, C') & \longrightarrow & H^{i+1}(G, A')
\end{array}
$$

coming from the functorial property and the long exact sequences commute for all $i \geq 0$.

3.2 Explicit resolutions

To calculate the groups $H^i(G, A)$ explicitly, one uses concrete projective resolutions. The most useful of these is the following one, inspired by simplicial constructions in topology.

Construction 3.2.1 (The standard resolution) Consider for each $i \geq 0$ the $\mathbf{Z}[G]$-module $\mathbf{Z}[G^{i+1}]$, where G^{i+1} is the $(i+1)$-fold direct power of G and the action of G is determined by $\sigma(\sigma_0, \ldots, \sigma_i) = (\sigma\sigma_0, \ldots, \sigma\sigma_i)$. These are projective (in fact, free) $\mathbf{Z}[G]$-modules, generated by the free $\mathbf{Z}[G]$-basis G^i. For $i > 0$ define G-homomorphisms $\delta^i : \mathbf{Z}[G^{i+1}] \to \mathbf{Z}[G^i]$ by $\delta^i = \sum_j (-1)^j s_j^i$, where $s_j^i : \mathbf{Z}[G^{i+1}] \to \mathbf{Z}[G^i]$ is the map determined by sending

$$(\sigma_0, \ldots, \sigma_i) \mapsto (\sigma_0, \ldots, \sigma_{j-1}, \sigma_{j+1}, \ldots, \sigma_i).$$

In this way, we get a projective resolution

$$\cdots \to \mathbf{Z}[G^3] \xrightarrow{\delta^2} \mathbf{Z}[G^2] \xrightarrow{\delta^1} \mathbf{Z}[G] \xrightarrow{\delta^0} \mathbf{Z} \to 0,$$

where δ^0 sends each σ_i to 1. This resolution is called the *standard resolution* of $\mathbf{Z}$. To see that the sequence is indeed exact, an immediate calculation shows first that $\delta^i \circ \delta^{i+1} = 0$ for all i. Then fix $\sigma \in G$ and define $h^i : \mathbf{Z}[G^{i+1}] \to \mathbf{Z}[G^{i+2}]$ by sending $(\sigma_0, \ldots, \sigma_i)$ to $(\sigma, \sigma_0, \ldots, \sigma_i)$. Another calculation shows $\delta^{i+1} \circ h^i + h^{i-1} \circ \delta^i = \mathrm{id}_{\mathbf{Z}[G^{i+1}]}$, which implies $\ker(\delta^i) = \mathrm{Im}\,(\delta^{i+1})$.

For a G-module A, one calls the elements of $\mathrm{Hom}_G(\mathbf{Z}[G^{i+1}], A)$ *i-cochains*, whereas those of $Z^{i+1}(\mathrm{Hom}_G(\mathbf{Z}[G^\bullet], A))$ and $B^{i+1}(\mathrm{Hom}_G(\mathbf{Z}[G^\bullet], A))$ *i-cocycles* and *i-coboundaries*, respectively. We shall denote these respective groups by $C^i(G, A)$, $Z^i(G, A)$ and $B^i(G, A)$. The cohomology groups $H^i(G, A)$ then arise as the groups $H^{i+1}(\mathrm{Hom}_G(\mathbf{Z}[G^\bullet], A))$. We shall see in the example below that for $i = 1$ we get back the notions of the previous chapter (in the commutative case).

For calculations, another expression is very useful.

Construction 3.2.2 (Inhomogeneous cochains) In $\mathbf{Z}[G^{i+1}]$ consider the particular basis elements

$$[\sigma_1, \ldots, \sigma_i] := (1, \sigma_1, \sigma_1\sigma_2, \ldots, \sigma_1 \ldots \sigma_i).$$

From the definition of the G-action on $\mathbf{Z}[G^{i+1}]$ we get that $\mathbf{Z}[G^{i+1}]$ is none but the free $\mathbf{Z}[G]$-module generated by the elements $[\sigma_1, \ldots, \sigma_i]$. A calculation shows that on these elements the differentials δ^i are expressed by

$$\delta^i([\sigma_1, \ldots, \sigma_i]) = \sigma_1[\sigma_2, \ldots, \sigma_i] + \sum_{j=1}^{i} (-1)^j [\sigma_1, \ldots, \sigma_j\sigma_{j+1}, \ldots, \sigma_i] +$$

$$+ (-1)^{i+1}[\sigma_1, \ldots, \sigma_{i-1}]. \tag{3.2}$$

Therefore we may identify i-cochains with functions $[\sigma_1, \ldots, \sigma_i] \to a_{\sigma_1, \ldots, \sigma_i}$ and compute the maps $\delta_{i-1}^* : C^{i-1}(G, A) \to C^i(G, A)$ by the formula

$$a_{\sigma_1, \ldots, \sigma_{i-1}} \mapsto \sigma_1 a_{\sigma_2, \ldots, \sigma_i} + \sum_{j=1}^{i} (-1)^j a_{\sigma_1, \ldots, \sigma_j\sigma_{j+1}, \ldots, \sigma_i} + (-1)^{i+1} a_{\sigma_1, \ldots, \sigma_{i-1}}.$$

The functions $a_{\sigma_1, \ldots, \sigma_i}$ are called *inhomogeneous cochains*.

Here is how to calculate the groups $H^i(G, A)$ in low dimensions by means of inhomogeneous cochains.

Examples 3.2.3

1. A 1-cocycle is given by a function $\sigma \mapsto a_\sigma$ satisfying $a_{\sigma_1 \sigma_2} = \sigma_1 a_{\sigma_2} + a_{\sigma_1}$. It is a 1-coboundary if and only if it is of the form $\sigma \mapsto \sigma a - a$ for some $a \in A$. We thus get back the first cohomology group defined in the noncommutative situation in the previous chapter. Note that in the special case when G acts trivially on A, i.e. $\sigma(a) = a$ for all $a \in A$, we have $Z^1(G, A) = \text{Hom}(G, A)$ and $B^1(G, A) = 0$, so finally $H^1(G, A) = \text{Hom}(G, A)$.
2. A 2-cocycle is given by a function $(\sigma_1, \sigma_2) \mapsto a_{\sigma_1, \sigma_2}$ satisfying

$$\sigma_1 a_{\sigma_2, \sigma_3} - a_{\sigma_1 \sigma_2, \sigma_3} + a_{\sigma_1, \sigma_2 \sigma_3} - a_{\sigma_1, \sigma_2} = 0.$$

It is a 2-coboundary, i.e. an element of $\text{Im}(\partial^{1*})$ if it is of the form $\sigma_1 a_{\sigma_2} - a_{\sigma_1 \sigma_2} + a_{\sigma_1}$ for some 1-cochain $\sigma \mapsto a_\sigma$. These are the relations of Definition 2.7.5, written additively.

Remark 3.2.4 Using the above description via cocycles one also gets explicit formulae for the coboundary maps $\delta^i : H^i(G, C) \to H^{i+1}(G, A)$ in long exact cohomology sequences. In particular, in the case $i = 0$ we get the same answer as in the noncommutative situation (Proposition 2.7.1): given $c \in C^G$, we lift it to an element $b \in B$, and $\delta^0(c)$ is represented by the map $\sigma \mapsto \sigma b - b$, which is readily seen to be a 1-cocycle with values in A. Similarly, for $i = 1$ we obtain the coboundary map constructed in the proof of Proposition 2.7.6.

Example 3.2.5 For some questions (e.g. as in the example of group extensions below) it is convenient to work with *normalized cochains*. These are obtained by considering the free resolution

$$\cdots \to L_2 \xrightarrow{\delta_n^2} L_1 \xrightarrow{\delta_n^1} L_0 \xrightarrow{\delta_n^0} \mathbf{Z} \to 0,$$

where L_i is the free G-submodule of $\mathbf{Z}[G^{i+1}]$ generated by those $[\sigma_1, \ldots \sigma_i]$ where none of the σ_j is 1. The morphisms δ_n^i are defined by the same formulae as for the δ^i in (3.2), except that if we happen to have $\sigma_j \sigma_{j+1} = 1$ for some j in $[\sigma_1, \ldots \sigma_i]$, we set the term involving $\sigma_j \sigma_{j+1}$ on the right-hand side to 0. This indeed defines a map $L_i \to L_{i-1}$, and a calculation shows that we again have $\ker(\delta_n^i) = \text{Im}(\delta_n^{i+1})$. So we have obtained a free resolution of $\mathbf{Z}$ and may use it for computing the cohomology of a G-module A. Elements in $\text{Hom}_G(L_i, A)$ may be identified with inhomogeneous i-cochains $a_{\sigma_1, \ldots, \sigma_i}$ which have the value 0 whenever one of the σ_j equals 1.

Example 3.2.6 (Group extensions) An important example of 2-cocycles arising 'in nature' comes from the theory of group extensions. Consider an exact sequence of groups $0 \to A \to E \to G \overset{\pi}{\to} 1$, with A abelian. The conjugation action of E on A passes to the quotient in G and gives A the structure of a G-module. Now associate with E a 2-cocycle as follows. Choose a *normalized set-theoretic section* of π, i.e. a map $s : G \to E$ with $s(1) = 1$ and $\pi \circ s = \mathrm{id}_G$. For elements $\sigma_1, \sigma_2 \in G$ the element $a_{\sigma_1,\sigma_2} := s(\sigma_1)s(\sigma_2)s(\sigma_1\sigma_2)^{-1}$ maps to 1 in G, and therefore defines an element of A. An immediate calculation shows that $(\sigma_1, \sigma_2) \mapsto a_{\sigma_1,\sigma_2}$ is a 2-cocycle of G with values in A, which is in fact normalized, i.e. satisfies $a_{1,\sigma} = a_{\sigma,1} = 1$ for all $\sigma \in G$. Another calculation shows that replacing s by another set-theoretic section yields a 2-cocycle with the same class in $H^2(G, A)$. In this way one associates with E a class $c(E) \in H^2(G, A)$. Furthermore, we see that in the case when there is a section s which is a group homomorphism, i.e. the extension E splits as a semi-direct product of G by A, then $c(E) = 0$.

In fact, once we fix a G-action on A, we may consider the set $\mathrm{Ext}(G, A)$ of equivalence classes of extensions E of G by A inducing the given action of G on A modulo the following equivalence relation: two extensions E and E' are called equivalent if there is an isomorphism $\lambda : E \overset{\sim}{\to} E'$ inducing a commutative diagram

$$
\begin{array}{ccccccccc}
0 & \longrightarrow & A & \longrightarrow & E & \longrightarrow & G & \longrightarrow & 1 \\
& & \downarrow{\scriptstyle \mathrm{id}} & & \downarrow{\scriptstyle \lambda} & & \downarrow{\scriptstyle \mathrm{id}} & & \\
0 & \longrightarrow & A & \longrightarrow & E' & \longrightarrow & G & \longrightarrow & 1.
\end{array}
$$

The map $E \mapsto c(E)$ is easily seen to preserve this equivalence relation, and in fact induces a bijection between $\mathrm{Ext}(G, A) \to H^2(G, A)$. The inverse is constructed as follows: one represents a class in $H^2(G, A)$ by a *normalized* cocycle a_{σ_1,σ_2} and defines a group E with underlying set $A \times G$ and group law $(a_1, \sigma_1) \cdot (a_2, \sigma_2) := (a_1 + \sigma_1(a_2) + a_{\sigma_1,\sigma_2}, \sigma_1\sigma_2)$. The cocycle relation implies that this product is associative, and the fact that a_{σ_1,σ_2} is normalized implies that $(0, 1)$ is a unit element. The element $(-\sigma^{-1}(a) - \sigma^{-1}(a_{\sigma,\sigma^{-1}}), \sigma^{-1})$ yields an inverse for (a, σ), therefore E is indeed a group and one checks that it is an extension of G by A with $c(E) = [a_{\sigma_1,\sigma_2}]$. All this is verified by straightforward calculations which we leave to the readers to carry out or to look up e.g. in Weibel [1], Section 6.6.

Remark 3.2.7 Given a homomorphism $\phi : A \to B$ of G-modules, the natural map $\phi_* : H^2(G, A) \to H^2(G, B)$ induced on cohomology has the following interpretation in terms of group extensions: the class $c(E)$ of an extension $0 \to A \xrightarrow{\iota} E \to G \to 1$ is mapped to that of the *pushforward* extension $\phi_*(E)$ defined as the quotient of $B \times E$ by the normal subgroup of elements of the form $(\phi(a), \iota(a)^{-1})$ for $a \in A$. One verifies that $\phi_*(E)$ is indeed an extension of G by B, and that $c(\phi_*(E)) = \phi_*(c(E))$ by the explicit description of the cocycle class $c(E)$ given above.

For special groups other projective resolutions may be useful for computing cohomology, as the examples of cyclic groups show.

Example 3.2.8 Let $G = \mathbf{Z}$. Then the sequence

$$0 \to \mathbf{Z}[\mathbf{Z}] \to \mathbf{Z}[\mathbf{Z}] \to \mathbf{Z} \to 0$$

gives a projective resolution of the trivial $\mathbf{Z}$-module $\mathbf{Z}$, where the second map is given by multiplication by $\sigma - 1$ for a generator σ of $\mathbf{Z}$ considered a cyclic group, and the third one is induced by mapping σ to 1. It is immediate to check the exactness of the sequence, and for a $\mathbf{Z}[\mathbf{Z}]$-module A we get

$$H^0(\mathbf{Z}, A) = A^\sigma, \quad H^1(\mathbf{Z}, A) = A/(\sigma - 1)A \quad \text{and}$$
$$H^i(\mathbf{Z}, A) = 0 \quad \text{for} \quad i > 1.$$

Example 3.2.9 Let now G be a finite cyclic group of order n, generated by an element σ. Consider the maps $\mathbf{Z}[G] \to \mathbf{Z}[G]$ defined by

$$N : a \mapsto \sum_{i=0}^{n-1} \sigma^i a \quad \text{and} \quad \sigma - 1 : a \mapsto \sigma a - a.$$

One checks easily that $\ker(N) = \operatorname{Im}(\sigma - 1)$ and $\operatorname{Im}(N) = \ker(\sigma - 1)$. Hence we obtain a free resolution

$$\cdots \xrightarrow{N} \mathbf{Z}[G] \xrightarrow{\sigma - 1} \mathbf{Z}[G] \xrightarrow{N} \mathbf{Z}[G] \xrightarrow{\sigma - 1} \mathbf{Z}[G] \to \mathbf{Z} \to 0,$$

the last map being induced by $\sigma \mapsto 1$.

For a G-module A, define maps $N : A \to A$ and $\sigma - 1 : A \to A$ by the same formulae as above and put $_N A := \ker(N)$. Using the above resolution, one finds

$$H^0(G, A) = A^G, \quad H^{2i+1}(G, A) = {_N A}/(\sigma - 1)A \text{ and}$$
$$H^{2i+2}(G, A) = A^G/NA \tag{3.3}$$

for $i > 0$.

Remark 3.2.10 If $K|k$ is a finite Galois extension with cyclic Galois group G as above, the above calculation shows $H^1(G, K^\times) = {}_N K^\times / (\sigma - 1) K^\times$. The first group is trivial by Hilbert's Theorem 90 and we get back the original form of the theorem, as established in Example 2.3.4 of the previous chapter.

3.3 Relation to subgroups

Let H be a subgroup of G and A an H-module. Then $\mathbf{Z}[G]$ with its canonical G-action is an H-module as well, and we can associate with A the G-module

$$M_H^G(A) := \mathrm{Hom}_H(\mathbf{Z}[G], A)$$

where the action of G on an H-homomorphism $\phi : \mathbf{Z}[G] \to A$ is given by $(\sigma \phi)(g) := \phi(g \sigma)$ for a basis element g of $\mathbf{Z}[G]$. One sees that $\sigma \phi$ is indeed an H-homomorphism.

Lemma 3.3.1 *Assume moreover given a G-module M. We have a canonical isomorphism*

$$\mathrm{Hom}_G(M, \mathrm{Hom}_H(\mathbf{Z}[G], A)) \xrightarrow{\sim} \mathrm{Hom}_H(M, A)$$

induced by mapping a G-homomorphism $m \to \phi_m$ in the left-hand side group to the H-homomorphism $m \mapsto \phi_m(1)$.

Proof Given an H-homomorphism $\lambda : M \to A$, consider the map $m \mapsto \lambda_m$, where $\lambda_m \in \mathrm{Hom}_H(\mathbf{Z}[G], A)$ is the map determined by $g \mapsto \lambda(gm)$. The kind reader will check that we get an element of $\mathrm{Hom}_G(M, \mathrm{Hom}_H(\mathbf{Z}[G], A))$ in this way, and that the two constructions are inverse to each other. $\square$

Now apply the lemma to the terms of a projective $\mathbf{Z}[G]$-resolution $P_\bullet$ of $\mathbf{Z}$. Note that this is also a resolution by projective H-modules because $\mathbf{Z}[G]$ is free as a $\mathbf{Z}[H]$-module (a system of coset representatives yields a basis). Passing to cohomology groups, we get:

Corollary 3.3.2 (Shapiro's lemma) *Given a subgroup H of G and an H-module A, there are canonical isomorphisms*

$$H^i(G, M_H^G(A)) \xrightarrow{\sim} H^i(H, A)$$

for all $i \geq 0$.

The case when $H = \{1\}$ is particularly important. In this case an H-module A is just an abelian group; we denote $M_H^G(A)$ simply by $M^G(A)$ and call it the *co-induced* module associated with A.

Corollary 3.3.3 *The group $H^i(G, M^G(A))$ is trivial for all $i > 0$.*

Proof In this case the right-hand side in Shapiro's lemma is trivial (e.g. because $0 \to \mathbf{Z} \to \mathbf{Z} \to 0$ gives a projective resolution of $\mathbf{Z}$). □

Remarks 3.3.4

1. It is important to note that the construction of co-induced modules is func-torial in the sense that every homomorphism $A \to B$ of abelian groups induces a G-homomorphism $M^G(A) \to M^G(B)$. Of course, a similar property holds for the modules $M_H^G(A)$.
2. For a G-module A there is a natural injective map $A \to M^G(A)$ given by assigning to $a \in A$ the homomorphism $\mathbf{Z}[G] \to A$ of abelian groups induced by the mapping $\sigma \mapsto \sigma a$.
3. If G is finite, the choice of a $\mathbf{Z}$-basis of $\mathbf{Z}[G]$ induces a *noncanonical* isomorphism $M^G(A) \cong A \otimes_{\mathbf{Z}} \mathbf{Z}[G]$ for all abelian groups A.

Using Shapiro's lemma we may define two basic maps relating the cohomology of a group to that of a subgroup.

Construction 3.3.5 (Restriction maps) Let G be a group, A a G-module and H a subgroup of G. There are natural maps of G-modules

$$A \xrightarrow{\sim} \mathrm{Hom}_G(\mathbf{Z}[G], A) \to \mathrm{Hom}_H(\mathbf{Z}[G], A) = M_H^G(A),$$

the first one given by mapping $a \in A$ to the unique G-homomorphism sending 1 to a and the second by considering a G-homomorphism as an H-homomorphism. Taking cohomology and applying Shapiro's lemma we thus get maps

$$\mathrm{Res} : H^i(G, A) \to H^i(H, A)$$

for all $i \geq 0$, called *restriction maps*. One sees that for $i = 0$ we get the natural inclusion $A^G \to A^H$.

When the subgroup H has finite index, there is a natural map in the opposite direction.

Construction 3.3.6 (Corestriction maps)   Let H be a subgroup of G of finite index n and let A be a G-module.

Given an H-homomorphism $\phi : \mathbf{Z}[G] \to A$, define a new map $\mathbf{Z}[G] \to A$ by the assignment

$$\phi_H^G : x \mapsto \sum_{j=1}^n \rho_j \phi(\rho_j^{-1} x),$$

where $\rho_1, \ldots, \rho_n$ is a system of left coset representatives for H in G. This is manifestly a group homomorphism which does not depend on the choice of the ρ_j; indeed, if we replace the system of representatives (ρ_j) by another system $(\rho_j \tau_j)$ with some $\tau_j \in H$, we get $\rho_j \tau_j \phi(\tau_j^{-1} \rho_j^{-1} x) = \rho_j \phi(\rho_j^{-1} x)$ for all j, the map ϕ being an H-homomorphism. Furthermore, the map ϕ_H^G is also a G-homomorphism, because we have for all $\sigma \in G$

$$\sum_{j=1}^n \rho_j \phi(\rho_j^{-1} \sigma x) = \sigma \left(\sum_{j=1}^n (\sigma^{-1} \rho_j) \phi((\sigma^{-1} \rho_j)^{-1} x) \right)$$

$$= \sigma \left(\sum_{j=1}^n \rho_j \phi(\rho_j^{-1} x) \right),$$

as the $\sigma^{-1} \rho_j$ form another system of left coset representatives.

The assignment $\phi \mapsto \phi_H^G$ thus defines a well-defined map

$$\mathrm{Hom}_H(\mathbf{Z}[G], A) \to \mathrm{Hom}_G(\mathbf{Z}[G], A) \cong A,$$

so by taking cohomology and applying Shapiro's lemma we get maps

$$\mathrm{Cor} : H^i(H, A) \to H^i(G, A)$$

for all $i \geq 0$, called *corestriction maps*.

An immediate consequence of the preceding constructions is the following basic fact.

Proposition 3.3.7   *Let G be a group, H a subgroup of finite index n in G and A a G-module. Then the composite maps*

$$\mathrm{Cor} \circ \mathrm{Res} : H^i(G, A) \to H^i(G, A)$$

are given by multiplication by n for all $i \geq 0$.

Proof   Indeed, if $\phi : \mathbf{Z}[G] \to A$ is a G-homomorphism, then for all $x \in \mathbf{Z}[G]$ we have $\phi_H^G(x) = \sum \rho_j \phi(\rho_j^{-1} x) = \sum \rho_j \rho_j^{-1} \phi(x) = n\phi(x)$.   □

In the case $H = \{1\}$ we get:

Corollary 3.3.8 *Let G be a finite group of order n. Then the elements of $H^i(G, A)$ have finite order dividing n for all G-modules A and integers $i > 0$.*

In some important cases the above torsion property implies vanishing of the higher cohomology groups. Recall that an abelian group A is called *uniquely n-divisible* for some integer $n > 0$ if the multiplication-by-n map on A is bijective. Examples of such groups are $\mathbf{Q}$-vector spaces and torsion groups whose elements have order prime to n.

Corollary 3.3.9 *Assume moreover that A is uniquely n-divisible. Then $H^i(G, A) = 0$ for all $i > 0$.*

Proof Indeed, in this case multiplication by n induces an isomorphism $A \xrightarrow{\sim} A$, hence also isomorphisms $H^i(G, A) \xrightarrow{\sim} H^i(G, A)$ for all i. But $nH^i(G, A) = 0$ for $i > 0$ by the previous corollary. □

Remark 3.3.10 When A is a G-module and $H \subset G$ is a normal subgroup of finite index in G, we may construct an isomorphism

$$\mathrm{Hom}_H(\mathbf{Z}[G], A) \xrightarrow{\sim} A \otimes_{\mathbf{Z}} \mathbf{Z}[G/H]$$

of G-modules by the assignment

$$\phi \mapsto \sum_{\bar{g} \in G/H} g\phi(g^{-1}) \otimes \bar{g}.$$

Here g denotes an arbitrary preimage of $\bar{g}$ in G; since ϕ is an H-homomorphism, the right-hand side is well defined. One readily verifies that the above map is bijective and compatible with the action of G if we make G act on $\mathbf{Z}[G/H]$ by the action induced from that on $\mathbf{Z}[G]$ and on $A \otimes_{\mathbf{Z}} \mathbf{Z}[G/H]$ via the rule $\sigma(a \otimes \bar{g}) = \sigma(a) \otimes \sigma(\bar{g})$.

Furthermore, the maps $A \to M_H^G(A)$ and $M_H^G(A) \to A$ used in the construction of restriction and corestriction maps identify with the maps

$$a \mapsto \sum_{\bar{g} \in G/H} a \otimes \bar{g} \quad \text{and} \quad a \otimes \bar{g} \mapsto a.$$

Another basic construction is the following one.

Construction 3.3.11 (Inflation maps) Let G be a group, and H a normal subgroup. Then for a G-module A the submodule A^H of fixed elements under

H is stable under the action of G (indeed, for $\sigma \in G, \tau \in H$ and $a \in A^H$ one has $\tau \sigma a = \sigma(\sigma^{-1}\tau\sigma)a = \sigma a$). Thus A^H carries a natural structure of a G/H-module.

Now take a projective resolution $P_\bullet$ of $\mathbf{Z}$ as a trivial G-module and a projective resolution $Q_\bullet$ of $\mathbf{Z}$ as a trivial G/H-module. Each Q_i can be considered as a G-module via the projection $G \to G/H$, so applying Lemma 3.1.7 with $R = \mathbf{Z}[G]$, $B^\bullet = Q_\bullet$ and $\alpha = \mathrm{id}_\mathbf{Z}$ we get a morphism $P_\bullet \to Q_\bullet$ of complexes of G-modules, whence also a map $\mathrm{Hom}_G(Q_\bullet, A^H) \to \mathrm{Hom}_G(P_\bullet, A^H)$. Now since $\mathrm{Hom}_G(Q_i, A^H) = \mathrm{Hom}_{G/H}(Q_i, A^H)$ for all i, the former complex equals $\mathrm{Hom}_{G/H}(Q_\bullet, A^H)$, so by taking cohomology we get maps $H^i(G/H, A^H) \to H^i(G, A^H)$ which do not depend on the choices of $P_\bullet$ and $Q_\bullet$ by the same argument as in the proof of Proposition 3.1.9. Composing with the natural map induced by the G-homomorphism $A^H \to A$ we finally get maps

$$\mathrm{Inf} : H^i(G/H, A^H) \to H^i(G, A),$$

for all $i \geq 0$, called *inflation maps*.

Remark 3.3.12 Calculating the inflation maps in terms of the standard resolution of $\mathbf{Z}$, we see that inflating an i-cocycle $\mathbf{Z}[(G/H)^{i+1}] \to A^H$ amounts to taking the lifting $\mathbf{Z}[G^{i+1}] \to A^H$ induced by the projection $G \to G/H$.

Similarly, one checks that the restriction of a cocycle $\mathbf{Z}[G^{i+1}] \to A$ to a subgroup H is given by restricting it to a map $\mathbf{Z}[H^{i+1}] \to A$.

Remark 3.3.13 Given a normal subgroup H in G and a G-module A, with trivial H-action, the inflation map $\mathrm{Inf} : H^2(G/H, A) \to H^2(G, A)$ has the following interpretation in terms of group extensions: given an extension $0 \to A \xrightarrow{\pi} E \to (G/H) \to 1$, its class $c(E)$ satisfies $\mathrm{Inf}(c(E)) = c(\rho^*(E))$, where $\rho : G \to G/H$ is the natural projection, and $\rho^*(E)$ is the *pullback extension* $\rho^*(E)$ defined as the subgroup of $E \times G$ given by elements (e, g) satisfying $\pi(e) = \rho(g)$. One verifies that $\rho^*(E)$ is indeed an extension of G by A, and the relation $c(\rho^*(E)) = \mathrm{Inf}(c(E))$ holds by the construction of inflation maps and that of the class $c(E)$ in Example 3.2.6.

We now turn to the last basic construction relative to subgroups.

Construction 3.3.14 (Conjugation) Let P and A be G-modules and H a *normal* subgroup of G. For each $\sigma \in G$ we define a map

$$\sigma_* : \mathrm{Hom}_H(P, A) \to \mathrm{Hom}_H(P, A)$$

by setting $\sigma_*(\phi)(p) := \sigma^{-1}\phi(\sigma(p))$ for each $p \in P$ and $\phi \in \mathrm{Hom}_H(P,A)$. To see that $\sigma_*(\phi)$ indeed lies in $\mathrm{Hom}_H(P,A)$, we compute for $\tau \in H$

$$\sigma_*(\phi)(\tau(p)) = \sigma^{-1}\phi(\sigma\tau(p)) = \sigma^{-1}\phi(\sigma\tau\sigma^{-1}\sigma(p))$$
$$= \sigma^{-1}\sigma\tau\sigma^{-1}\phi(\sigma(p)) = \tau\sigma_*(\phi)(p),$$

where we have used the normality of H in the penultimate step. As σ_*^{-1} is obviously an inverse for σ_*, we get an automorphism of the group $\mathrm{Hom}_H(P,A)$. It follows from the definition that σ_* is the identity for $\sigma \in H$.

Now we apply the above to a projective resolution $P_\bullet$ of the trivial G-module **Z**. Note that this is also a resolution by projective H-modules, because $\mathbf{Z}[G]$ is free as a $\mathbf{Z}[H]$-module (a system of coset representatives yields a basis). The construction yields an automorphism σ_* of the complex $\mathrm{Hom}_H(P_\bullet, A)$, i.e. an automorphism in each term compatible with the G-maps in the resolution. Taking cohomology we thus get automorphisms $\sigma_*^i : H^i(H,A) \to H^i(H,A)$ in each degree $i \geq 0$, and the same method as in Proposition 3.1.9 implies that they do not depend on the choice of $P_\bullet$. These automorphisms are trivial for $\sigma \in H$, so we get an action of the quotient G/H on the groups $H^i(H,A)$, called the *conjugation action*.

It is worthwhile to record an explicit consequence of this construction.

Lemma 3.3.15 *Let*

$$0 \to A \to B \to C \to 0$$

be a short exact sequence of G-modules, and H a normal subgroup in G. The long exact sequence

$$0 \to H^0(H,A) \to H^0(H,B) \to H^0(H,C) \to H^1(H,A) \to H^1(H,B) \to \dots$$

is an exact sequence of G/H-modules, where the groups $H^i(H,A)$ are equipped with the conjugation action defined above.

Proof This follows immediately from the fact that the conjugation action as defined above induces an isomorphism of the exact sequence of complexes

$$0 \to \mathrm{Hom}_H(P_\bullet, A) \to \mathrm{Hom}_H(P_\bullet, B) \to \mathrm{Hom}_H(P_\bullet, C) \to 0$$

onto itself. □

This lemma will be handy for establishing the following fundamental exact sequence involving inflation and restriction maps.

Proposition 3.3.16 *Let G be a group, H a normal subgroup and A a G-module. There is a natural map $\tau : H^1(H, A)^{G/H} \to H^2(G/H, A^H)$ fitting into an exact sequence*

$$0 \to H^1(G/H, A^H) \xrightarrow{\text{Inf}} H^1(G, A) \xrightarrow{\text{Res}} H^1(H, A)^{G/H} \xrightarrow{\tau}$$
$$\to H^2(G/H, A^H) \xrightarrow{\text{Inf}} H^2(G, A).$$

We begin the proof by the following equally useful lemma.

Lemma 3.3.17 *In the situation of the proposition we have*

$$M^G(A)^H \cong M^{G/H}(A) \quad and \quad H^j(H, M^G(A)) = 0 \ for \ all \ j > 0.$$

Proof   The first statement follows from the chain of isomorphisms

$$M^G(A)^H = \text{Hom}(\mathbf{Z}[G], A)^H \cong \text{Hom}(\mathbf{Z}[G/H], A) = M^{G/H}(A).$$

As for the second, the already used fact that $\mathbf{Z}[G]$ is free as a $\mathbf{Z}[H]$-module implies that $M^G(A)$ is isomorphic to a direct sum of copies of $M^H(A)$. But it follows from the definition of cohomology that $H^j(H, \bigoplus M^H(A)) \cong \bigoplus H^j(H, M^H(A))$, which is 0 by Corollary 3.3.3. $\square$

Proof of Proposition 3.3.16   Define C as the G-module fitting into the exact sequence

$$0 \to A \to M^G(A) \to C \to 0. \tag{3.4}$$

This is also an exact sequence of H-modules, so we get a long exact sequence

$$0 \to A^H \to M^G(A)^H \to C^H \to H^1(H, A) \to H^1(H, M^G(A)),$$

where the last group is trivial by the second statement of Lemma 3.3.17 and Corollary 3.3.3. Hence we may split up the sequence into two short exact sequences

$$0 \to A^H \to M^G(A)^H \to B \to 0, \tag{3.5}$$
$$0 \to B \to C^H \to H^1(H, A) \to 0. \tag{3.6}$$

Using Lemma 3.3.15 we see that these are exact sequences of G/H-modules. Taking the long exact sequence in G/H-cohomology coming from (3.5) we get

$$0 \to A^G \to M^G(A)^G \to B^{G/H} \to H^1(G/H, A^H) \to H^1(G/H, M^G(A)^H),$$

where the last group is trivial by Lemma 3.3.17. So we have a commutative diagram with exact rows

$$
\begin{array}{ccccccccc}
&&&&& 0 &&& \\
&&&&& \downarrow &&& \\
0 & \longrightarrow & A^G & \longrightarrow & M^G(A)^G & \longrightarrow & B^{G/H} & \longrightarrow & H^1(G/H, A^H) \to 0 \\
&& \downarrow{\scriptstyle \text{id}} && \downarrow{\scriptstyle \text{id}} && \downarrow && \\
0 & \longrightarrow & A^G & \longrightarrow & M^G(A)^G & \longrightarrow & C^G & \longrightarrow & H^1(G, A) \to 0 \\
&&&&& \downarrow && \\
&&&&& H^1(H, A)^{G/H} && \\
&&&&& \downarrow && \\
&&&&& H^1(G/H, B) &&
\end{array}
$$

where the second row comes from the long exact G-cohomology sequence of (3.4), and the column from the long exact sequence of (3.6). A diagram chase shows that we obtain from the diagram above an exact sequence

$$ 0 \to H^1(G/H, A^H) \xrightarrow{\alpha} H^1(G, A) \xrightarrow{\beta} H^1(H, A)^{G/H} \to H^1(G/H, B). $$

Here we have to identify the maps α and β with inflation and restriction maps, respectively. For α, this follows by viewing A^H and B as G-modules via the projection $G \to G/H$ and considering the commutative diagram

$$
\begin{array}{ccccc}
B^{G/H} & \xrightarrow{\text{id}} & B^G & \longrightarrow & C^G \\
\downarrow && \downarrow && \downarrow \\
H^1(G/H, A^H) & \xrightarrow{\lambda} & H^1(G, A^H) & \longrightarrow & H^1(G, A)
\end{array}
$$

where the composite of the maps in the lower row is by definition the inflation map. Here λ is simply given by viewing a 1-cocycle $G/H \to A^H$ as a 1-cocycle $G \to A^H$, and the diagram commutes by the functoriality of the long exact cohomology sequence. As for β, its identification with a restriction map follows from the commutative diagram

$$
\begin{array}{ccc}
C^G & \longrightarrow & H^1(G, A) \\
\downarrow && \downarrow{\scriptstyle \text{Res}} \\
C^H & \longrightarrow & H^1(H, A)
\end{array}
$$

where the left vertical map is the natural inclusion.

Now the remaining part of the required exact sequence comes from the commutative diagram

$$H^1(H,A)^{G/H} \longrightarrow H^1(G/H,B) \longrightarrow H^1(G/H,C^H) \xrightarrow{\text{Inf}} H^1(G,C)$$

$$\left\downarrow{\cong}\right. \qquad\qquad\qquad\qquad\qquad\qquad \left\downarrow{\cong}\right.$$

$$H^2(G/H,A^H) \xrightarrow{\hspace{1cm}\text{Inf}\hspace{1cm}} H^2(G,A)$$

where the top row, coming from (3.6), is exact at $H^1(G/H,B)$, and the vertical isomorphisms are induced by the long exact sequences coming from (3.5) and (3.4), using again that $M^G(A)$ and $M^G(A)^H$ have trivial cohomology. Commutativity of the diagram relies on a compatibility between inflation maps and long exact sequences, which is proven in the same way as the one we have just considered for H^1. Finally, the exactness of the sequence of the proposition at $H^2(G/H,B^H)$ comes from the exactness of the row in the above diagram, together with the injectivity of the inflation map $H^1(G/H,C^H) \to H^1(G,C)$ that we have already proven (for A in place of C). □

Remark 3.3.18 The map τ of the proposition is called the *transgression map*. For an explicit description of τ in terms of cocycles, see Neukirch–Schmidt–Wingberg [1], Proposition 1.6.5.

The proposition has variants for higher cohomology groups as well. Here is a useful one.

Proposition 3.3.19 *In the situation of the previous proposition, let $i > 1$ be an integer and assume moreover that the groups $H^j(H,A)$ are trivial for $1 \le j \le i - 1$. Then there is a natural map*

$$\tau_{i,A} : H^i(H,A)^{G/H} \to H^{i+1}(G/H,A^H)$$

fitting into an exact sequence

$$0 \to H^i(G/H,A^H) \xrightarrow{\text{Inf}} H^i(G,A) \xrightarrow{\text{Res}} H^i(H,A)^{G/H} \xrightarrow{\tau_{i,A}}$$
$$\to H^{i+1}(G/H,A^H) \xrightarrow{\text{Inf}} H^{i+1}(G,A).$$

Proof Embed A into the co-induced module $M^G(A)$ and let C_A be the cokernel of this embedding. The G-module $M^G(A)$ is an H-module in particular, and the assumption that $H^1(H,A)$ vanishes implies the exactness of the sequence $0 \to A^H \to M^G(A)^H \to C_A^H \to 0$ by the long exact cohomology sequence. This is a short exact sequence of G/H-modules, so taking the associated long exact sequence yields the first and fourth vertical maps in the commutative diagram

$$0 \longrightarrow H^{j-1}(G/H, C_A^H) \xrightarrow{\text{Inf}} H^{j-1}(G, C_A) \xrightarrow{\text{Res}} H^{j-1}(H, C_A)^{G/H} \rightarrow$$

$$\downarrow \qquad\qquad \downarrow \qquad\qquad \downarrow$$

$$0 \longrightarrow H^j(G/H, A^H) \xrightarrow{\text{Inf}} H^j(G, A) \xrightarrow{\text{Res}} H^j(H, A)^{G/H} \rightarrow$$

$$\xrightarrow{\tau_{j-1, C_A}} H^j(G/H, C_A^H) \xrightarrow{\text{Inf}} H^j(G, C_A)$$

$$\downarrow \qquad\qquad\qquad \downarrow$$

$$\xrightarrow{\tau_{j, A}} H^{j+1}(G/H, A^H) \xrightarrow{\text{Inf}} H^{j+1}(G, A)$$

where the other vertical maps come from long exact sequences associated with $0 \rightarrow A \rightarrow M^G(A) \rightarrow C_A \rightarrow 0$, and the maps $\tau_{j,A}$ and τ_{j-1,C_A} are yet to be defined. The second and fifth vertical maps are isomorphisms because $H^j(G, M^G(A)) = 0$ for $j > 0$ according to Corollary 3.3.3. Moreover, Lemma 3.3.17 shows that the groups $H^j(G/H, M^G(A)^H)$ and $H^j(H, M^G(A))$ are also trivial for $j > 0$, hence the first and fourth vertical maps and the map $H^{j-1}(H, C_A) \rightarrow H^j(H, A)$ inducing the third vertical map are isomorphisms as well. In particular, the assumption yields that $H^j(H, C_A) = 0$ for all $1 \leq j < i - 1$. By induction starting from the case $i = 1$ proven in the previous proposition, we may thus assume that the map τ_{i-1, C_A} has been defined and the upper row is exact for $j = i$. We may then define $\tau_{i,A}$ by identifying it to τ_{i-1,C_A} via the isomorphisms in the diagram, and from this obtain an exact lower row. □

Another, easier consequence of Proposition 3.3.16 is the following.

Proposition 3.3.20 *In the situation of Proposition 3.3.16 assume moreover that H has finite index n in G and that A is uniquely n-divisible. Then the restriction maps $H^i(G, A) \rightarrow H^i(H, A)$ induce isomorphisms*

$$H^i(G, A) \xrightarrow{\sim} H^i(H, A)^{G/H}$$

for all $i \geq 0$.

Proof The case $i = 0$ is obvious and the case $i = 1$ is an immediate consequence of Proposition 3.3.16 and Corollary 3.3.9. To treat the case $i > 1$, consider again the short exact sequence

$$0 \rightarrow A \rightarrow M^G(A) \rightarrow C_A \rightarrow 0.$$

As in the previous proof, it induces isomorphisms $H^{i-1}(G, C_A) \cong H^i(G, A)$ and $H^{i-1}(H, C_A) \cong H^i(H, A)$ for $i > 1$, the latter being compatible with the action of G/H by Lemma 3.3.19. Now observe that since A is uniquely n-divisible, so is $M^G(A) = \mathrm{Hom}_{\mathbf{Z}}(\mathbf{Z}[G], A)$. Moreover, the map $\phi \mapsto \phi(1)$ gives a splitting of the embedding $A \hookrightarrow \mathrm{Hom}_{\mathbf{Z}}(\mathbf{Z}[G], A)$ as a $\mathbf{Z}$-homomorphism and therefore $M^G(A) \cong A \oplus C_A$ as an abelian group. It follows that C_A is n-divisible as well, and therefore the map $H^{i-1}(G, C_A) \to H^{i-1}(H, C_A)^{G/H}$ is an isomorphism by induction on i. Hence so is the map $H^i(G, A) \to H^i(H, A)^{G/H}$. $\qquad\square$

Remarks 3.3.21

1. The two last propositions are easy to establish using the *Hochschild–Serre spectral sequence* for group extensions (see e.g. Shatz [1] or Weibel [1]).
2. The arguments proving Propositions 3.3.19 and 3.3.20 are examples of a very useful technique called *dimension shifting*, which consists of proving statements about cohomology groups by embedding G-modules into co-induced modules and then using induction in long exact sequences. For other examples where this technique can be applied, see the exercises.

3.4 Cup-products

In this section we construct an associative product operation

$$H^i(G, A) \times H^j(G, B) \to H^{i+j}(G, A \otimes B), \qquad (a, b) \mapsto a \cup b$$

which is *graded-commutative*, i.e. it satisfies

$$a \cup b = (-1)^{ij}(b \cup a). \tag{3.7}$$

Here $A \otimes B$ is the tensor product of A and B over $\mathbf{Z}$, equipped with the G-module structure given by $\sigma(a \otimes b) = \sigma(a) \otimes \sigma(b)$. Note that in general this is different from the tensor product of A and B over $\mathbf{Z}[G]$. Also, in equality (3.7) we have tacitly identified $H^{i+j}(G, A \otimes B)$ with $H^{i+j}(G, B \otimes A)$ along the isomorphism $A \otimes B \cong B \otimes A$ obtained by permuting the terms.

We begin the construction with general considerations on complexes. We restrict to the case of abelian groups, the only one we shall need.

Construction 3.4.1   Let $A^\bullet$ and $B^\bullet$ be complexes of abelian groups. We define the *tensor product* complex $A^\bullet \otimes B^\bullet$ by first considering the *double complex*

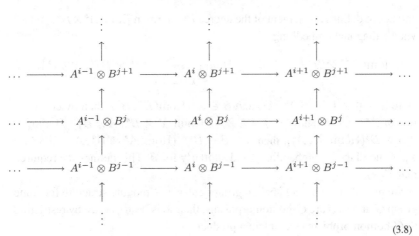

$$(3.8)$$

where the horizontal maps $\partial_{ij}^h : A^i \otimes B^j \to A^{i+1} \otimes B^j$ are given by $\partial_A^i \otimes \mathrm{id}$ and the vertical maps $\partial_{ij}^v : A^i \otimes B^j \to A^i \otimes B^{j+1}$ by $\mathrm{id} \otimes (-1)^i \partial_B^j$. In this way, the squares *anticommute*, i.e. one has

$$\partial_{i,j+1}^h \circ \partial_{ij}^v = -\partial_{i+1,j}^v \circ \partial_{ij}^h.$$

Now take the *total complex* associated with this double complex. By definition, this is the complex $T^\bullet$ with

$$T^n = \bigoplus_{i+j=n} A^i \otimes B^j$$

and $\partial^n : T^n \to T^{n+1}$ given on the component $A^i \otimes B^j$ by $\partial_{ij}^h + \partial_{ij}^v$. The above anticommutativity then implies $\partial^{n+1} \circ \partial^n = 0$, i.e. that $T^\bullet$ is a complex. We define $T^\bullet$ to be the tensor product of $A^\bullet$ and $B^\bullet$ and denote it by $A^\bullet \otimes B^\bullet$.

We now proceed to the second step of the construction.

Construction 3.4.2 In the situation of the above construction, assume further given abelian groups A and B. Consider the complexes $\mathrm{Hom}(A^\bullet, A)$ and $\mathrm{Hom}(B^\bullet, B)$ whose degree i terms are $\mathrm{Hom}(A^{-i}, A)$ and $\mathrm{Hom}(B^{-i}, B)$, respectively, with differentials induced by those of $A^\bullet$ and $B^\bullet$. We construct a product operation

$$H^i(\mathrm{Hom}(A^\bullet, A)) \times H^j(\mathrm{Hom}(B^\bullet, B)) \to H^{i+j}(\mathrm{Hom}(A^\bullet \otimes B^\bullet, A \otimes B))$$
$$(3.9)$$

as follows. Given homomorphisms $\alpha : A^{-i} \to A$ and $\beta : B^{-j} \to B$ with $i + j = n$, the tensor product $\alpha \otimes \beta$ is a homomorphism $A^{-i} \otimes B^{-j} \to A \otimes B$, and hence defines an element of the degree $i+j$ term in $\mathrm{Hom}(A^\bullet \otimes B^\bullet, A \otimes B)$ via the diagonal embedding

$$\mathrm{Hom}(A^{-i} \otimes B^{-j}, A \otimes B) \to \mathrm{Hom}\left(\bigoplus_{k+l=i+j} A^{-k} \otimes B^{-l}, A \otimes B \right).$$

Here if $\alpha \in Z^i(\mathrm{Hom}(A^\bullet, A))$ and $\beta \in Z^j(\mathrm{Hom}(B^\bullet, B))$, then by construction of $A^\bullet \otimes B^\bullet$ we have $\alpha \otimes \beta \in Z^{i+j}(\mathrm{Hom}(A^\bullet \otimes B^\bullet, A \otimes B))$. Moreover, if $\alpha \in B^i(\mathrm{Hom}(A^\bullet, A))$, then $\alpha \otimes \beta \in B^{i+j}(\mathrm{Hom}(A^\bullet \otimes B^\bullet, A \otimes B))$ (use again the diagonal embedding), and similarly for β. This defines the required map (3.9).

We note that if here all abelian groups carry a G-module structure for some group G and α, β are G-homomorphisms, then so is $\alpha \otimes \beta$, hence by restricting to G-homomorphisms we obtain a product

$$H^i(\mathrm{Hom}_G(A^\bullet, A)) \times H^j(\mathrm{Hom}_G(B^\bullet, B)) \to H^{i+j}(\mathrm{Hom}_G(A^\bullet \otimes B^\bullet, A \otimes B)),$$

where $A \otimes B$ and $A^\bullet \otimes B^\bullet$ are endowed with the G-module structure defined at the beginning of this section.

The next step is the following key proposition. Recall that the lower numbering in a projective resolution $P_\bullet$ is defined by $P_i := P^{-i}$.

Proposition 3.4.3 *Let G be a group, and let $P_\bullet$ be a complex of G-modules which is a projective resolution of the trivial G-module $\mathbf{Z}$. Then $P_\bullet \otimes P_\bullet$ is a projective resolution of the trivial $\mathbf{Z}[G \times G]$-module $\mathbf{Z}$.*

Here the terms of $P_\bullet \otimes P_\bullet$ are endowed by a $G \times G$-action coming from

$$(\sigma_1, \sigma_2)(p_1 \otimes p_2) = \sigma_1(p_1) \otimes \sigma_2(p_2).$$

The proof is based on the following lemma.

Lemma 3.4.4 *If $A^\bullet$ and $B^\bullet$ are acyclic complexes of free abelian groups, then so is the complex $A^\bullet \otimes B^\bullet$.*

Similarly, if $A^\bullet$ and $B^\bullet$ are complexes of free abelian groups concentrated in nonpositive degrees, acyclic in negative degrees and having a free abelian group as 0-th cohomology, then so is the complex $A^\bullet \otimes B^\bullet$. Moreover, we have $H^0(A^\bullet \otimes B^\bullet) \cong H^0(A^\bullet) \otimes H^0(B^\bullet)$.

Proof As tensor products and direct sums of free abelian groups are again free, we get that the terms of $A^\bullet \otimes B^\bullet$ are free. The proof of acyclicity is based on the fact that a subgroup of a free abelian group is again free. This implies that for all i the subgroups $B^i(A^\bullet)$ are free, and in particular projective. Consider for all i the short exact sequences

$$0 \to Z^i(A^\bullet) \to A^i \to B^{i+1}(A^\bullet) \to 0.$$

The terms here are free abelian groups, so the sequence splits. Moreover, we have $Z^i(A^\bullet) = B^i(A^\bullet)$ by the acyclicity of $A^\bullet$, therefore we may rewrite the above exact sequence as

$$0 \to B^i(A^\bullet) \xrightarrow{\text{id}} B^i(A^\bullet) \oplus B^{i+1}(A^\bullet) \xrightarrow{(0,\text{id})} B^{i+1}(A^\bullet) \to 0.$$

Hence the complex $A^\bullet$ decomposes as an infinite direct sum of complexes of the shape

$$\cdots \to 0 \to 0 \to A \xrightarrow{\text{id}} A \to 0 \to 0 \to \ldots,$$

and similarly, the complex $B^\bullet$ decomposes as a direct sum of complexes

$$\cdots \to 0 \to 0 \to B \xrightarrow{\text{id}} B \to 0 \to 0 \to \ldots.$$

As the construction of tensor products of complexes manifestly commutes with arbitrary direct sums, we are reduced to check acyclicity for the tensor product of complexes of this type. But by definition, these are complexes of the form

$$\cdots \to 0 \to 0 \to A{\otimes}B \xrightarrow{(\text{id},\text{id})} (A{\otimes}B){\oplus}(A{\otimes}B) \xrightarrow{\text{id}-\text{id}} A{\otimes}B \to 0 \to 0 \to \ldots,$$

or similar ones with the second identity map replaced by $-\text{id}$. The first statement is then obvious. The second one is proven by the same argument, and the description of the 0-th cohomology follows from right exactness of the tensor product. □

Proof of Proposition 3.4.3 By definition, the P_i are direct summands in some free G-module, which is in particular a free abelian group, so they are also free abelian groups. Hence the second statement of the lemma applies. Therefore the corollary is proven if we show that the terms of $P_\bullet \otimes P_\bullet$ are projective as $\mathbf{Z}[G \times G]$-modules. For this, notice first the canonical isomorphism $\mathbf{Z}[G \times G] \cong \mathbf{Z}[G] \otimes_{\mathbf{Z}} \mathbf{Z}[G]$: indeed, both abelian groups are free on a basis corresponding to pairs of elements in G. Taking direct sums we get that tensor products of free $\mathbf{Z}[G]$-modules are free $\mathbf{Z}[G \times G]$-modules with the above $G \times G$-action. Finally, if P_i (resp. P_j) are projective $\mathbf{Z}[G]$-modules with direct complement Q_i (resp. Q_j) in some free $\mathbf{Z}[G]$-module, the isomorphism

$$(P_i \oplus Q_i) \otimes (P_j \oplus Q_j) \cong (P_i \otimes P_j) \oplus (P_i \otimes Q_j) \oplus (Q_i \otimes P_j) \oplus (Q_i \otimes Q_j)$$

shows that $P_i \otimes P_j$ is a direct summand in a free $\mathbf{Z}[G \times G]$-module, and hence it is projective. Projectivity of the terms of $P_\bullet \otimes P_\bullet$ follows. □

Putting everything together, we can finally construct the cup-product.

Construction 3.4.5 Let A and B be G-modules, and $P_\bullet$ a projective resolution of the trivial G-module $\mathbf{Z}$. Applying Construction 3.4.2 with $A^\bullet = B^\bullet = P_\bullet$ we get maps

$$H^i(\mathrm{Hom}(P_\bullet, A)) \times H^j(\mathrm{Hom}(P_\bullet, B)) \to H^{i+j}(\mathrm{Hom}(P_\bullet \otimes P_\bullet, A \otimes B)).$$

By the proposition above, the complex $P_\bullet \otimes P_\bullet$ is a projective resolution of $\mathbf{Z}$ as a $G \times G$-module, so by definition of group cohomology we may rewrite the above as

$$H^i(G, A) \times H^j(G, B) \to H^{i+j}(G \times G, A \otimes B).$$

On the other hand, the diagonal embedding $G \to G \times G$ induces a restriction map

$$\mathrm{Res}: H^{i+j}(G \times G, A \otimes B) \to H^{i+j}(G, A \otimes B).$$

Composing the two, we finally get an operation

$$H^i(G, A) \times H^j(G, B) \to H^{i+j}(G, A \otimes B),$$

which we call the *cup-product* map. We denote the image of two elements $a \in H^i(G, A)$ and $b \in H^j(G, B)$ by $a \cup b$. The kind reader will check that this construction does not depend on the chosen projective resolution $P_\bullet$.

Remarks 3.4.6

1. The construction is functorial in the sense that for a morphism $A \to A'$ of G-modules the diagram

$$
\begin{array}{ccc}
H^i(G, A) \times H^j(G, B) & \longrightarrow & H^{i+j}(G, A \otimes B) \\
\downarrow & & \downarrow \\
H^i(G, A') \times H^j(G, B) & \longrightarrow & H^{i+j}(G, A' \otimes B)
\end{array}
$$

   commutes, and similarly in the second variable.
2. Given a morphism of G-modules $A \times B \to C$, we get pairings

$$H^i(G, A) \times H^j(G, B) \to H^{i+j}(G, C)$$

   by composing the cup-product with the natural map

$$H^{i+j}(G, A \otimes B) \to H^{i+j}(G, C).$$

   We shall also refer to these more general pairings as cup-products.

3. It follows from the construction that for $i = j = 0$ the cup-product

$$H^0(G, A) \times H^0(G, B) \to H^0(G, A \otimes B)$$

is just the natural map $A^G \otimes B^G \to (A \otimes B)^G$.

Proposition 3.4.7 *The cup-product is associative and graded-commutative, i.e. it satisfies the relation (3.7).*

Proof One checks associativity by carefully following the construction. It ultimately boils down to the associativity of the tensor product; we leave the details to the reader. For graded-commutativity, we first work on the level of tensor products of complexes and compare the images of the obvious maps

$$A^i \otimes B^j \to \bigoplus_{k+l=i+j} A^k \otimes B^l \quad \text{and} \quad B^j \otimes A^i \to \bigoplus_{k+l=i+j} B^l \otimes A^k$$

in the complexes $A^\bullet \otimes B^\bullet$ and $B^\bullet \otimes A^\bullet$, respectively. Given $a \otimes b \in A^i \otimes B^j$, the differential in $A^\bullet \otimes B^\bullet$ acts on it by $\partial_A^i \otimes \mathrm{id}_B + (-1)^i \mathrm{id}_A \otimes \partial_B^j$, whereas the differential of $B^\bullet \otimes A^\bullet$ acts on $b \otimes a$ by $\partial_B^j \otimes \mathrm{id}_A + (-1)^j \mathrm{id}_B \otimes \partial_A^i$. Therefore mapping $a \otimes b$ to $(-1)^{ij}(b \otimes a)$ induces an isomorphism of complexes

$$A^\bullet \otimes B^\bullet \xrightarrow{\sim} B^\bullet \otimes A^\bullet.$$

Applying this with $A^\bullet = B^\bullet = P_\bullet$ and performing the rest of the construction of the cup-product, we get that via the above isomorphism the elements $a \cup b$ and $(-1)^{ij}(b \cup a)$ get mapped to the same element in $H^{i+j}(G, A \otimes B)$. □

The cup-product enjoys the following exactness property.

Proposition 3.4.8 *Given an exact sequence*

$$0 \to A_1 \to A_2 \to A_3 \to 0 \tag{3.10}$$

of G-modules such that the tensor product over $\mathbf{Z}$

$$0 \to A_1 \otimes B \to A_2 \otimes B \to A_3 \otimes B \to 0 \tag{3.11}$$

with a G-module B remains exact, we have for all elements $a \in H^i(G, A_3)$ and $b \in H^j(G, B)$ the relation

$$\delta(a) \cup b = \delta(a \cup b)$$

in $H^{i+j+1}(G, A_1 \otimes B)$, where the δ are the connecting maps in the associated long exact sequences.

Similarly, if

$$0 \to B_1 \to B_2 \to B_3 \to 0$$

is an exact sequence of G-modules such that the tensor product over $\mathbf{Z}$

$$0 \to A \otimes B_1 \to A \otimes B_2 \to A \otimes B_3 \to 0$$

with a G-module A remains exact, we have for all elements $a \in H^i(G, A)$ *and* $b \in H^j(G, B_3)$ *the relation*

$$a \cup \delta(b) = (-1)^i \delta(a \cup b)$$

in $H^{i+j+1}(G, A \otimes B_1)$.

Proof For the first statement, fix an element $b \in H^j(G, B)$. Take a projective resolution $P_\bullet$ of the trivial G-module $\mathbf{Z}$ and consider the sequences

$$0 \to \operatorname{Hom}(P_\bullet, A_1) \to \operatorname{Hom}(P_\bullet, A_2) \to \operatorname{Hom}(P_\bullet, A_3) \to 0 \qquad (3.12)$$

and

$$0 \to \operatorname{Hom}(P_\bullet \otimes P_\bullet, A_1 \otimes B) \to \operatorname{Hom}(P_\bullet \otimes P_\bullet, A_2 \otimes B) \to \operatorname{Hom}(P_\bullet \otimes P_\bullet, A_3 \otimes B) \to 0.$$

These are exact sequences of complexes by virtue of the projectivity of the P_i and the exactness of sequences (3.10) and (3.11). Lifting b to an element $\beta \in \operatorname{Hom}(P_j, B)$, tensor product with β yields maps

$$\operatorname{Hom}(P_i, A_k) \to \operatorname{Hom}(P_i \otimes P_j, A_k \otimes B)$$

for $k = 1, 2, 3$. Hence proceeding as in Construction 3.4.2 we obtain maps from the terms in the first sequence to those of the second (increasing degrees by j), giving rise to a commutative diagram by functoriality of the cup-product construction. The connecting maps δ are obtained by applying the snake lemma to the above sequences as in Proposition 3.1.1, and one gets the first statement from the aforementioned commutativity by following the image of the element $a \in H^i(G, A)$. The proof of the second statement is similar, except that one has to replace the differentials in the complexes $\operatorname{Hom}^\bullet(P_\bullet, B_\lambda)$ by their multiples by $(-1)^i$ in order to get a commutative diagram, by virtue of the sign convention we have taken in Construction 3.4.1. □

We shall also need another exactness property of the cup-product.

Proposition 3.4.9 *Assume given exact sequences*

$$0 \to A_1 \to A_2 \to A_3 \to 0 \quad \text{and} \quad 0 \to B_1 \to B_2 \to B_3 \to 0$$

of G-modules and a **Z**-*bilinear pairing* $A_2 \times B_2 \to C$ *into some G-module C, compatible with the action of G. Assume further that the restriction of this pairing to* $A_1 \times B_1$ *is trivial. Then it induces pairings*

$$A_1 \times B_3 \to C \quad \text{and} \quad A_3 \times B_1 \to C$$

such that the induced cup-products satisfy the compatibility

$$\delta_A(\alpha) \cup \beta = (-1)^{i+1} \alpha \cup \delta_B(\beta)$$

for $\alpha \in H^i(G, A_3)$ *and* $\beta \in H^j(G, B_3)$, *where* $\delta_A : H^i(G, A_3) \to H^{i+1}(G, A_1)$ *and* $\delta_B : H^j(G, B_3) \to H^{j+1}(G, B_1)$ *are boundary maps coming from the above short exact sequences.*

Proof Take again a projective resolution $P_\bullet$ of the trivial G-module **Z**, giving rise to an exact sequence of the form (3.12) and a similar one with the B_i. These are linked by a pairing

$$\text{Hom}(P_\bullet, A_2) \times \text{Hom}(P_\bullet, B_2) \to \text{Hom}(P_\bullet \otimes P_\bullet, C)$$

trivial on $\text{Hom}(P_\bullet, A_1) \times \text{Hom}(P_\bullet, B_1)$. Represent α and β by cocycles $\alpha_3 \in Z^i(\text{Hom}(P_\bullet, A_3))$ and $\beta_3 \in Z^j(\text{Hom}(P_\bullet, B_3))$, respectively. Recall from the proof of Proposition 3.1.1 that the class $\delta_A(\alpha)$ is constructed as follows. We first lift α_3 to an element $\alpha_2 \in \text{Hom}(P_i, A_2)$, and then take $\partial_A^i(\alpha_2)$ in $B^{i+1}(\text{Hom}(P_\bullet, A_2))$. This is an element mapping to 0 in $Z^{i+1}(\text{Hom}(P_\bullet, A_3))$ and hence coming from some $\alpha_1 \in Z^{i+1}(\text{Hom}(P_\bullet, A_1))$, and we define $\delta_A(\alpha)$ to be its class in $H^{i+1}(\text{Hom}(P_\bullet, A_1))$. By definition of our pairing, $\delta_A(\alpha) \cup \beta$ is constructed by lifting β_3 to some $\beta_2 \in \text{Hom}(P_j, B_2)$ and then taking the image of $\partial_A^i(\alpha_2) \otimes \beta_2$ in $\text{Hom}(P_{i+1} \otimes P_j, C)$. Since α_2 comes from $Z^{i+1}(\text{Hom}(P_\bullet, A_1))$, this does not depend on the choice of the lifting β_2, and moreover it yields a cocycle in $Z^{i+j+1}(\text{Hom}(P_\bullet \otimes P_\bullet, C))$. In a similar way, one represents $\alpha \cup \delta_B(\beta)$ by the image of $\alpha_2 \otimes \partial_B^j(\beta_2)$ in $Z^{i+j+1}(\text{Hom}(P_\bullet \otimes P_\bullet, C))$. Now viewing $\partial_A^i(\alpha_2) \otimes \beta_2 + (-1)^i \alpha_2 \otimes \partial_B^j(\beta_2)$ as an element in $Z^{i+j+1}(\text{Hom}(P_\bullet \otimes P_\bullet, C))$, we see that it is none but $\partial^{i+j}(\alpha_2 \otimes \beta_2)$, where ∂^{i+j} is the total differential of the complex. Hence it becomes 0 in $H^{i+j+1}(\text{Hom}(P_\bullet \otimes P_\bullet, C))$, which yields the required formula. $\square$

Finally, given a subgroup H of G (normal or of finite index if needed), the cup-product satisfies the following compatibility relations with the associated restriction, inflation and corestriction maps.

Proposition 3.4.10 *Given G-modules A and B, the following relations hold.*

1. *For $a \in H^i(G, A)$ and $b \in H^j(G, B)$ we have*

$$\text{Res}(a \cup b) = \text{Res}(a) \cup \text{Res}(b).$$

2. *Assume H is normal in G. Then we have for $a \in H^i(G/H, A^H)$ and $b \in H^j(G/H, B^H)$*

$$\text{Inf}(a \cup b) = \text{Inf}(a) \cup \text{Inf}(b).$$

3. **(Projection Formula)** *Assume that H is of finite index in G. Then for $a \in H^i(H, A)$ and $b \in H^j(G, B)$ we have*

$$\text{Cor}(a \cup \text{Res}(b)) = \text{Cor}(a) \cup b.$$

4. *Assume H is normal in G. Then for all $\sigma \in G/H$, $a \in H^i(H, A)$ and $b \in H^j(H, B)$ we have*

$$\sigma_*(a \cup b) = \sigma_*(a) \cup \sigma_*(b).$$

Proof According to the definition of restriction maps, the first statement follows by performing the cup-product construction for the modules $M_H^G(A) = \text{Hom}_H(\mathbf{Z}[G], A)$ and $M_H^G(B) = \text{Hom}_H(\mathbf{Z}[G], B)$, and using functoriality of the construction for the natural maps $A \to M_H^G(A)$ and $B \to M_H^G(B)$. Similarly, the second statement follows by performing the cup-product construction simultaneously for the projective resolutions $P_\bullet$ and $Q_\bullet$ considered in the definition of inflation maps, and using functoriality. For the projection formula consider the diagram

$$\text{Hom}_H(\mathbf{Z}[G], A) \times \text{Hom}_H(\mathbf{Z}[G], B) \to \text{Hom}_{H \times H}(\mathbf{Z}[G \times G], A \otimes B)$$

$$\downarrow \qquad\qquad\qquad \uparrow \qquad\qquad\qquad\qquad\qquad \downarrow$$

$$\text{Hom}_G(\mathbf{Z}[G], A) \times \text{Hom}_G(\mathbf{Z}[G], B) \to \text{Hom}_{G \times G}(\mathbf{Z}[G \times G], A \otimes B),$$

where the horizontal maps are induced by the tensor product, the middle vertical map is the one inducing the restriction and the two others are those inducing the corestriction maps. The diagram is commutative in the sense that starting from elements in $\text{Hom}_H(\mathbf{Z}[G], A)$ and $\text{Hom}_G(\mathbf{Z}[G], B)$ we get the same elements in $\text{Hom}_{G \times G}(\mathbf{Z}[G \times G], A \otimes B)$ by going through the diagram in the two possible ways; this follows from the definition of the maps. The claim then again follows by performing the cup-product construction for the pairings in the two rows of the diagram and using functoriality. Finally, part (4) follows from the fact that the action of σ on $\operatorname{Hom}_H(P^\bullet, A)$ for a projective resolution $P^\bullet$ of the trivial G-module $\mathbf{Z}$ defined in the construction of the map σ_* is compatible with taking tensor products of resolutions. □

We close this section with an important compatibility relation which complements the calculation of the cohomology of finite cyclic groups in Example 3.2.9.

Proposition 3.4.11 *Let G be a finite cyclic group of order n. Fix a generator $\sigma \in G$, and let χ be the element of the group $H^1(G, \mathbf{Z}/n\mathbf{Z}) \cong \operatorname{Hom}(G, \mathbf{Z}/n\mathbf{Z})$ corresponding to the homomorphism given by $\sigma \mapsto 1$.*

1. *Denote by $\delta : H^1(G, \mathbf{Z}/n\mathbf{Z}) \to H^2(G, \mathbf{Z})$ the boundary map coming from the short exact sequence*

$$0 \to \mathbf{Z} \xrightarrow{n} \mathbf{Z} \to \mathbf{Z}/n\mathbf{Z} \to 0. \tag{3.13}$$

The element $\delta(\chi)$ is a generator of the cyclic group $H^2(G, \mathbf{Z})$.

2. *If A is a G-module, the isomorphisms*

$$H^i(G, A) \cong H^{i+2}(G, A)$$

of Example 3.2.9 are induced by cup-product with $\delta(\chi)$ for all $i > 0$.

3. *The isomorphism*

$$A^G/NA \cong H^2(G, A)$$

is induced by mapping an element of $A^G = H^0(G, A)$ to its cup-product with $\delta(\chi)$.

Proof Recall the free resolution

$$\cdots \to \mathbf{Z}[G] \xrightarrow{\sigma-1} \mathbf{Z}[G] \xrightarrow{N} \mathbf{Z}[G] \xrightarrow{\sigma-1} \mathbf{Z}[G] \to \mathbf{Z} \to 0 \tag{3.14}$$

used to calculate the cohomology of G. To prove the first statement, it will suffice to check that the element $\delta(\chi) \in H^2(G, \mathbf{Z})$ is represented by the homomorphism $\bar{\chi} : \mathbf{Z}[G] \to \mathbf{Z}$ given by sending the generator σ of G to 1, with $\mathbf{Z}[G]$ placed in degree -2 in the above resolution. This is done by carefully going through the construction of δ, given by applying Proposition 3.1.1 to the short exact sequence of complexes arising from homomorphisms of the above resolution to the sequence (3.13). It yields the following: first we lift χ to the element $\psi \in \operatorname{Hom}(\mathbf{Z}[G], \mathbf{Z})$ sending σ to 1. We then compose ψ by $N : \mathbf{Z}[G] \to \mathbf{Z}[G]$ to get a homomorphism with values in $n\mathbf{Z}$. The class $\delta(\chi)$

is then represented by any map $\lambda : \mathbf{Z}[G] \to \mathbf{Z}$ satisfying $n\lambda = \psi \circ N$; the map $\lambda = \bar{\chi}$ manifestly has this property.

This being said, the calculation of the cup-product with $\delta(\chi)$ is shown by the diagram

$$
\begin{array}{ccccc}
\mathrm{Hom}(\mathbf{Z}[G], A) & \xrightarrow{\;N_*\;} & \mathrm{Hom}(\mathbf{Z}[G], A) & \xrightarrow{(\sigma-1)*} & \mathrm{Hom}(\mathbf{Z}[G], A) \to \dots \\
\otimes\bar{\chi}\downarrow & & \otimes\bar{\chi}\downarrow & & \otimes\bar{\chi}\downarrow \\
\mathrm{Hom}(\mathbf{Z}[G \times G], A) & \longrightarrow & \mathrm{Hom}(\mathbf{Z}[G \times G], A) & \longrightarrow & \mathrm{Hom}(\mathbf{Z}[G \times G], A) \to \dots \\
\mathrm{Res}\downarrow & & \mathrm{Res}\downarrow & & \mathrm{Res}\downarrow \\
\mathrm{Hom}(\mathbf{Z}[G], A) & \xrightarrow{\;N_*\;} & \mathrm{Hom}(\mathbf{Z}[G], A) & \xrightarrow{(\sigma-1)*} & \mathrm{Hom}(\mathbf{Z}[G], A) \to \dots
\end{array}
$$

where the maps in the bottom line are the same as in the top one *except that the whole complex is shifted by degree 2*. But the resolution (3.14) is periodic by 2, whence the second statement.

The last statement is proven by a similar argument: here we represent $a \in H^0(G, A)$ by the homomorphism $\bar{a} : \mathbf{Z}[G] \to A$ sending σ to a, with $\mathbf{Z}[G]$ placed in degree 0 this time. Then it remains to observe that tensoring with $\bar{a}$ and taking restriction along the diagonal yields the natural diagram

$$
\begin{array}{ccccc}
\mathrm{Hom}(\mathbf{Z}[G], \mathbf{Z}) & \xrightarrow{\;N_*\;} & \mathrm{Hom}(\mathbf{Z}[G], \mathbf{Z}) & \xrightarrow{(\sigma-1)*} & \mathrm{Hom}(\mathbf{Z}[G], \mathbf{Z}) \to \dots \\
\downarrow & & \downarrow & & \downarrow \\
\mathrm{Hom}(\mathbf{Z}[G], A) & \xrightarrow{\;N_*\;} & \mathrm{Hom}(\mathbf{Z}[G], A) & \xrightarrow{(\sigma-1)*} & \mathrm{Hom}(\mathbf{Z}[G], A) \to \dots
\end{array}
$$

corresponding to the map $\mathbf{Z} \to A$ given by sending 1 to a.   □

EXERCISES

1. Let $\phi : G_1 \to G_2$ be a homomorphism of groups, and equip each G_2-module A with the G_1-action induced by ϕ. Show that there exists a unique family of homomorphisms

$$\phi_A^i : H^i(G_2, A) \to H^i(G_1, A)$$

   for each $i \geq 0$ such that ϕ_A^0 is the natural inclusion map $A^{G_2} \to A^{G_1}$, and moreover for every short exact sequence

$$0 \to A \to B \to C \to 0$$

   of G_2-modules the arising diagrams

$$H^i(G_2, A) \longrightarrow H^i(G_2, B) \longrightarrow H^i(G_2, C) \longrightarrow H^{i+1}(G_2, A)$$

$$\phi_A^i \downarrow \qquad\qquad \phi_B^i \downarrow \qquad\qquad \phi_C^i \downarrow \qquad\qquad \phi_A^{i+1} \downarrow$$

$$H^i(G_1, A) \longrightarrow H^i(G_1, B) \longrightarrow H^i(G_1, C) \longrightarrow H^{i+1}(G_1, A)$$

commute. [*Note:* This gives in particular another construction of restriction and inflation maps.]

2. Let H be a subgroup of G of finite index n, and let $\rho_1, \ldots, \rho_n$ be a system of left coset representatives.

   (a) Check that the map $\mathrm{Cor}^0 : A^H \to A^G$ given by $x \mapsto \sum_j \rho_j x$ does not depend on the choice of the ρ_j.
   (b) Show that the corestriction maps $H^i(H, A) \to H^i(G, A)$ are the only maps which coincide with the above Cor^0 for $i = 0$ and satisfy a property analogous to that of the maps ϕ_A^i of the previous exercise.

3. With notation as in the previous exercise, assume moreover that H is *normal* in G. Define for all $i \geq 0$ *norm maps* $N_{G/H} : H^i(H, A) \to H^i(H, A)$ by the formula $N_{G/H} = \sum_{j=1}^n \rho_{j*}$.

   (a) Check that the above definition does not depend on the choice of the ρ_j.
   (b) Verify the formula $\mathrm{Res} \circ \mathrm{Cor} = N_{G/H}$.

4. In the situation of the previous exercise assume moreover that A is a G-module uniquely divisible by $[G : H]$.

   (a) Verify that in this case the assignment $\alpha \mapsto (1/[G : H]) \mathrm{Cor}(\alpha)$ gives a well-defined map $\lambda : H^i(H, A) \to H^i(G, A)$ such that $\lambda \circ \mathrm{Res}$ is the identity map of $H^i(G, A)$.
   (b) Give another proof of Proposition 3.3.20 using part (a) and the previous exercise.

5. Show that using the standard resolution one can give the following explicit description of the cup-product using cocycles: if $a \in H^i(G, A)$ is represented by an i-cocycle $(\sigma_1, \ldots, \sigma_i) \mapsto a_{\sigma_1, \ldots, \sigma_i}$ and $b \in H^i(G, B)$ is represented by a j-cocycle $(\sigma_1, \ldots, \sigma_j) \mapsto b_{\sigma_1, \ldots, \sigma_j}$, then $a \cup b \in H^{i+j}(G, A \otimes B)$ is represented by the $(i + j)$-cocycle $(\sigma_1, \ldots, \sigma_{i+j}) \mapsto a_{\sigma_1, \ldots, \sigma_i} \otimes \sigma_1 \ldots \sigma_i (b_{\sigma_{i+1}, \ldots, \sigma_{i+j}})$.

6. Give an explicit interpretation of the piece

$$H^1(H, A)^{G/H} \to H^2(G/H, A^H) \xrightarrow{\mathrm{Inf}} H^2(G, A)$$

of the exact sequence of Proposition 3.3.16 in terms of classes of group extensions. (You may assume for simplicity that H acts trivially on A.)

7. Let G be a finite cyclic group generated by $\sigma \in G$ and let A, B be G-modules.

   (a) Describe directly the pairing

$$(A^G / NA) \times (_N B / (\sigma - 1)B) \to {}_N (A \otimes B) / (\sigma - 1)(A \otimes B)$$

   induced by the cup-product

$$H^{2i+2}(G, A) \times H^{2j+1}(G, B) \to H^{2i+2j+1}(G, A \otimes B)$$

   via the formulae of Example 3.2.9.

(b) Similar questions for the pairings

$$(A^G/NA) \times (B^G/NB) \to (A \otimes B)^G/N(A \otimes B)$$

and

$$({}_N A/(\sigma - 1)A) \times ({}_N B/(\sigma - 1)B) \to (A \otimes B)^G/N(A \otimes B).$$

4

The cohomological Brauer group

We now apply the cohomology theory of the previous chapter to the study of the Brauer group. However, we shall have to use a slightly modified construction which takes into account the fact that the absolute Galois group of a field is determined by its finite quotients. This is the cohomology theory of profinite groups, which we develop first. As a fruit of our labours, we identify the Brauer group of a field with a second, this time commutative, cohomology group of the absolute Galois group. As applications, we obtain easy proofs of some basic facts concerning the Brauer group, and give an important characterization of the index of a central simple algebra. Last but not least, one of the main objects of study in this book makes its appearance: the Galois symbol.

The cohomology theory of profinite groups was introduced in the late 1950s by John Tate, motivated by sheaf-theoretic considerations of Alexander Grothendieck. His original aim was to find the appropriate formalism for developing class field theory. Tate himself never published his work, which thus became accessible to the larger mathematical community through the famous account of Serre [4], which also contains many original contributions. It was Brauer himself who described the Brauer group as a second cohomology group, using his language of factor systems. We owe to Serre the insight that descent theory can be used to give a more conceptual proof. The Galois symbol was defined by Tate in connection with the algebraic theory of power residue symbols, a topic extensively studied in the 1960s by Bass, Milnor, Moore, Serre and others.

4.1 Profinite groups and Galois groups

It can be no surprise that the main application of the cohomological techniques of the previous chapter will be in the case when G is the Galois group of a finite Galois extension. However, it will be convenient to consider the case of infinite Galois extensions as well, and first and foremost that of the extension $k_s|k$, where k_s is a separable closure of k.

Recall that a (possibly infinite) algebraic field extension $K|k$ is a Galois extension if it is separable (i.e. the minimal polynomials of all elements of K have distinct roots in an algebraic closure) and if for each element $x \in K \setminus k$ there exists a field automorphism σ of K fixing k elementwise such that $\sigma(x) \neq x$. We denote the group of k-automorphisms of K by $\mathrm{Gal}\,(K|k)$ as in the finite case, and call it the Galois group of $K|k$. A basic example of an infinite Galois extension is given by a separable closure k_s of k. Its Galois group is called (somewhat abusively) the *absolute Galois group* of k.

A Galois extension $K|k$ is a union of finite Galois extensions, because we may embed each simple extension $k(\alpha) \subset K$ into the splitting field of the minimal polynomial of α, which is a finite Galois extension contained in K. This fact has a crucial consequence for the Galois group $\mathrm{Gal}\,(K|k)$, namely that it is determined by its finite quotients. We shall prove this in Proposition 4.1.3 below, in a more precise form. To motivate its formulation, consider a tower of finite Galois subextensions $M|L|k$ contained in an infinite Galois extension $K|k$. The main theorem of Galois theory provides us with a canonical surjection $\phi_{ML} : \mathrm{Gal}\,(M|k) \to \mathrm{Gal}\,(L|k)$. Moreover, if $N|k$ is yet another finite Galois extension containing M, we have $\phi_{NL} = \phi_{ML} \circ \phi_{NM}$. Thus one expects that if we somehow 'pass to the limit in M', then $\mathrm{Gal}\,(L|k)$ will actually become a quotient of the infinite Galois group $\mathrm{Gal}\,(K|k)$ itself. This is achieved by the following construction.

Construction 4.1.1   A *(filtered) inverse system* of groups $(G_\alpha, \phi_{\alpha\beta})$ consists of:

- a partially ordered set $(\Lambda, \leq)$ which is directed in the sense that for all $(\alpha, \beta) \in \Lambda$ there is some $\gamma \in \Lambda$ with $\alpha \leq \gamma$, $\beta \leq \gamma$;
- for each $\alpha \in \Lambda$ a group G_α;
- for each $\alpha \leq \beta$ a homomorphism $\phi_{\alpha\beta} : G_\beta \to G_\alpha$ such that we have equalities $\phi_{\alpha\gamma} = \phi_{\alpha\beta} \circ \phi_{\beta\gamma}$ for $\alpha \leq \beta \leq \gamma$.

The *inverse limit* of the system is defined as the subgroup of the direct product $\prod_{\alpha \in \Lambda} G_\alpha$ consisting of sequences (g_α) such that $\phi_{\alpha\beta}(g_\beta) = g_\alpha$ for all $\alpha \leq \beta$. It is denoted by $\varprojlim G_\alpha$; we shall not specify the inverse system in the notation when it is clear from the context. Also, we shall often say loosely that $\varprojlim G_\alpha$ is the inverse limit of the groups G_α, without special reference to the inverse system.

Plainly, this notion is not specific to the category of groups and one can define the inverse limit of sets, rings, modules, even of topological spaces in an analogous way.

We can now define a *profinite group* as an inverse limit of a system of finite groups. For a prime number p, a *pro-p group* is an inverse limit of finite p-groups.

Examples 4.1.2

1. A finite group is profinite; indeed, it is the inverse limit of the system $(G_\alpha, \phi_{\alpha\beta})$ for any directed index set Λ, with $G_\alpha = G$ and $\phi_{\alpha\beta} = \mathrm{id}_G$.

2. Given a group G, the set of its finite quotients can be turned into an inverse system as follows. Let Λ be the index set formed by the normal subgroups of finite index partially ordered by the following relation: $U_\alpha \leq U_\beta$ if $U_\alpha \supset U_\beta$. Then if $U_\alpha \leq U_\beta$ are such normal subgroups, we have a quotient map $\phi_{\alpha\beta} : G/U_\beta \to G/U_\alpha$. The inverse limit of this system is called the *profinite completion* of G, customarily denoted by $\widehat{G}$. There is a canonical homomorphism $G \to \widehat{G}$.

3. Take $G = \mathbf{Z}$ in the previous example. Then Λ is just the set $\mathbf{Z}_{>0}$, since each subgroup of finite index is generated by some positive integer m. The partial order is induced by the divisibility relation: $m|n$ if $m\mathbf{Z} \supset n\mathbf{Z}$. The completion $\widehat{\mathbf{Z}}$ is usually called *zed hat* (or *zee hat* in the US).

4. In the previous example, taking only powers of some prime p in place of m we get a subsystem of the inverse system considered there; in fact it is more convenient to index it by the exponent of p. With this convention the partial order becomes the usual (total) order of $\mathbf{Z}_{>0}$. The inverse limit is $\mathbf{Z}_p$, *the additive group of p-adic integers*. This is a commutative pro-p-group. The Chinese Remainder Theorem implies that the direct product of the groups $\mathbf{Z}_p$ for all primes p is isomorphic to $\widehat{\mathbf{Z}}$.

Now we come to the main example, that of Galois groups.

Proposition 4.1.3 *Let $K|k$ be a Galois extension of fields. Then the Galois groups of finite Galois subextensions of $K|k$ together with the homomorphisms $\phi_{ML} : \mathrm{Gal}\,(M|k) \to \mathrm{Gal}\,(L|k)$ form an inverse system whose inverse limit is isomorphic to $\mathrm{Gal}\,(K|k)$. In particular, $\mathrm{Gal}\,(K|k)$ is a profinite group.*

Proof Only the isomorphism statement needs a proof. For this, define a group homomorphism $\phi : \mathrm{Gal}\,(K|k) \to \prod \mathrm{Gal}\,(L|k)$ (where the product is over all finite Galois subextensions $L|k$) by sending a k-automorphism σ of K to the direct product of its restrictions to the various subfields L indexing the product. This map is injective, since if an automorphism σ does not fix an element α of k_s, then its restriction to a finite Galois subextension containing $k(\alpha)$ is nontrivial (as we have already remarked, such an extension always exists). On the other hand, the main theorem of Galois theory assures that the image of ϕ is contained in $\varprojlim \mathrm{Gal}\,(L|k)$. It is actually all of $\varprojlim \mathrm{Gal}\,(L|k)$, which is seen as follows: take an element (σ_L) of $\varprojlim \mathrm{Gal}\,(L|k)$ and define a k-automorphism σ of K by putting $\sigma(\alpha) = \sigma_L(\alpha)$ with some finite Galois L containing $k(\alpha)$. The fact that σ is well defined follows from the fact that by hypothesis the σ_L form a compatible system of automorphisms; finally, σ maps to $(\sigma_L) \in \varprojlim \mathrm{Gal}\,(L|k)$ by construction. $\square$

Corollary 4.1.4 *Projection to the components of the inverse limit of the proposition yields natural surjections* $\mathrm{Gal}\,(K|k) \to \mathrm{Gal}\,(L|k)$ *for all finite Galois subextensions* $L|k$ *contained in* K.

Example 4.1.5 (Finite fields) Let F be a finite field and F_s a separable closure of F. It is well known that for each integer $n > 0$ the extension $F_s|F$ has a unique subextension $F_n|F$ with $[F_n : F] = n$. Moreover, the extension $F_n|F$ is Galois with group $\mathrm{Gal}\,(F_n|F) \cong \mathbf{Z}/n\mathbf{Z}$, and via this isomorphism the natural projections $\mathrm{Gal}\,(F_{mn}|F) \to \mathrm{Gal}\,(F_n|F)$ correspond to the projections $\mathbf{Z}/mn\mathbf{Z} \to \mathbf{Z}/n\mathbf{Z}$. It follows that $\mathrm{Gal}\,(F_s|F) \cong \widehat{\mathbf{Z}}$.

Example 4.1.6 (Laurent series fields) As another example of a field with absolute Galois group $\widehat{\mathbf{Z}}$, we may consider the formal Laurent series field $k((t))$ over an algebraically closed field k of characteristic 0.

Here is a sketch of the proof of this fact. Take a finite extension $L|k((t))$ of degree n. As we are in characteristic 0, we may write $L = k((t))(\alpha)$ with some $\alpha \in L$. Multiplying α by a suitable element of $k[[t]]$ we may assume α satisfies an irreducible monic polynomial equation $f(\alpha) = \alpha^n + a_{n-1}\alpha^{n-1} + \cdots + a_0 = 0$ with $a_i \in k[[t]]$ and $f'(\alpha) \neq 0$. Then by the implicit function theorem for formal power series (which can be easily proven by Newton's method) we may express t as a formal power series $t = b_0 + b_1\alpha + b_2\alpha^2 + \cdots \in k[[\alpha]]$. In particular, plugging this expression into the power series expansions of the a_i and using the above equation for α we get that $b_j = 0$ for $j < n$. Next, we may find a formal power series $\tau = c_1\alpha + c_2\alpha^2 + \cdots \in k[[\alpha]]$ with $\tau^n = t$. Indeed, comparing power series expansions we get $c_1^n = b_1$, $nc_1^{n-1}c_2 = b_2$ and so on, from which we may determine the c_i inductively. Finally, we may also express α as a power series $\alpha = d_1\tau + d_2\tau^2 + \cdots \in k[[\tau]]$, with $d_1 = c_1^{-1}, d_2 = -c_2c_1^{-3}$ and so on. Hence we may embed L into the Laurent series field $k((\tau))$, but this field is none but the degree n cyclic Galois extension $k((t))(\tau)$ of $k((t))$. We conclude as in the previous example.

Profinite groups are endowed with a natural topology as follows: if G is the inverse limit of a system of finite groups $(G_\alpha, \phi_{\alpha\beta})$, endow the G_α with the discrete topology, their product with the product topology and the subgroup $G \subset \prod G_\alpha$ with the subspace topology.

Proposition 4.1.7 *The natural projection maps $G \to G_\alpha$ are continuous and their kernels U_α form a basis of open neighbourhoods of 1 in G.*

Proof The first statement holds by construction. For the second, note that the image of each element $g \neq 1$ of G must have nontrivial image in some G_α, by definition of the inverse limit. □

Remark 4.1.8 As a consequence of the proposition, we see that if we consider the inverse system of the quotients G/U indexed by the system of *all* open normal subgroups in G partially ordered by inclusion, the quotients G/U_α are cofinal in the inverse system and hence $\varprojlim G/U \cong G$. Thus in practice we may replace the system of the U_α by the system of all open normal subgroups in G.

To state other topological properties, we need a lemma.

Lemma 4.1.9 *Let $(G_\alpha, \phi_{\alpha\beta})$ be an inverse system of groups endowed with the discrete topology. Then the inverse limit $\varprojlim G_\alpha$ is a closed topological subgroup of the product $\prod G_\alpha$.*

Proof Take an element $g = (g_\alpha) \in \prod G_\alpha$. If $g \notin \varprojlim G_\alpha$, we have to show that it has an open neighbourhood which does not meet $\varprojlim G_\alpha$. By assumption for some α and β we must have $\phi_{\alpha\beta}(g_\beta) \neq g_\alpha$. Now take the subset of $\prod G_\alpha$ consisting of all elements with α-th component g_α and β-th component g_β. It is a suitable choice, being open (by the discreteness of the G_α and by the definition of the topological product) and containing g but avoiding $\varprojlim G_\alpha$. □

Corollary 4.1.10 *A profinite group is compact and totally disconnected (i.e. the only connected subsets are the one-element subsets). Moreover, the open subgroups are precisely the closed subgroups of finite index.*

Proof   Recall that finite groups are compact, and so is a product of compact groups, by Tikhonov's theorem. Compactness of the inverse limit then follows from the lemma, as closed subspaces of compact spaces are compact. Complete disconnectedness follows from the construction. For the second statement, note that each open subgroup U is closed since its complement is a disjoint union of cosets gU which are themselves open (the map $U \mapsto gU$ being a homeomorphism in a topological group); by compactness of G, these must be finite in number. Conversely, a closed subgroup of finite index is open, being the complement of the finite disjoint union of its cosets which are also closed.   □

Remark 4.1.11   In fact, one may characterize profinite groups as being those topological groups which are compact and totally disconnected. See e.g. Ribes–Zalesskii [1], Theorem 2.1.3 or Shatz [1], Theorem 2 for a proof.

We may now state and prove the main theorem of Galois theory for possibly infinite extensions. Observe first that if L is a subextension of a Galois extension $K|k$, then K is also a Galois extension of L and $\mathrm{Gal}\,(K|L)$ is naturally identified with a subgroup of $\mathrm{Gal}\,(K|k)$.

Theorem 4.1.12 (Krull)   *Let L be a subextension of the Galois extension $K|k$. Then $\mathrm{Gal}\,(K|L)$ is a closed subgroup of $\mathrm{Gal}\,(K|k)$. Moreover, in this way we get a bijection between subextensions of $K|k$ and closed subgroups of $\mathrm{Gal}\,(K|k)$, where open subgroups correspond to finite extensions of k contained in K.*

Proof   Take first a finite separable extension $L|k$ contained in K. Recall that we can embed it in a finite Galois extension $M|k$ contained in K (use the theorem of the primitive element to write $L = k(\alpha)$ and take the associated splitting field). Then $\mathrm{Gal}\,(M|k)$ is one of the standard finite quotients of $\mathrm{Gal}\,(K|k)$, and it contains $\mathrm{Gal}\,(M|L)$ as a subgroup. Let U_L be the inverse image of $\mathrm{Gal}\,(M|L)$ by the natural projection $\mathrm{Gal}\,(K|k) \to \mathrm{Gal}\,(M|k)$. Since the projection is continuous and $\mathrm{Gal}\,(M|k)$ has the discrete topology, U_L is open. It thus suffices to show $U_L = \mathrm{Gal}\,(K|L)$. We have $U_L \subset \mathrm{Gal}\,(K|L)$, for each element of U_L fixes L. On the other hand, the image of $\mathrm{Gal}\,(K|L)$ by the projection $\mathrm{Gal}\,(K|k) \to \mathrm{Gal}\,(M|k)$ is contained in $\mathrm{Gal}\,(M|L)$, whence the reverse inclusion. Now if $L|k$ is an arbitrary subextension of $K|k$, write it as a union of finite subextensions $L_\alpha|k$. By what we have just proven, each $\mathrm{Gal}\,(K|L_\alpha)$ is an open subgroup of $\mathrm{Gal}\,(K|k)$, hence it is also closed by Corollary 4.1.10. Their intersection is precisely $\mathrm{Gal}\,(K|L)$ which is thus a closed subgroup; its fixed field is exactly L, for K is Galois over L.

Conversely, given a closed subgroup $H \subset G$, it fixes some extension $L|k$ and is thus contained in $\mathrm{Gal}\,(K|L)$. To show equality, let σ be an element of $\mathrm{Gal}\,(K|L)$, and pick a fundamental open neighbourhood U_M of the identity in $\mathrm{Gal}\,(K|L)$, corresponding to a Galois extension $M|L$. Now $H \subset \mathrm{Gal}\,(K|L)$ surjects onto $\mathrm{Gal}\,(M|L)$ by the natural projection; indeed, otherwise its image in $\mathrm{Gal}\,(M|L)$ would fix a subfield of M strictly larger than L according to finite Galois theory, which would contradict our assumption that each element of $M \setminus L$ is moved by some element of H. In particular, some element of H must map to the same element in $\mathrm{Gal}\,(M|L)$ as σ. Hence H contains an element of the coset σU_M and, as U_M was chosen arbitrarily, this implies that σ is in the closure of H in $\mathrm{Gal}\,(K|L)$. But H is closed by assumption, whence the claim. Finally, the assertion about finite extensions follows from the above in view of Corollary 4.1.10. □

Remark 4.1.13 The group $\mathrm{Gal}\,(K|k)$ contains many nonclosed subgroups if $K|k$ is an infinite extension. For instance, cyclic subgroups are usually nonclosed; as a concrete example, one may take the cyclic subgroup of $\widehat{\mathbf{Z}}$ generated by 1. In fact, a closed subgroup of a profinite group is itself profinite (see Lemma 4.2.8 below), but it can be shown that an infinite profinite group is always uncountable (see Ribes–Zalesskii [1], Proposition 2.3.1). Thus none of the countable subgroups in a profinite group are closed.

4.2 Cohomology of profinite groups

Let $G = \varprojlim G_\alpha$ be a profinite group. In this section we attach to G another system of cohomology groups, different from that of the previous chapter for infinite G, which reflects the profiniteness of G and which is more suitable for applications.

By a *(discrete) continuous G-module* we shall mean a G-module A such that the stabilizer of each $a \in A$ is open in G. Unless otherwise stated, we shall always regard A as equipped with the discrete topology; continuous G-modules are then precisely the ones for which the action of G (equipped with its profinite topology) is continuous. If $G_\alpha = G/U_\alpha$ is one of the standard quotients of G, the submodule A^{U_α} is naturally a G_α-module. The canonical surjection $\phi_{\alpha\beta} : G_\beta \to G_\alpha$ between two of the standard quotients induces inflation maps $\mathrm{Inf}_\alpha^\beta : H^i(G_\alpha, A^{U_\alpha}) \to H^i(G_\beta, A^{U_\beta})$ for all $i \geq 0$. Furthermore, the compatibility condition $\phi_{\alpha\gamma} = \phi_{\alpha\beta} \circ \phi_{\beta\gamma}$ implies that the groups $H^i(G_\alpha, A)$ together with the maps $\mathrm{Inf}_\alpha^\beta$ form a direct system in the following sense.

Construction 4.2.1 A *(filtered) direct system* of abelian groups $(B_\alpha, \psi_{\alpha\beta})$ consists of:

- a directed partially ordered set $(\Lambda, \leq)$;
- for each $\alpha \in \Lambda$ an abelian group B_α;
- for each $\alpha \leq \beta$ a homomorphism $\psi_{\alpha\beta} : B_\alpha \to B_\beta$ such that we have equalities $\psi_{\alpha\gamma} = \psi_{\beta\gamma} \circ \psi_{\alpha\beta}$ for $\alpha \leq \beta \leq \gamma$.

The *direct limit* of the system is defined as the quotient of the direct sum $\bigoplus_{\alpha \in \Lambda} B_\alpha$ by the subgroup generated by elements of the form $b_\beta - \psi_{\alpha\beta}(b_\alpha)$. It is denoted by $\varinjlim B_\alpha$. Direct limits of abelian groups with additional structure (e.g. rings or modules) are defined in an analogous way.

Also, given direct systems $(B_\alpha, \psi_{\alpha\beta})$ and $(C_\alpha, \rho_{\alpha\beta})$ indexed by the same directed set Λ, together with maps $\lambda_\alpha : B_\alpha \to C_\alpha$ satisfying $\lambda_\beta \circ \psi_{\alpha\beta} = \rho_{\alpha\beta} \circ \lambda_\alpha$ for all $\alpha \leq \beta$, we have an induced map $\lambda : \varinjlim B_\alpha \to \varinjlim C_\alpha$, called the *direct limit of the maps* λ_α.

We can now define:

Definition 4.2.2 Let $G = \varprojlim G_\alpha$ be a profinite group, and A a continuous G-module. For all integers $i \geq 0$, we define the *i-th continuous cohomology group* $H^i_{\mathrm{cont}}(G, A)$ as the direct limit of the direct system $(H^i(G_\alpha, A^{U_\alpha}), \mathrm{Inf}^\beta_\alpha)$ constructed above. In the case when $G = \mathrm{Gal}\,(k_s|k)$ for some separable closure k_s of a field k, we also denote $H^i_{\mathrm{cont}}(G, A)$ by $H^i(k, A)$, and call it the *i-th Galois cohomology group of k with values in A*.

Example 4.2.3 Consider $\mathbf{Z}$ with trivial action by a profinite group G. Then $H^1_{\mathrm{cont}}(G, \mathbf{Z}) = 0$. Indeed, by definition this is the direct limit of the groups $H^1(G/U, \mathbf{Z}) = \mathrm{Hom}(G/U, \mathbf{Z})$ for U open and normal in G, which are trivial, as the G/U are finite and $\mathbf{Z}$ is a torsion free abelian group.

Remark 4.2.4 It follows from the definition that $H^0_{\mathrm{cont}}(G, A) = H^0(G, A)$ for all continuous G-modules A, and that $H^i_{\mathrm{cont}}(G, A) = H^i(G, A)$ if G is finite.

However, for $i > 0$ and G infinite the two groups are different in general. Take, for instance, $i = 1$, $G = \widehat{\mathbf{Z}}$ and $A = \mathbf{Q}$ with trivial $\widehat{\mathbf{Z}}$-action. Then $H^1_{\mathrm{cont}}(\widehat{\mathbf{Z}}, \mathbf{Q}) = \varinjlim \mathrm{Hom}(\mathbf{Z}/n\mathbf{Z}, \mathbf{Q}) = 0$, because $\mathbf{Q}$ is torsion free.

On the other hand, $H^1(\widehat{\mathbf{Z}}, \mathbf{Q})$ is the group of $\mathbf{Z}$-module homomorphisms $\widehat{\mathbf{Z}} \to \mathbf{Q}$. But as $\mathbf{Q}$ is a divisible abelian group (i.e. the equation $nx = y$ is solvable in $\mathbf{Q}$ for all $n \in \mathbf{Z}$), one knows that a homomorphism $C \to \mathbf{Q}$ from a subgroup C of an abelian group B extends to a homomorphism $B \to \mathbf{Q}$ (see e.g. Weibel [1], p. 39; note that the proof of this fact uses Zorn's lemma). Applying this with $C = \mathbf{Z}$, $B = \widehat{\mathbf{Z}}$ and the natural inclusion $\mathbf{Z} \to \mathbf{Q}$ we get a nontrivial homomorphism $\widehat{\mathbf{Z}} \to \mathbf{Q}$.

Convention 4.2.5 *From now on, all cohomology groups of a profinite group will be understood to be continuous, and we drop the subscript* cont *from the notation.*

We now come to a basic property of the cohomology of profinite groups.

Proposition 4.2.6 *For a profinite group G and a continuous G-module A the groups $H^i(G, A)$ are torsion abelian groups for all $i > 0$. Moreover, if G is a pro-p-group, then they are p-primary torsion groups.*

Proof This follows from the definition together with Corollary 3.3.8. □

Corollary 4.2.7 *Let V be a $\mathbf{Q}$-vector space equipped with a continuous action by a profinite group G. Then $H^i(G, V) = 0$ for all $i > 0$.*

Proof It follows from the construction of cohomology that in this case the groups $H^i(G, V)$ are $\mathbf{Q}$-vector spaces; since for $i > 0$ they are also torsion groups, they must be trivial. □

Recall that Corollary 3.3.8 was obtained as a consequence of a statement about restriction and corestriction maps. We now adapt these to the profinite situation. To do so, we need a lemma on profinite groups.

Lemma 4.2.8 *Let G be a profinite group, and (U_α) the system of open normal subgroups in G.*

1. *Given a closed subgroup $H \subset G$, there is a canonical isomorphism*

$$H \cong \varprojlim H/(H \cap U_\alpha)$$

of topological groups. Consequently, the group H is profinite and its profinite topology is the same as its subgroup topology.
2. *If moreover H is normal, the natural map*

$$\varprojlim G/HU_\alpha \to G/H$$

is an isomorphism of topological groups as well, and therefore G/H is a profinite group.

Proof We only prove the first statement, the proof of the second one being similar. The quotients $H/(H \cap U_\alpha)$ form an inverse system of finite groups as subgroups of the quotients G/U_α, and hence their inverse limit identifies with a subgroup of G. Thus the inclusion map $H \hookrightarrow G$ factors as a composite $H \to \varprojlim H/(H \cap U_\alpha) \hookrightarrow G$. Each element g of the open complement of H in G has an open neighbourhood of the form gU_α not meeting H, and therefore its class in G/U_α does not come from $H/(H \cap U_\alpha)$. Hence each element of $\varprojlim H/(H \cap U_\alpha)$ comes from H, and the homomorphism $H \to \varprojlim H/(H \cap U_\alpha)$ is a continuous bijection. Since its source and target are compact, it is an isomorphism of topological groups. $\square$

Construction 4.2.9 Let G be a profinite group, H a *closed* subgroup and A a continuous G-module. Define continuous restriction maps

$$\mathrm{Res}: H^i(G, A) \to H^i(H, A)$$

as the direct limit of the system of composite maps

$$H^i(G/U_\alpha, A^{U_\alpha}) \xrightarrow{\mathrm{Res}} H^i(H/(H \cap U_\alpha), A^{U_\alpha}) \to H^i(H/H \cap U_\alpha, A^{H \cap U_\alpha}),$$

where the U_α are the open normal subgroups of G. Lemma 4.2.8 (1) ensures that the target of Res is indeed the group $H^i(H, A)$.

In the case when H is *open* in G, one defines continuous corestriction maps $\mathrm{Cor}: H^i(H, A) \to H^i(G, A)$ in a similar way. Finally, when H is a closed normal subgroup in G, one defines inflation maps

$$\mathrm{Inf}: H^i(G/H, A^H) \to H^i(G, A)$$

as the direct limit of the system of inflation maps

$$H^i(G/HU_\alpha, A^{HU_\alpha}) \to H^i(G/U_\alpha, A^{U_\alpha}),$$

noting that the groups G/HU_α have inverse limit G/H by Lemma 4.2.8 (2).

Manifestly, in the case of a finite G we get back the previous restriction, corestriction and inflation maps.

Remark 4.2.10 In the above situation, one may define the module $M_H^G(A)$ to be the direct limit $\varinjlim \mathrm{Hom}_{H/(H \cap U_\alpha)}(\mathbf{Z}[G/U_\alpha], A^{U_\alpha})$, where the U_α are the standard open normal subgroups of G. We have a continuous G-action defined by $g(\phi_\alpha(x_\alpha)) = \phi_\alpha(x_\alpha g_\alpha)$, where g_α is the image of g in G/U_α; one checks that this action is well defined and continuous. (Note that in the spirit of the convention above we employ the notation $M_H^G(A)$ for another G-module as before; the one defined in Chapter 3 is not continuous in general.)

Then we have $M_G^G(A) \cong A$ and the Shapiro isomorphism $H^i(G, M_H^G(A)) \cong H^i(H, A)$ holds with a similar proof as in the noncontinuous case. In particular, one has the vanishing of the cohomology $H^i(G, M^G(A))$ of (continuous) co-induced modules for $i > 0$. One may then also define the continuous restriction and corestriction maps using this Shapiro isomorphism, by mimicking the construction of Chapter 3.

The construction of restriction maps implies the following basic property of the cohomology of profinite groups.

Proposition 4.2.11 *Let G be a profinite group, A a continuous G-module. Given a class $c \in H^i(G, A)$ for some $i > 0$, there exists an open subgroup $U_c \subset G$ such that c is annihilated by the restriction map $H^i(G, A) \to H^i(U_c, A)$.*

Proof Write $H^i(G, A)$ as the direct limit of the groups $H^i(G/U_\alpha, A^{U_\alpha})$ as in the definition of profinite group cohomology. The class c comes from a class $c_\alpha \in H^i(G/U_\alpha, A^{U_\alpha})$ for suitable α. Set $U_c := U_\alpha$ for the above α. By definition, the restriction map $H^i(G, A) \to H^i(U_c, A)$ is the direct limit of the restriction maps $\mathrm{Res}_\beta : H^i(G/U_\beta, A^{U_\beta}) \to H^i(U_c/(U_c \cap U_\beta), A^{U_\beta})$. But for $\beta = \alpha$ we have $U_\beta = U_\alpha = U_c$ and hence $\mathrm{Res}_\alpha(c_\alpha) = 0$. $\square$

Next, as in the noncontinuous case, we have:

Proposition 4.2.12 *Let G be a profinite group, H an open subgroup of index n and A a continuous G-module. Then the composite maps*

$$\mathrm{Cor} \circ \mathrm{Res} : H^i(G, A) \to H^i(G, A)$$

are given by multiplication by n for all $i > 0$. Consequently, the restriction $H^i(G, A) \to H^i(H, A)$ is injective on the prime-to-n torsion part of $H^i(G, A)$.

Proof Each element of $H^i(G, A)$ comes from some $H^i(G/U_\alpha, A^{U_\alpha})$, and Proposition 3.3.7 applies. The second statement follows because the multiplication-by-n map is injective on the subgroup of elements of order prime to n. $\square$

A refined version of the last statement is the following.

Corollary 4.2.13 *Let G be a profinite group, p a prime number and H a closed subgroup such that the image of H in each finite quotient of G has order prime to p. Then for each continuous G-module A the restriction map $H^i(G, A) \to H^i(H, A)$ is injective on the p-primary torsion part of $H^i(G, A)$.*

Proof Assume that an element of $H^i(G, A)$ of p-power order maps to 0 in $H^i(H, A)$. It comes from an element of some $H^i(G_\alpha, A^{U_\alpha})$ of which we may assume, up to replacing U_α by a smaller subgroup, that it maps to 0 in $H^i(H/(H \cap U_\alpha), A^{U_\alpha})$. By the proposition (applied to the finite group G/U_α) it must then be 0.   □

The main application of the above corollary will be to *pro-p-Sylow subgroups* of a profinite group G. By definition, these are subgroups of G which are pro-p-groups for some prime number p and whose images in each finite quotient of G are of index prime to p.

Proposition 4.2.14 *A profinite group G possesses pro-p-Sylow subgroups for each prime number p, and any two of these are conjugate in G.*

The proof uses the following well-known lemma.

Lemma 4.2.15 *An inverse limit of nonempty finite sets is nonempty.*

Proof The proof works more generally for compact topological spaces. Given an inverse system $(X_\alpha, \phi_{\alpha\beta})$ of nonempty compact spaces, consider the subsets $X_{\lambda\mu} \subset \prod X_\alpha$ consisting of the sequences (x_α) satisfying $\phi_{\lambda\mu}(x_\mu) = x_\lambda$ for a fixed pair $\lambda \le \mu$. These are closed subsets of the product, and their intersection is precisely $\varprojlim X_\alpha$. Furthermore, the directedness of the index set implies that finite intersections of the $X_{\lambda\mu}$ are nonempty. Since $\prod X_\alpha$ is compact by Tikhonov's theorem, it ensues that $\varprojlim X_\alpha$ is nonempty.   □

Proof of Proposition 4.2.14 Write G as an inverse limit of a system of finite groups G_α. For each G_α, denote by S_α the set of its p-Sylow subgroups (for the classical Sylow theorems, see e.g. Lang [3]). These form an inverse system of finite sets, hence by the lemma we may find an element S in the limit $\varprojlim S_\alpha$. This S corresponds to an inverse limit of p-Sylow subgroups of the G_α and hence gives a pro-p-Sylow subgroup of G. If P and Q are two pro-p-Sylow subgroups of G, their images in each G_α are p-Sylow subgroups there and hence are conjugate by some $x_\alpha \in G_\alpha$ by the finite Sylow theorem. Writing X_α for the set of possible x_α's, we get again an inverse system of finite sets, whose nonempty inverse limit contains an element x with $x^{-1}Px = Q$. □

Corollary 4.2.13 now implies:

Corollary 4.2.16 *If P is a pro-p-Sylow subgroup of a profinite group G, the restriction maps* Res : $H^i(G, A) \to H^i(P, A)$ *are injective on the p-primary torsion part of $H^i(G, A)$ for all $i > 0$ and continuous G-modules A.*

To conclude this section, we mention another construction from the previous chapter which carries over without considerable difficulty to the profinite case, that of cup-products.

Construction 4.2.17 Given a profinite group G and continuous G-modules A and B, define the tensor product $A \otimes B$ as the tensor product of A and B over $\mathbf{Z}$ equipped with the continuous G-action induced by $\sigma(a \otimes b) = \sigma(a) \otimes \sigma(b)$. In the previous chapter we have constructed for all $i, j \geq 0$ and all open subgroups U of G cup-product maps

$$H^i(G/U, A^U) \times H^j(G/U, B^U) \to H^{i+j}(G/U, A^U \otimes B^U), \quad (a, b) \mapsto a \cup b$$

satisfying the relation

$$\mathrm{Inf}(a \cup b) = \mathrm{Inf}(a) \cup \mathrm{Inf}(b)$$

for the inflation map arising from the quotient map $G/V \to G/U$ for an open inclusion $V \subset U$. Note that by the above definition of the G-action we have a natural map $A^U \otimes B^U \to (A \otimes B)^U$, so by passing to the limit over all inflation maps of the above type we obtain cup-product maps

$$H^i(G, A) \times H^j(G, B) \to H^{i+j}(G, A \otimes B), \quad (a, b) \mapsto a \cup b$$

for continuous cohomology.

It follows immediately from the noncontinuous case that this cup-product is also associative and graded-commutative, and that moreover it satisfies compatibility formulae with restriction, corestriction and inflation maps as in Proposition 3.4.10. It also satisfies the exactness property of Proposition 3.4.8, but for this we have to establish first the long exact cohomology sequence in the profinite setting. We treat this question in the next section.

4.3 The cohomology exact sequence

We now show that the analogues of exact sequences established for usual cohomology groups also hold for continuous ones. We begin with the long exact cohomology sequence.

Proposition 4.3.1 *Given a profinite group G and a short exact sequence*

$$0 \to A \to B \to C \to 0$$

of continuous G-modules, there is a long exact sequence of abelian groups

$$\cdots \to H^i(G, A) \to H^i(G, B) \to H^i(C, C) \to H^{i+1}(G, A) \to \cdots$$

starting from $H^0(G, A)$.

For the proof we need two formal statements about direct limits.

Lemma 4.3.2 *Let $(A_\alpha, \phi_{\alpha\beta})$, $(B_\alpha, \psi_{\alpha\beta})$ and $(C_\alpha, \rho_{\alpha\beta})$ be three direct systems indexed by the same directed set Λ. Assume moreover given exact sequences*

$$A_\alpha \xrightarrow{\lambda_\alpha} B_\alpha \xrightarrow{\mu_\alpha} C_\alpha$$

for each $\alpha \in \Lambda$ such that the diagrams

$$
\begin{array}{ccc}
A_\alpha & \xrightarrow{\lambda_\alpha} & B_\alpha & \xrightarrow{\mu_\alpha} & C_\alpha \\
\phi_{\alpha\beta} \downarrow & & \psi_{\alpha\beta} \downarrow & & \rho_{\alpha\beta} \downarrow \\
A_\beta & \xrightarrow{\lambda_\beta} & B_\beta & \xrightarrow{\mu_\beta} & C_\beta
\end{array}
$$

commute for all $\alpha \leq \beta$. Then the limit sequence

$$\varinjlim A_\alpha \xrightarrow{\lambda} \varinjlim B_\alpha \xrightarrow{\mu} \varinjlim C_\alpha$$

is exact as well.

Proof An element of $\ker(\mu)$ is represented by some $b_\alpha \in B_\alpha$ with the property that $\rho_{\alpha\beta}(\mu_\alpha(b_\alpha)) = \mu_\beta(\psi_{\alpha\beta}(b_\alpha)) = 0$ for some $\beta \geq \alpha$. But then there is some $a_\beta \in A_\beta$ with $\lambda_\beta(a_\beta) = \psi_{\alpha\beta}(b_\alpha)$. $\square$

Lemma 4.3.3 *Consider a profinite group G and a direct system $(A_\alpha, \phi_{\alpha\beta})$ of continuous G-modules (in particular, the $\phi_{\alpha\beta}$ are G-homomorphisms). Then* the G-module $\varinjlim A_\alpha$ is also continuous, the groups $H^i(G, A_\alpha)$ with the induced maps form a direct system, and there exist canonical isomorphisms

$$\varinjlim H^i(G, A_\alpha) \xrightarrow{\sim} H^i(G, \varinjlim A_\alpha)$$

for all $i \geq 0$.

Proof The first statement follows from the construction of direct limits, which also shows that for each open subgroup U the G/U-module $(\varinjlim A_\alpha)^U$ is the direct limit of the G/U-modules A_α^U. Hence it suffices to show the isomorphism statement for the cohomology of the latter. Taking a projective resolution $P^\bullet$ of the trivial G/U-module $\mathbf{Z}$, we are reduced to establishing isomorphisms of the form

$$\varinjlim \mathrm{Hom}(P^i, A_\alpha) \xrightarrow{\sim} \mathrm{Hom}(P^i, \varinjlim A_\alpha)$$

compatible with coboundary maps. Such isomorphisms again follow from the construction of direct limits (for instance, one may observe that the canonical isomorphisms $\oplus \mathrm{Hom}(P^i, A_\alpha) \cong \mathrm{Hom}(P^i, \oplus A_\alpha)$ preserve the relations defining the direct limit). $\qquad\square$

Proof of Proposition 4.3.1 The homomorphisms $H^i(G, A) \to H^i(G, B)$ and $H^i(G, B) \to H^i(G, C)$ arise from the finite case by passing to the limit. To define the connecting homomorphism $\partial : H^i(G, C) \to H^{i+1}(G, A)$, consider first an open subgroup U of G and define K_U as the cokernel of the map $B^U \to C^U$ (this is a nontrivial group in general). As the map $B \to C$ is surjective, we get that the direct limit of the groups K_U with U running over all open subgroups is trivial. Therefore the last term in the sequence

$$\varinjlim H^i(G/U, (B^U/A^U)) \to \varinjlim H^i(G/U, C^U) \to \varinjlim H^i(G/U, K_U)$$

is trivial by Lemma 4.3.3 (note that each K_U is also a G-module via the natural projection). The sequence is exact by Lemma 4.3.2, hence we may always lift an element γ of the middle term, which is none but $H^i(G, C)$, to an element in some $H^i(G/U, (B^U/A^U))$. The usual long exact sequence coming from the sequence of G/U-modules

$$0 \to A^U \to B^U \to B^U/A^U \to 0 \tag{4.1}$$

then yields an element in $H^{i+1}(G/U, A^U)$, and hence in $H^{i+1}(G, A)$, which we define to be $\partial(\gamma)$. This definition manifestly does not depend on the choice of U and, furthermore, the long exact sequence coming from

$$0 \to B^U/A^U \to C^U \to K_U \to 0$$

shows that any two liftings of γ into $H^i(G/U, (B^U/A^U))$ differ by an element of $H^{i-1}(G/U, K_U)$, which then maps to 0 in $H^{i-1}(G/V, K_V)$ for some $V \subset U$. This shows that the map ∂ is well defined. Exactness of the sequence at the terms $H^i(G, A)$ and $H^i(G, C)$ now follows from that of the long exact sequence associated with (4.1) and exactness at the terms $H^i(G, B)$ follows from Lemma 4.3.2. $\square$

Remarks 4.3.4

1. A more elegant way for establishing the above proposition is by constructing continuous cohomology groups directly as Ext-groups in the category of continuous G-modules; the long exact sequence then becomes a formal consequence, just as in the previous chapter (see e.g. Weibel [1], Section 6.11). We have chosen the above more pedestrian presentation in order to emphasize the viewpoint that all basic facts for the cohomology of profinite groups follow from the finite case by passing to the limit.

2. It is important to note that if the exact sequence

$$0 \to A \to B \xrightarrow{p} C \to 0$$

has a splitting, i.e. a map of G-modules $i : C \to B$ with $p \circ i = \mathrm{id}_C$, then the induced maps $p_* : H^i(G, B) \to H^i(G, C)$ and $i_* : H^i(G, C) \to H^i(G, B)$ also satisfy $p_* \circ i_* = \mathrm{id}$ by the functoriality of cohomology. Therefore the long exact sequence splits up into a collection of (split) short exact sequences

$$0 \to H^i(G, A) \to H^i(G, B) \to H^i(G, C) \to 0,$$

a fact we shall use many times later.

The inflation-restriction sequences of the last chapter also carry over to the profinite setting:

Corollary 4.3.5 *Let G be a profinite group, H a closed normal subgroup. Then the statements of Propositions 3.3.16, 3.3.19 and 3.3.20 remain valid in the continuous cohomology.*

Proof This follows from *loc. cit.* via Lemma 4.3.2. $\square$

We conclude this section with a first application of the cohomology of profinite groups which will be invaluable for the sequel.

Proposition 4.3.6 (Kummer Theory) *Let k be a field, and $m > 0$ an integer prime to the characteristic of k. Denote by μ_m the group of m-th roots of unity in a fixed separable closure of k, equipped with its Galois action. There exists a canonical isomorphism*

$$k^\times / k^{\times m} \xrightarrow{\sim} H^1(k, \mu_m)$$

induced by sending an element $a \in k^\times$ to the class of the 1-cocycle $\sigma \mapsto \sigma(\alpha)\alpha^{-1}$, where α is an m-th root of a.

For the proof we need the continuous version of Hilbert's Theorem 90:

Lemma 4.3.7 *The Galois cohomology group $H^1(k, k_s^\times)$ is trivial.*

Proof This follows from Example 2.3.4 after passing to the limit. □

Proof of Proposition 4.3.6 Consider the exact sequence of $\mathrm{Gal}\,(k_s|k)$-modules

$$1 \to \mu_m \to k_s^\times \xrightarrow{m} k_s^\times \to 1, \tag{4.2}$$

where the third map is given by raising elements to the m-th power. This map is surjective because the polynomial $x^m - a$ is separable for all $a \in k_s$, in view of the assumption on m. A piece of the associated long exact sequence reads

$$H^0(k, k_s^\times) \to H^0(k, k_s^\times) \to H^1(k, \mu_m) \to H^1(k, k_s^\times),$$

where the last group is trivial by the lemma. Noting that $H^0(k, k_s^\times) = k^\times$ and that the first map is multiplication by m, by construction of cohomology, we obtain the required isomorphism. Its explicit description follows from the construction of the coboundary map in cohomology (see Remark 3.2.4). □

Remark 4.3.8 Note that it was crucial here to work with k_s and Galois cohomology, for we do not dispose of the analogue of exact sequence (4.2) at a finite level.

The proposition has the following consequence (which is the original form of Kummer's theorem):

Corollary 4.3.9 *For k and m as above, assume moreover that k contains a primitive m-th root of unity ω. Then every finite Galois extension of k with Galois group isomorphic to $\mathbf{Z}/m\mathbf{Z}$ is of the form $k(\alpha)|k$ with some $\alpha \in k_s^\times$ satisfying $\alpha^m \in k^\times$.*

Proof The Galois group of an extension as in the corollary is a quotient of $\mathrm{Gal}\,(k_s|k)$ isomorphic to $\mathbf{Z}/m\mathbf{Z}$, and thus corresponds to a surjection $\chi : \mathrm{Gal}\,(k_s|k) \to \mathbf{Z}/m\mathbf{Z}$. But since by assumption $\mu_m \subset k$, we have isomorphisms $\mathrm{Hom}(\mathrm{Gal}\,(k_s|k), \mathbf{Z}/m\mathbf{Z}) \cong H^1(k, \mathbf{Z}/m\mathbf{Z}) \cong H^1(k, \mu_m)$ (the second one depending on the choice of ω). By the proposition χ corresponds to the class of some $a \in k^\times$ modulo $k^{\times m}$, and moreover the kernel of χ is precisely $\mathrm{Gal}\,(k(\alpha)|k)$, where α is an m-th root of a. $\qquad\square$

In positive characteristic we have the following complement to Kummer theory.

Proposition 4.3.10 (Artin–Schreier Theory) *Let k be a field of characteristic $p > 0$. Denote by $\wp : k \to k$ the endomorphism mapping $x \in k$ to $x^p - x$. Then there exists a canonical isomorphism*

$$k/\wp(k) \xrightarrow{\sim} H^1(k, \mathbf{Z}/p\mathbf{Z})$$

induced by mapping $a \in k$ to the cocycle $\sigma \mapsto \sigma(\alpha) - \alpha$, where α is a root of the equation $x^p - x = a$.

The proof is based on the following lemma. It is sometimes called the additive version of Hilbert's Theorem 90, as it concerns the additive group of k_s viewed as a $\mathrm{Gal}\,(k_s|k)$-module instead of the multiplicative group.

Lemma 4.3.11 *For an arbitrary field k the groups $H^i(k, k_s)$ are trivial for all $i > 0$.*

Proof We prove the triviality of $H^i(G, K)$ for all finite Galois extensions $K|k$ with group G and all $i > 0$. According to the normal basis theorem of Galois theory (see e.g. Lang [3], Chapter VI, Theorem 13.1), we may find an element $x \in K$ such that $\sigma_1(x), \ldots, \sigma_n(x)$ form a basis of the k-vector space K, where $1 = \sigma_1, \ldots, \sigma_n$ are the elements of G. This means that K is isomorphic to $k \otimes_{\mathbf{Z}} \mathbf{Z}[G]$ as a G-module. The latter is a co-induced G-module by Remark 3.3.4 (3), so its cohomology is trivial by Corollary 3.3.3. $\qquad\square$

Remark 4.3.12 In characteristic 0 the lemma is easy to prove: the coefficient module k_s is a $\mathbf{Q}$-vector space, hence Corollary 4.2.7 applies. However, we are about to apply the positive characteristic case.

Proof of Proposition 4.3.10 The endomorphism $\wp$ extends to the separable closure k_s with the same definition. Its kernel is the prime field $\mathbf{F}_p$, which is isomorphic to the trivial $\mathrm{Gal}\,(k_s|k)$-module $\mathbf{Z}/p\mathbf{Z}$ as a $\mathrm{Gal}\,(k_s|k)$-module. Moreover, the map $\wp\,:\,k_s\,\to\,k_s$ is surjective, because for each $a\,\in\,k_s$ the polynomial x^p-x-a is separable. We thus have an exact sequence of $\mathrm{Gal}\,(k_s|k)$-modules

$$0 \to \mathbf{Z}/p\mathbf{Z} \to k_s \xrightarrow{\ \wp\ } k_s \to 0, \tag{4.3}$$

from which we conclude as in the proof of Proposition 4.3.6, using Lemma 4.3.11 in place of Hilbert's Theorem 90. □

Remarks 4.3.13

1. In a similar way as in Corollary 4.3.9 above, one derives from the proposition that every finite Galois extension of k with Galois group $\mathbf{Z}/p\mathbf{Z}$ is generated by a root of some polynomial x^p-x-a, with $a\in k$.

2. There is a generalization of Artin–Schreier theory to powers of the prime p due to Witt. It uses the ring $W_r(k_s)$ of r-truncated Witt vectors briefly described at the end of Section A.8 of the Appendix. The subgroup of fixed elements of the Frobenius operator $F\,:\,W_r(k_s)\,\to\,W_r(k_s)$ identifies with $W_r(\mathbf{F}_p) = \mathbf{Z}/p^r\mathbf{Z}$, whence an exact sequence

$$0 \to \mathbf{Z}/p^r\mathbf{Z} \to W_r(k_s) \xrightarrow{F-\mathrm{id}} W_r(k_s) \to 0$$

generalizing (4.3). It carries a $\mathrm{Gal}\,(k_s|k)$-action induced from that on k_s, and one deduces from Lemma 4.3.11 the vanishing of the groups $H^i(k, W_r(k_s))$ for all $i, r > 0$ by induction on r using the exact sequences

$$0 \to W_r(k_s) \xrightarrow{V} W_{r+1}(k_s) \to k_s \to 0.$$

Thus a direct generalization of the above proof gives rise to an isomorphism

$$H^1(k, \mathbf{Z}/p^r\mathbf{Z}) \cong W_r(k)/(F - \mathrm{id})W_r(k).$$

4.4 The Brauer group revisited

The main goal of this section is to identify the Brauer group of a field k with the Galois cohomology group $H^2(k, k_s^\times)$, which is more tractable than the group $H^1(k, \mathrm{PGL}_\infty)$ encountered in Chapter 2. To this aim, recall first that in Proposition 2.7.9 we have constructed an injective group homomorphism

$$\delta_\infty\,:\,H^1(G, \mathrm{PGL}_\infty) \to H^2(G, K^\times)$$

for a finite Galois extension $K|k$ with group G. Having group cohomological techniques at our disposal, we can now prove:

Theorem 4.4.1 *The map δ_∞ induces an isomorphism*

$$H^1(G, \mathrm{PGL}_\infty) \xrightarrow{\sim} H^2(G, K^\times)$$

of abelian groups.

Proof It remains to prove surjectivity of the map δ_∞. We show much more, namely that the map δ_n is surjective, where n is the order of G. For this, consider $K \otimes_k K$ as a K-vector space. Multiplication by an invertible element of $K \otimes_k K$ is a K-linear automorphism $K \otimes_k K \to K \otimes_k K$. In this way we get a group homomorphism $(K \otimes_k K)^\times \to \mathrm{GL}_n(K)$ which we may insert into a commutative diagram with exact rows

$$
\begin{array}{ccccccccc}
1 & \longrightarrow & K^\times & \longrightarrow & (K \otimes_k K)^\times & \longrightarrow & (K \otimes_k K)^\times / K^\times & \longrightarrow & 1 \\
 & & \mathrm{id}\downarrow & & \downarrow & & \downarrow & & \\
1 & \longrightarrow & K^\times & \longrightarrow & \mathrm{GL}_n(K) & \longrightarrow & \mathrm{PGL}_n(K) & \longrightarrow & 1
\end{array}
$$

where all maps are compatible with the action of G if we make G act on $K \otimes_k K$ via the right factor and on the other terms by the standard action. Hence by taking cohomology we get a commutative diagram

$$
\begin{array}{ccc}
H^1(G, (K \otimes_k K)^\times / K^\times) & \xrightarrow{\alpha} H^2(G, K^\times) \longrightarrow H^2(G, (K \otimes_k K)^\times) \\
\downarrow & \quad\quad \mathrm{id}\downarrow \\
H^1(G, \mathrm{PGL}_n(K)) & \xrightarrow{\delta_n} H^2(G, K^\times)
\end{array}
$$

where the upper row is exact. Recall now the G-isomorphism $K \otimes_k K \cong \bigoplus K e_i$ explained before the proof of Lemma 2.3.8. In other words, it says that $K \otimes_k K$ is isomorphic as a G-module to $K \otimes_{\mathbf{Z}} \mathbf{Z}[G]$, which implies that $(K \otimes_k K)^\times$ is isomorphic to the G-module $K^\times \otimes_{\mathbf{Z}} \mathbf{Z}[G]$, because the invertible elements in $\bigoplus K e_i$ are exactly those with coefficients in $K^\times$. Now by Remark 3.3.4 (3) the G-module $K^\times \otimes_{\mathbf{Z}} \mathbf{Z}[G]$ is co-induced, hence the group $H^2(G, (K \otimes_k K)^\times)$ is trivial. This yields the surjectivity of the map α in the diagram, and hence also that of δ_n by commutativity of the diagram. $\quad\square$

The above proof shows much more than the assertion of the theorem. Namely, the fact that we have at our disposal both the injectivity of δ_m for all m and the surjectivity of δ_n has the following remarkable consequences.

Corollary 4.4.2 *Let $K|k$ be a finite Galois extension of degree n and group G. Then the maps $\lambda_{nm} : H^1(G, \mathrm{PGL}_n(K)) \to H^1(G, \mathrm{PGL}_{nm}(K))$ of Lemma 2.4.4 are bijective for all m.*

Furthermore, the pointed set $H^1(G, \mathrm{PGL}_n(K))$ is equipped with a group structure via

$$H^1(G, \mathrm{PGL}_n(K)) \times H^1(G, \mathrm{PGL}_n(K)) \to H^1(G, \mathrm{PGL}_{n^2}(K))$$
$$\xleftarrow{\sim} H^1(G, \mathrm{PGL}_n(K))$$

and the map $\delta_n : H^1(G, \mathrm{PGL}_n(K)) \to H^2(G, K^\times)$ is an isomorphism of abelian groups.

Proof In the first assertion only surjectivity requires a proof, and this follows from the surjectivity of δ_n together with Lemma 2.7.8. The second assertion then follows from the theorem. □

Combining Theorem 4.4.1 with Proposition 2.4.9 we get:

Theorem 4.4.3 *Let k be a field, $K|k$ a finite Galois extension and k_s a separable closure of k. There exist natural isomorphisms of abelian groups*

$$\mathrm{Br}\,(K|k) \cong H^2(G, K^\times) \quad and \quad \mathrm{Br}\,(k) \cong H^2(k, k_s^\times).$$

The theorem has a number of corollaries. Here is a first one which is quite cumbersome to establish in the context of central simple algebras (see Remark 4.4.8 below) but is almost trivial once one disposes of cohomological techniques.

Corollary 4.4.4 *Let $K|k$ be a Galois extension of degree n. Then each element of the relative Brauer group $\mathrm{Br}\,(K|k)$ has order dividing n.*

Proof This follows from the theorem together with Corollary 3.3.8. □

Note that the above corollary gives another proof of the fact that $\mathrm{Br}\,(k)$ is a torsion abelian group, first proven in Corollary 2.8.5. Also, together with Theorem 2.2.7 we obtain another proof of the fact that the period of a Brauer class divides its index (Theorem 2.8.7(1)).

One also has the following cohomological interpretation of the m-torsion part ${}_m\mathrm{Br}\,(k)$ of the Brauer group.

Corollary 4.4.5 *For each positive integer m prime to the characteristic of k we have a canonical isomorphism*

$$_m\mathrm{Br}\,(k) \cong H^2(k, \mu_m).$$

Recall that μ_m denotes the group of m-th roots of unity in k_s equipped with its canonical Galois action.

Proof   We again exploit the exact sequence (4.2). A piece of the associated long exact sequence is

$$H^1(k, k_s^\times) \to H^2(k, \mu_m) \to H^2(k, k_s^\times) \to H^2(k, k_s^\times),$$

where the first group is trivial by Hilbert's Theorem 90 (Lemma 4.3.7). The corollary follows by noting that the last map is multiplication by m. $\square$

As another corollary, we have a nice description of the relative Brauer group in the case of a cyclic extension.

Corollary 4.4.6   *For a cyclic Galois extension $K|k$ there is an isomorphism*

$$\mathrm{Br}\,(K|k) \cong k^\times / N_{K|k}(K^\times).$$

Proof   This follows from the theorem in view of the calculation of the cohomology of cyclic groups (Example 3.2.9). $\square$

Finally, we also record the following inflation-restriction sequence:

Corollary 4.4.7   *For a finite Galois extension $K|k$ there is an exact sequence*

$$0 \to \mathrm{Br}\,(K|k) \xrightarrow{\mathrm{Inf}} \mathrm{Br}\,(k) \xrightarrow{\mathrm{Res}} \mathrm{Br}\,(K).$$

Proof   This follows from the theorem and Corollary 4.3.5 in degree 2 (which applies to the $\mathrm{Gal}\,(k_s|k)$-module $k_s^\times$ in view of Hilbert's Theorem 90). $\square$

Remark 4.4.8   It is straightforward to verify that in the above exact sequence the restriction map is induced by the base change map $A \mapsto A \otimes_k K$ on central simple algebras; this in fact holds for an arbitrary finite separable extension $K|k$. The cohomological theory also yields a corestriction map $\mathrm{Br}\,(K) \to \mathrm{Br}\,(k)$ for finite separable extensions. This map is, however, much more difficult to define on the level of algebras. A construction is given in Tignol [2] (see also the exposition in Rowen [2]): he associates with a central simple K-algebra A a central simple k-algebra $\mathrm{Cor}_{K|k}(A)$ representing the class of $\mathrm{Cor}([A])$ in the Brauer group, such that moreover for a central simple k-algebra B one has $\mathrm{Cor}_{K|k}(B \otimes_k K) \cong B^{\otimes[K:k]}$. In the case of a quadratic extension the construction is much simpler; it can be found in the books Knus–Merkurjev–Rost–Tignol [1] or Scharlau [2].

4.5 Another characterization of the index

Recall that we have given two proofs of the fact that the Brauer group is torsion: in Corollary 2.8.5 we proved that the class of a central simple algebra A is annihilated by the index $\mathrm{ind}(A)$, and in Corollary 4.4.4 we proved that it is annihilated by the degree of some splitting field. This hints at a possible relation between the index and the degrees of splitting fields. Indeed, we have:

Proposition 4.5.1 *Let A be a central simple k-algebra. The index $\mathrm{ind}(A)$ is the greatest common divisor of the degrees of finite separable field extensions $K|k$ that split A.*

Before proving the proposition we derive some consequences. First, combining with Theorem 2.2.7 we get:

Corollary 4.5.2 *The index $\mathrm{ind}(A)$ is the smallest among the degrees of finite separable field extensions $K|k$ that split A.*

Here are some other easy corollaries.

Corollary 4.5.3 *Let A and B be central simple k-algebras that generate the same subgroup in $\mathrm{Br}\,(k)$. Then $\mathrm{ind}(A) = \mathrm{ind}(B)$.*

Proof The proposition implies that for all i we have $\mathrm{ind}(A^{\otimes i}) \mid \mathrm{ind}(A)$. But for suitable i and j we have $[A^{\otimes i}] = [B]$ and $[B^{\otimes j}] = [A]$ in $\mathrm{Br}\,(k)$ by assumption, so the result follows, taking Remark 2.8.2 (2) into account. □

Corollary 4.5.4 *Let $K|k$ be a finite separable field extension.*

1. We have the divisibility relations

$$\mathrm{ind}_K(A \otimes_k K) \mid \mathrm{ind}_k(A) \mid [K : k]\,\mathrm{ind}_K(A \otimes_k K).$$

2. If $\mathrm{ind}_k(A)$ is prime to $[K : k]$, then $\mathrm{ind}_k(A) = \mathrm{ind}_K(A \otimes_k K)$. In particular, if A is a division algebra, then so is $A \otimes_k K$.

Proof It is enough to prove the first statement. The divisibility relation $\mathrm{ind}_K(A \otimes_k K) \mid \mathrm{ind}_k(A)$ is immediate from the proposition. For the second one, use Theorem 2.2.7 to find a finite separable field extension $K'|K$ splitting $A \otimes_k K$ with $[K' : K] = \mathrm{ind}_K(A \otimes_k K)$. Then K' is also a splitting field of A, so Proposition 4.5.1 shows $\mathrm{ind}_k(A) \mid [K' : k] = \mathrm{ind}_K(A \otimes_k K)[K : k]$. □

The proof of Proposition 4.5.1 will be based on the following refinement of Theorem 4.4.1, interesting in its own right.

Proposition 4.5.5   *Let $K|k$ be a separable field extension of degree n. Then the boundary map $\delta_n : H^1(k, \mathrm{PGL}_n(k_s)) \to \mathrm{Br}\,(k)$ induces a bijection*

$$\ker(H^1(k, \mathrm{PGL}_n(k_s)) \to H^1(K, \mathrm{PGL}_n(k_s))) \xrightarrow{\sim} \mathrm{Br}\,(K|k).$$

We need a lemma from Galois theory.

Lemma 4.5.6   *Let $\widetilde{K}$ be the Galois closure of K, and denote the Galois groups $\mathrm{Gal}\,(\widetilde{K}|k)$ and $\mathrm{Gal}\,(\widetilde{K}|K)$ by G and H, respectively. Making G act on the tensor product $K \otimes_k \widetilde{K}$ via the second factor, we have an isomorphism of G-modules*

$$(K \otimes_k \widetilde{K})^\times \cong M_H^G(\widetilde{K}^\times).$$

Proof   According to the theorem of the primitive element, we may write $K = k(\alpha)$ for some $\alpha \in K$ with minimal polynomial $f \in k[x]$, so that $\widetilde{K}$ is the splitting field of f. By Galois theory, if $1 = \sigma_1, \ldots, \sigma_n$ is a system of left coset representatives for H in G, the roots of f in $\widetilde{K}$ are exactly the $\sigma_i(\alpha)$ for $1 \leq i \leq n$. So we get, just like before the proof of Speiser's lemma in Chapter 2, a chain of isomorphisms

$$K \otimes_k \widetilde{K} \cong \widetilde{K}[x]/\prod_{i=1}^n (x - \sigma_i(\alpha)) \cong \mathrm{Hom}_H(\mathbf{Z}[G], \widetilde{K}) = M_H^G(\widetilde{K}).$$

The lemma follows by restricting to invertible elements.   □

Proof of Proposition 4.5.5   We have already shown in the proof of Theorem 4.4.1 the injectivity of δ_n (even of δ_∞), so it suffices to see surjectivity. With the notations of the lemma above, consider the short exact sequence of G-modules

$$1 \to \widetilde{K}^\times \to (K \otimes_k \widetilde{K})^\times \to (K \otimes_k \widetilde{K})^\times/\widetilde{K}^\times \to 1,$$

where G acts on $K \otimes_k \widetilde{K}$ via the second factor. Part of the associated long exact sequence reads

$$H^1(G, (K \otimes_k \widetilde{K})^\times/\widetilde{K}^\times) \to H^2(G, \widetilde{K}^\times) \to H^2(G, (K \otimes_k \widetilde{K})^\times). \quad (4.4)$$

Using the previous lemma, Shapiro's lemma and Theorem 4.4.1, we get a chain of isomorphisms

$$H^2(G, (K \otimes_k \widetilde{K})^\times) \cong H^2(G, M_H^G(\widetilde{K})) \cong H^2(H, \widetilde{K}) \cong \mathrm{Br}\,(\widetilde{K}|K).$$

We also have $H^2(G, \widetilde{K}^\times) \cong \mathrm{Br}\,(\widetilde{K}|k)$, so all in all we get from exact sequence (4.4) a surjection

$$\widetilde{\alpha} : H^1(G, (K \otimes_k \widetilde{K})^\times / \widetilde{K}^\times) \to \mathrm{Br}\,(K|k).$$

On the other hand, the choice of a k-basis of K provides an embedding $K \hookrightarrow M_n(k)$, whence a G-equivariant map $K \otimes_k \widetilde{K} \to M_n(\widetilde{K})$, and finally a map $(K \otimes_k \widetilde{K})^\times \to \mathrm{GL}_n(\widetilde{K})$. Arguing as in the proof of Theorem 4.4.1, we get a commutative diagram:

$$
\begin{array}{ccc}
H^1(G, (K \otimes_k \widetilde{K})^\times / K^\times) & \xrightarrow{\;\;\widetilde{\alpha}\;\;} & H^2(G, \widetilde{K}^\times) \\[1em]
\downarrow & & \downarrow{\scriptstyle \mathrm{id}} \\[1em]
H^1(G, \mathrm{PGL}_n(\widetilde{K})) & \xrightarrow{\;\;\delta_n\;\;} & H^2(G, \widetilde{K}^\times).
\end{array}
$$

Therefore by the surjectivity of $\widetilde{\alpha}$ each element of $\mathrm{Br}\,(K|k) \subset H^2(G, \widetilde{K}^\times)$ comes from some element in $H^1(G, \mathrm{PGL}_n(\widetilde{K}))$. By the injectivity of δ_n and its obvious compatibility with restriction maps, this element restricts to 1 in $H^1(H, \mathrm{PGL}_n(\widetilde{K}))$, as required.   □

Proof of Proposition 4.5.1   In view of Theorem 2.2.7 it is enough to show that if a finite separable extension $K|k$ of degree n splits A, then $\mathrm{ind}(A)$ divides n. For such a K, the class of A in $\mathrm{Br}\,(K|k)$ comes from a class in $H^1(k, \mathrm{PGL}_n(k))$ according to Proposition 4.5.5. By Theorem 2.4.3 this class is also represented by some central simple k-algebra B of degree n, hence of index dividing n. But $\mathrm{ind}(A)$ equals $\mathrm{ind}(B)$ by Remark 2.8.2 (2).   □

We conclude this section by showing how Proposition 4.5.1 and its consequences allow one to give quick proofs of the main results of Section 2.8.

Remark 4.5.7   We give another proof of Theorem 2.8.7 (2) according to which the index and period of a central simple algebra A have the same prime factors. Given a prime number p that does not divide $\mathrm{per}(A)$, Proposition 4.5.1 implies that it suffices to find a finite separable splitting field K of A with $[K : k]$ prime to p. To do so, let $L|k$ be a finite Galois extension that splits A, and let K be the fixed field of a p-Sylow subgroup P of $\mathrm{Gal}\,(L|k)$. Then $\mathrm{Br}\,(L|K)$ is a p-primary torsion group by Corollary 4.4.4, so the assumption implies that the image of $[A]$ by the restriction map $\mathrm{Br}\,(L|k) \to \mathrm{Br}\,(L|K)$ is trivial. This means that A is split by K.

Remark 4.5.8   We finally give another proof of the primary decomposition of a division algebra D (Proposition 2.8.13). As in the first proof, we decompose the class of D in Br (k) as the sum of classes of division algebras D_i of prime power index $p_i^{r_i}$, and show that the equality $\mathrm{ind}(D) = \prod_i \mathrm{ind}(D_i)$ holds. A repeated application of Theorem 2.2.7 shows that for fixed i one may find a finite separable extension $K_i|k$ of degree prime to p_i that splits all the D_j for $j \neq i$. Then $D \otimes_k K_i$ and $D_i \otimes_k K_i$ have the same class in Br (K_i), and thus $\mathrm{ind}_{K_i}(D_i \otimes_k K_i) \mid \mathrm{ind}(D)$ by Corollary 4.5.4 (1). The algebras $D_i \otimes_k K_i$ are still division algebras of index $\mathrm{ind}(D_i)$ over K_i by Corollary 4.5.4 (2). To sum up, we have proven that $\mathrm{ind}(D_i)$ divides $\mathrm{ind}(D)$ for all i, hence so does their product $\prod_i \mathrm{ind}(D_i)$. Divisibility in the other direction follows from Theorem 2.8.7 (2) as in the first proof.

4.6 The Galois symbol

It is time to introduce one of the main protagonists of this book, the Galois symbol. To construct it, consider an integer $m > 0$ and a field k of characteristic prime to m. Recall that μ_m denotes the group of m-th roots of unity in a fixed separable closure k_s of k, equipped with its canonical action by $G = \mathrm{Gal}\,(k_s|k)$. Kummer theory (Proposition 4.3.6) then defines a map

$$\partial : k^\times \to H^1(k, \mu_m),$$

which is surjective with kernel $k^{\times m}$. On the other hand, for an integer $n > 0$ we may take n copies of $H^1(k, \mu_m)$ and consider the cup-product

$$H^1(k, \mu_m) \otimes \cdots \otimes H^1(k, \mu_m) \to H^n(k, \mu_m^{\otimes n}),$$

where according to the convention taken in Chapter 3 the G-module $\mu_m^{\otimes n}$ is the tensor product over $\mathbf{Z}$ of n copies of μ_m, equipped with the Galois action defined by $\sigma(\omega_1 \otimes \cdots \otimes \omega_n) = \sigma(\omega_1) \otimes \cdots \otimes \sigma(\omega_n)$.

Putting the two together, we obtain a homomorphism from the n-fold tensor product

$$\partial^n : k^\times \otimes_{\mathbf{Z}} \cdots \otimes_{\mathbf{Z}} k^\times \to H^n(k, \mu_m^{\otimes n}).$$

We now have the following basic fact due to Tate.

Proposition 4.6.1   *Assume that $a_1, \ldots, a_n \in k^\times$ is a sequence of elements such that $a_i + a_j = 1$ for some $1 \leq i < j \leq n$. Then*

$$\partial^n(a_1 \otimes \cdots \otimes a_n) = 0.$$

The proof uses some simple compatibility statements for the Kummer map.

Lemma 4.6.2 *Let $K|k$ be a finite separable field extension. Then the diagrams*

$$
\begin{array}{ccc}
k^\times & \xrightarrow{\partial_k} & H^1(k, \mu_m) \\
\downarrow & & \downarrow {\scriptstyle \mathrm{Res}} \\
K^\times & \xrightarrow{\partial_K} & H^1(K, \mu_m)
\end{array}
\qquad and \qquad
\begin{array}{ccc}
K^\times & \xrightarrow{\partial_K} & H^1(K, \mu_m) \\
\downarrow {\scriptstyle N_{K|k}} & & \downarrow {\scriptstyle \mathrm{Cor}} \\
k^\times & \xrightarrow{\partial_k} & H^1(k, \mu_m)
\end{array}
$$

commute, where in the first diagram the left vertical map is the natural inclusion.

Proof It follows from the construction of restriction and corestriction maps and Remark 3.1.10 (3) that they are compatible with the boundary maps on cohomology. It therefore remains to see that the maps $\mathrm{Res} : H^0(k, k_s^\times) \to H^0(K, k_s^\times)$ and $\mathrm{Cor} : H^0(K, k_s^\times) \to H^0(k, k_s^\times)$ are given by the inclusion $k^\times \to K^\times$ and the norm $N_{K|k} : K^\times \to k^\times$, respectively. The first of these statements is obvious, and the second comes from the fact that if we embed $K|k$ into a finite Galois extension $L|k$, the norm of an element $\alpha \in K^\times$ is given by the product $\prod \sigma_i(\alpha)$, where $1 = \sigma_1, \dots, \sigma_l$ is a system of left coset representatives of $\mathrm{Gal}\,(L|k)$ modulo $\mathrm{Gal}\,(L|K)$. $\square$

Proof of Proposition 4.6.1 By graded-commutativity and associativity of the cup-product we may assume $i = 1$ and $j = n = 2$, and use the notation $a_1 = a$, $a_2 = 1 - a$. Take an irreducible factorization

$$
x^m - a = \prod_l f_l
$$

in the polynomial ring $k[x]$, for each l let α_l be a root of f_l in k_s and define $K_l = k(\alpha_l)$. We then have

$$
1 - a = \prod_l f_l(1) = \prod_l N_{K_l|k}(1 - \alpha_l)
$$

by definition of the field norm. Therefore, as ∂^2 is a group homomorphism,

$$
\partial^2(a \otimes (1 - a)) = \sum_l \partial^2(a \otimes N_{K_l|k}(1 - \alpha_l)).
$$

Here

$$
\partial^2(a \otimes N_{K_l|k}(1 - \alpha_l)) = \partial(a) \cup \partial(N_{K_l|k}(1 - \alpha_l)) =
$$
$$
= \partial(a) \cup \mathrm{Cor}_k^{K_l}(\partial(1 - \alpha_l)) = \mathrm{Cor}_k^{K_l}(\mathrm{Res}_k^{K_l}(\partial(a)) \cup \partial(1 - \alpha_l)),
$$

where we have used the definition of ∂^2, the above lemma and the projection formula (Proposition 3.4.10 (3)), respectively. But (again using the lemma)

$$\mathrm{Res}_k^{K_l}(\partial(a)) = \partial_{K_l}(a) = 0,$$

because by definition we have $a = \alpha_l^m$ in K_l, and so a lies in $K_l^{\times m}$, which is the kernel of ∂_{K_l}. This proves the proposition. $\qquad\qquad\square$

The proposition prompts the following definition.

Definition 4.6.3 Let k be a field. For $n > 1$ we define the *n-th Milnor K-group* $K_n^M(k)$ to be the quotient of the n-fold tensor product $k^\times \otimes_{\mathbf{Z}} \cdots \otimes_{\mathbf{Z}} k^\times$ by the subgroup generated by those elements $a_1 \otimes \cdots \otimes a_n$ with $a_i + a_j = 1$ for some $1 \le i < j \le n$. By convention, we put $K_0(k) := \mathbf{Z}$ and $K_1(k) := k^\times$.

For elements $a_1, \ldots, a_n \in k^\times$, we denote the class of $a_1 \otimes \cdots \otimes a_n$ in $K_n^M(k)$ by $\{a_1, \ldots, a_n\}$. We usually call these classes *symbols*.

By the proposition, the map ∂^n factors through $K_n^M(k)$ and yields a map

$$h_{k,m}^n : K_n^M(k) \to H^n(k, \mu_m^{\otimes n}),$$

which makes sense even for $n = 0$.

Definition 4.6.4 The above map $h_{k,m}^n$ is called the *Galois symbol*.

We now have the following fundamental theorem, formerly known as the *Bloch–Kato conjecture*.

Theorem 4.6.5 (Voevodsky, Rost) *The Galois symbol yields an isomorphism*

$$K_n^M(k)/m \xrightarrow{\sim} H^n(k, \mu_m^{\otimes n})$$

for all $n \ge 0$, all fields k and all m prime to the characteristic of k.

The case when m is a power of 2 had been usually referred to as *Milnor's Conjecture*; it was proven in Voevodsky [1]. The general case was settled in Voevodsky [2], using important contributions due to Markus Rost.

For $n = 0$ the statement of the theorem is trivial, and for $n = 1$ it is none but Kummer theory (Proposition 4.3.6). The case $n = 2$ was proven in 1982 by Merkurjev and Suslin [1]; this theorem will occupy us in Chapter 8 of this book.

Theorem 4.6.6 (Merkurjev–Suslin) *The Bloch–Kato conjecture is true for $n = 2$ and all m invertible in k.*

We shall explain in the next section the relation of this statement to the one given in Chapter 2, Section 2.5.

4.7 Cyclic algebras and symbols

Continuing the discussion of the previous section, let us now focus on the case $n = 2$. Assume first that k has characteristic prime to m and contains a primitive m-th root of unity ω. The symbol h_k^2 then has as target $H^2(k, \mu_m^{\otimes 2})$, but choosing an isomorphism $\mu_m \cong \mathbf{Z}/m\mathbf{Z}$ by sending ω to 1 we get isomorphisms

$$H^2(k, \mu_m^{\otimes 2}) \cong H^2(k, \mathbf{Z}/m\mathbf{Z}) \cong H^2(k, \mu_m) \cong {}_m\mathrm{Br}\,(k),$$

the last one by Corollary 4.4.5. We emphasize that this chain of isomorphisms depends on the choice of ω. We then have:

Proposition 4.7.1 *Let $a, b \in k^\times$. Then under the above isomorphisms the element $h_k^2(\{a, b\}) \in H^2(k, \mu_m^{\otimes 2})$ goes to the Brauer class of the cyclic algebra $(a, b)_\omega$ defined in Chapter 2, Section 2.5.*

Remarks 4.7.2

1. The statement makes sense because we have seen in Chapter 2, Section 2.5 that the algebra $(a, b)_\omega$ is split by an extension of degree m, therefore it has period dividing m.
2. The proposition implies that the form of the Merkurjev–Suslin theorem stated in Theorem 2.5.7 is equivalent to the surjectivity of h_k^2 under the assumption $\omega \in k$. Henceforth, by 'Merkurjev–Suslin theorem' we shall mean this most general form, i.e. Theorem 4.6.6.

Before embarking on the proof, recall from the construction of the Kummer map $\partial : k^\times \to H^1(k, \mu_m)$ that under the identification

$$H^1(k, \mu_m) \cong H^1(k, \mathbf{Z}/m\mathbf{Z}) = \mathrm{Hom}(\mathrm{Gal}\,(k_s|k), \mathbf{Z}/m\mathbf{Z})$$

induced by sending ω to 1 the element $\partial(a)$ is mapped to the character sending the automorphism $\sigma : \sqrt[m]{a} \mapsto \omega \sqrt[m]{a}$ to 1, where $\sqrt[m]{a}$ is an m-th root of a in k_s. The kernel of this character fixes the cyclic Galois extension $K = k(\sqrt[m]{a})$ of k, whence an isomorphism $\chi : \mathrm{Gal}\,(K|k) \to \mathbf{Z}/m\mathbf{Z}$. In Corollary 2.5.5 we have shown that $(a, b)_\omega$ is isomorphic to the more general cyclic algebra (χ, b) introduced in Construction 2.5.1 of *loc. cit.*, the isomorphism depending, as always, on the choice of ω.

We shall derive Proposition 4.7.1 from the following more general one which is valid without assuming m prime to the characteristic of k.

Proposition 4.7.3   *Let k be a field and $m > 0$ an integer. Assume given a degree m cyclic Galois extension $K|k$ with group G, and let $\chi : G \xrightarrow{\sim} \mathbf{Z}/m\mathbf{Z}$ be an isomorphism. Take a lifting $\widetilde{\chi}$ of χ to a character $\mathrm{Gal}\,(k_s|k) \to \mathbf{Z}/m\mathbf{Z}$, and fix an element $b \in k^\times$. Denoting by δ the coboundary map $H^1(k, \mathbf{Z}/m\mathbf{Z}) \to H^2(k, \mathbf{Z})$ coming from the exact sequence $0 \to \mathbf{Z} \xrightarrow{m} \mathbf{Z} \to \mathbf{Z}/m\mathbf{Z} \to 0$, the cup-product map*

$$H^2(k, \mathbf{Z}) \times H^0(k, k_s^\times) \to H^2(k, k_s^\times) \cong \mathrm{Br}\,(k)$$

sends the element $\delta(\widetilde{\chi}) \cup b$ to the Brauer class of the cyclic algebra (χ, b).

Proof   Recall from Chapter 2, Section 2.5 that we have constructed the algebra (χ, b) via Galois descent, by twisting the standard Galois action on the matrix algebra $M_m(k)$ by the 1-cocycle $z(b) : G \to \mathrm{PGL}_m(K)$ given by applying first χ and then sending 1 to the class $F(b)$ of the invertible matrix

$$\widetilde{F}(b) = \begin{bmatrix} 0 & 0 & \cdots & 0 & b \\ 1 & 0 & \cdots & 0 & 0 \\ 0 & 1 & \cdots & 0 & 0 \\ \vdots & & \ddots & & \vdots \\ 0 & 0 & \cdots & 1 & 0 \end{bmatrix}$$

in $\mathrm{PGL}_m(K)$. Recall also that $\widetilde{F}(b)^m = b \cdot I_m$.

Consider now the commutative diagram of G-groups

$$
\begin{array}{ccccccccc}
1 & \longrightarrow & \mathbf{Z} & \xrightarrow{\ m\ } & \mathbf{Z} & \longrightarrow & \mathbf{Z}/m\mathbf{Z} & \longrightarrow & 1 \\
& & \downarrow{\scriptstyle b} & & \downarrow{\scriptstyle \widetilde{F}(b)} & & \downarrow{\scriptstyle F(b)} & & \\
1 & \longrightarrow & K^\times & \longrightarrow & \mathrm{GL}_m(K) & \longrightarrow & \mathrm{PGL}_m(K) & \longrightarrow & 1,
\end{array}
$$

where the maps denoted by $b, F(b), \widetilde{F}(b)$ mean the map induced by sending 1 to the corresponding element. The commutativity of the left square follows from the equality $\widetilde{F}(b)^m = b \cdot I_m$, and that of the right one is straightforward. Taking cohomology we obtain the commutative diagram

$$
\begin{array}{ccc}
H^1(G, \mathbf{Z}/m\mathbf{Z}) & \xrightarrow{\ \delta\ } & H^2(G, \mathbf{Z}) \\
\downarrow{\scriptstyle (F(b))_*} & & \downarrow{\scriptstyle b_*} \\
H^1(G, \mathrm{PGL}_m(K)) & \xrightarrow{\ \delta_m\ } & H^2(G, K^\times),
\end{array}
$$

where the horizontal arrows are boundary maps. The character χ is naturally an element of $\mathrm{Hom}(G, \mathbf{Z}/m\mathbf{Z}) = H^1(G, \mathbf{Z}/m\mathbf{Z})$ and, as explained above, it is mapped by $F(b)_*$ to the class of the 1-cocycle $z(b)$. By definition, we have therefore $\delta_m((F(b))_* \chi) = [(\chi, b)]$, so by commutativity of the diagram $[(\chi, b)] = b_*(\delta(\chi))$. But one checks from the definition of cup-products that the map b_* is given by cup-product with the class of b in $H^0(G, K^\times)$, so we get $[(\chi, b)] = \delta(\chi) \cup b$. Moreover, we defined the character $\tilde{\chi}$ as the image of χ by the inflation map $H^1(G, \mathbf{Z}/m\mathbf{Z}) \to H^1(k, \mathbf{Z}/m\mathbf{Z})$, and the inflation map $H^0(G, K^\times) \to H^0(k, k_s^\times)$ is obviously the identity. We finally get $[(\chi, b)] = \delta(\tilde{\chi}) \cup b$ by compatibility of the cup-product with inflations. $\qquad\square$

Proof of Proposition 4.7.1 In view of Proposition 4.7.3 and the discussion preceding it, all that remains to be seen is the equality

$$\delta(\tilde{\chi}) \cup b = \tilde{\chi} \cup \partial(b),$$

where $\delta(\tilde{\chi}) \cup b \in H^2(k, k_s^\times)$ is the element considered in Proposition 4.7.3, the map $\partial : k^\times \to H^1(k, \mu_m)$ is the Kummer coboundary and the cup-product on the left is that between $H^1(k, \mathbf{Z}/m\mathbf{Z})$ and $H^1(k, \mu_m)$. This follows from (the profinite version of) Proposition 3.4.9, with the exact sequences

$$0 \to \mathbf{Z} \xrightarrow{m} \mathbf{Z} \to \mathbf{Z}/m\mathbf{Z} \to 0, \quad 1 \to \mu_m \to k_s^\times \to k_s^\times \to 1,$$

the pairing $\mathbf{Z} \times k_s^\times \to k_s^\times$ (which is trivial on $m\mathbf{Z} \times \mu_m$), and the elements $\tilde{\chi} \in H^1(k, \mathbf{Z})$ and $b \in H^0(k, k_s^\times)$. $\qquad\square$

We now derive some interesting consequences from Proposition 4.7.1.

Corollary 4.7.4 *For $K|k$, G and χ as above, the isomorphism*

$$H^2(G, K^\times) \cong k^\times / N_{K|k}(K^\times) \qquad (4.5)$$

of Corollary 4.4.6 is induced by the map $k^\times \to H^2(G, K^\times)$ sending an element $b \in k^\times$ to the class of the cyclic algebra (χ, b).

Proof By Proposition 3.4.11, the isomorphism (4.5) is induced (from right to left) by cup-product with $\delta(\chi)$, as it is a generator of the group $H^2(G, \mathbf{Z}) \cong \mathbf{Z}/m\mathbf{Z}$. On the other hand, the previous proposition implies that $\delta(\chi) \cup b$ is exactly the class of (χ, b) in $\mathrm{Br}\,(K|k) \cong H^2(G, K^\times)$. $\qquad\square$

This immediately implies the following key property of cyclic algebras that will serve many times in what follows.

Corollary 4.7.5 *With notation as in the previous corollary, the period of the cyclic algebra* (χ,b) *equals the order of the element* $b \in k^{\times}$ *in the torsion group* $k^{\times}/N_{K|k}(K^{\times})$.

In particular, the cyclic algebra (χ, b) *is split if and only if b is a norm from the extension* $K|k$.

Remark 4.7.6 The period of a cyclic division algebra may be strictly smaller than its index; see Rowen [1], Theorem 4 for examples in degree p^2 for an odd prime p. Given integers $s > r$, one may also construct examples of cyclic algebras of index p^s and period p^r using the theorem of Rehmann–Tikhonov–Yanchevskiĭ stated in Remark 5.5.2 (2) below.

Another consequence of Proposition 4.7.1 is the following characterization of cyclic algebras.

Corollary 4.7.7 *Let* $K|k$, *m and G be as above, and let A be a central simple k-algebra split by* K.

1. *There exist an isomorphism* $\chi : G \xrightarrow{\sim} \mathbf{Z}/m\mathbf{Z}$ *and an element* $b \in k$ *such that the cyclic algebra* (χ, b) *is Brauer equivalent to A.*
2. *If moreover A has degree m, then we have actually* $(\chi, b) \cong A$.

Proof The first statement is an immediate consequence of Corollary 4.7.4. The second follows from the first, since we are then dealing with Brauer equivalent algebras of the same degree. □

Putting together Proposition 4.7.1 and Corollary 4.7.5 we get:

Corollary 4.7.8 *Assume k contains a primitive m-th root of unity* ω, *and let* $a, b \in k^{\times}$. *Then the following statements are equivalent.*

1. *The symbol* $h_k^2(\{a, b\})$ *is trivial.*
2. *The cyclic algebra* $(a, b)_{\omega}$ *is split.*
3. *The element b is a norm from the extension* $k(\sqrt[m]{a})|k$.

Note that the equivalence (2) $\Leftrightarrow$ (3) generalizes the equivalence (1) $\Leftrightarrow$ (4) in Proposition 1.1.7.

Remark 4.7.9 Since the first two conditions of the corollary are symmetric in a and b, we get that b *is a norm from the extension* $k(\sqrt[m]{a})|k$ *if and only if* a

is a norm from the extension $k(\sqrt[m]{b})|k$. This type of statement is usually called a *reciprocity law* in arithmetic.

EXERCISES

1. Show that in the correspondence of Theorem 4.1.12 Galois extensions $L|k$ contained in K correspond to closed normal subgroups of $\mathrm{Gal}\,(K|k)$.

2. (Continuous cochains) Let G be a profinite group and A a continuous G-module. Define the group $C^i_{\mathrm{cont}}(G, A)$ of *continuous i-cochains* as the subgroup of those maps in $\mathrm{Hom}_G(\mathbf{Z}[G^{i+1}], A)$ whose restriction to G^{i+1} is continuous when G^{i+1} is equipped with the product topology. Show that the boundary maps δ^{i*} of the complex $C^\bullet(G, A)$ introduced in Construction 3.2.1 map $C^i_{\mathrm{cont}}(G, A)$ into $C^{i+1}_{\mathrm{cont}}(G, A)$, and that the cohomology groups of the arising complex $C^\bullet_{\mathrm{cont}}(G, A)$ are isomorphic to the continuous cohomology groups $H^i_{\mathrm{cont}}(G, A)$.

3. Let $m > 0$ be an integer, and k a field containing a primitive m^2-th root of unity. Consider a degree m cyclic extension $K = k(\sqrt[m]{a})|k$ with Galois group G.

   (a) Show that the group $(K^\times/K^{\times m})^G$ is generated by $k^\times$ and $\sqrt[m]{a}$. [*Hint:* Use Proposition 3.3.16.]

   (b) Determine explicitly the cokernel of the map

   $$k^\times/k^{\times m} \to (K^\times/K^{\times m})^G.$$

4. (Theorem of Frobenius) Prove that $\mathrm{Br}\,(\mathbf{R}) \cong \mathbf{Z}/2\mathbf{Z}$, the nontrivial class being that of the Hamilton quaternions.

5. Let k be a field of characteristic 0 such that $\mathrm{Gal}\,(\bar{k}|k) \cong \mathbf{Z}/p\mathbf{Z}$ for some prime number p.

   (a) Show that $\mathrm{Br}\,(k) \cong \mathrm{Br}\,(k)/p\,\mathrm{Br}\,(k) \cong k^\times/k^{\times p}$. [*Hint:* Use the Kummer sequence and the periodicity of the cohomology of cyclic groups.]

   (b) By computing $\mathrm{Br}\,(k)$ in a different way, show that $N_{\bar{k}|k}(\bar{k}^\times) = k^{\times p}$.

   (c) Conclude that the above is only possible for $p = 2$ and $\bar{k} = k(\sqrt{-1})$.

   (d) Show that moreover in the above case k may be equipped with an ordered field structure. [*Hint:* Declare the squares to be the positive elements.]

6. Using the previous exercise, prove the following theorem of E. Artin and O. Schreier: If k is a field of characteristic 0 whose absolute Galois group is a nontrivial finite group, then $\bar{k} = k(\sqrt{-1})$ and k has an ordered field structure. [*Hint:* Begin by taking a p-Sylow subgroup in the Galois group and recall that p-groups are solvable.]

   [*Remark:* In fact, Artin and Schreier also showed that in positive characteristic the absolute Galois group is either trivial or infinite.]

7. (Wang's theorem in the general case) Let D be a central division algebra over k.

   (a) Consider the primary decomposition

   $$D \cong D_1 \otimes_k D_2 \otimes_k \cdots \otimes_k D_r$$

of Proposition 2.8.13. Show that

$$SK_1(D) \cong \bigoplus_{i=1}^{r} SK_1(D_i).$$

   (b) Assume that $\operatorname{ind}(D)$ is squarefree, i.e. a product of distinct primes. Show that $SK_1(D) = 0$. [*Hint:* Reduce to Theorem 2.10.12.]
8. (a) Verify the relations $(\chi, b_1) \otimes (\chi, b_2) \cong (\chi, b_1 b_2)$ and $(\chi, b)^{\mathrm{op}} \cong (\chi, b^{-1})$ for cyclic algebras.
   (b) Deduce that the cyclic algebras (χ, b_1) and (χ, b_2) are isomorphic if and only if $b_1 b_2^{-1}$ is a norm from the cyclic extension of the base field determined by χ.

9. Let n be a positive integer, and let k be a field containing a primitive n-th root of unity ω. Consider a purely transcendental extension $k(x, y)$ of dimension 2. Given integers i, j prime to n, show that the cyclic $k(x, y)$-algebras $(x, y)_{\omega^i}$ and $(x, y)_{\omega^j}$ are isomorphic if and only if $i - j$ is divisible by n.

5

Severi–Brauer varieties

In Chapter 1 we associated with each quaternion algebra a conic with the property that the conic has a k-point if and only if the algebra splits over k. We now generalize this correspondence to arbitrary dimension: with each central simple algebra A of degree n over an arbitrary field k we associate a projective k-variety X of dimension $n - 1$ which has a k-point if and only if A splits. Both objects will correspond to a class in $H^1(G, \mathrm{PGL}_n(K))$, where K is a Galois splitting field for A with group G. The varieties X arising in this way are called Severi–Brauer varieties; they are characterized by the property that they become isomorphic to some projective space over the algebraic closure. This interpretation will enable us to give another, geometric construction of the Brauer group. Another central result of this chapter is a theorem of Amitsur which states that for a Severi–Brauer variety X with function field $k(X)$ the kernel of the natural map $\mathrm{Br}\,(k) \to \mathrm{Br}\,(k(X))$ is a cyclic group generated by the class of X. This seemingly technical statement (which generalizes Witt's theorem proven in Chapter 1) has very fruitful algebraic applications. At the end of the chapter we shall present one such application, due to Saltman, which shows that all central simple algebras of fixed degree n over a field k containing the n-th roots of unity can be made cyclic via base change to some large field extension of k.

Severi–Brauer varieties were introduced in the pioneering paper of Châtelet [1], under the name 'variétés de Brauer'. Practically all results in the first half of the present chapter stem from this work. The term 'Severi–Brauer variety' was coined by Beniamino Segre in his note [1], who expressed his discontent that Châtelet had ignored previous work by Severi in the area. Indeed, in the paper of Severi [1] Severi–Brauer varieties are studied in a classical geometric context, and what is known today as Châtelet's theorem is proven in some cases. As an amusing feature, we may mention that Severi calls the varieties in question 'varietà di Segre', but beware, this does not refer to Beniamino but to his second uncle Corrado Segre. The groundbreaking paper by Amitsur [1] was the first to emphasize the importance of the birational viewpoint on Severi–Brauer varieties in the study of central simple algebras. This observation was a milestone on the road leading to the proof of the Merkurjev–Suslin theorem.

5.1 Basic properties

In Chapter 2 we have seen that as a consequence of Wedderburn's theorem one may define a central simple algebra over a field k as a finite-dimensional k-algebra that becomes isomorphic to some full matrix algebra $M_n(K)$ over a finite extension $K|k$ of the base field. As a consequence of descent theory, we have seen that when $K|k$ is Galois, the central simple k-algebras split by K can be described by means of the automorphism group $\mathrm{PGL}_n(K)$ of $M_n(K)$. But $\mathrm{PGL}_n(K)$ is also the automorphism group of projective $(n-1)$-space $\mathbf{P}_K^{n-1}$ (considered as an algebraic variety), which motivates the following definition.

Definition 5.1.1   A Severi–Brauer variety over a field k is a projective algebraic variety X over k such that the base extension $X_K := X \times_k K$ becomes isomorphic to $\mathbf{P}_K^{n-1}$ for some finite field extension $K|k$. The field K is called a *splitting field* for X.

Remarks 5.1.2

1. A k-variety X is a Severi–Brauer variety if and only if $X_{\bar{k}} \cong \mathbf{P}_{\bar{k}}^{n-1}$ for an algebraic closure $\bar{k}$ of k. Indeed, necessity is obvious and sufficiency follows from the fact that the coefficients of the finitely many polynomials defining an isomorphism $X_{\bar{k}} \cong \mathbf{P}_{\bar{k}}^{n-1}$ are all contained in a finite extension of k.
2. It follows from general considerations in algebraic geometry that a Severi–Brauer variety is necessarily smooth. Also, the assumption that X be projective is also superfluous: it can be shown that an algebraic variety (i.e. separated scheme of finite type) over k that becomes isomorphic to a projective variety over a finite extension of k is itself projective.

As examples of Severi–Brauer varieties we may cite the projective plane conics encountered in Chapter 1. The next section will describe a general method for constructing examples.

We now come to the fundamental result about Severi–Brauer varieties. Before stating it, let us introduce some (nonstandard) terminology: we say that a closed subvariety $Y \to X$ defined over k is a *twisted-linear* subvariety of X

if Y is a Severi–Brauer variety and moreover over $\bar{k}$ the inclusion $Y_{\bar{k}} \subset X_{\bar{k}}$ becomes isomorphic to the inclusion of a linear subvariety of $\mathbf{P}_{\bar{k}}^{n-1}$.

Theorem 5.1.3 (Châtelet) *Let X be a Severi–Brauer variety of dimension $n - 1$ over the field k. The following are equivalent:*

1. *X is isomorphic to projective space $\mathbf{P}_k^{n-1}$ over k.*
2. *X is birationally isomorphic to projective space $\mathbf{P}_k^{n-1}$ over k.*
3. *X has a k-rational point.*
4. *X contains a twisted-linear subvariety D of codimension 1.*

It is usually the equivalence of statements (1) and (3) that is referred to as Châtelet's theorem. The only implication which is not straightforward to establish is (3) $\Rightarrow$ (4); we owe the beautiful proof given below to Endre Szabó. This proof uses some elementary notions from algebraic geometry; however, for the less geometrically minded, we note that in Section 5.3 another proof will be given, under the assumption that X has a Galois splitting field. We shall see in Corollary 5.1.5 below that this condition is always satisfied.

Proof The implication (1) $\Rightarrow$ (2) is obvious. If (2) holds, then X and $\mathbf{P}_k^{n-1}$ have k-isomorphic Zariski open subsets. If k is infinite, a Zariski open subset of $\mathbf{P}_k^{n-1}$ contains a k-rational point, whence (3). On the other hand, over a finite field (3) always holds; see Remark 5.1.6 below. Next we prove (4) $\Rightarrow$ (1). The subvariety D whose existence is postulated by (4) is a divisor, so we may consider the associated complete linear system $|D|$ (see the Appendix) which defines a rational map ϕ_D into some projective space. Over $\bar{k}$ the divisor D becomes a hyperplane by assumption, so the rational map it defines is in fact an isomorphism with projective $(n - 1)$-space $\mathbf{P}_{\bar{k}}^{n-1}$. Hence the target of ϕ_D must be $\mathbf{P}_k^{n-1}$ and it must be an everywhere defined isomorphism.

It remains to prove the implication (3) $\Rightarrow$ (4). Let P be a k-rational point and denote by $\pi : Y \to X$ the blow-up of X at P (see Appendix, Example A.2.3). As X (and in particular P) is smooth, the exceptional divisor E is isomorphic to $\mathbf{P}_k^{n-2}$. Pick a hyperplane $L \subset E$. Over the algebraic closure $\bar{k}$ our $Y_{\bar{k}}$ is isomorphic to the blow-up of $\mathbf{P}_{\bar{k}}^{n-1}$ in P, hence it is a subvariety of $\mathbf{P}_{\bar{k}}^{n-1} \times \mathbf{P}_{\bar{k}}^{n-2}$. The second projection induces a morphism $\psi_{\bar{k}} : Y_{\bar{k}} \to \mathbf{P}_{\bar{k}}^{n-2}$, mapping $E_{\bar{k}}$ isomorphically onto $\mathbf{P}_{\bar{k}}^{n-2}$. As the fibres of $\psi_{\bar{k}}$ are projective lines and $\pi_{\bar{k}}$ is an isomorphism outside P, we see that the subvariety $D_{\bar{k}} := \pi_{\bar{k}}(\psi_{\bar{k}}^{-1}(\psi_{\bar{k}}(L_{\bar{k}}))) \subset X_{\bar{k}}$ is a hyperplane in $\mathbf{P}_{\bar{k}}^{n-1}$. We want to define this structure over k, i.e. we are looking for a morphism $\psi : Y \to Z$ defined over k which becomes $\psi_{\bar{k}}$ after base extension to $\bar{k}$.

Let $A \subset X$ be an ample divisor, and let d denote the degree of $A_{\bar{k}}$ in the projective space $X_{\bar{k}} \cong \mathbf{P}_{\bar{k}}^{n-1}$. The divisor $(\pi^* A - dE)_{\bar{k}}$ has degree 0 on the fibres of $\psi_{\bar{k}}$ and has degree d on $E_{\bar{k}}$. Hence the morphism $Y_{\bar{k}} \to \mathbf{P}_{\bar{k}}^N$ associated with the corresponding linear system factors as the composite

$$Y_{\bar{k}} \xrightarrow{\psi_{\bar{k}}} Z_{\bar{k}} \xrightarrow{\phi_d} \mathbf{P}_{\bar{k}}^N,$$

where ϕ_d is the d-uple embedding. The linear system $|\pi^* A - dE|$ defines (over k) a rational map $\psi : Y \to \mathbf{P}_k^N$. By construction this ψ becomes the above $\psi_{\bar{k}}$ after base extension to $\bar{k}$, hence it is actually an everywhere defined morphism $Y \to Z$, where $Z := \psi(Y)$. Then the subvariety $D := \pi(\psi^{-1}(\psi(L))) \subset X$ is defined over k, and becomes $D_{\bar{k}}$ after extension to $\bar{k}$. This is the D we were looking for. $\qquad \square$

Corollary 5.1.4 *A Severi–Brauer variety X always splits over a finite separable extension of the base field k.*

Proof By an argument similar to that in the proof of Theorem 2.2.1, it is enough to show that X becomes isomorphic to projective space over a separable closure k_s of k. This follows from the theorem, for X_{k_s} always has a rational point over k_s (see Appendix, Proposition A.1.2). $\qquad \square$

By embedding a separable splitting field into its Galois closure, we get:

Corollary 5.1.5 *A Severi–Brauer variety X always splits over a finite Galois extension of the base field k.*

Remark 5.1.6 Based on the above corollary, we shall construct in Theorem 5.2.1 below a 1–1 correspondence between isomorphism classes of Severi–Brauer varieties of dimension $n - 1$ and classes in $H^1(k, \mathrm{PGL}_n)$. When k is finite, the corollary is straightforward and $H^1(k, \mathrm{PGL}_n)$ is trivial (see Remark 2.2.8). This means that over a finite field every Severi–Brauer variety is split, and therefore in this case Theorem 5.1.3 is obvious.

5.2 Classification by Galois cohomology

Let X be a Severi–Brauer variety of dimension $n - 1$ over a field k, and let $K|k$ be a finite Galois extension with group G which is a splitting field of X. We now associate with X a 1-cohomology class of G with values in $\mathrm{PGL}_n(K)$ by a construction analogous to that in the theory of central simple algebras.

First some conventions. Given quasi-projective varieties Y, Z over k, denote by Y_K, Z_K the varieties obtained by base extension to K, and make G act on the set of morphisms $Y_K \to Z_K$ by $\phi \mapsto \sigma(\phi) := \phi \circ \sigma^{-1}$. In particular, for $Y = Z$ we obtain a left action of G on the K-automorphism group $\mathrm{Aut}_K(Y)$.

Given a K-isomorphism $\phi : \mathbf{P}_K^{n-1} \xrightarrow{\sim} X_K$, define for each element $\sigma \in G$ a K-automorphism $a_\sigma \in \mathrm{Aut}_K(\mathbf{P}_K^{n-1})$ by

$$a_\sigma := \phi^{-1} \circ \sigma(\phi).$$

Exactly the same computations as in Chapter 2, Section 2.3 show that the map $\sigma \mapsto a_\sigma$ is a 1-cocycle of G with values in $\mathrm{Aut}_K(\mathbf{P}_K^{n-1})$, and that changing ϕ amounts to changing a_σ by a 1-coboundary. Therefore we have assigned to X a class $[a_\sigma]$ in $H^1(G, \mathrm{Aut}_K(\mathbf{P}_K^{n-1}))$. Fixing an isomorphism $\mathrm{Aut}_K(\mathbf{P}_K^{n-1}) \cong \mathrm{PGL}_n(K)$ (see Appendix, Example A.2.2), we may consider it as a class in $H^1(G, \mathrm{PGL}_n(K))$. Using the boundary map $H^1(G, \mathrm{PGL}_n(K)) \to \mathrm{Br}\,(K|k)$ we can also assign to X a class $[X]$ in $\mathrm{Br}\,(K|k)$.

Denote by $SB_n(k)$ the pointed set of isomorphism classes of Severi–Brauer varieties of dimension $n - 1$ over k, the base point being the class of $\mathbf{P}_k^{n-1}$.

Theorem 5.2.1 *The map $SB_n(k) \to H^1(k, \mathrm{PGL}_n)$ given by $X \mapsto [a_\sigma]$ is a base point preserving bijection.*

Combining the theorem with Theorem 2.4.3 we thus get a base point preserving bijection

$$CSA_n(k) \leftrightarrow SB_n(k).$$

Given a central simple k-algebra A of degree n, we shall call (somewhat abusively) a Severi–Brauer variety X whose isomorphism class corresponds to that of A by the bijection above a *Severi–Brauer variety associated with A*.

We now prove Theorem 5.2.1 using a construction due to Kang [1]. The proof will at the same time yield the following important property.

Theorem 5.2.2 *Let X be a Severi–Brauer variety of dimension $n - 1$ over k, and let d be the period of X, i.e. the order of $[X]$ in the Brauer group $\mathrm{Br}\,(K|k)$. Then there exists a projective embedding*

$$\rho : X \hookrightarrow \mathbf{P}_k^{N-1}, \quad N = \binom{n + d - 1}{d}$$

such that $\rho_K : X_K \hookrightarrow \mathbf{P}_K^{N-1}$ is isomorphic to the d-uple embedding ϕ_d.

Proof of Theorems 5.2.1 and 5.2.2 We begin by proving the injectivity of the map $SB_n(k) \to H^1(k, \mathrm{PGL}_n)$. Let X and Y be Severi–Brauer varieties split by K and having the same class in $H^1(G, \mathrm{PGL}_n)$. Take trivialization isomorphisms $\phi : \mathbf{P}_K^{n-1} \xrightarrow{\sim} X_K$ and $\psi : \mathbf{P}_K^{n-1} \xrightarrow{\sim} Y_K$. Our assumption that the cocycles $\phi^{-1} \circ \sigma(\phi)$ and $\psi^{-1} \circ \sigma(\psi)$ have the same class in $H^1(G, \mathrm{PGL}_n(K))$ means that there exists $h \in \mathrm{PGL}_n(K)$ such that

$$\phi^{-1} \circ \sigma(\phi) = h^{-1} \circ \psi^{-1} \circ \sigma(\psi) \circ \sigma(h)$$

for all $\sigma \in G$. We then have

$$\psi \circ h \circ \phi^{-1} = \sigma(\psi \circ h \circ \phi^{-1}) \in \mathrm{Hom}_K(X_K, Y_K),$$

so the K-isomorphism $\psi \circ h \circ \phi^{-1} : X_K \to Y_K$ is G-equivariant. It follows that $\psi \circ h \circ \phi^{-1}$ is defined over k, and hence yields a k-isomorphism between X and Y.

Let now $\alpha = [a_\sigma]$ be a class in $H^1(G, \mathrm{PGL}_n(K))$. We show that α can be realized as the cohomology class of a Severi–Brauer variety of dimension $n - 1$ which becomes isomorphic over K to $\phi_d(\mathbf{P}_K^{n-1})$. This will prove the surjectivity statement in Theorem 5.2.1 and at the same time Theorem 5.2.2.

Consider the boundary map $\delta : H^1(G, \mathrm{PGL}_n(K)) \to H^2(G, K^\times)$. By definition, a 2-cocycle representing $\delta(\alpha)$ is obtained by lifting the elements $a_\sigma \in \mathrm{PGL}_n(K)$ to elements $\widetilde{a}_\sigma \in \mathrm{GL}_n(K)$ and setting

$$b_{\sigma,\tau} = \widetilde{a}_\sigma\, \sigma(\widetilde{a}_\tau)\, \widetilde{a}_{\sigma\tau}^{-1}. \tag{5.1}$$

Let d be the order of $\delta(\alpha)$ in $\mathrm{Br}\,(K|k)$ (which is a torsion group by Corollary 2.8.5). In terms of the cocycle $b_{\sigma,\tau}$ this means that $(b_{\sigma,\tau})^d$ is a 2-coboundary, i.e. there is a 1-cochain $\sigma \mapsto c_\sigma$ with values in $K^\times$ such that

$$(b_{\sigma,\tau})^d = c_\sigma\, \sigma(c_\tau)\, c_{\sigma\tau}^{-1}. \tag{5.2}$$

Now consider the natural left action of the group $\mathrm{GL}_n(K)$ on $V := K^n$ which is compatible with that of G in the sense of Construction 2.3.6. This action extends to the symmetric powers $V_i := V^{\otimes i}/S_i$ by setting

$$\widetilde{a}(v_1 \otimes \cdots \otimes v_i) = \widetilde{a}(v_1) \otimes \cdots \otimes \widetilde{a}(v_i).$$

Note that V_i is none but the space of homogeneous polynomials of degree i over K. We extend the action of G to the V_i in a similar way. Now consider the case $i = d$, and for each $\sigma \in G$ define an element $\nu_\sigma \in \mathrm{Aut}_K(V_d)$ by

$$\nu_\sigma := c_\sigma^{-1}\widetilde{a}_\sigma,$$

where c_σ^{-1} acts as constant multiplication and $\tilde{a}_\sigma$ via the action of $\mathrm{GL}_n(K)$ on V_d described above. We contend that the map $\sigma \to \nu_\sigma$ is a 1-cocycle. Indeed, for $\sigma, \tau \in G$, we compute using (5.1)

$$\nu_{\sigma\tau} = c_{\sigma\tau}^{-1} \tilde{a}_{\sigma\tau} = c_{\sigma\tau}^{-1} (b_{\sigma,\tau}^{-1} \tilde{a}_\sigma \, \sigma(\tilde{a}_\tau)) = (c_{\sigma\tau}^{-1} b_{\sigma,\tau}^{-d})(\tilde{a}_\sigma \sigma(\tilde{a}_\tau)),$$

where in the second step we considered $b_{\sigma\tau}$ as a scalar matrix in $\mathrm{GL}_n(K)$ and in the third just as a scalar. Hence using (5.2) we get

$$\nu_{\sigma\tau} = c_{\sigma\tau}^{-1} (c_{\sigma\tau} \sigma(c_\tau)^{-1} c_\sigma^{-1})(\tilde{a}_\sigma \sigma(\tilde{a}_\tau)) = \left(c_\sigma \, \sigma(c_\tau)\right)^{-1} \tilde{a}_\sigma \sigma(\tilde{a}_\tau) = \nu_\sigma \sigma(\nu_\tau),$$

so that we have indeed defined a 1-cocycle.

Now equip V_d with the twisted G-action defined by ν_σ (see Construction 2.3.6 for the definition). Let $W := \left({}_\nu V_d\right)^G$ be the invariant subspace under this twisted action. By Speiser's lemma (Lemma 2.3.8), this is a k-vector space such that $W \otimes_k K \cong V_d$. Let $k[X]$ be the graded k-subalgebra of $K[x_0, \dots, x_{n-1}] \cong \bigoplus_i V_i$ generated by W. Choosing a k-basis $v_0, \dots, v_{N-1}$ of W we get a natural surjection of graded k-algebras $k[x_0, \dots, x_{N-1}] \to k[X]$ induced by sending x_i to v_i. The kernel of this surjection is a homogeneous ideal in $k[x_0, \dots, x_{N-1}]$ defining a closed subset $X \subset \mathbf{P}_k^{N-1}$. Moreover, the isomorphism $W \otimes_k K \cong V_d$ implies that $k[X] \otimes_k K$ becomes isomorphic to the graded K-subalgebra of $K[x_0, \dots, x_{n-1}]$ generated by V_d. But this is none but the homogeneous coordinate ring of $\phi_d(\mathbf{P}_K^{n-1})$ (see Appendix, Example A.2.1), hence $X \otimes_k K \cong \phi_d(\mathbf{P}_K^{n-1})$. This shows that X is a Severi–Brauer variety, and at the same time that Theorem 5.2.2 holds. By construction the class in $H^1(G, \mathrm{PGL}_n(K))$ associated with X is indeed α. $\square$

Remarks 5.2.3

1. It should be noted that the embedding $X \hookrightarrow \mathbf{P}_k^{N-1}$ constructed in the above proof is not canonical, but depends on the choice of the cochain $\sigma \mapsto c_\sigma$. Two choices differ by a 1-cocycle $G \to K^\times$ which is of the form $\sigma \mapsto c\sigma(c)^{-1}$ with some $c \in K^\times$ by Hilbert's Theorem 90.
2. In geometric terms, we have shown in the above proof that if d is the period of X, then the linear system of degree d hypersurfaces on $\mathbf{P}_K^{n-1}$ descends to a linear system on X, and moreover it induces a closed embedding in a projective space over k. It can be shown that the period is exactly the smallest positive integer d with this property, which yields a geometric definition of the period (see Kollár [1], §5).

Example 5.2.4 Theorem 5.2.1 shows that the 1-dimensional Severi–Brauer varieties are exactly the smooth projective conics. Indeed, such a variety X defines a class in $H^1(k, \mathrm{PGL}_2)$ which also corresponds to a central simple algebra of degree 2 by Theorem 2.4.3. By the results of Chapter 1, this must be a quaternion algebra (a, b), whose class has always order 2 in the Brauer group. By Theorem 5.2.2, we can embed X as a smooth subvariety in $\mathbf{P}_k^2$ which is isomorphic to the conic $x_1^2 = x_0 x_2$ over the algebraic closure. It is a well-known fact from algebraic geometry that then X itself is a conic. In Section 5.4 we shall prove that X is in fact the conic $C(a, b)$ associated with the quaternion algebra (a, b) in Chapter 1.

Remarks 5.2.5

1. In the literature one finds other approaches to the construction of Severi–Brauer varieties. The classical approach, going back to Châtelet, is to construct the Severi–Brauer variety associated with a degree n algebra A as the variety of left ideals of dimension n in A, by embedding it as a closed subvariety into the Grassmannian $Gr(n, n^2)$ (see e.g. Saltman [3] or Knus–Merkurjev–Rost–Tignol [1]). This construction has the advantage of being canonical, but the projective embedding it gives is far from being the most 'economical' one. For instance, Severi–Brauer varieties of dimension 1 are realized not as plane conics, but as curves in $\mathbf{P}^5$ defined by 31 (non-independent) equations; see Jacobson [3], p. 113. Another approach is that of Grothendieck, which is based on general techniques in descent theory. It has similar advantages and disadvantages: it is more conceptual than the one above and works in a much more general situation, but it does not give explicit information on the projective embedding. See Jahnel [1] for a very detailed exposition of Grothendieck's construction.

2. It is also possible to go in the reverse direction and associate a central simple algebra to a Severi–Brauer variety via a geometric construction. This method uses vector bundles. If X is Severi–Brauer variety over k such that $X \times_k \bar{k} \cong \mathbf{P}_{\bar{k}}^{n-1}$, one can show that the n-fold direct power $\mathcal{O}(1)^{\oplus n}$ of Serre's twisting bundle $\mathcal{O}(1)$ on $\mathbf{P}_{\bar{k}}^{n-1}$ comes from a vector bundle $\mathcal{F}_X$ on X by base change to $X \times_k \bar{k}$. In (Quillen [1], §8.4) the existence of $\mathcal{F}_X$ is established by flat descent theory; in (Kollár [1], §1) by showing that up to scaling by an element of $k^\times$ it is the unique nonsplit extension of the tangent bundle of X by a trivial line bundle. One then defines the k-algebra A_X as the opposite algebra to the endomorphism algebra of the vector bundle $\mathcal{F}_X$. By construction, after base change to $\bar{k}$ it becomes isomorphic to $\operatorname{End}_{\bar{k}}(\mathcal{O}(1)^{\oplus n}) \cong M_n(\bar{k})$ and therefore it is a central simple algebra.

5.3 Geometric Brauer equivalence

In the previous section we have shown that isomorphism classes of Severi–Brauer varieties of dimension $n - 1$ correspond bijectively to elements in the pointed set $H^1(k, \mathrm{PGL}_n)$. They therefore have a class in $\mathrm{Br}\,(k)$. Defining this class involves, however, the consideration of an equivalence relation on the disjoint union of the sets $H^1(k, \mathrm{PGL}_n)$ for all n, which corresponds to Brauer equivalence on central simple algebras. We now show that Brauer equivalence is quite easy to define geometrically on Severi–Brauer varieties, using closed embeddings of twisted-linear subvarieties.

The first step in this direction is:

Proposition 5.3.1 *Let X be a Severi–Brauer variety and Y a twisted-linear subvariety of X. Then X and Y have the same class in $\mathrm{Br}\,(k)$.*

Proof Let $n - 1$ be the dimension of X, and $d - 1$ that of Y. Let $V_d \subset k^n$ be the linear subspace generated by the first d standard basis vectors, and $\widetilde{P}_d(k)$ the subgroup of $\mathrm{GL}_n(k)$ consisting of elements leaving $V_d \subset k^n$ invariant. In other words, $\widetilde{P}_d(k)$ is the subgroup

$$\begin{bmatrix} \mathrm{GL}_d(k) & * \\ 0 & \mathrm{GL}_{n-d}(k) \end{bmatrix} \subset \mathrm{GL}_n(k).$$

We denote by $P_d(k)$ its image in $\mathrm{PGL}_n(k)$. Note that restriction to the subspace V_d yields a natural map $\widetilde{P}_d(k) \to \mathrm{GL}_d(k)$, and hence a map $P_d(k) \to \mathrm{PGL}_d(k)$.

Now by definition of a twisted-linear subvariety (taking Corollary 5.1.5 into account), there exists a finite Galois extension $K|k$ of group G and a commutative diagram of trivializations

$$
\begin{array}{ccc}
X \times_k K & \xleftarrow[\sim]{\;\phi\;} & \mathbf{P}^{n-1} \times_k K \\
\uparrow & & \uparrow \\
Y \times_k K & \xleftarrow[\sim]{\;\psi\;} & \mathbf{P}^{d-1} \times_k K,
\end{array}
$$

where the right vertical map is the inclusion of a projective linear subspace. Since $\mathrm{PGL}_n(k)$ acts transitively on the $(d - 1)$-dimensional projective linear subspaces of $\mathbf{P}_k^{n-1}$, we may assume that this map actually is the projectivization of the inclusion map $V_d \otimes_k K \to K^n$. In this case the cocycle $a_\sigma : \sigma \mapsto \phi \circ \sigma(\phi^{-1})$ defining the class of X in $H^1(G, \mathrm{PGL}_n(K))$ takes its values in the subgroup $P_d(K) \subset \mathrm{PGL}_n(K)$, and the class of $[Y]$ is in the image of the map $H^1(G, P_d(K)) \to H^1(G, \mathrm{PGL}_n(K))$. The commutative diagram

$$
\begin{array}{ccccccccc}
1 & \longrightarrow & K^\times & \longrightarrow & \mathrm{GL}_n(K) & \longrightarrow & \mathrm{PGL}_n(K) & \longrightarrow & 1 \\
& & \mathrm{id}\uparrow & & \uparrow & & \uparrow & & \\
1 & \longrightarrow & K^\times & \longrightarrow & \widetilde{P}_d(K) & \longrightarrow & P_d(K) & \longrightarrow & 1 \\
& & \mathrm{id}\downarrow & & \downarrow & & \downarrow & & \\
1 & \longrightarrow & K^\times & \longrightarrow & \mathrm{GL}_d(K) & \longrightarrow & \mathrm{PGL}_d(K) & \longrightarrow & 1
\end{array}
$$

yields the commutative diagram of boundary maps

$$
\begin{array}{ccc}
[X] \in H^1(G, \mathrm{PGL}_n(K)) & \longrightarrow & H^2(G, K^\times) \\
\uparrow & & \mathrm{id}\uparrow \\
[a_\sigma] \in \ \ H^1(G, P_d(K)) & \longrightarrow & H^2(G, K^\times) \\
\downarrow & & \mathrm{id}\downarrow \\
[Y] \in H^1(G, \mathrm{PGL}_d(K)) & \longrightarrow & H^2(G, K^\times).
\end{array}
\tag{5.3}
$$

Here the commutativity of the upper square is obvious and that of the lower one is proven by an argument similar to that of Lemma 2.7.8. We conclude that $[Y] = [X] \in \mathrm{Br}\,(k)$. $\qquad\square$

Proposition 5.3.2 *Let B be a central simple algebra, and let $A = M_r(B)$ for some $r > 0$. Denote by X and Y the Severi–Brauer varieties associated with A and B, respectively. Then Y can be embedded as a twisted-linear subvariety into X.*

The proof below is due to Michael Artin.

Proof We keep the notations from the proof of Proposition 5.3.1; in particular, let d be the degree of B, and $n = rd$ that of A. It will be enough to show that the class of B in $H^1(G, \mathrm{PGL}_d(K))$ lies in the image of the natural map

$$
H^1(G, P_d(K)) \to H^1(G, \mathrm{PGL}_d(K)).
$$

Indeed, then diagram (5.3) shows that $[X]$ and $[Y]$ are both images of the same class in $H^1(G, P_d(K))$ because the horizontal maps in the diagram are injective by Proposition 2.7.9. Therefore the construction of Severi–Brauer varieties out of 1-cocycles given in the previous chapter implies that Y embeds as a twisted-linear subvariety into X. To see this, consider the natural projection $\pi_d : K^n \to K^d$ given by mapping the last $n - d$ basis elements to 0. In the construction of the varieties X and Y we twisted the G-action on a symmetric power of these vector spaces by the action of PGL_n, resp. PGL_d. These twisted actions are compatible with each other under the maps $\mathrm{PGL}_n \leftarrow P_d \to \mathrm{PGL}_d$. The induced map on G-invariants is surjective, because so is the map given by base change to K, which is just the above π_d by Speiser's lemma. The construction of X and Y then shows that this surjection of k-vector spaces induces a surjection of homogeneous coordinate rings $k[X] \to k[Y]$, which corresponds to a closed embedding of Y into X. This embedding is twisted linear because X and Y have the same period, and therefore we have worked with the same homogeneous component of the coordinate ring in the construction.

Now to prove our claim about the class $[B]$, consider the commutative diagram with exact rows and columns

$$
\begin{array}{ccccccccc}
 & & 1 & & & & 1 & & \\
 & & \downarrow & & & & \downarrow & & \\
1 \to K^\times & \xrightarrow{\Delta} & K^\times \times K^\times & \xrightarrow{(+,-)} & & & K^\times & & \to 1 \\
\text{id}\downarrow & & \downarrow & & & & \downarrow & & \\
1 \to K^\times & \longrightarrow & \mathrm{GL}_d(K) \times \mathrm{GL}_{n-d}(K) & \longrightarrow & & & (\mathrm{GL}_d(K) \times \mathrm{GL}_{n-d}(K))/K^\times & & \to 1 \\
 & & \downarrow & & & & \downarrow & & \\
 & & \mathrm{PGL}_d(K) \times \mathrm{PGL}_{n-d}(K) & \xrightarrow{\text{id}} & & & \mathrm{PGL}_d(K) \times \mathrm{PGL}_{n-d}(K) & & \\
 & & \downarrow & & & & \downarrow & & \\
 & & 1 & & & & 1 & &
\end{array}
$$

where Δ is the diagonal map, and $(+, -)$ is the map $(a, b) \mapsto ab^{-1}$. This is a diagram of groups equipped with a G-action, so by taking cohomology we get a commutative diagram of pointed sets with exact columns

$$H^1(G, \mathrm{PGL}_d(K)) \times H^1(G, \mathrm{PGL}_{n-d}(K)) \to H^1(G, \mathrm{GL}_d(K) \times \mathrm{GL}_{n-d}(K))/K^\times)$$

$$\downarrow \qquad\qquad\qquad\qquad\qquad\qquad\qquad\qquad \downarrow$$

$$H^1(G, \mathrm{PGL}_d(K)) \times H^1(G, \mathrm{PGL}_{n-d}(K)) \xrightarrow{\text{id}} H^1(G, \mathrm{PGL}_d(K)) \times H^1(G, \mathrm{PGL}_{n-d}(K))$$

$$\downarrow \qquad\qquad\qquad\qquad\qquad\qquad\qquad\qquad \downarrow$$

$$H^2(G, K^\times) \times H^2(G, K^\times) \xrightarrow{(+,-)} H^2(G, K^\times).$$

We have $n - d = (r - 1)d$, so $M_{r-1}(B)$ is a central simple algebra of degree $n - d$ satisfying

$$[B] - [M_{r-1}(B)] = [B] - [B] = 0 \in H^2(G, K^\times).$$

The diagram then shows that the pair $([B], [M_{r-1}(B)])$ defines an element of $H^1(G, \mathrm{PGL}_d(K)) \times H^1(G, \mathrm{PGL}_{n-d}(K))$ which is in the image of the map

$$H^1(G, (\mathrm{GL}_d(K) \times \mathrm{GL}_{n-d}(K))/K^\times) \to H^1(G, \mathrm{PGL}_d(K)) \times H^1(G, \mathrm{PGL}_{n-d}(K)).$$

In particular, the class $[B]$ is in the image of the map

$$\lambda : H^1(G, (\mathrm{GL}_d(K) \times \mathrm{GL}_{n-d}(K))/K^\times) \to H^1(G, \mathrm{PGL}_d(K))$$

obtained from the previous one by composing with the natural projection. Now observe that the natural surjection

$$\alpha : P_d(K) \to (\mathrm{GL}_d(K) \times \mathrm{GL}_{n-d}(K))/K^\times$$

induced by the mapping

$$\begin{bmatrix} \mathrm{GL}_d(K) & * \\ 0 & \mathrm{GL}_{n-d}(K) \end{bmatrix} \longrightarrow \begin{bmatrix} \mathrm{GL}_d(K) & 0 \\ 0 & \mathrm{GL}_{n-d}(K) \end{bmatrix}$$

has a section $\beta : (\mathrm{GL}_d(K) \times \mathrm{GL}_{n-d}(K))/K^\times \to P_d(K)$ satisfying $\alpha \circ \beta = \mathrm{id}$, which is induced by the obvious map in the reverse direction. Consequently, the natural map

$$H^1(G, P_d(K)) \xrightarrow{\alpha_*} H^1(G, (\mathrm{GL}_d(K) \times \mathrm{GL}_{n-d}(K))/K^\times)$$

induced on cohomology is surjective, so we conclude that the class $[B]$ lies in the image of the composite map

$$\lambda \circ \alpha_* : H^1(G, P_d(K)) \to H^1(G, \mathrm{PGL}_d(K)),$$

as was to be shown. $\qquad\square$

We can sum up the two previous propositions in the following statement.

Theorem 5.3.3 (Châtelet) *Two Severi–Brauer varieties X and Y over k have the same class in $\mathrm{Br}(k)$ if and only if there exists a Severi–Brauer variety Z over k into which both X and Y can be embedded as twisted-linear subvarieties.*

Remark 5.3.4 Châtelet formulated this statement in a different but equivalent way: he stated that two Severi–Brauer varieties are Brauer equivalent if and only if they have isomorphic twisted-linear subvarieties. (Indeed, the corresponding central simple algebras are then matrix algebras over the same division algebra, by Wedderburn's theorem.)

The theorem has several interesting consequences. First some terminology: we call a Severi–Brauer variety *minimal* if it has no proper twisted-linear subvarieties.

Corollary 5.3.5 *A central simple algebra A is a division algebra if and only if the associated Severi–Brauer variety is minimal.*

Proof This follows from the theorem and the fact that division algebras are the central simple algebras of lowest dimension in their Brauer class. □

Next recall that we have defined the index of a central simple algebra A to be the degree of the division algebra D for which $A \cong M_r(D)$ according to Wedderburn's theorem. In other words, the index $\mathrm{ind}(A)$ is the degree of the unique division algebra in the Brauer class of A. Hence:

Corollary 5.3.6 (Châtelet) *Let A be a central simple algebra, and let X be the Severi–Brauer variety associated with A. Then all minimal twisted-linear subvarieties of X have the same dimension d, satisfying the equality*

$$d = \mathrm{ind}(A) - 1.$$

We thus get a geometric definition of the index.

Remarks 5.3.7

1. Recall that A is split if and only if it has index 1. According to the proposition, this happens if and only if the minimal twisted-linear subvarieties have dimension 0. The subvarieties of dimension 0 defined over k are precisely the k-rational points, and conversely these are trivially twisted-linear subvarieties (if they exist). We thus get another proof of Châtelet's theorem (assuming the existence of a separable splitting field, which was used in the proof of Theorem 5.3.3).
2. It is possible to define the group structure on Brauer equivalence classes via geometric constructions on Severi–Brauer varieties. See Kollár [1], §4.

5.4 Amitsur's theorem

Let V be a variety over a field k. The natural inclusion $k \subset k(V)$ induces a map

$$r_V : \mathrm{Br}\,(k) \to \mathrm{Br}\,(k(V))$$

given by mapping the class of a Severi–Brauer variety X over k to the class of the variety $X_{k(V)}$ obtained by base extension. In particular, this applies to $V = X$. In this case, the base extension $X_{k(X)}$ has a $k(X)$-rational point coming from the generic point of X. Hence by Châtelet's theorem the class of X in $\mathrm{Br}\,(k)$ lies in the kernel of the map r_X. The following famous theorem shows that this construction already describes the kernel.

Theorem 5.4.1 (Amitsur)   *Let X be a Severi–Brauer variety defined over a field k. Then the kernel of the restriction map $r_X : \mathrm{Br}\,(k) \to \mathrm{Br}\,(k(X))$ is a cyclic group generated by the class of X in $\mathrm{Br}\,(k)$.*

An immediate corollary is:

Corollary 5.4.2   *Let X and Y be Severi–Brauer varieties that are birational over k. Then their classes $[X]$ and $[Y]$ generate the same subgroup in $\mathrm{Br}\,(k)$.*

Remark 5.4.3   *Amitsur's conjecture* predicts that the converse to the above corollary should be true: if $[X]$ and $[Y]$ generate the same subgroup in $\mathrm{Br}\,(k)$ and they have the same dimension, then X should be birational to Y over k. See Roquette [1] and Tregub [1] for partial results in this direction, as well as Tabuada–van den Bergh [1] for an interesting reformulation.

Note, however, that a weaker result is quite easy to prove: *If $[X]$ and $[Y]$ generate the same subgroup in $\mathrm{Br}\,(k)$, then X and Y are* stably birational *over k, i.e. there exist positive integers m, n such that $X \times_k \mathbf{P}^m$ is birational to $Y \times_k \mathbf{P}^n$ over k.* Indeed, the assumption implies that $X \times_k k(Y)$ and $Y \times_k k(Y)$ generate the same subgroup in $\mathrm{Br}\,(k(Y))$. But $Y \times_k k(Y)$ has a $k(Y)$-rational point (coming from the generic point of Y), so by Châtelet's theorem its class in $\mathrm{Br}\,(k(Y))$ is trivial. Hence so is that of $X \times_k k(Y)$, which means that $X \times_k k(Y) \cong \mathbf{P}^n \times_k k(Y)$. In particular, these varieties have the same function field, which by definition equals $k(X \times_k Y)$ for the left-hand side and $k(\mathbf{P}^n \times_k Y)$ for the right-hand side. Thus $X \times_k Y$ is birational to $\mathbf{P}^n \times_k Y$, and the claim follows by symmetry.

The main ingredient in the proof of Amitsur's theorem is the following proposition. Before stating it, we recall from Proposition A.4.4 (2) of the

Appendix that the Picard group of projective space $\mathbf{P}_K^n$ over a field K is isomorphic to $\mathbf{Z}$, generated by the class of a K-rational hyperplane. We call the map realizing the isomorphism $\mathrm{Pic}\,(\mathbf{P}_K^n) \cong \mathbf{Z}$ the *degree map*, and define the degree of a divisor on $\mathbf{P}_K^n$ to be the image of its class by the degree map. This map is not to be confused with the degree map defined for curves.

Proposition 5.4.4 *Let $K|k$ be a finite Galois extension with group G that is a splitting field for X. There is an exact sequence*

$$0 \to \mathrm{Pic}\,(X) \overset{\mathrm{deg}}{\to} \mathbf{Z} \overset{\delta}{\to} H^2(G, K^\times) \to H^2(G, K(X)^\times),$$

where the map deg *is given by composing the natural map* $\mathrm{Pic}\,(X) \to \mathrm{Pic}\,(X_K)$ *with the degree map.*

We first prove a lemma.

Lemma 5.4.5 *We have an isomorphism of G-modules*

$$\mathrm{Div}(X_K) \cong \bigoplus_{P \in X^1} M_{G_P}^G(\mathbf{Z})$$

where X^1 (resp. X_K^1) denotes the set of irreducible codimension 1 subvarieties of X (resp. X_K), and G_P denotes the stabilizer in G of a fixed element of X_K^1 above P.

Proof Note first that for each $P \in X^1$ the group G permutes the finite set $\{Q \mapsto P\}$ of elements of X_K^1 lying above P. We thus get a direct sum decomposition

$$\mathrm{Div}(X_K) \cong \bigoplus_{P \in X^1} \left(\bigoplus_{Q \mapsto P} \mathbf{Z} \right).$$

By definition of $M_{G_P}^G(\mathbf{Z})$, we thus have to construct an isomorphism

$$\mathrm{Hom}_{G_P}(\mathbf{Z}[G], \mathbf{Z}) \overset{\sim}{\to} \bigoplus_{Q \mapsto P} \mathbf{Z}.$$

For this, choose a system of left coset representatives $1 = \sigma_1, \ldots, \sigma_r$ of G modulo G_P. The map $\phi \mapsto \phi(\sigma_1), \ldots, \phi(\sigma_r)$ induces an isomorphism

$$\mathrm{Hom}_{G_P}(\mathbf{Z}[G], \mathbf{Z}) \overset{\sim}{\to} \bigoplus_{i=1}^{r} \mathbf{Z},$$

which does not depend on the choice of the system $\{\sigma_1, \ldots, \sigma_r\}$, as ϕ is a G_P-homomorphism. So it will be enough to identify the right-hand side with the sum indexed by the set $\{Q \mapsto P\}$. For this, note that G acts transitively on $\{Q \mapsto P\}$. Indeed, each Q gives rise to a discrete valuation v_Q on $K(X)$ lying above the valuation v_P defined by P on $k(X)$ (see Appendix, Section A.4), and for $\sigma \in G$ we have $v_Q \circ \sigma = v_{\sigma(Q)}$. As G acts transitively on the discrete valuations above v_P (Appendix, Proposition A.6.3 (1)), it acts transitively on the set $\{Q \mapsto P\}$. Fixing Q_0 above P with stabilizer G_P, we may thus identify $\{Q \mapsto P\}$ with the set $\{\sigma_1(Q_0), \ldots, \sigma_r(Q_0)\}$. $\qquad\square$

Proof of Proposition 5.4.4 By definition of the Picard group, we have an exact sequence of G-modules

$$0 \to K(X)^\times / K^\times \to \mathrm{Div}(X_K) \to \mathrm{Pic}\,(X_K) \to 0. \qquad (5.4)$$

The beginning of the associated long exact cohomology sequence reads

$$0 \to (K(X)^\times / K^\times)^G \to \mathrm{Div}(X_K)^G \to \mathrm{Pic}\,(X_K)^G \to$$

$$\to H^1(G, K(X)^\times / K^\times) \to H^1(G, \mathrm{Div}(X_K)).$$

Here we have

$$H^1(G, \mathrm{Div}(X_K)) \cong \bigoplus_P H^1(G_P, \mathbf{Z}) = 0$$

by Lemma 5.4.5, Shapiro's lemma and the fact that $H^1(G_P, \mathbf{Z}) = \mathrm{Hom}(G_P, \mathbf{Z}) = 0$ for the finite group G_P.

Next, a piece of the long exact sequence coming from the sequence of G-modules

$$0 \to K^\times \to K(X)^\times \to K(X)^\times / K^\times \to 0 \qquad (5.5)$$

reads

$$H^1(G, K(X)^\times) \to H^1(G, K(X)^\times / K^\times) \to H^2(G, K^\times) \to H^2(G, K(X)^\times).$$

Here the group $H^1(G, K(X)^\times)$ is trivial by Hilbert's Theorem 90 (applied to the extension $K(X)|k(X)$). Therefore by splicing the two long exact sequences together we get

$$0 \to (K(X)^\times / K^\times)^G \to \mathrm{Div}(X_K)^G \to \mathrm{Pic}\,(X_K)^G \to$$

$$\to H^2(G, K^\times) \to H^2(G, K(X)^\times).$$

To identify this sequence with that of the proposition, we make the following observations. First, by Lemma 5.4.5 we have $(\mathrm{Div}(X_K))^G = \mathrm{Div}(X)$. Next, the beginning of the long exact sequence associated with (5.5) and the vanishing of $H^1(G, K^\times)$ (again by Hilbert's Theorem 90) yields the isomorphism

$k(X)^{\times}/k^{\times} \cong (K(X)^{\times}/K^{\times})^{G}$. So we may replace the first two terms in the sequence above by Pic (X).

Finally, we have $X_K \cong \mathbf{P}_K^{n-1}$, whence an isomorphism Pic $(X_K) \cong \mathbf{Z}$ given by the degree map. To finish the proof, we have to show that Pic (X_K) is a *trivial* G-module. Indeed, the group G can only act on $\mathbf{Z}$ by sending 1 to 1 or -1. This action, however, comes from the action of G on line bundles on $\mathbf{P}_K^{n-1}$ and the line bundles in the class of -1 have no global sections, whereas those in the class of 1 do. (In terms of linear systems, the complete linear system associated with the class of 1 is that of hyperplanes in $\mathbf{P}_K^{n-1}$, whereas that associated with the class of -1 is empty.) This implies that 1 can only be fixed by G. □

Now it is easy to derive the following basic exact sequence.

Theorem 5.4.6 *There is an exact sequence*

$$0 \to \mathrm{Pic}\,(X) \xrightarrow{\mathrm{deg}} \mathbf{Z} \xrightarrow{\delta} \mathrm{Br}\,(k) \to \mathrm{Br}\,(k(X)),$$

with deg $:$ Pic $(X) \to \mathbf{Z}$ *the same map as above.*

For the proof of the theorem we need the following lemma.

Lemma 5.4.7 *Let V be a k-variety having a smooth k-rational point. Then the restriction map* Br $(k) \to$ Br $(k(V))$ *is injective.*

Proof If P is a smooth k-point on V, the local ring $\mathcal{O}_{X,P}$ embeds into the formal power series ring $k[[t_1, \ldots, t_n]]$, where n is the dimension of V (see Appendix, Theorem A.5.4). Passing to quotient fields we get an injection $k(V) \subset k((t_1, \ldots, t_n))$. This field in turn can be embedded into the iterated Laurent series field $k((t_1))((t_2)) \ldots ((t_n))$. All in all, we have an induced map Br $(k(V)) \to$ Br $(k((t_1)) \ldots ((t_n)))$. We show injectivity of the composite map $r :$ Br $(k) \to$ Br $(k((t_1)) \ldots ((t_n)))$. For this it will be enough to treat the case $n = 1$, i.e. the injectivity of $r :$ Br $(k) \to$ Br $(k((t)))$, as the general case then follows by a straightforward induction.

Represent a class in the kernel of r by a Severi–Brauer variety X defined over k. Regarding it as a variety defined over $k((t))$, Châtelet's theorem implies that it has a $k((t))$-rational point. If X is embedded into projective space $\mathbf{P}^N$, this point has homogeneous coordinates $(x_0, \ldots, x_N)$. Viewing $k((t))$ as the quotient field of the ring $k[[t]]$, we may assume that each x_i lies in $k[[t]]$ and not all of them are divisible by t. Setting $t = 0$ then defines a rational point of

X over k, and we conclude by Châtelet's theorem that the class of X in $\mathrm{Br}\,(k)$ is trivial.   □

Proof of Theorem 5.4.6   By Theorem 4.4.3 we have isomorphisms

$$H^2(G, K^\times) \cong \mathrm{Br}\,(K|k) \quad \text{and} \quad H^2(G, K(X)^\times) \cong \mathrm{Br}\,(K(X)|k(X)).$$

Now the definition of relative Brauer groups gives a commutative diagram with exact rows:

$$
\begin{array}{ccccccc}
0 & \longrightarrow & \mathrm{Br}\,(K|k) & \longrightarrow & \mathrm{Br}\,(k) & \longrightarrow & \mathrm{Br}\,(K) \\
& & \downarrow & & \downarrow & & \downarrow \\
0 & \longrightarrow & \mathrm{Br}\,(K(X)|k(X)) & \longrightarrow & \mathrm{Br}\,(k(X)) & \longrightarrow & \mathrm{Br}\,(K(X)).
\end{array}
$$

Here the third vertical map is injective by the lemma above. Hence the snake lemma gives an isomorphism

$$\ker(\mathrm{Br}\,(K|k) \to \mathrm{Br}\,(K(X)|k(X))) \cong \ker(\mathrm{Br}\,(k) \to \mathrm{Br}\,(k(X)),$$

and the theorem results from the previous proposition.   □

Remark 5.4.8   The exact sequence of the theorem is easy to establish using the *Hochschild–Serre spectral sequence* in étale cohomology (see e.g. Milne [2]).

We can now prove Amitsur's theorem.

Proof of Theorem 5.4.1   The exact sequence of the theorem shows that the kernel of the map $r_X : \mathrm{Br}\,(k) \to \mathrm{Br}\,(k(X))$ is cyclic, so it is a finite cyclic group, because $\mathrm{Br}\,(k)$ is a torsion group. By the remarks at the beginning of this section, the class of X is contained in $\ker(r_X)$, so if d denotes the order of this class, we see that $\ker(r_X)$ has order divisible by d. On the other hand, by the exact sequence of the theorem the group $\ker(r_X)$ is the quotient of $\mathbf{Z}$ by the image of the map $\deg : \mathrm{Pic}\,(X) \to \mathbf{Z}$. Theorem 5.2.2 implies that there is a divisor class on X which becomes the d-th power of the class of a hyperplane over the algebraic closure. This means that $\mathrm{Im}\,(\deg) \subset \mathbf{Z}$ contains d, therefore $\ker(r_X)$ must have exact order d and the class of X must be a generator.   □

Our next goal is to show that Witt's theorem (Theorem 1.4.2) follows from that of Amitsur. First a corollary already announced before:

Corollary 5.4.9   *Assume that the base field k is not of characteristic 2. Let (a, b) be a quaternion algebra over k. Then the Severi–Brauer variety associated with (a, b) is the conic $C(a, b)$ introduced in Chapter 1.*

Proof The conic $C := C(a, b)$ is a Severi–Brauer variety of dimension 1, so it defines a class $[C]$ in $\mathrm{Br}\,(k)$. The conic has a point over some quadratic extension $L|k$, so by Châtelet's theorem $[C]$ restricts to the trivial class in $\mathrm{Br}\,(L)$. By the restriction-corestriction formula (Corollary 4.2.12) $[C]$ therefore lies in the 2-torsion of $\mathrm{Br}\,(k)$. By Amitsur's theorem, this 2-torsion class generates the kernel of the map $\mathrm{Br}\,(k) \rightarrow \mathrm{Br}\,(k(C))$. On the other hand, by Proposition 1.3.2 the algebra $(a, b) \otimes_k k(C)$ splits, so $[(a, b)] = [C]$ or $[(a, b)] = 0$. If (a, b) is split, C has a k-point by *loc. cit.* and Châtelet's theorem implies that $[C] = 0$ as required. In the other case, $[(a, b)]$ must be the nontrivial element in the kernel which is $[C]$, and we are done again. $\quad\square$

Remarks 5.4.10

1. Now we see how Witt's theorem follows from Amitsur's theorem above: by the above proof, for a quaternion division algebra (a, b) the only nontrivial class in the kernel of the map $\mathrm{Br}\,(k) \rightarrow \mathrm{Br}\,(k(C))$ is that of (a, b). But if two division algebras of the same degree have the same Brauer class, they are isomorphic by Wedderburn's theorem.
2. The corollary also holds in characteristic 2, for the quaternion algebras $[a, b)$ and the associated conics defined in Exercise 4 of Chapter 1.

We conclude this section by the following refinement of Theorem 5.4.6 which incorporates most of the results obtained so far.

Theorem 5.4.11 (Lichtenbaum) *Let X be a Severi–Brauer variety over a field k. In the exact sequence*

$$0 \rightarrow \mathrm{Pic}\,(X) \rightarrow \mathbf{Z} \xrightarrow{\delta} \mathrm{Br}\,(k) \rightarrow \mathrm{Br}\,(k(X))$$

the map δ is given by sending 1 to the class of X in $\mathrm{Br}\,(k)$.

Proof Let $K|k$ be a Galois extension splitting X. By the proof of Proposition 5.4.4, the map δ arises as the composition of the coboundary maps

$$\mathrm{Pic}\,(X_K)^G \rightarrow H^1(G, K(X)^\times / K^\times) \quad \text{and}$$
$$H^1(G, K(X)^\times / K^\times) \rightarrow H^2(G, K^\times)$$

coming from the short exact sequences (5.4) and (5.5), respectively. In view of the construction of these coboundary maps (see the proofs of Proposition 2.7.1 and Proposition 2.7.6), we can therefore describe $\delta(1)$ as follows. One takes first a divisor D representing the divisor class $1 \in \mathbf{Z} \cong \mathrm{Pic}\,(X_K)$. Here the divisor D is not G-invariant in general but its class is (see the end of the proof of Proposition 5.4.4), so one finds a function f_σ with $\mathrm{div}(f_\sigma) = \sigma(D) - D$. The $K(X)^\times$-valued map $\sigma \mapsto f_\sigma$ is the lifting of a 1-cocycle with values in $K(X)^\times/K^\times$, and the image of this cocycle by the second coboundary map is by definition the 2-cocycle $(\sigma, \tau) \mapsto f_\sigma \sigma(f_\tau) f_{\sigma\tau}^{-1}$. This is the 2-cocycle representing $\delta(1)$.

Now since D is of degree 1, the linear system $|D|$ defines an isomorphism $X_K \cong \mathbf{P}_K^{n-1}$, where $n - 1 = \dim X$. This isomorphism arises by taking the associated projective space to an isomorphism of vector spaces $L(D) \cong K^n$. Let $g_0, \ldots, g_{n-1}$ be a basis of the left-hand side mapping to the standard basis $e_0, \ldots, e_{n-1}$ of K^n. Denote by λ the inverse isomorphism sending e_i to g_i. In terms of linear systems, the map λ sends e_i to the positive divisor $(g_i) + D$. So for $\sigma \in G$, the isomorphism $\sigma(\lambda)$ sends e_i to the divisor $(\sigma(g_i)) + \sigma(D) = (\sigma(g_i)) + (f_\sigma) + D = (f_\sigma \sigma(g_i)) + D$. This last divisor is also an element of $|D|$, therefore $f_\sigma \sigma(g_i) \in L(D)$. We may therefore write

$$f_\sigma \sigma(g_i) = \sum a_{ij\sigma} g_i \tag{5.6}$$

with some $a_{ij\sigma} \in K$. The matrix $A_\sigma := [a_{ij\sigma}]$ is therefore the matrix of the K-automorphism $\lambda^{-1} \circ \sigma(\lambda)$. Comparing with the definition at the beginning of Section 5.2, we see that this is exactly the matrix defining the class of X in $H^1(G, \mathrm{PGL}_n(K))$, and the class in $\mathrm{Br}\,(k)$ is therefore given by the 2-cocycle $(\sigma, \tau) \mapsto A_\sigma \sigma(A_\tau) A_{\sigma\tau}^{-1}$.

To compare these two 2-cocycles, we perform the following computation in the function field $K(X)$:

$$\sigma\tau(g_i) = \sigma(\tau(g_i)) = \sigma(f_\tau^{-1} A_\tau g_i) = \sigma(f_\tau^{-1}) \sigma(A_\tau) \sigma(g_i)$$
$$= \sigma(f_\tau^{-1}) \sigma(A_\tau)(f_\sigma^{-1} A_\sigma g_i) = \sigma(f_\tau^{-1}) f_\sigma^{-1} \sigma(A_\tau) A_\sigma g_i.$$

Comparing with equation (5.6) applied to $\sigma\tau$ gives

$$g_i = f_{\sigma\tau} \sigma(f_\tau^{-1}) f_\sigma^{-1} A_{\sigma\tau}^{-1} \sigma(A_\tau) A_\sigma g_i$$

for all i, and therefore

$$f_\sigma \sigma(f_\tau) f_{\sigma\tau}^{-1} = (A_\sigma^{-1} \sigma(A_\tau^{-1}) A_{\sigma\tau})^{-1}.$$

It remains to observe that the 2-cocycle $(\sigma, \tau) \mapsto (A_\sigma^{-1} \sigma(A_\tau^{-1}) A_{\sigma\tau})$ represents the class $-[X]$ in $\mathrm{Br}\,(k)$. $\qquad\square$

Remarks 5.4.12

1. Lichtenbaum's theorem immediately implies Amitsur's, and therefore yields a proof which does not use the results of Section 2, just the construction of the Brauer class associated with X.

2. We also get a second (less explicit) proof of Theorem 5.2.2: if the class of X has order d in the Brauer group, then there exists a divisor class of degree d on X. The associated linear system defines the d-uple embedding over a splitting field K.

5.5 An application: making central simple algebras cyclic

We give now the following nice application of Amitsur's theorem, whose statement is purely algebraic and apparently does not involve Severi–Brauer varieties.

Theorem 5.5.1 (Saltman) *Assume that k contains a primitive n-th root of unity ω, and let A be a central simple algebra of degree n over k. There exists a field extension $F|k$ such that*

- *the algebra $A \otimes_k F$ is isomorphic to a cyclic algebra;*
- *the restriction map $\mathrm{Br}(k) \to \mathrm{Br}(F)$ is injective.*

Saltman himself did not publish this result (but see Berhuy–Frings [1], Theorem 4 for a slightly more general statement).

Remarks 5.5.2

1. An iterated application of the theorem (possibly infinitely many times) shows that there exists a field extension $K|k$ such that the map $\mathrm{Br}(k) \to \mathrm{Br}(K)$ is injective, and *all* central simple k-algebras of degree n become cyclic over K.
2. In the paper Rehmann–Tikhonov–Yanchevskiĭ [1] an even stronger statement is proven. Namely, the authors construct a field extension $F|k$ such that k is algebraically closed in F, every central simple F-algebra is cyclic and moreover the period and index of k-algebras are preserved after base change to F.

For the proof of Saltman's theorem we need the following lemma.

Lemma 5.5.3 *Consider a purely transcendental extension $k(x,y)|k$ generated by the independent variables x and y. The degree n cyclic algebra $(x,y)_\omega$ over $k(x,y)$ has period n.*

Proof We prove slightly more than required, namely that the algebra $(x,y)_\omega \otimes_{k(x,y)} K$ has period n, where K denotes the field $k((x))((y))$. The extension $L := K(\sqrt[n]{y})$ is cyclic of degree n and splits $(x,y)_\omega \otimes_{k(x,y)} K$.

By Corollary 4.7.4, the isomorphism $K^\times / N_{L|K}(L^\times) \cong \mathrm{Br}\,(L|K)$ is given by mapping $a \in K^\times$ to the class of the cyclic algebra $(a, y)_\omega$ over K, therefore the period of the K-algebra $(x, y)_\omega$ equals the order of $x \in K^\times$ in the group $K^\times / N_{L|K}(L^\times)$.

Denoting this order by e, we thus have by definition some $z \in L^\times$ with $x^e = N_{L|K}(z)$. By the general theory of formal power series, L is the formal Laurent series ring in one variable $\sqrt[n]{y}$ over $k((x))$. If z viewed as a Laurent series in $\sqrt[n]{y}$ had a nonzero term of negative degree, the same would be true of x^e viewed as a (constant) Laurent series in the variable y, which is not the case. Therefore $z \in k((x))[[\sqrt[n]{y}]]$, and taking its image by the natural map $k((x))[[\sqrt[n]{y}]] \to k((x))$ sending $\sqrt[n]{y}$ to 0 we get an element $\overline{z} \in k((x))$ satisfying $x^e = (\overline{z})^n$. Writing $\overline{z}$ as a power series in x, we see that n must divide e. On the other hand, e divides n, because quite generally the period divides the degree (even the index; see Proposition 2.8.7 (1)). Therefore $e = n$, and the lemma is proven. $\qquad\square$

Proof of Theorem 5.5.1 Define the field F to be the function field of a Severi–Brauer variety associated to the central simple algebra

$$B := (A \otimes_k k(x, y)) \otimes_{k(x,y)} (x, y)_\omega$$

defined over the field $k(x, y)$. By Châtelet's theorem (see the discussion before Theorem 5.4.1), the algebra

$$B \otimes_{k(x,y)} F \cong (A \otimes_k F) \otimes_F \left((x, y)_\omega \otimes_{k(x,y)} F \right)$$

splits. This implies that $A \otimes_k F$ and the opposite algebra of $(x, y)_\omega \otimes_k F$ have the same class in $\mathrm{Br}\,(F)$. As they both have degree n, they must be isomorphic. But the latter algebra is isomorphic to the F-algebra $(x, y^{-1})_\omega$, as one sees from their presentation. We conclude that $A \otimes_k F$ is isomorphic to a cyclic algebra.

We now show that $\mathrm{Br}\,(k)$ injects into $\mathrm{Br}\,(F)$. Let α be an element in the kernel of the map $\mathrm{Br}\,(k) \to \mathrm{Br}\,(F)$. For a field K containing k, we denote by α_K the image of α in $\mathrm{Br}\,(K)$. According to Amitsur's theorem, the group $\ker(\mathrm{Br}\,(k(x, y)) \to \mathrm{Br}\,(F))$ is the cyclic subgroup generated by the class of B, so there exists an integer $m > 0$ for which the equality

$$\alpha_{k(x,y)} = m\,[A \otimes_k k(x, y)] + m\,[(x, y)_\omega] \qquad (5.7)$$

holds in $\mathrm{Br}\,(k(x, y))$. By passing to the field $k_s(x, y)$ we obtain

$$0 = \alpha_{k_s(x,y)} = m\,[(x, y)_\omega] \in \mathrm{Br}\,(k_s(x, y)),$$

because A and α split over k_s. By Lemma 5.5.3, the $k_s(x, y)$-algebra $(x, y)_\omega$ has period n, so n divides m. But since both A and $(x, y)_\omega$ have degree n and the period divides the degree, we have $n[A] = 0$ in $\mathrm{Br}\,(k)$ and $n\,[(x, y)_\omega] = 0$ in $\mathrm{Br}\,(k(x, y))$. Therefore we get from the identity (5.7) that $\alpha_{k(x,y)} = 0$, whence $\alpha = 0$ by Lemma 5.4.7, as desired. $\square$

EXERCISES

1. Let k be a field containing a primitive n-th root ω of unity, and let $K = k(\sqrt[n]{a})$ be a cyclic extension of degree n. Given $b \in k^\times$, consider the closed subvariety Y_b of $\mathbf{A}_k^{n+1}$ defined by the equation

$$bx^n = N_{K/k}\left(\sum_{i=0}^{n-1} (\sqrt[n]{a})^i y_i\right),$$

   where we denoted the coordinates by $(x, y_0, y_1, \cdots, y_{n-1})$.

   (a) Verify that $Y_b(k) \neq \emptyset$ if and only if the cyclic algebra $(a, b)_\omega$ is split.
   (b) If $Y_b(k) \neq \emptyset$, show that Y is a k-rational variety.
   (c) Show that Y_b is *stably birational* to the Severi–Brauer variety associated to the cyclic algebra $(a, b)_\omega$. [*Hint:* Argue as in Remark 5.4.3.]

2. (Heuser) Let A be a central simple algebra of degree n over k, and let $e_1, \ldots, e_{n^2}$ be a k-basis of A. Consider the reduced characteristic polynomial $\mathrm{Nrd}_A(x - \sum e_i x_i)$ as a polynomial in the variables $x, x_1, \ldots, x_{n^2}$, and let $X \subset \mathbf{A}_k^{n^2+1}$ be the associated affine hypersurface. Moreover, let $Y \subset \mathbf{P}_k^{n^2-1}$ be the projective hypersurface associated to the homogeneous polynomial $\mathrm{Nrd}_A(\sum e_i x_i)$; it is called the *norm hypersurface* of A.

   (a) Show that the function field $k(X)$ of X is a splitting field of A. [*Hint:* Observe that $k(X)$ is a degree n extension of $k(x_1, \ldots, x_{n^2})$ that may be embedded into $A \otimes_k k(x_1, \ldots, x_{n^2})$.]
   (b) Show that the function field $k(Y)$ of Y is a splitting field of A. [*Hint:* Let $\widetilde{Y} \subset \mathbf{A}^{n^2}$ be the affine cone over Y, i.e. the affine hypersurface defined by $\mathrm{Nrd}_A(\sum e_i x_i) = 0$. Show that $k(X)|k(\widetilde{Y})$ and $k(\widetilde{Y})|k(Y)$ are purely transcendental extensions, and specialize.]

3. Let k be a field, and let A_1, A_2 be central simple algebras over k. Denote by X_1, resp. X_2 the associated Severi–Brauer varieties. Compute the kernel of the natural map $\mathrm{Br}\,(k) \to \mathrm{Br}\,(k(X_1 \times X_2))$.
   [*Hint:* Mimic the proof of Amitsur's theorem, and use the isomorphism

   $$\mathrm{Pic}\,(\mathbf{P}^n \times \mathbf{P}^m) \cong \mathbf{Z} \oplus \mathbf{Z}$$

   (Shafarevich [2], III.1.1, Example 3).]

4. Let k be a field of characteristic different from 2, and let C be the projective conic over k without k-rational points. Construct a field $F \supset k$ and a central simple algebra A over F such that $1 < \mathrm{ind}_{F(C)}(A \otimes_F F(C)) < \mathrm{ind}_F(A)$.

5. Let $\phi : C_1 \to C_2$ be a nonconstant morphism of projective conics defined over a field k. Denote by d the degree of ϕ, i.e. the degree of the induced extension $k(C_1)|\phi^* k(C_2)$ of function fields.

(a) Show that ϕ induces a commutative diagram

$$
\begin{array}{ccc}
\mathrm{Pic}\,(\overline{C}_2)^G & \xrightarrow{\ \delta\ } & \mathrm{Br}\,(k) \\
{\scriptstyle d}\big\downarrow & & \big\downarrow{\scriptstyle \mathrm{id}} \\
\mathrm{Pic}\,(\overline{C}_1)^G & \xrightarrow{\ \delta\ } & \mathrm{Br}\,(k),
\end{array}
$$

where G is the absolute Galois group of k, and δ is the map of Theorem 5.4.6.

(b) Conclude that if d is even, then C_2 has a k-rational point. [*Hint:* Use Lichtenbaum's theorem.]

6. Let k be a field of characteristic 0, A a central simple algebra over k and X the associated Severi–Brauer variety. Denote by $N_X(k) \subset k^\times$ the subgroup of $k^\times$ generated by the subgroups $N_{K/k}(K^\times) \subset k^\times$ for those finite field extensions $K|k$ for which $X(K) \neq \emptyset$. Prove that $\mathrm{Nrd}(A^\times) = N_X(k)$.

[*Remark:* The group $N_X(k)$ is called the *norm group* of X.]

7. Let V be an n-dimensional vector space over a field k, and assume that there is a product of fields $K = k_1 \times \cdots \times k_r$ contained as an n-dimensional k-subalgebra in $\mathrm{End}_k(V)$. Show that there exists an isomorphism of k-vector spaces $V \xrightarrow{\sim} K$ under which the inclusion map $K \hookrightarrow \mathrm{End}_k(V)$ becomes identified with the map $K \to \mathrm{End}_k(K)$ sending $x \in K$ to the k-linear map $y \mapsto xy$.

8. Let A be a central simple algebra of degree n over a field k, and assume that there is a product of fields $K = k_1 \times \cdots \times k_r$ contained as an n-dimensional k-subalgebra in A. Show that $N_{K|k}(x) = \mathrm{Nrd}_A(x)$ for all $x \in K$.

[*Hint:* Consider the base change of A to the function field of its Severi–Brauer variety and then use the previous exercise. This exercise generalizes Proposition 2.6.3 (2).]

6

Residue maps

Residue maps constitute a fundamental technical tool for the study of the cohomological symbol. Their definition is not particularly enlightening at a first glance, but the reader will see that they emerge naturally during the computation of Brauer groups of function fields or complete discretely valued fields. When one determines these, a natural idea is to pass to a field extension having trivial Brauer group. There are three famous classes of such fields: finite fields, function fields of curves over an algebraically closed field and complete discretely valued fields with algebraically closed residue field. Once we know that the Brauer groups of these fields vanish, we are able to compute the Brauer groups of function fields over an arbitrary perfect field. The central result here is Faddeev's exact sequence for the Brauer group of a rational function field. We give two important applications of this theory: one to the class field theory of curves over finite fields, the other to constructing counterexamples to the rationality of the field of invariants of a finite group acting on some linear space. Following this ample motivation, we finally attack residue maps with finite coefficients, thereby preparing the ground for the next two chapters.

Residue maps for the Brauer group first appeared in the work of the German school on class field theory; the names of Artin, Hasse and F. K. Schmidt are the most important to be mentioned here. It was apparently Witt who first noticed the significance of residue maps over arbitrary discretely valued fields. Residue maps with finite coefficients came into the foreground in the 1960s in the context of étale cohomology; another source for their emergence in Galois cohomology is work by Arason [1] on quadratic forms.

6.1 Cohomological dimension

Before embarking on the study of fields with vanishing Brauer group it is convenient to discuss the relevant cohomological background: this is the theory of cohomological dimension for profinite groups, introduced by Tate.

163

Recall that for an abelian group B and a prime number p, the notation $B\{p\}$ stands for the p-primary torsion subgroup of B, i.e. the subgroup of elements of p-power order.

Definition 6.1.1 Let G be a profinite group, p a prime number. We say that G has *p-cohomological dimension* $\leq n$ if $H^i(G, A)\{p\} = 0$ for all $i > n$ and all continuous *torsion G-modules* A. We define the *p-cohomological dimension* $\mathrm{cd}_p(G)$ to be the smallest positive integer n for which G has cohomological dimension $\leq n$ if such an n exists, and set $\mathrm{cd}_p(G) = \infty$ otherwise.

One may wonder why we restrict to torsion G-modules in the definition and why not take all G-modules. This is solely for technical convenience; the analogous notion defined using all G-modules is called the *strict p-cohomological dimension* of G in the literature. In fact, there is not much difference between the two concepts, as the following proposition shows.

Proposition 6.1.2 *Assume that $\mathrm{cd}_p(G) \leq n$. Then $H^i(G, A)\{p\} = 0$ for all $i > n + 1$ and all continuous G-modules A.*

Proof Let A be a continuous G-module, and consider the multiplication-by-p map $p : A \to A$. Its kernel $_pA$ and cokernel A/pA are torsion G-modules fitting into the exact sequence

$$0 \to {}_pA \to A \xrightarrow{p} A \to A/pA \to 0,$$

which may be split into two short exact sequences

$$0 \to {}_pA \to A \xrightarrow{p} C \to 0 \quad \text{and} \quad 0 \to C \to A \to A/pA \to 0,$$

with $C := \mathrm{Im}\,(p)$. By assumption, the groups $H^i(G, {}_pA)$ and $H^i(G, A/pA)$ vanish for $i > n$, so the associated long exact sequences induce isomorphisms

$$H^i(G, A) \cong H^i(G, C) \quad \text{and} \quad H^{i+1}(G, C) \cong H^{i+1}(G, A)$$

for $i > n$. Thus for $i > n + 1$ the induced map $p_* : H^i(G, A) \to H^i(G, A)$ is an isomorphism. But by the construction of cohomology, the map p_* is also given by multiplication by p, so if it is an isomorphism, then the group $H^i(G, A)$ cannot have p-primary torsion. The claim follows. □

As a first example, we have:

Proposition 6.1.3 *We have $\mathrm{cd}_p(\hat{\mathbf{Z}}) = 1$ for all primes p.*

Proof Note first that $\mathrm{cd}_p(\hat{\mathbf{Z}}) \neq 0$, because

$$H^1(\hat{\mathbf{Z}}, \mathbf{Z}/p\mathbf{Z}) = \varinjlim \mathrm{Hom}(\mathbf{Z}/n\mathbf{Z}, \mathbf{Z}/p\mathbf{Z}) \cong \mathbf{Z}/p\mathbf{Z}.$$

Next we show the vanishing of $H^2(\hat{\mathbf{Z}}, A)$ for all torsion $\hat{\mathbf{Z}}$-modules A. By definition, this group is the direct limit of the groups $H^2(\mathbf{Z}/n\mathbf{Z}, A^{n\hat{\mathbf{Z}}})$ via the inflation maps $\mathrm{Inf} : H^2(\mathbf{Z}/n\mathbf{Z}, A^{n\hat{\mathbf{Z}}}) \to H^2(\mathbf{Z}/mn\mathbf{Z}, A^{mn\hat{\mathbf{Z}}})$, which by construction are induced by the natural map between the projective resolutions of $\mathbf{Z}$ considered as a trivial $(\mathbf{Z}/mn\mathbf{Z})$- and $(\mathbf{Z}/n\mathbf{Z})$-module, respectively. A calculation shows that on the low degree terms of the special projective resolution of Example 3.2.9 this map is given by the commutative diagram

$$
\begin{array}{ccccccc}
\mathbf{Z}[\mathbf{Z}/mn\mathbf{Z}] & \xrightarrow{N} & \mathbf{Z}[\mathbf{Z}/mn\mathbf{Z}] & \xrightarrow{\sigma-1} & \mathbf{Z}[\mathbf{Z}/mn\mathbf{Z}] & \xrightarrow{\sigma\mapsto 1} & \mathbf{Z} \\
\downarrow{\scriptstyle\sigma\mapsto m\sigma^m} & & \downarrow{\scriptstyle\sigma\mapsto\sigma^m} & & \downarrow{\scriptstyle\sigma\mapsto\sigma^m} & & \downarrow{\scriptstyle\mathrm{id}} \\
\mathbf{Z}[\mathbf{Z}/n\mathbf{Z}] & \xrightarrow{N} & \mathbf{Z}[\mathbf{Z}/n\mathbf{Z}] & \xrightarrow{\sigma^m-1} & \mathbf{Z}[\mathbf{Z}/n\mathbf{Z}] & \xrightarrow{\sigma^m\mapsto 1} & \mathbf{Z}
\end{array}
$$

where σ and σ^m denote generators of $\mathbf{Z}/mn\mathbf{Z}$ and $\mathbf{Z}/n\mathbf{Z}$, respectively. It follows that in degree 2 the inflation map is induced by multiplication by m on A, and as such annihilates all m-torsion elements of $H^2(\mathbf{Z}/n\mathbf{Z}, A^{n\hat{\mathbf{Z}}})$. This implies the claim because m was arbitrary here.

Finally, we prove the vanishing of $H^i(\hat{\mathbf{Z}}, A)$ for $i > 2$ by dimension shifting as follows. Given a continuous torsion $\hat{\mathbf{Z}}$-module A, we may embed it to the co-induced module $M^G(A)$ which is torsion by construction (see Remark 4.2.10). Hence so is the quotient $M^G(A)/A$, and so Corollary 4.3.1 gives $H^i(\hat{\mathbf{Z}}, M^G(A)/A) \cong H^{i+1}(\hat{\mathbf{Z}}, A)$, which is trivial for $i > 1$ by induction, starting from the case $i = 2$ treated above. □

Next a general lemma about cohomological dimension.

Lemma 6.1.4 *Let G and p be as above, and let H be a closed subgroup of G. Then $\mathrm{cd}_p(H) \leq \mathrm{cd}_p(G)$. Here equality holds in the case when the image of H in all finite quotients of G has index prime to p. In particular, $\mathrm{cd}_p(G) = \mathrm{cd}_p(G_p)$ for a pro-p-Sylow subgroup G_p of G.*

Proof Let B be a continuous torsion H-module. Then the continuous G-module $M_H^G(B)$ introduced in Remark 4.2.10 is also torsion and satisfies $H^i(H, B) = H^i(G, M_H^G(B))$ for all $i \geq 0$ by Shapiro's lemma, whence the inequality $\mathrm{cd}_p(H) \leq \mathrm{cd}_p(G)$. The opposite inequality in the case when H satisfies the prime-to-p condition of the lemma follows from Corollary 4.2.13. □

In the case of pro-p-groups there is a very useful criterion for determining the p-cohomological dimension.

Proposition 6.1.5 *Let G be a pro-p-group for some prime number p. Then $\mathrm{cd}_p(G) \le n$ if and only if $H^{n+1}(G, \mathbf{Z}/p\mathbf{Z}) = 0$.*

For the proof we need the following lemma from module theory.

Lemma 6.1.6 *The only finite simple G-module of p-power order is $\mathbf{Z}/p\mathbf{Z}$ with trivial action.*

Proof If A is a finite G-module of p-power order, then A^G must be a nontrivial G-submodule. Indeed, the complement $A \setminus A^G$ is the disjoint union of G-orbits each of which has p-power order and thus A^G cannot consist of the unit element only. Now if moreover we assume A to be simple, we must have $A = A^G$, i.e. triviality of the G-action. But then A must be $\mathbf{Z}/p\mathbf{Z}$, because a subgroup of a trivial G-module is a G-submodule. $\square$

Proof of Proposition 6.1.5 Necessity of the condition is obvious. For sufficiency, note first that $H^j(G, A\{p\}) = H^j(G, A)$ for all torsion G-modules A and $j > 0$; indeed, decomposing A into the direct sum of its p-primary components, we see that for a prime $\ell \ne p$ the group $H^j(G, A\{\ell\})$ is both ℓ-primary torsion (by definition of cohomology) and p-primary torsion (by Proposition 4.2.6), hence trivial. Thus we may restrict to p-primary torsion modules. Next observe that it is enough to prove $H^{n+1}(G, A) = 0$ for all p-primary torsion G-modules A, by a similar dimension shifting argument as at the end of the proof of Proposition 6.1.3. By the definition of continuous cohomology we may assume that G is finite. Writing A as the direct limit of its finitely generated G-submodules, we may assume using Lemma 4.3.3 that A is finitely generated, hence finite of p-power order. Then by general module theory A has a composition series whose successive quotients are simple G-modules. The long exact cohomology sequence and induction on the length of the composition series implies that it is enough to consider these. We have arrived at the situation of the above lemma, and may conclude from the assumption. $\square$

Now we come to the cohomological dimension of fields.

Definition 6.1.7 The *p-cohomological dimension* $\mathrm{cd}_p(k)$ of a field k is the p-cohomological dimension of the absolute Galois group $\mathrm{Gal}\,(k_s|k)$ for some separable closure k_s. Its *cohomological dimension* $\mathrm{cd}(k)$ is defined as the supremum of the $\mathrm{cd}_p(k)$ for all primes p.

For us the most interesting case is that of fields of p-cohomological dimension 1, for this is a property that can be characterized using the Brauer group.

Theorem 6.1.8 *Let k be a field, and p a prime number different from the characteristic of k. Then the following statements are equivalent:*

1. *The p-cohomological dimension of k is ≤ 1.*
2. *For all separable algebraic extensions $K|k$ we have $\mathrm{Br}\,(K)\{p\} = 0$.*
3. *The norm map $N_{L|K} : L^{\times} \to K^{\times}$ is surjective for all separable algebraic extensions $K|k$ and all Galois extensions $L|K$ with $\mathrm{Gal}\,(L|K) \cong \mathbf{Z}/p\mathbf{Z}$.*

Proof For the implication (1) $\Rightarrow$ (2), choose a separable closure k_s of k containing K. Then $\mathrm{Gal}\,(k_s|K)$ identifies with a closed subgroup of $\mathrm{Gal}\,(k_s|k)$, and hence we have $\mathrm{cd}_p(K) \leq \mathrm{cd}_p(k) \leq 1$ using Lemma 6.1.4. In particular, the group $H^2(K, \mu_{p^i})$ is trivial for all $i > 0$, but this group is none but the p^i-torsion part of $\mathrm{Br}\,(K)$ according to Corollary 4.4.5. For (2) $\Rightarrow$ (3), note first that for $L|K$ as in (3) we have $\mathrm{Br}\,(L|K) \cong K^{\times}/N_{L|K}(L^{\times})$ thanks to Corollary 4.4.6. But $\mathrm{Gal}\,(L|K) \cong \mathbf{Z}/p\mathbf{Z}$ also implies that $\mathrm{Br}\,(L|K)$ is annihilated by p, so $\mathrm{Br}\,(L|K) \subset \mathrm{Br}\,(K)\{p\} = 0$, whence the claim.

Finally, for (3) $\Rightarrow$ (1) let G_p be a pro-p-Sylow subgroup of $\mathrm{Gal}\,(k_s|k)$. Lemma 6.1.4 implies that it is enough to prove $\mathrm{cd}_p(G_p) \leq 1$, and moreover for this it is enough to show $H^2(G_p, \mathbf{Z}/p\mathbf{Z}) = 0$ by Proposition 6.1.5. As the extension $k(\mu_p)|k$ has degree $p - 1$, the fixed field k_p of G_p contains the p-th roots of unity, hence we have a chain of isomorphisms $H^2(G_p, \mathbf{Z}/p\mathbf{Z}) \cong H^2(k_p, \mu_p) \cong {}_p\mathrm{Br}\,(k_p)$. Let $K_p|k_p$ be a finite extension contained in k_s and denote by P the Galois group $\mathrm{Gal}\,(K_p|k_p)$. As $\mathrm{Br}\,(K_p|k_p)$ injects into $\mathrm{Br}\,(k_p)$, we are reduced to showing ${}_p\mathrm{Br}\,(K_p|k_p) = 0$. The group P, being a finite p-group, is solvable, i.e. there exists a finite chain

$$P = P_0 \supset P_1 \supset \cdots \supset P_n = \{1\}$$

of normal subgroups such that $P_i/P_{i+1} \cong \mathbf{Z}/p\mathbf{Z}$. These subgroups correspond to field extensions

$$k_p = K_0 \subset K_1 \subset \cdots \subset K_n = K_p$$

such that $\mathrm{Gal}\,(K_i|k_p) \cong P/P_i$. We now show ${}_p\mathrm{Br}\,(K_i|k_p) = 0$ by induction on i, the case $i = 0$ being trivial. Assuming the statement for $i - 1$, consider the exact sequence

$$0 \to H^2(P/P_{i-1}, K_{i-1}^{\times}) \to H^2(P/P_i, K_i^{\times}) \to H^2(P_{i-1}/P_i, K_i^{\times})$$

coming from Proposition 3.3.19 applied with $G = P/P_i$, $H = P_{i-1}/P_i$ and $A = K_i^\times$, noting that $H^1(P_i/P_{i-1}, K_i^\times) = 0$ thanks to Hilbert's Theorem 90. Restricting to p-torsion subgroups, we get

$$0 \to {}_p\mathrm{Br}\,(K_{i-1}|k_p) \to {}_p\mathrm{Br}\,(K_i|k_p) \to {}_p\mathrm{Br}\,(K_i|K_{i-1}).$$

Here the right-hand side group is trivial by (3) applied with $K = K_{i-1}$ and $L = K_i$ (and noting Corollary 4.4.6 again), and the left-hand side group is trivial by induction. Hence so is the middle one, which completes the proof of the inductive step. □

Remark 6.1.9 With notation as in the above proof, given a class α in ${}_p\mathrm{Br}\,(K_p|k_p)$, one may find a finite subextension $\tilde{k}_p$ of $k_p|k$ and a finite Galois extension $\tilde{K}_p|\tilde{k}_p$ of p-power degree contained in K_p such that α comes by base change to k_p from a class in ${}_p\mathrm{Br}\,(\tilde{K}_p|\tilde{k}_p)$. This observation shows that in the statement of Theorem 6.1.8 one may replace conditions (2) and (3) with similar conditions requiring only that K is a finite separable extension of k.

We have the following complement to Theorem 6.1.8:

Proposition 6.1.10 *Let k be a field of characteristic $p > 0$. Then $\mathrm{cd}_p(k) \le 1$.*

Proof By Lemma 6.1.4, we may replace k by the fixed field of some pro-p-Sylow subgroup of $\mathrm{Gal}\,(k_s|k)$. Hence we may assume that k is a field of characteristic p whose absolute Galois group is a pro-p-group. By Proposition 6.1.5, it suffices therefore to establish the vanishing of $H^2(k, \mathbf{Z}/p\mathbf{Z})$. For this, recall the exact sequence

$$0 \to \mathbf{Z}/p\mathbf{Z} \to k_s \xrightarrow{\wp} k_s \to 0$$

from the proof of Proposition 4.3.10, where $\wp : k_s \to k_s$ is given by $\wp(x) = x^p - x$. Part of the associated long exact sequence reads

$$H^1(k, k_s) \to H^2(k, \mathbf{Z}/p\mathbf{Z}) \to H^2(k, k_s),$$

from which we get the required vanishing, the two extremal terms being trivial by Lemma 4.3.11. □

Remark 6.1.11 According to the proposition, the higher Galois cohomology groups with p-torsion coefficients are trivial invariants for fields of characteristic $p > 0$. In the study of these other cohomology theories have been helpful. One approach, proposed by Milne [1] and Kato [2], is to use the modules $\nu(n)$ of logarithmic differentials that we shall discuss later in Section 9.5, and consider the groups $H_p^{n+1}(k) := H^1(k, \nu(n)_{k_s})$ for $n \geq 1$. As we shall see in Section 9.2, for $n = 1$ one has $H_p^2(k) \cong {}_p\mathrm{Br}\,(k)$, which is a nontrivial group in general for nonperfect k, in contrast to the situation of Theorem 6.1.8.

This phenomenon is related to the problem of defining the 'right' notion of p-cohomological dimension for fields of characteristic p. In Serre [2], §II.3 such a field k is defined to be of p-*dimension* ≤ 1 if ${}_p\mathrm{Br}\,(K) = 0$ for all finite extensions $K|k$. In Kato [2] and Kato–Kuzumaki [1] a definition is proposed for all n: they say that k is of p-dimension n if n is the smallest integer with $[k : k^p] \leq p^n$ and $H_p^{n+1}(K) = 0$ for all finite extensions $K|k$.

We conclude this chapter by examples of fields of cohomological dimension ≤ 1.

Examples 6.1.12 Finite fields and Laurent series fields over an algebraically closed field of characteristic 0 have absolute Galois group isomorphic to $\hat{\mathbf{Z}}$, by Examples 4.1.5 and 4.1.6, respectively. They therefore have cohomological dimension 1 by Proposition 6.1.3.

Another class of examples arises from the work done in Chapter 2, Section 2.9.

Example 6.1.13 If K is a complete discretely valued field with perfect residue field, then its maximal unramified extension K_{nr} has cohomological dimension ≤ 1.

To see this, it suffices in view of Proposition 6.1.10 to verify $\mathrm{cd}_p(K_{nr}) \leq 1$ for p invertible in K. We apply Proposition 6.1.8 and show that $\mathrm{Br}\,(L) = 0$ for a separable algebraic extension $L|K_{nr}$. We may assume $L|K_{nr}$ is a finite extension contained in K_s, in which case $L = K_{nr}(\alpha)$ for some $\alpha \in L$. We contend that L is the maximal unramified extension of $K(\alpha)$. To see this, we may assume that the extension $K(\alpha)|K$ has no nontrivial unramified subextension after replacing K by the unramified subfield given by Proposition A.6.8 (3) of the Appendix. But then the same proposition implies that for a finite unramified extension $M|K(\alpha)$ contained in K_s we have $M = (M \cap K_{nr})(\alpha)$ which is a subfield of L. We conclude by applying Corollary 2.9.4 to the complete discretely valued field $K(\alpha)$.

Remarks 6.1.14

1. For $K = k((t))$ with k of characteristic 0 there is an easier proof of $\mathrm{cd}(K_{nr}) \leq 1$ because in that case $\mathrm{Gal}\,(K_s|K_{nr}) \cong \hat{\mathbf{Z}}$ by the same arguments as in Example 4.1.6.

2. One can also prove $\mathrm{cd}(K_{nr}) \leq 1$ by verifying condition (3) of Proposition 6.1.8 (see Serre [2], Chapter X, §7).

6.2 C_1-fields

The pertinence of the following condition to our subject matter has been first observed by Emil Artin.

Definition 6.2.1 A field k is said to satisfy condition C_1 if every homogeneous polynomial $f \in k[x_1, \ldots, x_n]$ of degree $d < n$ has a nontrivial zero in k^n.

We briefly call such fields C_1-fields.

Remarks 6.2.2

1. More generally, a field k satisfies condition C_r for an integer $r > 0$ if every homogeneous polynomial $f \in k[x_1, \ldots, x_n]$ of degree d with $d^r < n$ has a nontrivial zero in k^n. This condition was introduced and first studied by Lang [1].
2. Even more generally, a field k is said to satisfy condition C'_r for some integer $r > 0$ if each finite system $f_1, \ldots, f_m \in k[x_1, \ldots, x_n]$ of homogeneous polynomials of respective degrees $d_1, \ldots, d_m$ has a nontrivial common zero in k^n, provided that $d_1^r + \cdots + d_m^r < n$. For more on this property, see the book of Pfister [1].

Artin himself called C_1-fields quasi-algebraically closed, because they have the property that there is no nontrivial finite-dimensional central division algebra over them. In fact, one has:

Proposition 6.2.3 *Let k be a C_1-field. Then $\mathrm{cd}(k) \leq 1$, and $\mathrm{Br}(L) = 0$ for every finite extension $L|k$.*

Note that the extension $L|k$ is not assumed to be separable. We prove first the following lemma which will be also useful later.

Lemma 6.2.4 *If K is a C_1-field, then so is every finite extension $L|K$.*

Proof Let $f \in L[x_1, \ldots, x_n]$ be a homogeneous polynomial of degree $d < n$, and let $v_1, \ldots, v_m$ be a basis of the K-vector space L. Introduce new variables x_{ij} ($1 \leq i \leq n, 1 \leq j \leq m$) satisfying $x_{i1}v_1 + \cdots + x_{im}v_m = x_i$, and consider the equation $N_{L|K}(f(x_1, \ldots, x_n)) = 0$. This is then a homogeneous equation over K of degree md in the mn variables x_{ij}, so by assumption it has a nontrivial zero $(\alpha_{11}, \ldots, \alpha_{mn})$ in K^{mn}, since $md < mn$. Whence a nontrivial element $(\alpha_1, \ldots, \alpha_n) \in L^n$ satisfying $N_{L|K}(f(\alpha_1, \ldots, \alpha_n)) = 0$, which holds if and only if $f(\alpha_1, \ldots, \alpha_n) = 0$. $\square$

Proof of Proposition 6.2.3   If our C_1-field k has positive characteristic p, we have $\mathrm{cd}_p(k) \leq 1$ by the general Proposition 6.1.10. So as far as cohomological dimension is concerned, we may concentrate on the other primes and conclude from Theorem 6.1.8 and Lemma 6.2.4 that it is enough to show the second statement in the case $L = k$, i.e. that a C_1-field has trivial Brauer group.

So consider a division algebra D of degree n over a C_1-field k, and denote by Nrd : $D \to k$ the associated reduced norm. Choosing a k-basis $v_1, \ldots, v_{n^2}$ of D considered as a k-vector space, we see from the construction of Nrd in Chapter 2 that $f(x_1, \ldots, x_{n^2}) := \mathrm{Nrd}(x_1 v_1 + \ldots x_{n^2} v_{n^2})$ is a homogeneous polynomial of degree n in the n^2 variables $x_1, \ldots, x_{n^2}$. If here $n > 1$, then by the C_1 property f has a nontrivial zero in k^n. But this contradicts the assumption that D is a division algebra, by Proposition 2.6.2. Therefore $n = 1$, and $D = k$ is the trivial division algebra over k. $\square$

Remark 6.2.5   The question arises whether the converse of the proposition holds true. The answer is no: Ax [1] has constructed a field of cohomological dimension 1 (and of characteristic 0) which is not a C_1-field. See also the book of Shatz [1] for details.

Here are the first nontrivial examples of C_1-fields.

Theorem 6.2.6 (Chevalley)   *Finite fields satisfy the C_1 property.*

Proof   Let $\mathbf{F}_q$ be the field with q elements, where q is some power of a prime number p. Following Warning, we prove more, namely that the number of solutions in $\mathbf{F}_q^n$ of a polynomial equation $f(x_1, \ldots, x_n) = 0$ of degree $d < n$ is divisible by p. If f is moreover homogeneous, it already has the trivial solution, whence the claim.

For a polynomial $g \in k[x_1, \ldots, x_n]$ denote by $N(g)$ the number of its zeroes in $\mathbf{F}_q^n$, and introduce the notation

$$\Sigma(g) := \sum_{(\alpha_1, \ldots, \alpha_n) \in \mathbf{F}_q^n} (g(\alpha_1, \ldots, \alpha_n))^{q-1}.$$

As $\alpha^{q-1} = 1$ for each nonzero $\alpha \in \mathbf{F}_q$, we see that the element $\Sigma(g) \in \mathbf{F}_q$ actually lies in $\mathbf{F}_p \subset \mathbf{F}_q$, and moreover

$$q^n - \Sigma(g) = \sum_{(\alpha_1,\ldots,\alpha_n) \in \mathbf{F}_q^n} (1 - (g(\alpha_1,\ldots,\alpha_n))^{q-1}) \equiv N(g) \mod p.$$

Therefore it is enough to show that $\Sigma(f) = 0$ in $\mathbf{F}_q$ for our particular f above. For this, write $f(x_1,\ldots,x_n)^{q-1}$ as a linear combination of monomials $x_1^{r_1} \ldots x_n^{r_n}$. We show that $\Sigma(x_1^{r_1} \ldots x_n^{r_n}) = 0$ in $\mathbf{F}_q$ for all occurring monomials $x_1^{r_1} \ldots x_n^{r_n}$. This is obvious if one of the r_i is 0, so we may assume this is not the case. Then, as f has degree less than n by assumption, we may assume that one of the r_i, say r_1, is smaller than $q - 1$. Then fixing $(\alpha_2,\ldots,\alpha_n) \in \mathbf{F}_q^{n-1}$ and taking a generator ω of the cyclic group $\mathbf{F}_q^\times$ we get

$$\sum_{\alpha \in \mathbf{F}_q} \alpha^{r_1} \alpha_2^{r_2} \ldots \alpha_n^{r_n} = \alpha_2^{r_2} \ldots \alpha_n^{r_n} \sum_{i=0}^{q-2} \omega^{i r_1} = (\alpha_2^{r_2} \ldots \alpha_n^{r_n}) \frac{(\omega^{r_1})^{q-1} - 1}{\omega^{r_1} - 1},$$

which equals 0 in $\mathbf{F}_q$. We conclude by making $(\alpha_2,\ldots,\alpha_n)$ run over $\mathbf{F}_q^{n-1}$. $\square$

Remark 6.2.7 Together with the previous proposition, the theorem gives another proof of the fact that finite fields have cohomological dimension 1. Moreover, we also get that finite fields have trivial Brauer group (up to now, we only knew that the prime-to-p part is trivial, by Theorem 6.1.8). In other words, we have proven a theorem of Wedderburn: *A finite-dimensional division algebra over a finite field is a field.*

Other classic examples of C_1-fields are given by the following theorem.

Theorem 6.2.8 (Tsen) *Let k be an algebraically closed field, and let K be the function field of an algebraic curve over k. Then K is a C_1-field.*

Proof Using Lemma 6.2.4 we may assume K is a simple transcendental extension $k(t)$ of k. Given a homogeneous polynomial $f \in k(t)[x_1,\ldots,x_n]$ of degree $d < n$, we may also assume the coefficients to be in $k[t]$, and we may look for solutions in $k[t]^n$. Choose an integer $N > 0$ and look for the x_i in the form

$$x_i = \sum_{j=0}^{N} a_{ij} t^j,$$

with the $a_{ij} \in k$ to be determined. Plugging this expression into f and regrouping according to powers of t, we get a decomposition

$$0 = f(x_1, \ldots, x_n) = \sum_{l=0}^{dN+r} f_l(a_{10}, \ldots, a_{nN})t^l,$$

where r is the maximal degree of the coefficients of f and the f_l are homogeneous polynomials in the a_{ij}, all of which should equal 0. Since $d < n$ by assumption, for N sufficiently large the number $dN + r + 1$ of the polynomials f_l is smaller than the number $n(N+1)$ of the indeterminates a_{ij}, so they define a nonempty Zariski closed subset in projective $(nN + n - 1)$-space $\mathbf{P}^{nN+n-1}$ (see Appendix, Corollary A.3.3). As k is algebraically closed, this closed set has a point in $\mathbf{P}^{nN+n-1}(k)$, whence the a_{ij} we were looking for. $\square$

Remarks 6.2.9

1. The theorems of Chevalley and Tsen can be sharpened in the sense that finite fields as well as function fields of curves over algebraically closed fields (or even C_1-fields) satisfy the C_1' property of Remark 6.2.2 (2). The proofs are similar to the ones given above and are left as an exercise.
2. Tsen's theorem has the following geometric interpretation. Let C be a smooth projective curve with function field K. The homogeneous polynomial $f \in K[x_1, \ldots, x_n]$ defines an $(n - 1)$-dimensional projective variety equipped with a surjective morphism $X \to C$. A nontrivial solution of $f(x_1, \ldots, x_n) = 0$ in K^n defines a *section* of p, i.e. a morphism $s : C \to X$ with $p \circ s = \mathrm{id}_C$. In particular, $s(C) \subset X$ is a closed subvariety of dimension 1 mapped isomorphically onto C by p.

   As a particular example, consider a degree 2 homogeneous polynomial in three variables with coefficients in $k[t]$. It defines a surface fibred in conics over the projective line. By Tsen's theorem, there is a curve on the surface intersecting each fibre in exactly one point. For remarkable recent generalizations of this fact, see Graber-Harris-Starr [1] and de Jong-Starr [1].

Before moving over to other classes of C_1-fields, we point out the following interesting corollary to Tsen's theorem.

Corollary 6.2.10 *Let C be a smooth projective geometrically irreducible curve over a finite field $\mathbf{F}$. Every central simple algebra over the function field $\mathbf{F}(C)$ is split by a cyclic field extension, and hence it is Brauer equivalent to a cyclic algebra.*

Proof Let $\overline{\mathbf{F}}$ be an algebraic closure of $\mathbf{F}$. We have $\mathrm{Br}\,(\overline{\mathbf{F}}(C)) = 0$ by Tsen's theorem, so every central simple algebra over $\mathbf{F}(C)$ is split by $\mathbf{F}'(C)$ for some finite extension $\mathbf{F}'|\mathbf{F}$. This is necessarily a cyclic extension as $\mathbf{F}$ is finite. The second statement follows from Proposition 4.7.7.   □

The third famous class of C_1-fields is that of fields complete with respect to a discrete valuation with algebraically closed residue field. The C_1 property for these and for some of their dense subfields was established by Serge Lang in his thesis (Lang [1]). In order to illustrate some of the methods involved in the proof, we now discuss a special case of Lang's theorem.

Let k be a field of characteristic 0, and let $k((t))$ be the field of formal Laurent series over k. It is a complete discretely valued field, and we may consider its maximal unramified extension $k((t))_{nr}$ inside a fixed separable closure. It can be described as the union of the fields $k'((t))$ for all finite extensions $k'|k$ inside a separable closure k_s of k.

Theorem 6.2.11 (Lang) *For a field k of characteristic 0 the field $k((t))_{nr}$ is a C_1-field. In particular, if k is algebraically closed, then $k((t))$ itself is a C_1-field.*

We shall deduce the theorem above from Tsen's theorem using an approximation method taken from Greenberg [1]. The crucial statement is:

Theorem 6.2.12 (Greenberg) *Let k be a field of characteristic 0, and let $S = \{f_1, \ldots, f_m\}$ be a system of polynomials in $k[[t]][x_1, \ldots, x_n]$. The f_i have a common nontrivial zero $(a_1, \ldots, a_n) \in k[[t]]^n$ if and only if for all $N > 0$ the congruences*

$$f_i(x_1, \ldots, x_n) = 0 \bmod (t^N), \quad i = 1, \ldots, m$$

have a common nontrivial solution $(a_1^{(N)}, \ldots, a_n^{(N)})$.

We first show that *Theorem 6.2.12 implies Theorem 6.2.11.* Consider a homogeneous polynomial $f \in k((t))_{nr}[x_1, \ldots, x_n]$ of degree $d < n$. To prove that f has a zero in $k((t))_{nr}^n$, after multiplying with a common denominator we may assume that f has coefficients in $k'[[t]]$ for a finite extension $k'|k$. Since the rings $k'[[t]]/(t^N)$ and $k'[t]/(t^N)$ are isomorphic for all $N > 0$, we may find for each N a degree d homogeneous polynomial $f^{(N)} \in k'[t][x_1, \ldots, x_n]$ with $f^{(N)} = f \bmod (t^N)$. By Tsen's theorem, after replacing k' by a finite extension we see that $f^{(N)}$ has a zero $(a_1^{(N)}, \ldots, a_n^{(N)}) \in k'(t)^n$, where we may assume the $a_j^{(N)}$ to lie in $k'[t]$ by homogeneity of $f^{(N)}$. Reducing modulo

(t^N) thus yields a zero of f modulo (t^N), and so the case $m = 1$ of Theorem 6.2.12 applies. $\qquad\square$

Proof of Theorem 6.2.12 We prove a stronger statement: there exist constants c, ν depending only on S such that for N sufficiently large every common nontrivial zero $(a_1^{(N)}, \dots, a_n^{(N)})$ of the f_i modulo (t^N) may be approximated by a common nontrivial zero $(a_1, \dots, a_n) \in k[[t]]^n$ of the f_i such that each $a_i^{(N)}$ is congruent to a_i modulo $(t^{N/c-\nu})$.

Consider the affine closed subset $V \subset \mathbf{A}^n$ defined as the locus of common zeroes of the $f_i \in S$. We prove the above statement by induction on the dimension d of V, starting from the obvious case $d = -1$, i.e. $V = \emptyset$.

We first make a reduction to the case when V is a subvariety of $\mathbf{A}^n$. For this, let J be the ideal in $k((t))[x_1, \dots, x_n]$ generated by the $f_i \in S$. Let g be a polynomial with $g \notin J$ but $g^r \in J$ for some $r > 1$. Then if $(a_1, \dots, a_n) \in k[[t]]^n$ satisfies $f_i(a_1, \dots, a_n) = 0 \bmod t^N$ for all i, we conclude the same for g^r, and hence we get $g(a_1, \dots, a_n) = 0 \bmod t^s$ for all integers $0 < s \le N/r$. Applying this to a system of generators $T = \{g_1, \dots, g_M\}$ of the radical of J, we see that if the theorem holds for the system T, it also holds for the system S with possibly different c, ν. So we may assume J equals its own radical, and hence is an intersection of finitely many prime ideals $P_1, \dots, P_r$. Now if $g_j \in P_j$ are such that $g_1 \dots g_r \in J$, we see as above that $f_i(a_1, \dots, a_n) = 0$ mod t^N for all i implies that there is some j with $g_j(a_1, \dots, a_n) = 0$ mod (t^s) for all $0 < s \le N/r$. Reasoning as above, we therefore conclude that it is enough to prove the theorem for the P_j. In this case the associated V is an affine variety in the sense of the Appendix since k has characteristic 0 (see Remark A.1.1).

Now for each subset $I \subset \{1, \dots, m\}$ of cardinality $n - d$ consider the closed subset V_I defined in $\mathbf{A}^n$ by the system $S_I = \{f_i \in S : i \in I\}$, and let $V_I^+ \subset V_I$ be the union of the d-dimensional $k((t))$-irreducible components different from V. The sets V_I^+ are defined as the locus of zeroes of some finite system $S_I^+ \supset S_I$ of polynomials. Consider also the singular locus $W \subset V$ of V. As remarked in Section A.1 of the Appendix, it is a proper closed subset of V, and as such has dimension $< d$. Furthermore, it is defined by a system S_W of polynomials obtained by adding some equations (namely the $(n - d) \times (n - d)$ minors of the Jacobian of the f_i) to S. Finally, let $P = (a_1, \dots, a_n) \in k[[t]]^n$ be a point satisfying $f_i(a_1, \dots, a_n) = 0 \bmod (t^N)$ for all $f_i \in S$, with some N to be determined later. If P also happens to satisfy all the other equations in S_W modulo (t^N), then up to changing c and ν we conclude by the inductive hypothesis that there is some point in $W \subset V$ congruent to P modulo $(t^{N/c-\nu})$ if N is sufficiently large, and we are finished.

Similarly, if P also satisfies the equations in some S_I^+ modulo (t^N), then by the inductive hypothesis applied to the proper closed subset $V \cap V_I^+$ we get a point in $V \cap V_I^+$ congruent to P modulo $(t^{N/c-\nu})$ for N sufficiently large, again after changing c and ν if necessary. So we may assume we are not in the above cases. Then if P is congruent to a $k[[t]]$-valued point of V modulo (t^N), it must be a smooth point not contained in any of the other components of the V_I, so it will be enough to assure that P is congruent to some smooth point in V_I.

We may assume $I = \{1, \ldots, n - d\}$, and we may enlarge the system S_I by adding the linear polynomials $x_{n-d+1} - a_{n-d+1}, \ldots, x_n - a_n$. For ease of notation we denote this new system again by S. Let J_S be the Jacobian matrix of the system S, and let h be its determinant evaluated at P. Observe that since the j-th partial derivatives of the $x_i - a_i$ equal 0 for $i \neq j$ and 1 for $i = j$, the subdeterminant formed by the first $n - d$ columns in the Jacobian of the system S_I at P also equals h. Hence h is nontrivial modulo (t^N) by assumption; denote by ν the highest power of t dividing h, and take N so large that $N > 2\nu$. Under this assumption a refined form of Hensel's lemma (cf. Appendix, Proposition A.5.6) implies that there is a point of V_I over $k[[t]]$ congruent to P modulo $(t^{N-\nu})$, and we are done.   □

Remarks 6.2.13

1. An examination of the above proof shows that even if we only need the case $m = 1$ for the application to Lang's theorem, in order to prove this special case we still have to consider systems of polynomials to make the induction work. Thus working with several polynomials is often more advantageous than with a single one; in particular, the C_1' property can be more handy than just C_1. In fact, assuming the C_1' analogue of Tsen's theorem (Remark 6.2.9 (1)), we get from the above proof that the fields $k((t))_{nr}$ are actually C_1'-fields.

2. Greenberg's theorem is much more general than the form proven above, and is very useful for many applications. It states that given a discrete valuation ring R for which Hensel's lemma holds and a system of equations with coefficients in R, then under a separability assumption one may approximate solutions over the completion $\widehat{R}$ by solutions over R arbitrarily closely in the topology of $\widehat{R}$. For example, this more general statement works for the subring $R \subset k[[t]]$ formed by power series algebraic over $k(t)$. In the characteristic 0 case there is also a constructive method for finding a good approximation (Kneser [1]).

6.3 Cohomology of complete discretely valued fields

In the next section we shall apply Tsen's theorem to study the cohomology of function fields of curves. As a preparation, we examine a local situation first. Let K be a field complete with respect to a discrete valuation v with residue field κ. Fix a separable closure K_s of K, and let K_{nr} be the maximal unramified extension of K, i.e. the composite of all finite unramified extensions of K inside K_s. The valuation v extends uniquely to a discrete valuation of K_{nr} with residue field κ_s, a separable closure of κ. The extension $K_{nr}|K$ is Galois, and the Galois group $G := \mathrm{Gal}\,(K_{nr}|K)$ is canonically isomorphic to $\mathrm{Gal}\,(\kappa_s|\kappa)$.

Denoting by U_{nr} the multiplicative group of units in K_{nr}, we get an exact sequence of G-modules

$$1 \to U_{nr} \to K_{nr}^{\times} \xrightarrow{v} \mathbf{Z} \to 0 \tag{6.1}$$

which is split by the map $\mathbf{Z} \to K_{nr}^{\times}$ sending 1 to a local parameter π of v. Hence for each $i \geq 0$ we have a split exact sequence of cohomology groups

$$0 \to H^i(G, U_{nr}) \to H^i(G, K_{nr}^{\times}) \to H^i(G, \mathbf{Z}) \to 0$$

by Remark 4.3.4 (2). For $i = 0$ this is just the analogue of exact sequence (6.1) for K instead of K_{nr}, and for $i = 1$ it is uninteresting because of Hilbert's Theorem 90. For $i \geq 2$, we may use the exact sequence

$$0 \to \mathbf{Z} \to \mathbf{Q} \to \mathbf{Q}/\mathbf{Z} \to 0$$

to obtain isomorphisms $H^i(G, \mathbf{Z}) \cong H^{i-1}(G, \mathbf{Q}/\mathbf{Z})$, as $H^i(G, \mathbf{Q}) = 0$ for $i > 0$ by Corollary 4.2.7. Hence we may rewrite the above sequence as

$$0 \to H^i(G, U_{nr}) \to H^i(G, K_{nr}^{\times}) \xrightarrow{r_v^i} H^{i-1}(G, \mathbf{Q}/\mathbf{Z}) \to 0.$$

The map r_v^i is called the *residue map* associated to v.

As regards the kernel of the residue map, we have:

Proposition 6.3.1 *The natural reduction map $U_{nr} \to \kappa_s^{\times}$ induces isomorphisms*

$$H^i(G, U_{nr}) \cong H^i(G, \kappa_s^{\times})$$

for all $i > 0$. Therefore we have split exact sequences

$$0 \to H^i(G, \kappa_s^{\times}) \to H^i(G, K_{nr}^{\times}) \xrightarrow{r_v^i} H^{i-1}(G, \mathbf{Q}/\mathbf{Z}) \to 0$$

for $i \geq 2$.

For the proof we need a formal lemma.

Lemma 6.3.2 *Let Γ be a finite group, and (A_j, ϕ_j) an inverse system of Γ-modules indexed by the set $\mathbf{Z}_+$ of positive integers, with surjective transition maps ϕ_j. If $i > 0$ is an integer such that $H^i(\Gamma, A_1) = H^i(\Gamma, \ker(\phi_j)) = 0$ for all j, then $H^i(\Gamma, \varprojlim A_j) = 0$.*

Here the inverse limit is equipped, as before, with the discrete topology.

Proof Choose a projective resolution $P_\bullet$ of $\mathbf{Z}$, and represent an element of $H^i(\Gamma, \varprojlim A_j)$ by a collection of homomorphisms $\lambda_j : P_i \to A_j$ each of which are mapped to 0 by δ_*^i. By induction on j we construct an element $\mu \in \operatorname{Hom}(P_{i-1}, \varprojlim A_j)$ represented by homomorphisms $\mu_j : P_{i-1} \to A_j$ with $\delta_*^{i-1}(\mu_j) = \lambda_j$. The existence of μ_1 follows from $H^i(\Gamma, A_1) = 0$. Assuming μ_j has been constructed, lift it to a homomorphism $\mu'_{j+1} : P_{i-1} \to A_{j+1}$ using the surjectivity of $\phi_j : A_{j+1} \to A_j$. Then $\lambda_{j+1} - \delta_*^i(\mu'_{j+1})$ is a map $P^i \to \ker(\phi_j)$ mapped to 0 by δ_*^i, and hence of the form $\delta_*^{i-1}(\nu_{j+1})$ with some $\nu_{j+1} : P_{i-1} \to \ker(\phi_j)$ by the assumption $H^i(\Gamma, \ker(\phi_j)) = 0$. Setting $\mu_{j+1} = \mu'_{j+1} + \nu_{j+1}$ completes the inductive step. $\square$

Proof of Proposition 6.3.1 In view of the discussion preceding the proposition it will be enough to prove the first statement. For this it will be enough to consider a finite unramified Galois extension $L|K$ with group Γ and residual extension $\lambda|\kappa$, and establish isomorphisms $H^i(\Gamma, L^\times) \cong H^i(\Gamma, \lambda^\times)$. Consider for all $j > 0$ the multiplicative subgroups

$$U_L^j := \{x \in L : v(x - 1) \geq i\}$$

in the group of units U_L of L. The groups U_L^j form a decreasing filtration of U_L^1 such that the natural map $U_L^1 \to \varprojlim U_L^1/U_L^j$ is an isomorphism. Furthermore, the reduction map $U_L \to \lambda^\times$ yields an exact sequence

$$1 \to U_L^1 \to U_L \to \lambda^\times \to 1$$

whose associated long exact sequence shows that the proposition follows if we show $H^i(\Gamma, U_L^1) = 0$ for all $i > 0$. For this, fix a local parameter π generating the maximal ideal of the valuation ring $A_v \subset L$ of v, and consider the maps $U_L^j \to \lambda$ sending $1 + a\pi^j$ to the image of $a \in A_v$ in λ. These maps are surjective group homomorphisms giving rise to exact sequences of Γ-modules

$$1 \to U_L^{j+1} \to U_L^j \to \lambda \to 0$$

for all j. Here we have $H^i(\Gamma, \lambda) = 0$ for $i > 0$ by Lemma 4.3.11, from which we infer $H^i(\Gamma, U_L^j/U_L^{j+1}) = 0$ for $i, j > 0$. By induction on j using the exact sequences

$$1 \to U_L^j/U_L^{j+1} \to U_L^1/U_L^{j+1} \to U_L^1/U_L^j \to 1$$

we obtain $H^i(\Gamma, U_L^1/U_L^j) = 0$ for all $i > 0$ and $j > 0$. We conclude by applying the above lemma to the inverse system of Γ-modules formed by the quotients U_L^1/U_L^j.   □

Remark 6.3.3 If κ has characteristic 0, one can give a simpler proof of the proposition by remarking that the union U_{nr}^1 of the groups U_L^1 in K_{nr} is a uniquely divisible abelian group. This fact can be proven using Hensel's lemma (see Appendix, Proposition A.5.5). When κ has characteristic $p > 0$, the group U_{nr}^1 is only divisible by integers prime to p.

For $i = 2$ we get the Brauer group of κ as the left term in the exact sequence of the proposition. In fact, in this case the middle term of the sequence is none but the Brauer group of K, if we assume moreover that κ is perfect. To show this, consider the inflation maps induced by the surjection $\mathrm{Gal}\,(K_s|K) \to G$.

Proposition 6.3.4 *Assume moreover that κ is perfect. Then the inflation maps*

$$\mathrm{Inf} : H^i(G, K_{nr}^\times) \to H^i(K, K_s^\times)$$

are isomorphisms for all $i > 0$.

Proof By Example 6.1.13, the field K_{nr} has cohomological dimension ≤ 1 and trivial Brauer group. This implies the vanishing of the groups $H^i(K_{nr}, K_s^\times)$ for $i > 1$ in view of Proposition 6.1.2; it also holds for $i = 1$ by Hilbert's Theorem 90. Therefore the condition for the exactness of the inflation-restriction sequence (Proposition 3.3.19 completed by Corollary 4.3.5) is satisfied, so we have for each $i > 0$ an exact sequence

$$0 \to H^i(G, K_{nr}^\times) \xrightarrow{\mathrm{Inf}} H^i(K, K_s^\times) \xrightarrow{\mathrm{Res}} H^i(K_{nr}, K_s^\times).$$

But here the last group vanishes for $i > 0$, and the proposition follows.   □

Thus in the exact sequence of Proposition 6.3.1 we may replace the middle term by $H^i(K, K_s^\times)$. As already indicated, the most important case is when $i = 2$, and we record it separately.

Corollary 6.3.5 (Witt) *For a complete discretely valued field K with perfect residue field κ there is a split exact sequence*

$$0 \to \mathrm{Br}\,(\kappa) \to \mathrm{Br}\,(K) \to \mathrm{Hom}_{\mathrm{cont}}(G, \mathbf{Q}/\mathbf{Z}) \to 0 \qquad (6.2)$$

where $G = \mathrm{Gal}\,(\kappa_s|\kappa)$.

Proof The identification with Brauer groups follows from Theorem 4.4.3, and the isomorphism $H^1(G, \mathbf{Q}/\mathbf{Z}) \cong \mathrm{Hom}_{\mathrm{cont}}(G, \mathbf{Q}/\mathbf{Z})$ follows from Example 3.2.3 (1) by passing to the limit. $\qquad\square$

Remark 6.3.6 In the case when $K = k((t))$ for a perfect field k, the last corollary may be restated in terms of central simple algebras as follows: *Every central simple algebra over $k((t))$ is Brauer equivalent to a tensor product of the form $(A \otimes_k k((t))) \otimes_{k((t))} (\chi, t)$, where A is a central simple algebra over k, and (χ, t) is a cyclic algebra over $k((t))$ for some character $\chi : G \to \mathbf{Q}/\mathbf{Z}$.* This statement follows from the corollary above and the observation that the section of exact sequence (6.2) coming from the splitting $\mathbf{Z} \to K_{nr}^{\times}, 1 \mapsto t$ of the valuation map is given by $\chi \mapsto (\chi, t)$.

We now focus on the important special case where the residue field is finite. Such complete discretely valued fields are usually called local fields in the literature; they are Laurent series fields over a finite field or finite extensions of the field $\mathbf{Q}_p$ of p-adic numbers for some p (see e.g. Neukirch [1], Chapter II, Proposition 5.2).

Proposition 6.3.7 (Hasse) *For a complete discretely valued field K with finite residue field we have a canonical isomorphism*

$$\mathrm{Br}(K) \cong \mathbf{Q}/\mathbf{Z}.$$

Moreover, for a finite separable extension $L|K$ there are commutative diagrams

$$
\begin{array}{ccc}
\mathrm{Br}(L) & \xrightarrow{\cong} & \mathbf{Q}/\mathbf{Z} \\
{\scriptstyle\mathrm{Cor}}\downarrow & & \downarrow{\scriptstyle\mathrm{id}} \\
\mathrm{Br}(K) & \xrightarrow{\cong} & \mathbf{Q}/\mathbf{Z}
\end{array}
\quad and \quad
\begin{array}{ccc}
\mathrm{Br}(K) & \xrightarrow{\cong} & \mathbf{Q}/\mathbf{Z} \\
{\scriptstyle\mathrm{Res}}\downarrow & & \downarrow{\scriptstyle[L:K]} \\
\mathrm{Br}(L) & \xrightarrow{\cong} & \mathbf{Q}/\mathbf{Z},
\end{array}
$$

where the right vertical map in the second diagram is multiplication by the degree $[L : K]$.

The map inducing the isomorphism $\mathrm{Br}(K) \cong \mathbf{Q}/\mathbf{Z}$ is classically called the *Hasse invariant map*.

Proof The first statement results from Corollary 6.3.5, taking into account that the Brauer group of a finite field is trivial (Example 6.1.12), and that $\mathrm{Hom}_{\mathrm{cont}}(\hat{\mathbf{Z}}, \mathbf{Q}/\mathbf{Z}) \cong \mathbf{Q}/\mathbf{Z}$. For the second statement, note first that it is enough to verify the commutativity of the first diagram, in view of the formula

Cor∘Res = $[L : K]$ (Proposition 4.2.12). Next, we apply Proposition A.6.8 (3) of the Appendix to find an intermediate extension $L \supset M \supset K$ unramified over K and with residue field λ equal to that of L. It will then be enough to treat the case of the extensions $M|K$ and $L|M$ separately. This follows from the fact that the composition of corestriction maps is again a corestriction map, as one sees directly from the definition. For the extension $M|K$, the commutativity of the second diagram follows from that of the diagram

$$
\begin{array}{ccc}
H^1(\lambda, \mathbf{Q}/\mathbf{Z}) & \xrightarrow{\ \cong\ } & \mathbf{Q}/\mathbf{Z} \\
{\scriptstyle \text{Cor}}\downarrow & & \downarrow{\scriptstyle \text{id}} \\
H^1(\kappa, \mathbf{Q}/\mathbf{Z}) & \xrightarrow{\ \cong\ } & \mathbf{Q}/\mathbf{Z}
\end{array}
$$

whose commutativity results from the definition of corestriction maps for procyclic groups. In the case of the extension $L|M$, the corestriction map on Brauer groups induces the identity on $H^1(\lambda, \mathbf{Q}/\mathbf{Z})$, whence the required commutativity. □

We finally describe central simple algebras over complete discretely valued fields with finite residue field.

Proposition 6.3.8 *If K is as in the previous proposition, every central simple algebra A over K is isomorphic to a cyclic algebra, and its period equals its index.*

Proof As before, let G be the Galois group of the residue field κ of K. Under the isomorphism Br $(K) \cong \mathrm{Hom}_{\mathrm{cont}}(G, \mathbf{Q}/\mathbf{Z})$ established during the proof of Proposition 6.3.7 the class $[A] \in$ Br (K) of A corresponds to a character $\chi : G \to \mathbf{Q}/\mathbf{Z}$ whose order d in $\mathrm{Hom}_{\mathrm{cont}}(G, \mathbf{Q}/\mathbf{Z}) \cong \mathbf{Q}/\mathbf{Z}$ is the period of A. On the other hand, if we view G as the Galois group of $K_{nr}|K$, the kernel of χ fixes an unramified extension $L|K$ of degree d such that the restriction map Br $(K) \to$ Br (L) annihilates α by construction. Thus A is split by an extension of degree d, which by virtue of Proposition 4.5.1 means that its index divides d. Hence it must equal d by Proposition 2.8.7 (1), which shows the equality of period and index. For cyclicity, note that since Gal $(K_{nr}|K) \cong \hat{\mathbf{Z}}$, we may embed the above L in a cyclic extension $M|K$ whose degree equals that of A. It then remains to apply Corollary 4.7.7 (2). □

6.4 Cohomology of function fields of curves

Let k again be a *perfect* field and C a smooth projective curve over k with function field K. We choose an algebraic closure $\bar{k}$ of k, and denote by G the Galois group $\mathrm{Gal}\,(\bar{k}|k)$. We denote the curve $C \times_k \bar{k}$ (which is assumed to be connected) by $\overline{C}$; its function field is by definition the composite $K\bar{k}$.

We shall investigate the cohomology of G with values in the multiplicative group $(K\bar{k})^\times$. As in Chapter 5, Section 5.4, the key tool for this will be the exact sequence of G-modules

$$0 \to \bar{k}^\times \to (K\bar{k})^\times \xrightarrow{\mathrm{div}} \mathrm{Div}(\overline{C}) \to \mathrm{Pic}\,(\overline{C}) \to 0, \qquad (6.3)$$

but we shall go further in the associated long exact sequences this time. There is a similar exact sequence over k:

$$0 \to k^\times \to K^\times \xrightarrow{\mathrm{div}} \mathrm{Div}(C) \to \mathrm{Pic}\,(C) \to 0. \qquad (6.4)$$

This sequence exists in arbitrary dimension, but our assumption that C is a curve makes it possible to define a *degree map*

$$\deg\,:\,\mathrm{Div}(C) \to \mathbf{Z}$$

associating to a divisor $\sum_P m_P P$ the integer $\sum_P m_P[\kappa(P) : k]$ (not to be confused with the degree map used in Chapter 5, Section 5.4). It is a fundamental fact (see Appendix, Proposition A.4.6) that the image of the divisor map div: $K^\times \to \mathrm{Div}(C)$ is contained in the kernel $\mathrm{Div}^0(C)$ of the degree map, so we have an induced map $\deg\,:\,\mathrm{Pic}\,(C) \to \mathbf{Z}$. We denote its kernel by $\mathrm{Pic}^0(C)$.

For each closed point P, the group G permutes the closed points lying over P. Therefore we get a direct sum decomposition

$$\mathrm{Div}(\overline{C}) = \bigoplus_{P \in C_0} \left(\bigoplus_{Q \mapsto P} \mathbf{Z} \right), \qquad (6.5)$$

where C_0 denotes the set of closed points of C, and the notation $Q \mapsto P$ stands for the closed points Q of $\overline{C}$ lying over P.

Hence for each $i \geq 0$ the divisor map induces maps

$$H^i(G, (K\bar{k})^\times) \to H^i(G, \mathrm{Div}(\overline{C})) \xrightarrow{\sim} \bigoplus_{P \in C_0} H^i\left(G, \bigoplus_{Q \mapsto P} \mathbf{Z}\right), \qquad (6.6)$$

as cohomology commutes with direct sums (more generally, with direct limits; see Lemma 4.3.3 and its proof).

Fix a preimage Q_0 of P in $\overline{C}$, and denote by G_P the stabilizer of Q_0 in G; it is an open subgroup of G depending on Q_0 only up to conjugation. Exactly as in the proof of Lemma 5.4.5, we have an isomorphism of G-modules

$$M_{G_P}^G(\mathbf{Z}) \cong \bigoplus_{Q \mapsto P} \mathbf{Z}. \tag{6.7}$$

Using Shapiro's lemma (Remark 4.2.10), we may thus rewrite the maps (6.6) as

$$H^i(G, (K\bar{k})^\times) \to \bigoplus_{P \in C_0} H^i(G_P, \mathbf{Z}).$$

Furthermore, the exact sequence $0 \to \mathbf{Z} \to \mathbf{Q} \to \mathbf{Q}/\mathbf{Z} \to 0$ induces isomorphisms $H^i(G_P, \mathbf{Z}) \cong H^{i-1}(G_P, \mathbf{Q}/\mathbf{Z})$ for $i \geq 2$, as in the previous section. So finally we get for each $i \geq 2$ and $P \in C_0$ a map

$$r_P^i : H^i(G, (K\bar{k})^\times) \to H^{i-1}(G_P, \mathbf{Q}/\mathbf{Z}),$$

called the *residue map associated with P*. By construction, for fixed i these maps are trivial for all but finitely many P. In order to get honest maps, we still have to prove:

Lemma 6.4.1 *The maps r_P^i depend only on P, and not on the closed point Q_0 lying above P used in the previous lemma.*

Proof As G acts transitively on the set $\{Q \mapsto P\}$, if we work with another point Q' instead of Q_0, we may find an element $\tau \in G$ with $Q' = \tau(Q_0)$. The stabilizer of Q' then will be $\tau G_P \tau^{-1}$. So an inspection of the previous construction reveals that is enough to see that the maps $H^{i-1}(G_P, \mathbf{Q}/\mathbf{Z}) \to H^{i-1}(\tau G_P \tau^{-1}, \mathbf{Q}/\mathbf{Z})$ induced by the natural map $\mathbf{Z}[G_P] \to \mathbf{Z}[\tau G_P \tau^{-1}]$ on cohomology became identity maps after identification of G_P with $\tau G_P \tau^{-1}$, which in turn is an immediate consequence of the construction of group cohomology. $\square$

The relation of the above residue maps with those of the previous section is as follows. As C is a smooth curve, the completion of the local ring of C at P is isomorphic to a formal power series ring $\kappa(P)[[t]]$ (see Appendix, Proposition A.5.3). By our assumption that k is perfect, here $\bar{k}$ is a separable closure of $\kappa(P)$, with $\mathrm{Gal}\,(\bar{k}|\kappa(P)) \cong G_P$. The construction of the previous section therefore yields residue maps

$$r_v^i : H^i(G_P, k((t))_{nr}^\times) \to H^{i-1}(G_P, \mathbf{Q}/\mathbf{Z}).$$

Proposition 6.4.2 *The diagram*

$$H^i(G, (K\bar{k})^\times) \xrightarrow{\;r_P^i\;} H^{i-1}(G_P, \mathbf{Q}/\mathbf{Z})$$

$$\text{Res} \downarrow \qquad\qquad\qquad \uparrow r_v^i$$

$$H^i(G_P, (K\bar{k})^\times) \longrightarrow H^i(G_P, k((t))_{nr}^\times)$$

commutes, where the bottom map is induced by the inclusion $K\bar{k} \hookrightarrow k((t))_{nr}$ coming from completing the local ring of a point of $\overline{C}$ above P.

Proof By Shapiro's lemma and the isomorphism (6.7) tensored by $(K\bar{k})^\times$, we have a chain of isomorphisms

$$H^i(G_P, (K\bar{k})^\times) \cong H^i(G, M_{G_P}^G (K\bar{k})^\times) \cong \bigoplus_{Q \mapsto P} H^i(G, (K\bar{k})^\times),$$

and the restriction map in the diagram is induced by taking a component of the direct sum corresponding to a point above P, say Q_0. The component of the divisor map associated with Q_0 is none but the discrete valuation $v_{Q_0} : (K\bar{k})^\times \to \mathbf{Z}$ corresponding to the local ring $\mathcal{O}_{\overline{C}, Q_0}$ of Q_0. The Proposition now follows from the isomorphism $G_P \cong \mathrm{Gal}\,(\bar{k}|\kappa(P))$ and the obvious fact that the discrete valuation induced by v_{Q_0} on the completion $\bar{k}[[t]]$ of $\mathcal{O}_{\overline{C}, Q}$ is none but the usual valuation v of the power series ring. $\square$

The basic fact concerning residue maps is:

Theorem 6.4.3 (Residue Theorem) *With notations as above, consider the corestriction maps*

$$\mathrm{Cor}_P : H^{i-1}(G_P, \mathbf{Q}/\mathbf{Z}) \to H^{i-1}(G, \mathbf{Q}/\mathbf{Z})$$

for each closed point P. The sequence of morphisms

$$H^i(G, (K\bar{k})^\times) \xrightarrow{\;\oplus r_P^i\;} \bigoplus_{P \in C_0} H^{i-1}(G_P, \mathbf{Q}/\mathbf{Z}) \xrightarrow{\;\Sigma\,\mathrm{Cor}_P\;} H^{i-1}(G, \mathbf{Q}/\mathbf{Z})$$

is a complex for all $i \geq 2$.

Proof The long exact sequence associated with the exact sequence of G-modules

$$0 \to (K\bar{k})^\times/\bar{k}^\times \to \mathrm{Div}(\overline{C}) \to \mathrm{Pic}\,(\overline{C}) \to 0 \qquad (6.8)$$

yields exact sequences

$$H^i(G, (K\bar{k})^\times/\bar{k}^\times) \to H^i(G, \mathrm{Div}(\overline{C})) \to H^i(G, \mathrm{Pic}\,(\overline{C}))$$

for each i. By construction, the direct sum of the maps r_P^i is obtained by composing the natural map $H^i(G, (K\bar{k})^\times) \to H^i(G, (K\bar{k})^\times / \bar{k}^\times)$ with the first map in the above sequence, and then applying the chain of isomorphisms

$$H^i(G, \mathrm{Div}(\overline{C})) \cong \bigoplus_{P \in C_0} H^i(G_P, \mathbf{Z}) \cong \bigoplus_{P \in C_0} H^{i-1}(G_P, \mathbf{Q}/\mathbf{Z}).$$

On the other hand, the degree map $\deg : \mathrm{Pic}\,(\overline{C}) \to \mathbf{Z}$ induces a map $H^i(G, \mathrm{Pic}\,(\overline{C})) \to H^i(G, \mathbf{Z})$. Therefore the theorem follows if we prove that the diagram

$$
\begin{array}{ccc}
H^i(G, \mathrm{Div}(\overline{C})) & \longrightarrow & H^i(G, \mathrm{Pic}\,(\overline{C})) \\
\cong \downarrow & & \downarrow \\
\bigoplus\limits_{P \in C_0} H^i(G_P, \mathbf{Z}) & \xrightarrow{\Sigma \mathrm{Cor}_P} & H^i(G, \mathbf{Z})
\end{array}
$$

commutes. As the degree map $\mathrm{Div}(\overline{C}) \to \mathbf{Z}$ factors though $\mathrm{Pic}\,(\overline{C})$, it will be enough to show that the composite

$$H^i(G, \mathrm{Div}(\overline{C})) \xrightarrow{\sim} \bigoplus_{P \in C_0} H^i(G_P, \mathbf{Z}) \xrightarrow{\Sigma \mathrm{Cor}_P} H^i(G, \mathbf{Z})$$

equals the map induced by $\deg : \mathrm{Div}(\overline{C}) \to \mathbf{Z}$, or else, by decomposing $\mathrm{Div}(\overline{C})$ as in (6.5), that the composite

$$H^i\left(G, \bigoplus_{Q \to P} \mathbf{Z}\right) \xrightarrow{\sim} H^i(G_P, \mathbf{Z}) \xrightarrow{\mathrm{Cor}_P} H^i(G, \mathbf{Z})$$

equals the map induced by $(m_1, \dots, m_r) \mapsto \sum m_i$. But using the isomorphism (6.7) we may rewrite the above composite map as

$$H^i(G, M_{G_P}^G(\mathbf{Z})) \to H^i(G, \mathbf{Z}),$$

the map being induced by summation according to the definition of corestriction maps. This finishes the verification of commutativity. $\quad\square$

In special cases we can say more. The most important of these is when C is the projective line. Then K is a rational function field $k(t)$, and we have the following stronger statement.

Theorem 6.4.4 (Faddeev) *Assume that C is the projective line. Then for each $i \geq 2$ the sequence*

$$0 \to H^i(G, \bar{k}^\times) \to H^i(G, (K\bar{k})^\times) \xrightarrow{\oplus r_P^i} \bigoplus_{P \in \mathbf{P}_0^1} H^{i-1}(G_P, \mathbf{Q}/\mathbf{Z})$$

$$\xrightarrow{\Sigma \operatorname{Cor}_P} H^{i-1}(G, \mathbf{Q}/\mathbf{Z}) \to 0$$

is exact.

Proof   For $\overline{C} = \mathbf{P}^1$ the degree map deg : $\operatorname{Pic}(\overline{C}) \to \mathbf{Z}$ is an isomorphism, hence exact sequence (6.8) takes the form

$$0 \to (K\bar{k})^\times/\bar{k}^\times \to \operatorname{Div}(\overline{C}) \to \mathbf{Z} \to 0.$$

Moreover, the choice of a rational point of C (say 0) defines a G-equivariant splitting $\mathbf{Z} \to \operatorname{Div}(\overline{C})$ of the above exact sequence, so that the sequence

$$0 \to H^i(G, (K\bar{k})^\times/\bar{k}^\times) \longrightarrow H^i(G, \operatorname{Div}(\overline{C})) \xrightarrow{\deg_*} H^i(G, \mathbf{Z}) \to 0 \quad (6.9)$$

is (split) exact for all i (see Remark 4.3.4 (2)). We have seen in the proof of Theorem 6.4.3 that here for $i \geq 2$ the map $\deg_*$ can be identified with the map

$$\bigoplus_{P \in C_0} H^{i-1}(G_P, \mathbf{Q}/\mathbf{Z}) \to H^{i-1}(G, \mathbf{Q}/\mathbf{Z})$$

given by the sum of corestrictions. Hence to conclude the proof it will be enough to establish exact sequences

$$0 \to H^i(G, \bar{k}^\times) \to H^i(G, (K\bar{k})^\times) \to H^i(G, (K\bar{k})^\times/\bar{k}^\times) \to 0 \quad (6.10)$$

for all $i > 0$. For this, consider the exact sequence

$$0 \to \bar{k}^\times \to (K\bar{k})^\times \to (K\bar{k})^\times/\bar{k}^\times \to 0$$

of G-modules. We claim that in the associated long exact sequence

$$\ldots \to H^i(G, \bar{k}^\times) \xrightarrow{\alpha_i} H^i(G, (K\bar{k})^\times) \to H^i(G, (K\bar{k})^\times/\bar{k}^\times) \to H^{i+1}(G, \bar{k}^\times) \to \ldots$$
$$(6.11)$$

the maps $\alpha_i : H^i(G, \bar{k}^\times) \to H^i(G, (K\bar{k})^\times)$ are injective for $i > 0$. Indeed, the completion of the local ring at a k-rational point of $\mathbf{P}_k^1$ (say 0) is isomorphic to $\bar{k}[[t]]$ as a G-module, whence a sequence of G-equivariant embeddings $\bar{k}^\times \to K\bar{k}^\times \to \bar{k}((t))^\times$, where the second map factorizes through $k((t))_{nr}^\times$. The composite of the induced maps

$$H^i(G, \bar{k}^\times) \xrightarrow{\alpha_i} H^i(G, K\bar{k}^\times) \to H^i(G, k((t))_{nr}^\times)$$

is injective by Proposition 6.3.1, hence so is the map α_i. By this injectivity property the long exact sequence splits up into a collection of short exact sequences (6.10), as desired. $\square$

The case $i = 2$ is of particular importance because of the relation with the Brauer group.

Corollary 6.4.5 (Faddeev) *The sequence*

$$0 \to \operatorname{Br}(k) \to \operatorname{Br}(K) \xrightarrow{\oplus r_P^i} \bigoplus_{P \in \mathbf{P}_0^1} H^1(G_P, \mathbf{Q}/\mathbf{Z}) \xrightarrow{\Sigma \operatorname{Cor}_P} H^1(G, \mathbf{Q}/\mathbf{Z}) \to 0$$

is exact.

Proof The corollary follows from the case $i = 2$ of the theorem, once we show that $H^2(G, (K\bar{k})^\times) \cong \operatorname{Br}(K)$. This is established in the same way as the isomorphism $H^2(G, k((t))_{nr}^\times) \cong \operatorname{Br}(k((t)))$ in Proposition 6.3.4, except that we use Tsen's theorem instead of Example 6.1.13. □

Remark 6.4.6 One may also derive Faddeev's exact sequence using methods of étale cohomology. See Milne [2], Example 2.22.

6.5 Application to class field theory

We now investigate the particular case when the base field is finite, and combine the techniques of the last section with some nontrivial facts from algebraic geometry in order to derive the main results in the class field theory of function fields over finite fields, first obtained by Hasse using a different method.

Throughout this section, $\mathbf{F}$ will denote a finite field, $G \cong \operatorname{Gal}(\overline{\mathbf{F}}|\mathbf{F})$ its absolute Galois group, and K the function field of a smooth projective curve C over $\mathbf{F}$. We shall continue to use some notation from the previous section.

Theorem 6.5.1 *The complex*

$$0 \to \operatorname{Br}(K) \xrightarrow{\oplus r_P^i} \bigoplus_{P \in C_0} H^1(G_P, \mathbf{Q}/\mathbf{Z}) \xrightarrow{\Sigma \operatorname{Cor}_P} H^1(G, \mathbf{Q}/\mathbf{Z}) \to 0$$

coming from Theorem 6.4.3 is exact. Furthermore, we have $H^i(G, (K\overline{\mathbf{F}})^\times) = 0$ for $i \geq 3$.

Facts 6.5.2 The proof will use the following facts about curves over finite fields which we quote from the literature:

For a smooth projective curve C over a finite field $\mathbf{F}$, the group $\operatorname{Pic}^0(\overline{C})$ is a torsion abelian group and the group $H^1(\mathbf{F}, \operatorname{Pic}^0(\overline{C}))$ vanishes.

The first claim follows from the fact that $\mathrm{Pic}^0(\overline{C})$ can be identified with the group $J(\overline{\mathbf{F}})$ of $\overline{\mathbf{F}}$-points of an abelian variety J defined over $\mathbf{F}$, the *Jacobian* of C (see e.g. Milne [4]). Being a projective variety, J has only a finite number of points over each finite extension $\mathbf{F}'|\mathbf{F}$ of the finite field $\mathbf{F}$, and the group $J(\overline{\mathbf{F}})$ is the union of the $J(\mathbf{F}')$, so it is a torsion abelian group. The second fact is a theorem of Lang [2]: for an abelian variety A (in fact, for any connected algebraic group) over a finite field $\mathbf{F}$ the group $H^1(\mathbf{F}, A(\overline{\mathbf{F}}))$ vanishes. This holds in particular for J.

Proof of Theorem 6.5.1 As $\mathrm{cd}(\mathbf{F}) = 1$, the second statement follows from Proposition 6.1.2. Granted the Facts 6.5.2 above, the proof of the case $i = 2$ is very similar to that of Theorem 6.4.4. The point is that the groups $H^i(G, \mathrm{Pic}^0(\overline{C}))$ are trivial for $i > 0$; for $i = 1$ this is just Lang's theorem, and for $i > 1$ it results from the fact that $\mathrm{Pic}^0(\overline{C})$ is torsion, in view of $\mathrm{cd}(\mathbf{F}) \leq 1$. Now consider the exact sequence of G-modules

$$0 \to \mathrm{Pic}^0(\overline{C}) \to \mathrm{Pic}(\overline{C}) \to \mathbf{Z} \to 0.$$

In the piece

$$H^i(G, \mathrm{Pic}^0(\overline{C})) \to H^i(G, \mathrm{Pic}(\overline{C})) \to H^i(G, \mathbf{Z}) \to H^{i+1}(G, \mathrm{Pic}^0(\overline{C}))$$

of the associated long exact cohomology sequence the first and last groups are trivial for $i > 0$ by what we have just said. Hence in the piece

$$H^1(G, \mathrm{Pic}(\overline{C})) \to H^2(G, (K\overline{\mathbf{F}})^\times / \overline{\mathbf{F}}^\times) \to H^2(G, \mathrm{Div}(\overline{C})) \xrightarrow{\beta_2} H^2(G, \mathrm{Pic}(\overline{C}))$$

of the long exact sequence associated with (6.8) we may replace the groups $H^i(G, \mathrm{Pic}(\overline{C}))$ by $H^i(G, \mathbf{Z})$ for $i = 1, 2$. The map β_2 is surjective as $\mathrm{cd}(\mathbf{F}) = 1$. Since moreover $H^1(G, \mathbf{Z}) = \mathrm{Hom}(G, \mathbf{Z}) = 0$, we obtain an exact sequence

$$0 \to H^2(G, (K\overline{\mathbf{F}})^\times / \overline{\mathbf{F}}^\times) \longrightarrow H^2(G, \mathrm{Div}(\overline{C})) \xrightarrow{\beta_2} H^2(G, \mathbf{Z}) \to 0.$$

As $H^i(G, \overline{\mathbf{F}}^\times) = 0$ for $i > 0$, the long exact sequence (6.11) yields an isomorphism $H^2(G, (K\overline{\mathbf{F}})^\times) \cong H^2(G, (K\overline{\mathbf{F}})^\times / \overline{\mathbf{F}}^\times)$, so we may replace $(K\overline{\mathbf{F}})^\times / \overline{\mathbf{F}}^\times$ by $(K\overline{\mathbf{F}})^\times$ in the left-hand side group. Hence we arrive at the Brauer group of K as in Corollary 6.4.5. To conclude the proof, one identifies the map β_2 with the sum of corestrictions $\bigoplus H^1(G_P, \mathbf{Q}/\mathbf{Z}) \to H^1(G, \mathbf{Q}/\mathbf{Z})$, in the same way as in the proof of the Residue Theorem. $\square$

One can obtain a more classical formulation of the theorem as follows. Take the completion of the local ring of C at a closed point P and denote by K_P

its fraction field. It is a Laurent series field over the residue field $\kappa(P)$, so Proposition 6.3.7 yields an isomorphism $\mathrm{Br}\,(K_P) \cong \mathbf{Q}/\mathbf{Z}$ induced by the Hasse invariant map, which we denote here by inv_P. The theorem then implies the following statement, which can be regarded as the main theorem in the class field theory of curves over finite fields.

Corollary 6.5.3 (Hasse) *With assumptions and notations as above, we have an exact sequence*

$$0 \to \mathrm{Br}\,(K) \to \bigoplus_{P \in C_0} \mathrm{Br}\,(K_P) \xrightarrow{\Sigma\,\mathrm{inv}_P} \mathbf{Q}/\mathbf{Z} \to 0.$$

Proof This follows from the theorem and the discussion above, noting the compatibility of Proposition 6.4.2 and the first commutative diagram of Proposition 6.3.7.   □

Remark 6.5.4 Using the above corollary and the function field analogue of the so-called Grunwald–Wang theorem, one proves, following Hasse, that a central simple algebra A over K is cyclic and its period equals its index (see Weil [3]). Note a subtle point here: though we know by Corollary 6.2.10 that A is split by a cyclic extension of the base field $\mathbf{F}$, in general the degree of such an extension is larger than the degree n of A. But in order to apply Proposition 4.7.7 one needs a cyclic splitting field of degree n. Therefore in general the required cyclic extension does not come from the base field, and our method based on Tsen's theorem does not apply.

Remark 6.5.5 According to the celebrated theorem of Albert, Brauer, Hasse and Noether, there is an exact sequence like the one in Corollary 6.5.3 above also in the case when K is a *number field*, i.e. a finite extension of $\mathbf{Q}$. Here the fields K_P run over all completions of K with respect to its (inequivalent) valuations. As opposed to the geometric case discussed above, these may be of two types. Either they are discrete valuations coming from some prime ideal in the ring of integers; in these cases an invariant map for the Brauer group similar to that in Proposition 6.3.7 can be constructed according to a theorem of Hasse. But there also exist so-called archimedean valuations, for which the completion is isomorphic to $\mathbf{R}$ or $\mathbf{C}$. The Brauer groups of these fields are respectively $\mathbf{Z}/2\mathbf{Z}$ and 0, and thus may be viewed as subgroups of $\mathbf{Q}/\mathbf{Z}$, yielding the 'archimedean invariant maps' in the sequence. The proof of this theorem is different from the one given above, and uses the main results of class field theory for number fields. See e.g. Tate [2] or Neukirch–Schmidt–Wingberg [1].

Using this result, Brauer, Hasse and Noether also proved that over a number field every central simple algebra is cyclic, and its period equals its index. See Kersten [1], Pierce [1] or Roquette [4] for recent accounts of the proof.

These famous theorems were found a few years earlier than the geometric statements we have discussed above, but from today's viewpoint the latter are easier to establish thanks to the geometric techniques which are unavailable in the arithmetic case.

6.6 Application to the rationality problem: the method

In this section we show how an application of Faddeev's exact sequence yields a simple answer to a long-standing problem in algebraic geometry. The question may be stated in purely algebraic terms as follows.

Problem 6.6.1 *Let k be a field, and $k(t_1, \ldots, t_n)$ a purely transcendental extension of k. Let $k \subset K \subset k(t_1, \ldots, t_n)$ be a subfield such that $k(t_1, \ldots, t_n)|K$ is a finite extension. Is it true that $K|k$ is a purely transcendental extension?*

Remarks 6.6.2

1. In the language of algebraic geometry, the problem may be rephrased as follows. A k-variety X of dimension n is called *rational (over k)* if it is birational over k to projective n-space $\mathbf{P}_k^n$; it is *unirational* if there exists a dominant rational map $\mathbf{P}_k^n \to X$ over k. So the question is: *is every unirational k-variety rational?*
2. *Positive results.* When $n = 1$, the answer is yes, by a classical theorem due to Lüroth (see e.g. van der Waerden [1], §73). For this reason, the problem is sometimes called the Lüroth problem in the literature. In the case $n = 2$ counterexamples can be given if the ground field k is not assumed to be algebraically closed (see Exercise 10). However, when k is algebraically closed of characteristic 0, it follows from a famous theorem of Castelnuovo in the classification of surfaces (see e.g. Beauville [1], Chapter V) that the answer is positive. Zariski showed that the answer is also positive for k algebraically closed of characteristic $p > 0$ if one assumes the extension $K|K_0$ to be separable.
3. *Negative results.* Castelnuovo's theorem dates back to 1894. However, after some false starts by Fano and Roth, the first counterexamples showing that the answer may be negative over $k = \mathbf{C}$ in dimension 3 have only

been found around 1970, by Clemens–Griffiths [1] and Iskovskih–Manin [1], independently. Immediately afterwards, Artin and Mumford [1] found counterexamples which could be explained by the nonvanishing of a certain *birational invariant* (i.e. an element of some group associated functorially to varieties and depending only on the birational isomorphism class of the variety) which is trivial for projective space. The group in question was the torsion part of the cohomology group $H^3(X, \mathbf{Z})$.

Still, even the counterexamples cited above did not rule out the possibility that the answer to the following weaker question might be positive. Observe that a purely transcendental field extension $k(t_1, \ldots, t_n)$ of k may be identified with the field of rational functions on an n-dimensional k-vector space V (by looking at V as affine n-space $\mathbf{A}^n$ over k, or by passing to the symmetric algebra of V). If a finite group G acts k-linearly on V, there is an induced action on the field $k(V)$. The action is called *faithful* if the homomorphism $G \to \mathrm{GL}(V)$ is injective. In this case the field extension $k(V)|k(V)^G$ is Galois with group G.

Problem 6.6.3 *Let k be an algebraically closed field of characteristic 0, and let V be a finite-dimensional vector space over k. Assume that a finite group G acts k-linearly and faithfully on V. Is it true that the field of invariants $k(V)^G$ is a purely transcendental extension of k?*

In his 1984 paper [2] Saltman showed that the answer to even this weaker question is negative in general. His approach, which was inspired by that of Artin and Mumford, but much more elementary, was developed further in works of Bogomolov ([1] and [2]). Our account below has been influenced by the notes of Colliot-Thélène and Sansuc [1].

The starting point for the construction of the counterexample is the consideration of the following invariant. Let $K|k$ be an extension of fields of characteristic 0, and $A \supset k$ a discrete valuation ring with fraction field K. The completion of K is isomorphic to a Laurent series field $\kappa((t))$, where κ is the residue field of A (see Appendix, Proposition A.5.3). Note that if the transcendence degree of $K|k$ is at least 2, then the extension $\kappa|k$ is transcendental. Let $\overline{\kappa}$ be an algebraic closure of κ. As in Section 6.3, we have a residue map

$$ r_A : H^2(\mathrm{Gal}\,(\overline{\kappa}|\kappa), \kappa((t))^\times_{nr}) \to H^1(\kappa, \mathbf{Q}/\mathbf{Z}) $$

induced by the valuation associated with A. As $\mathrm{Br}\,(\kappa((t))_{nr})$ is trivial by Example 6.1.13, the inflation-restriction sequence shows, as in the proof of Proposition 6.3.4, that we may identify the Brauer group of $\kappa((t))$ with $H^2(\mathrm{Gal}\,(\overline{\kappa}|\kappa), \kappa((t))^\times_{nr})$, so we get a composite map

$$\operatorname{Br}(K) \to \operatorname{Br}(\kappa((t))) \to H^1(\kappa, \mathbf{Q}/\mathbf{Z}),$$

which we also denote by r_A.

Definition 6.6.4 The intersection $\bigcap \ker(r_A) \subset \operatorname{Br}(K)$ of the groups $\ker(r_A)$ for all discrete valuation rings of $K|k$ is called the *unramified Brauer group* of $K|k$ and denoted by $\operatorname{Br}_{\mathrm{nr}}(K)$.

Though not reflected in the notation, one should bear in mind that $\operatorname{Br}_{\mathrm{nr}}(K)$ is an invariant which is relative to k. Of course, an analogous definition can be made for the higher cohomology groups of $\bar{k}^\times$; the proofs of the basic properties established below carry over without change.

In the case when K is the function field of a variety X defined over k, we may view $\operatorname{Br}_{\mathrm{nr}}(K)$ as an invariant attached to X. As it depends only on K, it is a *birational invariant* in the sense explained above. We now have the following functorial property.

Lemma 6.6.5 *Given a field extension $L|K$, the natural map $\operatorname{Br}(K) \to \operatorname{Br}(L)$ sends the subgroup $\operatorname{Br}_{\mathrm{nr}}(K)$ into $\operatorname{Br}_{\mathrm{nr}}(L)$.*

Proof Let B be a discrete valuation ring of $L|k$, with residue field κ_B. Its completion is isomorphic to $\kappa_B((t))$. If B contains K as a subfield, then we must have $K \subset \kappa_B$ in $\kappa_B((t))$ since the elements of K are units, and thus $K \subset \ker(r_B)$ by Corollary 6.3.5. Otherwise the intersection $A := B \cap K$ is a discrete valuation ring of $K|k$. Denoting by κ_A its residue field, we have a natural inclusion $\kappa_A \subset \kappa_B$. The associated valuations satisfy an equality $v_A = e \cdot v_B$ with some integer $e \geq 1$, for if t_A generates the maximal ideal of A, we have $t_A = u t^e$ for some unit u in B. The construction of residue maps then implies the commutativity of the diagram

$$
\begin{array}{ccc}
\operatorname{Br}(\kappa_B((t))) & \xrightarrow{\ r_B\ } & H^1(\kappa_B, \mathbf{Q}/\mathbf{Z}) \\[2mm]
{\scriptstyle \mathrm{Res}}\uparrow & & \uparrow{\scriptstyle e \cdot \mathrm{Res}} \\[2mm]
\operatorname{Br}(\kappa_A((t_A))) & \xrightarrow{\ r_A\ } & H^1(\kappa_A, \mathbf{Q}/\mathbf{Z}),
\end{array}
$$

whence $\ker(r_A) \subset \ker(r_B)$, and the lemma follows. $\qquad\square$

The following crucial proposition implies that purely transcendental extensions have trivial unramified Brauer group.

Proposition 6.6.6   *Let K be as above, and let $K(t)|K$ be a purely transcendental extension. Then the natural map $\mathrm{Br}_{\mathrm{nr}}(K) \to \mathrm{Br}_{\mathrm{nr}}(K(t))$ given by the previous lemma is an isomorphism.*

Proof   The map $\mathrm{Br}(K) \to \mathrm{Br}(K(t))$ is injective by Corollary 6.4.5, hence so is the map $\mathrm{Br}_{\mathrm{nr}}(K) \to \mathrm{Br}_{\mathrm{nr}}(K(t))$ of the previous lemma. Therefore it is enough to check surjectivity. For this take an $\alpha \in \mathrm{Br}_{\mathrm{nr}}(K(t))$. As α is in the kernel of all residue maps coming from valuations trivial on K, we have $\alpha \in \mathrm{Br}(K)$, again by Corollary 6.4.5. It therefore remains to be seen that $r_A(\alpha) = 0$ for each discrete valuation ring A of $K|k$. But for such an A one may find a discrete valuation ring B of $K(t)|k$ with $B \cap K = A$, by continuing the discrete valuation v_A to a discrete valuation v_B on $K(t)$ via setting $v_B(t) = 0$ (see Appendix, Proposition A.6.12). For the associated discrete valuations one has $e = 1$, and hence $\ker(r_B) \cap K \subset \ker(r_A)$ by the diagram of the previous proof. Since $\alpha \in \ker(r_B)$ by assumption, the claim follows. $\square$

We get by induction starting from the case $n = 1$ (Faddeev's theorem):

Corollary 6.6.7   *For a purely transcendental extension $k(t_1, \ldots, t_m)$ of k one has*

$$\mathrm{Br}_{\mathrm{nr}}\, k(t_1, \ldots, t_m) \cong \mathrm{Br}(k).$$

In particular, if k is algebraically closed, then $\mathrm{Br}_{\mathrm{nr}}\, k(t_1, \ldots, t_m) = 0$.

We now turn to the construction of Bogomolov and Saltman. *In the rest of this section we assume that the base field is algebraically closed of characteristic 0.*

As an appetizer, we prove the following classical result of Fischer [1] which shows that in the counterexample to Problem 6.6.3 the group G must be noncommutative.

Theorem 6.6.8 (Fischer)   *Assume that a finite abelian group A acts k-linearly and faithfully on a finite-dimensional k-vector space V. Then the field of invariants $k(V)^A$ is a purely transcendental extension of k.*

Proof   As we are in characteristic 0, the A-representation on V is semisimple, i.e. the $k[A]$-module V decomposes as a direct sum of 1-dimensional sub-$k[A]$-modules V_i, such that on V_i the A-action is given by $\sigma(v) = \chi_i(\sigma)v$ for some character $\chi_i : A \to k^\times$. Let v_i be a nonzero vector in V_i for each

i, and let X be the subgroup of $k(V)^\times$ generated by the v_i. As the v_i are linearly independent, X is a free abelian group. Now let $\widehat{A} = \text{Hom}(A, k^\times)$ be the character group of A, and consider the homomorphism $\phi : X \to \widehat{A}$ given by $v_i \mapsto \chi_i$. By construction, we have $\sigma(x) = (\phi(x)(\sigma))x$ for $x \in X$ and $\sigma \in A$. In particular, with the notation $Y := \ker(\phi)$ we get $Y \subset k(V)^A$. On the other hand, the index of Y in X is at most $|\widehat{A}| = |A|$, so the field index $[k(V) : k(Y)]$ is at most $|A|$. But $[k(V) : k(V)^A] = |A|$, as this extension is Galois with group A. Thus we conclude $k(Y) = k(V)^A$. Now Y is a free abelian group, being a subgroup of X, and therefore we have $k(Y) = k(y_1, \dots, y_m)$ for a basis $y_1, \dots, y_m$ of Y. This proves the theorem. □

Remark 6.6.9 An examination of the above proof reveals that the theorem is valid more generally for an arbitrary finite abelian group A of exponent e and any ground field F of characteristic prime to e and containing the e-th roots of unity. In the case of arbitrary F the field $F(V)^A$ will be the function field of an algebraic torus, not necessarily rational over F (see Voskresenskiĭ [2], §7.2).

This being said, Corollary 6.6.7 shows that in order to find a counterexample to Problem 6.6.3 it suffices to find a faithful representation V of a finite group G with $\text{Br}_{\text{nr}}\left(k(V)^G\right) \neq 0$. The key to this will be the following basic theorem characterizing the unramified Brauer group of invariant fields.

Theorem 6.6.10 (Bogomolov) *Let V be a finite-dimensional k-vector space, and let G be a finite group acting k-linearly and faithfully on V. Then*

$$\text{Br}_{\text{nr}}\left(k(V)^G\right) = \ker\left(\text{Br}\left(k(V)^G\right) \to \prod_{H \in \mathcal{B}} \text{Br}\left(k(V)^H\right)\right),$$

where $\mathcal{B}$ denotes the set of bicyclic subgroups of G.

Recall that a bicyclic group is just a direct product of two cyclic groups.

Proof By Lemma 6.6.5 the image of $\text{Br}_{\text{nr}}\left(k(V)^G\right)$ by each restriction map $\text{Br}\left(k(V)^G\right) \to \text{Br}\left(k(V)^H\right)$ lies in $\text{Br}_{\text{nr}}\left(k(V)^H\right)$. But since bicyclic groups are abelian, we have $\text{Br}_{\text{nr}}\left(k(V)^H\right) = 0$ by Fischer's theorem (Theorem 6.6.8) and Corollary 6.6.7. So we conclude that the left-hand side is contained in the right-hand side.

For the reverse inclusion, take an element $\alpha \in \text{Br}\left(k(V)^G\right)$ with $r_A(\alpha) \neq 0$ for some discrete valuation ring A of $k(V)^G|k$. Let B be one of the finitely many discrete valuation rings of $k(V)|k$ lying above A, let $D \subset G$ be the stabilizer of B under the action of G, and let κ_B and κ_A be the respective residue fields of B and A. As the finite extension $k(V)|k(V)^G$ is Galois, it is known (cf. Appendix, Proposition A.6.3 (2)) that the extension $\kappa_B|\kappa_A$ is a finite Galois extension as well, and there is a natural surjection $D \to \text{Gal}(\kappa_B|\kappa_A)$. Denote by I the kernel of this map. As κ_A is algebraically closed of characteristic 0, it is also known (see Appendix, Corollary A.6.11) that I is a central cyclic subgroup in D. If the image of α by the restriction map $\text{Res}_I : \text{Br}(k(V)^G) \to \text{Br}(k(V)^I)$ is nonzero, then so is its image by the map $\text{Br}(k(V)^G) \to \text{Br}(k(V)^H)$ for a bicyclic subgroup H containing the cyclic subgroup I, and we are done. So we may assume $\text{Res}_I(\alpha) = 0$. Now consider the commutative diagram with exact rows

$$
\begin{array}{ccccccc}
0 & \longrightarrow & \text{Br}(k(V)^I|k(V)^G) & \xrightarrow{\ \text{Inf}\ } & \text{Br}(k(V)^G) & \xrightarrow{\ \text{Res}\ } & \text{Br}(k(V)^I) \\
& & \downarrow & & \downarrow{\scriptstyle r_A} & & \downarrow{\scriptstyle r_C} \\
0 & \longrightarrow & H^1(\text{Gal}(\kappa_B|\kappa_A), \mathbf{Q}/\mathbf{Z}) & \xrightarrow{\ \text{Inf}\ } & H^1(\kappa_A, \mathbf{Q}/\mathbf{Z}) & \xrightarrow{\ \text{Res}\ } & H^1(\kappa_B, \mathbf{Q}/\mathbf{Z})
\end{array}
$$

in which the rows are restriction-inflation sequences (Corollary 4.3.5), and the map r_C is the residue map associated with the discrete valuation ring $C := B \cap k(V)^I$ which has the same residue field κ_B as B. The diagram shows that α comes from an element of $\text{Br}(k(V)^I|k(V)^G)$, and $r_A(\alpha)$ may be identified with a homomorphism $\phi_\alpha : D/I \to \mathbf{Q}/\mathbf{Z}$. Let $g \in D$ be an element whose image $\bar{g}$ in $\text{Gal}(\kappa_B|\kappa_A) \cong D/I$ satisfies $\phi_\alpha(\bar{g}) \neq 0$. As I is a central cyclic subgroup in D, the subgroup $H^g \subset D$ generated by g and I is bicyclic. We now show that the image of α by the restriction map $\text{Res}_{H^g} : \text{Br}(k(V)^G) \to \text{Br}(k(V)^{H^g})$ is nontrivial. Indeed, if we denote by B^g the discrete valuation ring $B \cap k(V)^{H^g}$, then the same argument as above with B^g in place of A shows that the image of $\text{Res}_{H^g}(\alpha)$ by the associated residue map r_{B^g} is a homomorphism $H^g/I \to \mathbf{Q}/\mathbf{Z}$. By construction, this homomorphism is none but the restriction of ϕ_α to the group H^g/I, and hence is nonzero. So $\text{Res}_{H^g}(\alpha)$ itself is nonzero, as required. $\square$

Remark 6.6.11 The above proof (together with Fischer's theorem) shows that instead of $\mathcal{B}$ one could take the set of all *abelian* subgroups of G.

As a consequence of the preceding theorem, Bogomolov was able to give a purely group-theoretic characterization of $\text{Br}_{\text{nr}}(k(V)^G)$.

Theorem 6.6.12 *Let $k(V)$, G and $\mathcal{B}$ be as in the theorem above. The group $\text{Br}_{\text{nr}}(k(V)^G)$ is isomorphic to the group*

$$
H^2_{\mathcal{B}}(G) := \ker\left(H^2(G, \mathbf{Q}/\mathbf{Z}) \xrightarrow{\ \text{Res}\ } \prod_{H \in \mathcal{B}} H^2(H, \mathbf{Q}/\mathbf{Z}) \right).
$$

Proof The proof is in three steps.

Step 1. We first establish an isomorphism

$$\mathrm{Br}_{\mathrm{nr}}\,(k(V)^G) = \ker\left(H^2(G, k(V)^\times) \to \prod_{H \in \mathcal{B}} H^2(H, k(V)^\times)\right).$$

For this, consider the exact sequence

$$0 \to \mathrm{Br}\,(k(V)|k(V)^G) \xrightarrow{\mathrm{Inf}} \mathrm{Br}\,(k(V)^G) \xrightarrow{\mathrm{Res}} \mathrm{Br}\,(k(V))$$

of Corollary 4.4.7. As $\mathrm{Br}_{\mathrm{nr}}\,(k(V)) = 0$ by Corollary 6.6.7, we see using Lemma 6.6.5 that each element of $\mathrm{Br}_{\mathrm{nr}}\,(k(V)^G)$ comes from $\mathrm{Br}\,(k(V)|k(V)^G)$. Using the fact that the composite map $\mathrm{Br}\,(k(V)|k(V)^G) \to \mathrm{Br}\,(k(V)^G) \to \mathrm{Br}\,(k(V)^H)$ factors through $\mathrm{Br}\,(k(V)|k(V)^H)$ and noting the isomorphism $\mathrm{Br}\,(k(V)|k(V)^G) \cong H^2(G, k(V)^\times)$, we may rewrite the formula of the previous theorem as stated above.

Step 2. We next show that we may replace the coefficient module $k(V)^\times$ by $k^\times$, i.e. we have an isomorphism

$$\mathrm{Br}_{\mathrm{nr}}\,(k(V)^G) = \ker\left(H^2(G, k^\times) \to \prod_{H \in \mathcal{B}} H^2(H, k^\times)\right).$$

For this we view $k(V)$ as the function field of affine n-space $\mathbf{A}_k^n$. As the Picard group of $\mathbf{A}_k^n$ is trivial (cf. Appendix, Proposition A.4.4 (1)), we have an exact sequence of G-modules

$$0 \to k^\times \to k(V)^\times \to \mathrm{Div}(\mathbf{A}_k^n) \to 0. \tag{6.12}$$

Denote by W the affine variety with coordinate ring $k[t_1, \ldots, t_n]^G$. By Lemma 5.4.5 we have a direct sum decomposition

$$\mathrm{Div}(\mathbf{A}_k^n) \cong \bigoplus_{P \in W^1} M_{G_P}^G(\mathbf{Z}),$$

where W^1 denotes the set of codimension 1 irreducible subvarieties of W and G_P is the stabilizer of an irreducible component lying over the codimension 1 subvariety P. Therefore using Shapiro's lemma we get from sequence (6.12) an exact sequence

$$\bigoplus_{P \in W^1} H^1(G_P, \mathbf{Z}) \to H^2(G, k^\times) \xrightarrow{\pi} H^2(G, k(V)^\times) \xrightarrow{\rho} \bigoplus_{P \in W^1} H^2(G_P, \mathbf{Z}).$$

Here the groups $H^1(G_P, \mathbf{Z}) = \mathrm{Hom}(G_P, \mathbf{Z})$ are trivial because the G_P are finite, so the map π is injective.

As for the groups $H^2(G_P, \mathbf{Z}) \cong H^1(G_P, \mathbf{Q}/\mathbf{Z}) = \mathrm{Hom}(G_P, \mathbf{Q}/\mathbf{Z})$, we obviously have injections $\iota_P \; : \; \mathrm{Hom}(G_P, \mathbf{Q}/\mathbf{Z}) \hookrightarrow \bigoplus \mathrm{Hom}(\langle g \rangle, \mathbf{Q}/\mathbf{Z})$, where the sum is over all cyclic subgroups $\langle g \rangle$ of G_P. Now if the restrictions of an element $\alpha \in H^2(G, k(V)^\times)$ to all bicyclic subgroups are trivial, the same must be true for all restrictions to cyclic subgroups, so all components of $\rho(\alpha) \in \bigoplus \mathrm{Hom}(G_P, \mathbf{Q}/\mathbf{Z})$ are sent to 0 by the various ι_P. Therefore $\rho(\alpha) = 0$ and α comes from $H^2(G, k^\times)$. The claim follows by noting that the maps $H^2(G, k^\times) \to H^2(H, k(V)^\times)$ factor through $H^2(H, k^\times)$.

Step 3. In view of the previous step, to prove the proposition it is enough to establish isomorphisms

$$H^2(G, k^\times) \cong H^2(G, \mathbf{Q}/\mathbf{Z}) \quad \text{and} \quad H^2(H, k^\times) \cong H^2(H, \mathbf{Q}/\mathbf{Z})$$

for all H. Now as k is algebraically closed of characteristic 0, the group $k^\times$ is divisible and its torsion subgroup (i.e. the group μ of all roots of unity in k) is isomorphic to $\mathbf{Q}/\mathbf{Z}$. Therefore the quotient $k^\times/(\mathbf{Q}/\mathbf{Z})$ is a $\mathbf{Q}$-vector space, so the long exact sequence coming from the exact sequence

$$0 \to \mathbf{Q}/\mathbf{Z} \to k^\times \to k^\times/(\mathbf{Q}/\mathbf{Z}) \to 0$$

of trivial G-modules yields the required isomorphisms in view of Corollary 4.2.7. $\qquad\square$

Remark 6.6.13 The proof above shows that the isomorphism of the theorem depends only on the choice of an isomorphism of abelian groups $\mu \cong \mathbf{Q}/\mathbf{Z}$. If one uses μ-coefficients instead of $\mathbf{Q}/\mathbf{Z}$-coefficients, the isomorphism becomes canonical.

6.7 Application to the rationality problem: the example

Keeping the assumptions and the notations of the previous section, we show at last:

Theorem 6.7.1 *There exists a finite group G for which $H^2_B(G) \neq 0$. Therefore G yields a counterexample to Problem 6.6.3 over k.*

The proof below is based on an idea of Shafarevich [1]. The following lemma from the theory of group extensions will be a basic tool.

Lemma 6.7.2 *Let A, B be two abelian groups. Regard A as a B-module with trivial action.*

1. There is a homomorphism

$$\rho_A : H^2(B, A) \to \mathrm{Hom}(\Lambda^2 B, A),$$

functorial in A, sending the class of an extension

$$0 \to A \to E \to B \to 0$$

to the map $\phi_E : b_1 \wedge b_2 \to [\tilde{b}_1, \tilde{b}_2]$, where the $\tilde{b}_i$ are arbitrary liftings of the b_i to E, and $[\tilde{b}_1, \tilde{b}_2]$ denotes their commutator. The kernel of ρ_A consists of extension classes with E commutative.

2. In terms of cocycles, the map ρ_A sends the class of a normalized 2-cocycle c_{b_1,b_2} to the alternating map $b_1 \wedge b_2 \mapsto c_{b_1,b_2} - c_{b_2,b_1}$.

3. Assume moreover that A and B are finite dimensional $\mathbf{F}_p$-vector spaces for a prime number $p > 2$. Then ρ_A has a canonical splitting.

Here $\Lambda^2 B$ denotes the quotient of $B \otimes_{\mathbf{Z}} B$ by the subgroup generated by the elements $b \otimes b$ for all $b \in B$. Part (1) of the lemma can be proven using the universal coefficient sequence for cohomology (Weibel [1], Theorem 3.6.5); we give here a direct argument.

Proof For (1), note first that since B acts trivially on A, the extension E is central, and therefore in the above definition $\phi_E(b_1, b_2) = [\tilde{b}_1, \tilde{b}_2]$ does not depend on the choice of the liftings $\tilde{b}_1, \tilde{b}_2$. Moreover, ϕ_E satisfies $\phi_E(b, b) = 0$ for all $b \in B$; let us check that it is also bilinear. For this, let $s : B \to E$ be a (set-theoretic) section of the projection $E \to B$ satisfying $s(1) = 1$. As in Example 3.2.6 this yields the normalized 2-cocycle $c_{b_1,b_2} = s(b_1)s(b_2) s(b_1 + b_2)^{-1}$ of B with values in A. Recall also the formula $[g_1 g_2, g_3] = g_1[g_2, g_3]g_1^{-1}[g_1, g_3]$ which holds in any group. Since A is central in E, we have

$$\phi_E(b_1 + b_2, b_3) = [s(b_1 + b_2), s(b_3)] = [c_{b_1,b_2}^{-1} s(b_1)s(b_2), s(b_3)]$$

$$= [s(b_1)s(b_2), s(b_3)] = s(b_1)[s(b_2), s(b_3)]s(b_1)^{-1} [s(b_1), s(b_3)]$$

$$= \phi_E(b_2, b_3) + \phi_E(b_1, b_3).$$

Similarly, $\phi_E(b_1, b_2 + b_3) = \phi_E(b_1, b_2) + \phi_E(b_1, b_3)$, so ϕ_E is a well-defined alternating bilinear map. To finish the proof of (1), it remains to check that ρ_A is a group homomorphism, because the second statement in (1) is then immediate from the definition of ρ_A. For this it is enough to establish (2), because the map $c_{b_1,b_2} \mapsto c_{b_1,b_2} - c_{b_2,b_1}$ is manifestly a homomorphism from the group of normalized 2-cocycles of B with values in A to the group $\mathrm{Hom}(\Lambda^2 B, A)$. But for the 2-cocycle c_{b_1,b_2} associated with E above we have

$$\phi_E(b_1, b_2) = s(b_1)s(b_2)s(b_1)^{-1}s(b_2)^{-1} = c_{b_1,b_2} s(b_1 + b_2)s(b_2 + b_1)^{-1} c_{b_2,b_1}^{-1},$$

which is indeed $c_{b_1,b_2} - c_{b_2,b_1}$ in the additive notation.

We finally turn to (3). It will be enough to construct a splitting in the case $A = \mathbf{F}_p$. Recall from linear algebra that the space of bilinear forms splits as the direct sum of the spaces of symmetric and alternating forms via the map $\lambda \mapsto (\lambda + \lambda^\tau, \lambda - \lambda^\tau)$, where λ^τ is the bilinear form obtained from λ by switching the entries. Thus given an alternating bilinear form $\phi : \Lambda^2 B \to \mathbf{F}_p$, there exists a bilinear form $\gamma : B \times B \to \mathbf{F}_p$ such that

$$\phi(b_1 \wedge b_2) = \gamma(b_1, b_2) - \gamma(b_2, b_1).$$

Notice that the map $(b_1, b_2) \mapsto \gamma(b_1, b_2)$ is a normalized 2-cocycle, as we have $\gamma(0, b) = \gamma(b, 0) = 0$ for all $b \in B$, and the cocycle relation holds by the calculation

$$\gamma(b_2, b_3) - \gamma(b_1 + b_2, b_3) + \gamma(b_1, b_2 + b_3) - \gamma(b_1, b_2)$$
$$= -\gamma(b_1, b_3) + \gamma(b_1, b_3) = 0.$$

The difference of two choices of γ is a symmetric bilinear form. But if γ is symmetric, then since $p > 2$, we may write

$$\gamma(b_1, b_2) = \frac{1}{2}\Big(\gamma(b_1 + b_2, b_1 + b_2) - \gamma(b_1, b_1) - \gamma(b_2, b_2)\Big) = (df)(b_1, b_2),$$

where f is the 1-cocycle $b \mapsto (1/2)\gamma(b, b)$ of B with values in the trivial B-module $\mathbf{F}_p$. Therefore the class $[\gamma]$ of the 2-cocycle $(b_1, b_2) \mapsto \gamma(b_1, b_2)$ in $H^2(B, \mathbf{F}_p)$ only depends on the alternating form ϕ, and we may define the map $\xi : \mathrm{Hom}(\Lambda^2 B, \mathbf{F}_p) \to H^2(B, \mathbf{F}_p)$ by sending ϕ to $[\gamma]$. By (2), the map ξ satisfies $\rho_{\mathbf{F}_p} \circ \xi = \mathrm{id}_{\mathbf{F}_p}$. $\qquad\square$

The lemma enables us to construct important examples of nilpotent groups.

Example 6.7.3 Let $p > 2$ be a prime number, and let V be an n-dimensional $\mathbf{F}_p$-vector space. Applying the canonical splitting constructed in part (3) of the above lemma for $B = V$, $A = \Lambda^2 V$, we get that the identity map of $\Lambda^2 V$ gives rise to an extension $\overline{G}_n$ of V by $\Lambda^2 V$. Here $\Lambda^2 V$ is both the centre and the commutator subgroup of the group $\overline{G}_n$, which is in particular nilpotent of class 2. It is the *universal nilpotent group of class 2 and exponent p on n generators*. Its elements can be written in the form $\prod_i a_i^{\alpha_i} \prod_{i<j} [a_i, a_j]^{\beta_{ij}}$, where $a_1, \ldots, a_n$ are liftings of a basis of V to G and $\alpha_i, \beta_{ij} \in \mathbf{F}_p$.

Proof of Theorem 6.7.1 Let $p > 2$ be a prime number, and consider the group $\overline{G}_n$ of the above example for $n \geq 4$. Let $a_1, \ldots, a_n$ be a system of generators as above, and look at the element $z = [a_1, a_2][a_3, a_4]$. Note that z lies in the centre $\Lambda^2 V$ of $\overline{G}_n$, and it has the property that the powers $z^r = [a_1, a_2]^r[a_3, a_4]^r$ cannot be expressed as commutators $[b_1, b_2]$ of elements $b_1, b_2 \in \overline{G}_n$ whose images in V are linearly independent over $\mathbf{F}_p$.

Indeed, in $\Lambda^2 V$ the z^r correspond to bivectors of the form $r(v_1 \wedge v_2) + r(v_1 \wedge v_3)$, which are either trivial or indecomposable bivectors, i.e. not of the form $w_1 \wedge w_2$ for independent w_i.

Now define G as the quotient of $\overline{G}_n$ by the central cyclic subgroup $\langle z \rangle$ generated by z. The conjugation action of G on $\langle z \rangle$ is trivial, hence it is isomorphic to $\mathbf{Z}/p\mathbf{Z}$ as a G-module. The extension

$$1 \to \langle z \rangle \to \overline{G}_n \to G \to 1 \qquad (6.13)$$

therefore defines a class $c(\overline{G}_n) \in H^2(G, \mathbf{Z}/p\mathbf{Z})$ by Example 3.2.6. We may send it to a class in $H^2(G, \mathbf{Q}/\mathbf{Z})$ via the map $\iota_* : H^2(G, \mathbf{Z}/p\mathbf{Z}) \to H^2(G, \mathbf{Q}/\mathbf{Z})$ induced by the inclusion $\iota : \mathbf{Z}/p\mathbf{Z} \to \mathbf{Q}/\mathbf{Z}$ sending 1 to $1/p$. Let us now show that $\iota_*(c(\overline{G}_n))$ lies in $H^2_B(G)$. For this it will be enough to see that the images of $c(\overline{G}_n)$ by the restriction maps $H^2(G, \mathbf{Z}/p\mathbf{Z}) \to H^2(H, \mathbf{Z}/p\mathbf{Z})$ are trivial for each bicyclic subgroup $H \subset G$. Such a subgroup necessarily meets the centre $Z(G)$ of G. Indeed, write $H = \langle h_1 \rangle \times \langle h_2 \rangle$ with some generators h_1, h_2. As the h_i commute, we have $[h_1, h_2] = 1$ in G and so $[b_1, b_2] = z^k$ in $\overline{G}_n$ for some $k \in \mathbf{F}_p$ and liftings b_i of the h_i in $\overline{G}_n$. By the choice of z made above, the images of the b_i should be linearly dependent in $V \cong G/Z(G)$, which means precisely that $H \cap Z(G) \neq \{1\}$. Now as $Z(G)$ is the image of $\Lambda^2 V$ in G, this implies that the inverse image $\overline{H}$ of H in $\overline{G}_n$ is cyclic modulo $\overline{H} \cap (\Lambda^2 V)$, and thus it is a commutative subgroup. Moreover, it is an $\mathbf{F}_p$-vector space, because $\overline{G}_n$ is of exponent p. But then the extension $1 \to \langle z \rangle \to \overline{H} \to H \to 1$ is an extension of $\mathbf{F}_p$-vector spaces and therefore a split extension. On the other hand, its class in $H^2(H, \mathbf{Z}/p\mathbf{Z})$ is precisely the image of $c(\overline{G}_n) \in H^2(G, \mathbf{Z}/p\mathbf{Z})$ by the restriction map to H, and we are done.

It remains to see that $\iota_*(c(\overline{G}_n))$ is a nontrivial class. Note that this class is represented by the extension

$$0 \to \mathbf{Q}/\mathbf{Z} \to E \to G \to 1$$

obtained as the pushout of the extension (6.13) by the map $\iota : \langle z \rangle \to \mathbf{Q}/\mathbf{Z}$. Were there a splitting $E \to \mathbf{Q}/\mathbf{Z}$, the composite map $\langle z \rangle \to \overline{G}_n \to E \to \mathbf{Q}/\mathbf{Z}$ would be 0, because z is in the commutator subgroup of $\overline{G}_n$ and $\mathbf{Q}/\mathbf{Z}$ is commutative. But this composite map is none but ι, a contradiction. $\square$

Remarks 6.7.4

1. For $n = 4$ the group G considered above was one of the first examples of Saltman [2]. Today we know that examples of groups G of order n with $H^2_B(G) \neq 0$ exist for all n divisible by p^5 for an odd prime p or by

2^6 (Hoshi–Kang–Kunyavskiĭ [1]) and that there are no smaller examples (Chu–Kang [1], Chu–Hu–Kang–Prokhorov [1]).

Bogomolov has also made a thorough study of the unramified Brauer group of invariant fields under actions of reductive algebraic groups, a topic which has interesting connections with geometric invariant theory. Besides the original papers (Bogomolov [1], [2]) one may profitably consult the survey of Colliot-Thélène and Sansuc [1].

2. One may ask whether the vanishing of the unramified Brauer group is a sufficient condition for the rationality of a variety. This is not the case: Colliot-Thélène and Ojanguren [1] gave examples of unirational but nonrational varieties with trivial unramified Brauer group. Moreover, Peyre [1] found a finite group G acting faithfully on a **C**-vector space V with $\mathrm{Br}_{\mathrm{nr}}\left(\mathbf{C}(V)^G\right) = 0$, but $\mathbf{C}(V)^G$ not purely transcendental over **C**.

In these examples, nonrationality is explained by the nonvanishing of an unramified cohomology group of degree 3. Using methods coming from the proof of the Milnor conjecture, Asok [1] has produced examples for all $i > 2$ of unirational but nonrational varieties where all unramified cohomology groups with $\mathbf{Z}/2\mathbf{Z}$-coefficients vanish up to degree i but not in degree $i + 1$.

3. A question closely related to the above is the famous *Noether problem*. The issue is the same as in Problem 6.6.3, except that the ground field is $k = \mathbf{Q}$. Emmy Noether's interest in the problem stemmed from its connection with the inverse Galois problem. Namely, it is a consequence of Hilbert's irreducibility theorem (see e.g. Serre [3], §10.1) that a positive answer to the problem for a given group G would yield an infinite family of Galois extensions of **Q** with group G obtained via specializations $t_i \mapsto a_i$. However, as opposed to the case of an algebraically closed ground field, the answer here may be negative even for cyclic G. Swan [1] and Voskresenskiĭ [1] found independently the first counterexample with $G = \mathbf{Z}/47\mathbf{Z}$; this is the smallest group of prime order yielding a counterexample. Later, Lenstra [1] found a counterexample with $G = \mathbf{Z}/8\mathbf{Z}$ and gave a necessary and sufficient condition for the answer to be positive in the case of a general commutative G. Saltman [1] found a new approach to the counterexample $G = \mathbf{Z}/8\mathbf{Z}$ by relating it to Wang's counterexample to the so-called Grunwald theorem in class field theory. See Swan [2] or Kersten [2] for nice surveys of the area including an account of Saltman's work. See also Garibaldi–Merkurjev–Serre [1] for a discussion from the point of view of cohomological invariants. Theorem 33.16 of this reference explains Saltman's approach by showing that in his counterexample a certain element in $\mathrm{Br}_{\mathrm{nr}}\left(\mathbf{Q}(V)^G\right)$ does not come from $\mathrm{Br}\left(\mathbf{Q}\right)$, and hence $\mathbf{Q}(V)^G$ cannot be purely transcendental by Corollary 6.6.7.

6.8 Residue maps with finite coefficients

This section and the next are of a technical nature; their results will be needed for our study of the cohomological symbol. Our purpose here is to define and study residue maps of the form

$$\partial_v^i : H^i(K, \mu_m^{\otimes j}) \to H^{i-1}(\kappa(v), \mu_m^{\otimes(j-1)}),$$

where K is a field equipped with a discrete valuation v with residue field $\kappa(v)$, m is an integer invertible in $\kappa(v)$ and i, j are positive integers (by convention we set $\mu_m^{\otimes 0} := \mathbf{Z}/m\mathbf{Z}$). This is a finite coefficient analogue of the residue map studied earlier, because for $j = 1$ we get maps $H^i(K, \mu_m) \to H^{i-1}(\kappa(v), \mathbf{Z}/m\mathbf{Z})$, i.e. instead of the multiplicative group we work with its m-torsion part.

The basis for our labours is the following construction in homological algebra.

Construction 6.8.1 Let G be a profinite group, and let H be a closed normal subgroup in G with $\mathrm{cd}(H) \leq 1$. We construct maps

$$\partial_i : H^i(G, A) \to H^{i-1}(G/H, H^1(H, A))$$

for all torsion G-modules A and all integers $i > 0$ as follows. Embed A into the co-induced module $M^G(A)$, and let C be the G-module fitting into the exact sequence

$$0 \to A \to M^G(A) \to C \to 0. \qquad (6.14)$$

Observe that here $H^j(H, C) = 0$ for all $j \geq 1$. Indeed, by Lemma 3.3.17 we have $H^j(H, M^G(A)) = 0$ for all $j \geq 1$, so that the long exact sequence in H-cohomology associated with (6.14) yields isomorphisms $H^j(H, C) \cong H^{j+1}(H, A)$ for all $j \geq 1$, but the latter groups are all trivial by assumption.

This shows that for $i > 2$ the assumptions of Proposition 3.3.19 (completed by Corollary 4.3.5) are satisfied, and therefore the inflation maps

$$\mathrm{Inf} : H^{i-1}(G/H, C^H) \to H^{i-1}(G, C)$$

are isomorphisms. We draw a similar conclusion for $i = 2$ from Proposition 3.3.16 of *loc. cit.* On the other hand, for $i \geq 2$ we get from the long exact sequence associated with (6.14) isomorphisms $H^{i-1}(G, C) \cong H^i(G, A)$, so finally isomorphisms

$$H^i(G, A) \cong H^{i-1}(G/H, C^H). \qquad (6.15)$$

But from the long exact sequence in H-cohomology coming from (6.14) we also obtain a map $C^H \to H^1(H, A)$, which is a morphism of G/H-modules by Lemma 3.3.15. Hence there are induced maps

$$H^{i-1}(G/H, C^H) \to H^{i-1}(G/H, H^1(H, A))$$

for all $i \geq 1$. Composing with the isomorphism (6.15) we thus obtain a construction of the maps ∂_i for $i > 1$. The case $i = 1$ was treated in Proposition 3.3.16 (in fact, it is just a restriction map).

If p is a fixed prime, and we only assume $cd_\ell(H) \leq 1$ for $\ell \neq p$, then the same construction works for prime-to-p torsion G-modules A.

A fundamental property of the maps ∂_i is the following.

Proposition 6.8.2 *The maps ∂_i fit into a functorial long exact sequence*

$$\ldots \to H^i(G/H, A^H) \overset{\text{Inf}}{\to} H^i(G, A) \overset{\partial_i}{\to} H^{i-1}(G/H, H^1(H, A)) \to H^{i+1}(G/H, A^H) \to \ldots$$

starting from $H^1(G, A)$.

Proof The beginning of the long exact sequence in H-cohomology coming from exact sequence (6.14) above reads

$$0 \to A^H \to M^G(A)^H \to C^H \to H^1(H, A) \to 0.$$

We may split this up into two short exact sequences

$$0 \to A^H \to M^G(A)^H \to T \to 0, \tag{6.16}$$
$$0 \to T \to C^H \to H^1(H, A) \to 0. \tag{6.17}$$

The long exact sequence in G/H-cohomology associated with (6.17) reads

$$\ldots \to H^{i-1}(G/H, T) \to H^{i-1}(G/H, C^H) \to H^{i-1}(G/H, H^1(H, A)) \to H^i(G, T) \to \ldots$$

Using Lemma 3.3.17 we see that $H^i(G/H, M^G(A)^H) = 0$ for all $i > 0$, hence the long exact sequence associated with (6.16) yields isomorphisms $H^i(G/H, A^H) \cong H^{i-1}(G/H, T)$ for all $i > 1$. Taking isomorphism (6.15) into account we may therefore identify the above long exact sequence with that of the proposition. The fact that the maps $H^i(G/H, A^H) \to H^i(G, A)$ are indeed the usual inflation maps follows from an easy compatibility between inflations and boundary maps in long exact sequences, which readers may check for themselves. $\square$

Remark 6.8.3 In the literature the maps ∂_i are usually obtained as edge morphisms of the Hochschild–Serre spectral sequence for group extensions, and the exact sequence of the above proposition results from the degeneration of the spectral sequence. It can be shown that the two constructions yield the same map.

The maps ∂_i enjoy the following compatibility property with respect to cup-products.

Lemma 6.8.4 *In the situation above the diagram*

$$
\begin{array}{ccccc}
H^p(G,A) & \times & H^q(G,B) & \xrightarrow{\ \cup\ } & H^{p+q}(G,A \otimes B) \\[1ex]
\downarrow \partial_p & & \uparrow \mathrm{Inf} & & \downarrow \partial_{p+q} \\[2ex]
H^{p-1}(G/H,H^1(H,A)) \times H^q(G/H,H^0(H,B)) & & \xrightarrow{\cup} & & H^{p+q-1}(G/H,H^1(H,A \otimes B))
\end{array}
$$

commutes for all continuous G-modules A and B. In other words, for $a \in H^p(G,A)$ and $b \in H^q(G/H, H^0(H,B))$ we have

$$\partial_{p+q}(a \cup \mathrm{Inf}(b)) = \partial_p(a) \cup b.$$

Proof Observe first that we have $M^G(A) \otimes_{\mathbf{Z}} B \cong M^G(A \otimes B)$. To establish this isomorphism, we may assume G finite, by compatibility of tensor products with direct limits. Then given $\phi : \mathbf{Z}[G] \to A$ sending $\sigma_i \in G$ to $a_i \in A$, we may define for each $b \in B$ a map $\phi^b : \mathbf{Z}[G] \to A \otimes B$ by sending σ_i to $a_i \otimes b$. This construction is bilinear, and defines the required isomorphism. Next, recall from the proof of Proposition 3.3.20 that for G finite the exact sequence

$$0 \to A \to M^G(A) \to C \to 0$$

is split as an exact sequence of abelian groups by the map $M^G(A) \to A$ given by $\phi \mapsto \phi(1)$. In the general case we also get a splitting by applying this fact to the finite quotients of G and passing to the direct limit.

Thus by tensoring with B over $\mathbf{Z}$ we obtain an exact sequence

$$0 \to A \otimes B \to M^G(A \otimes B) \to C \otimes B \to 0,$$

which we may use to construct the map ∂_{p+q} on the right as in Construction 6.8.1. The compatibility of cup-products with boundary maps (Proposition 3.4.8) and inflations (Proposition 3.4.10 (2)) then gives rise to the commutative diagram

$$H^p(G, A) \quad \times \quad H^q(G, B) \quad \xrightarrow{\cup} \quad H^{p+q}(G, A \otimes B)$$

$$\uparrow \cong \qquad\qquad \uparrow \text{Inf} \qquad\qquad \uparrow \cong$$

$$H^{p-1}(G/H, C^H) \quad \times \quad H^q(G/H, B^H) \quad \xrightarrow{\cup} \quad H^{p+q-1}(G/H, (C \otimes B)^H)$$

where the two unnamed vertical maps come from the isomorphism (6.15), itself defined as the composite of a boundary map and an inflation. The lemma now follows from the commutative diagram

$$H^{p-1}(G/H, C^H) \quad \times \quad H^q(G/H, B^H) \quad \xrightarrow{\cup} \quad H^{p+q-1}(G/H, (C \otimes B)^H)$$

$$\downarrow \qquad\qquad \downarrow \text{id} \qquad\qquad \downarrow$$

$$H^{p-1}(G/H, H^1(H, A)) \times H^q(G/H, H^0(H, B)) \xrightarrow{\cup} H^{p+q-1}(G/H, H^1(H, A \otimes B))$$

resulting from the functoriality of cup-products. $\qquad\qquad\square$

We now turn to the promised construction of residue maps.

Construction 6.8.5 Let K be a field complete with respect to a discrete valuation v, with residue field $\kappa(v)$. Choose a separable closure K_s of K, and write G and H for the Galois groups $\text{Gal}(K_s|K)$ and $\text{Gal}(K_s|K_{nr})$, respectively, where K_{nr} is the maximal unramified extension of K. Note that the ℓ-Sylow subgroups of H are isomorphic to $\mathbf{Z}_\ell$ for ℓ prime to $\text{char}(\kappa(v))$ by Proposition A.6.9 of the Appendix. Therefore $\text{cd}_\ell(H) \leq 1$ for such ℓ by Proposition 6.1.3 and Lemma 6.1.4, so for m prime to $\text{char}(\kappa(v))$ Construction 6.8.1 applied with $A = \mu_m^{\otimes j}$ and the above G and H yields maps

$$H^i(K, \mu_m^{\otimes j}) \to H^{i-1}(\kappa(v), H^1(K_{nr}, \mu_m^{\otimes j}))$$

for all $i, j > 0$. As H acts trivially on μ_m, we see that there is an isomorphism of G/H-modules

$$H^1(K_{nr}, \mu_m^{\otimes j}) \cong H^1(K_{nr}, \mu_m) \otimes \mu_m^{\otimes(j-1)}.$$

Now Kummer theory gives an isomorphism

$$H^1(K_{nr}, \mu_m) \cong K_{nr}^\times / K_{nr}^{\times m},$$

which is also G/H-equivariant according to Lemma 3.3.15. This may be composed with the (equally G/H-equivariant) valuation map

$$K_{nr}^\times / K_{nr}^{\times m} \to \mathbf{Z}/m\mathbf{Z}$$

sending t to 1. Putting the above together, we get a map

$$H^i(K, \mu_m^{\otimes j}) \to H^{i-1}(\kappa(v), \mu_m^{\otimes (j-1)}),$$

as required.

For a field K equipped with a discrete valuation v which is not necessarily complete, we first pass to the completion $\widehat{K}$ to which the above construction may be applied. For a separable closure $\widehat{K}_s$ of $\widehat{K}$ containing K_s the Galois group $\mathrm{Gal}(\widehat{K}_s|\widehat{K})$ may be identified with a subgroup of $\mathrm{Gal}(K_s|K)$ by Proposition A.6.4 (2) of the Appendix, whence a restriction map $H^i(K, \mu_m^{\otimes j}) \to H^i(\widehat{K}, \mu_m^{\otimes j})$. By composition we finally obtain a map

$$\partial_v^i : H^i(K, \mu_m^{\otimes j}) \to H^{i-1}(\kappa(v), \mu_m^{\otimes (j-1)})$$

as at the beginning of this section. This is the *residue map* with $\mu_m^{\otimes j}$-coefficients associated with v.

By means of the residue map we may define another useful map in Galois cohomology.

Construction 6.8.6 (Specialization maps) Consider first the case where K is complete with respect to a discrete valuation v, and let π be a local parameter. Denote by $(-\pi)$ the image of $-\pi$ by the Kummer map $K^\times \to H^1(K, \mu_m)$. Using the cup-product we may associate with each $a \in H^i(K, \mu_m^{\otimes j})$ the element $\partial_v^{i+1}((-\pi) \cup a)$ lying in $H^i(\kappa(v), \mu_m^{\otimes j})$. The choice of the minus sign may have an air of mystery at the moment, but will be justified in the next chapter (Remark 7.1.6 (1) and Corollary 7.5.2).

In this way we obtain a map

$$H^i(K, \mu_m^{\otimes j}) \to H^i(\kappa(v), \mu_m^{\otimes j}).$$

As in the construction of residue maps, when K is not assumed to be complete we first apply a restriction map to the completion, and then perform the preceding construction. We obtain a map

$$s_\pi^i : H^i(K, \mu_m^{\otimes j}) \to H^i(\kappa(v), \mu_m^{\otimes j}).$$

This is the i-th *specialization map* associated with π. It depends on the choice of the parameter π.

For complete fields K the specialization map enjoys the following crucial property.

Proposition 6.8.7 *For a complete discretely valued field K with residue field $\kappa(v)$ the composite maps*

$$H^i(\kappa(v), \mu_m^{\otimes j}) \xrightarrow{\text{Inf}} H^i(K, \mu_m^{\otimes j}) \xrightarrow{s_\pi^i} H^i(\kappa(v), \mu_m^{\otimes j})$$

are identity maps for all $i, j > 0$ and all choices of π.

Proof Apply Lemma 6.8.4 with G and H as in Construction 6.8.5, $A = \mu_m$, $B = \mu_m^{\otimes j}$, $p = 1$ and $q = i$. For $b \in H^i(\kappa(v), \mu_m^{\otimes j})$ and $a = (-\pi)$ the lemma then yields

$$s_\pi^i(\text{Inf}(b)) = \partial_v^1((-\pi)) \cup b.$$

But $\partial_v^1((-\pi)) = 1$, because ∂_v^1 becomes the mod m valuation map via the Kummer isomorphism and $-\pi$ has valuation 1. Hence cup-product with $\partial_v^1((-\pi))$ is the identity, and the proposition is proven. □

Corollary 6.8.8 *For K complete the sequences*

$$0 \to H^i(\kappa(v), \mu_m^{\otimes j}) \xrightarrow{\text{Inf}} H^i(K, \mu_m^{\otimes j}) \xrightarrow{\partial^i} H^{i-1}(\kappa(v), \mu_m^{\otimes(j-1)}) \to 0$$

are split exact for all $i, j > 0$.

Proof Apply Proposition 6.8.2 with G, H and A as in Construction 6.8.5. By the previous proposition, the maps s_π^i split up the resulting long exact sequence into a collection of short exact sequences as in the corollary. □

To close this section, we relate the residue maps defined above to the residue maps $r_v^i : H^i(K, K_s^\times) \to H^{i-1}(\kappa(v), \mathbf{Q}/\mathbf{Z})$ encountered in Section 6.3.

Proposition 6.8.9 *Assume that K is a complete discretely valued field with perfect residue field $\kappa(v)$. The residue maps ∂_v^i constructed in this section are compatible with those of Section 6.3 via the commutative diagram*

$$
\begin{array}{ccc}
H^i(K, \mu_m) & \xrightarrow{\partial_v^i} & H^{i-1}(\kappa(v), \mathbf{Z}/m\mathbf{Z}) \\
\downarrow & & \downarrow- \\
H^i(K, K_s^\times) & \xrightarrow{r_v^i} & H^{i-1}(\kappa(v), \mathbf{Q}/\mathbf{Z})
\end{array}
$$

where the left vertical map is induced by the natural inclusion $\mu_m \subset K_s^\times$ and the right one by the opposite *of the natural map $\mathbf{Z}/m\mathbf{Z} \to \mathbf{Q}/\mathbf{Z}$.*

Proof Recall from Section 6.3 that the map r_v^i has a splitting given by the composite

$$H^{i-1}(\kappa(v), \mathbf{Q}/\mathbf{Z}) \xrightarrow{\text{Inf}} H^{i-1}(K, \mathbf{Q}/\mathbf{Z}) \xrightarrow{\sim} H^i(K, \mathbf{Z}) \to H^i(K, K_s^\times)$$

where the last map is cup-product with a local parameter $\pi \in H^0(K, K_s^\times)$. On the other hand, the same argument as in the proof of Proposition 6.8.7 shows that ∂_v^i has a splitting given by $b \mapsto (\pi) \cup \text{Inf}(b)$, where (π) is the image of π by the Kummer coboundary map $H^0(K, K_s^\times) \to H^1(K, \mu_m)$. Thus it suffices to check the commutativity of the diagram

$$
\begin{array}{ccc}
H^i(K, \mu_m) & \xleftarrow{\ (\pi)\cup\text{Inf}\ } & H^{i-1}(\kappa(v), \mathbf{Z}/m\mathbf{Z}) \\
\downarrow & & \downarrow{\scriptstyle -\delta} \\
H^i(K, K_s^\times) & \xleftarrow{\ \pi\cup\text{Inf}\ } & H^i(\kappa(v), \mathbf{Z})
\end{array}
$$

where δ is a coboundary map coming from the long exact sequence associated with $0 \to \mathbf{Z} \xrightarrow{m} \mathbf{Z} \to \mathbf{Z}/m\mathbf{Z} \to 0$. This in turn follows from the compatibility between cup-products and coboundary maps proven in Proposition 3.4.9. □

6.9 The Faddeev sequence with finite coefficients

We now come to the main result concerning our freshly constructed residue maps, namely the analogue of Faddeev's theorem with finite coefficients.

Theorem 6.9.1 *Let k be a field, $\mathbf{P}^1$ the projective line over k and K its function field. For each $i, j > 0$ and m invertible in k the sequence*

$$0 \to H^i(k, \mu_m^{\otimes j}) \xrightarrow{\text{Inf}} H^i(K, \mu_m^{\otimes j}) \xrightarrow{\oplus \partial_P^i} \bigoplus_{P \in \mathbf{P}_0^1} H^{i-1}(\kappa(P), \mu_m^{\otimes(j-1)}) \xrightarrow{\Sigma \text{Cor}_P}$$

$$\to H^{i-1}(k, \mu_m^{\otimes(j-1)}) \to 0$$

is exact.

Here ∂_P^i is the residue map induced by the discrete valuation of K corresponding to the closed point P.

Remark 6.9.2 Note that in contrast to Theorem 6.4.4 we did not assume here that k is perfect. Therefore we have to explain what we mean by the corestriction maps Cor_P in characteristic $p > 0$. For a finite separable extension $F'|F$ of fields, we define the associated corestriction map as before, using

Galois theory. For a purely inseparable extension $F''|F$ of degree p^r we define the corestriction map to be multiplication by p^r. In the case of a general finite extension $F''|F$, we let $F'|F$ be the maximal separable subextension and define the corestriction to be the composite of the above two maps. This definition works for Galois cohomology with coefficients in torsion modules having no nontrivial elements of order p. In the presence of p-torsion much more sophisticated constructions should be used (or one should work with a different cohomology theory; compare Remark 6.1.11).

Proof of Theorem 6.9.1 Assume first that k is a perfect field. In this case the proof follows a pattern similar to that of Theorem 6.4.4, with some local differences. First, using the isomorphism $\mathrm{Pic}\,(\mathbf{P}_k^1) \cong \mathbf{Z}$ we consider the exact sequence of G-modules

$$0 \to (K\bar{k})^\times/\bar{k}^\times \longrightarrow \mathrm{Div}(\mathbf{P}_{\bar{k}}^1) \longrightarrow \mathbf{Z} \to 0,$$

which has a G-equivariant splitting coming from a k-rational point of $\mathbf{P}^1$. Therefore after tensoring with $\mu_m^{\otimes(j-1)}$ we still get a split exact sequence

$$0 \to \left((K\bar{k})^\times/\bar{k}^\times\right) \otimes \mu_m^{\otimes(j-1)} \longrightarrow \mathrm{Div}(\mathbf{P}_{\bar{k}}^1) \otimes \mu_m^{\otimes(j-1)} \longrightarrow \mu_m^{\otimes(j-1)} \to 0.$$

For each $i > 0$ this induces short exact sequences

$$0 \longrightarrow H^{i-1}(k, \left((K\bar{k})^\times/\bar{k}^\times\right) \otimes \mu_m^{\otimes(j-1)}) \longrightarrow$$

$$\longrightarrow H^{i-1}(k, \mathrm{Div}(\mathbf{P}_{\bar{k}}^1) \otimes \mu_m^{\otimes(j-1)}) \xrightarrow{\alpha} H^{i-1}(k, \mu_m^{\otimes(j-1)}) \longrightarrow 0.$$

Now in exactly the same way as in the proof of Theorem 6.4.3 we identify the map α to the map $\Sigma \mathrm{Cor}_P$ of the theorem. Furthermore, since $\bar{k}^\times$ is an m-divisible group, the tensor product $\bar{k}^\times \otimes \mu_m^{\otimes(j-1)}$ vanishes, so that tensoring the exact sequence

$$0 \to \bar{k}^\times \to (K\bar{k})^\times \to (K\bar{k})^\times/\bar{k}^\times \to 0$$

by $\mu_m^{\otimes(j-1)}$ yields an isomorphism

$$(K\bar{k})^\times \otimes \mu_m^{\otimes(j-1)} \cong \left((K\bar{k})^\times/k^\times\right) \otimes \mu_m^{\otimes(j-1)}.$$

We may therefore make this replacement in the exact sequence above and thus reduce to identifying the group $H^{i-1}(k, (K\bar{k})^\times \otimes \mu_m^{\otimes(j-1)})$ with the cokernel of the inflation map

$$\mathrm{Inf}\,:\, H^i(k, \mu_m^{\otimes j}) \to H^i(K, \mu_m^{\otimes j}). \tag{6.18}$$

To this end, we use the long exact sequence of Proposition 6.8.2 with $G = \mathrm{Gal}\,(\overline{K}|K)$, $H = \mathrm{Gal}\,(\overline{K}|K\overline{k})$ and $A = \mu_m^{\otimes j}$. Here $\mathrm{cd}(H) \leq 1$ by Tsen's theorem, so the proposition applies and yields a long exact sequence

$$\cdots \to H^i(k, \mu_m^{\otimes j}) \xrightarrow{\mathrm{Inf}} H^i(K, \mu_m^{\otimes j}) \to H^{i-1}(k, (K\overline{k})^\times \otimes \mu_m^{\otimes(j-1)}) \to \cdots$$

after making the identification $H^1(K\overline{k}, \mu_m^{\otimes j}) \cong (K\overline{k})^\times \otimes \mu_m^{\otimes(j-1)}$ as in Construction 6.8.5 above. Now just like in the proof of Theorem 6.4.4, the point is that the inflation maps (6.18) are injective for all $i > 0$. To see this, it is enough to show injectivity of the composite maps $H^i(k, \mu_m^{\otimes j}) \to H^i(k((t)), \mu_m^{\otimes j})$ obtained via the embedding $K \hookrightarrow k((t))$. But these maps are injective, because the specialization map yields a section for them by virtue of Proposition 6.8.7. Finally, the identification of the resulting maps

$$H^i(K, \mu_m^{\otimes j}) \to \bigoplus_{P \in \mathbf{P}_0^1} H^{i-1}(\kappa(P), \mu_m^{\otimes(j-1)})$$

with a direct sum of residue maps follows by an argument similar to that in Proposition 6.4.2.

It remains to reduce the case of a general base field k of characteristic $p > 0$ to the perfect case. To do so, consider the *perfect closure* k_{p^∞} of k (recall that this is the perfect field obtained by adjoining all p-power roots of elements in k). Given a separable closure k_s of k, the composite $k_{p^\infty} k_s$ is a separable closure of k_{p^∞}, as the extension $k_{p^\infty}|k$ is purely inseparable. In this way we may identify the absolute Galois group of k with that of k_{p^∞}, and similar considerations apply to the absolute Galois groups of $k(t)$ and $k_{p^\infty}(t)$. As $\mu_m \subset k_s$ for m prime to p, the action of these groups on the modules $\mu_m^{\otimes j}$ is the same, so we get natural isomorphisms on the corresponding Galois cohomology groups. Whence the isomorphic vertical maps in the commutative diagram

$$\begin{array}{ccc} H^i(k, \mu_m^{\otimes j}) & \xrightarrow{\ \mathrm{Inf}\ } & H^i(K, \mu_m^{\otimes j}) \\[2pt] \cong \Big\downarrow & & \Big\downarrow \cong \\[2pt] H^i(k_{p^\infty}, \mu_m^{\otimes j}) & \xrightarrow{\ \mathrm{Inf}\ } & H^i(Kk_{p^\infty}, \mu_m^{\otimes j}). \end{array}$$

Next, consider a closed point P of $\mathbf{P}_k^1 \setminus \{\infty\}$. It corresponds to an irreducible polynomial $f \in k[t]$, which becomes the p^r-th power of an irreducible polynomial in $k_{p^\infty}[t]$, where p^r is the inseparability degree of the extension $\kappa(P)|k$. This shows that there is a unique closed point P' of $\mathbf{P}_{k_{p^\infty}}^1$ lying above P', with

$$[\kappa(P') : k_{p^\infty}] = p^{-r}[\kappa(P) : k]. \tag{6.19}$$

Therefore we have a commutative diagram with isomorphic vertical maps

$$
\begin{array}{ccccc}
H^i(K,\mu_m^{\otimes j}) & \xrightarrow{\;\oplus\partial_P^i\;} & \displaystyle\bigoplus_{P\in\mathbf{P}^1_{k,0}} H^{i-1}(\kappa(P),\mu_m^{\otimes(j-1)}) & \xrightarrow{\;\Sigma\,\mathrm{Cor}_P\;} & H^{i-1}(k,\mu_m^{\otimes(j-1)}) \\
\cong\Big\downarrow & & \cong\Big\downarrow & & \cong\Big\downarrow \\
H^i(Kk_{p\infty},\mu_m^{\otimes j}) & \xrightarrow{\;\oplus\partial_P^i\;} & \displaystyle\bigoplus_{P\in\mathbf{P}^1_{k_{p\infty},0}} H^{i-1}(\kappa(P'),\mu_m^{\otimes(j-1)}) & \xrightarrow{\;\Sigma\,\mathrm{Cor}_P\;} & H^{i-1}(k_{p\infty},\mu_m^{\otimes(j-1)}).
\end{array}
$$

Here the left square commutes by the construction of residue maps and our remarks on the Galois groups of K and $Kk_{p\infty}$. Commutativity of the right square follows from our definition of corestriction maps in Remark 6.9.2 and the formula (6.19). This completes the identification of the exact sequence over k with that over $k_{p\infty}$. □

Observe that for the point at infinity ∞ of $\mathbf{P}^1_k$ we have $\kappa(\infty) = k$, and the corestriction map $\mathrm{Cor}_\infty : H^{i-1}(\kappa(\infty),\mu_m^{\otimes j}) \to H^{i-1}(k,\mu_m^{\otimes j})$ is the identity map. Hence we get:

Corollary 6.9.3 *In the situation of the theorem there is an exact sequence*

$$
0 \to H^i(k,\mu_m^{\otimes j}) \xrightarrow{\;\mathrm{Inf}\;} H^i(K,\mu_m^{\otimes j}) \xrightarrow{\;\oplus\partial_P^i\;} \bigoplus_{P\in\mathbf{P}^1_0\setminus\{\infty\}} H^{i-1}(\kappa(P),\mu_m^{\otimes(j-1)}) \to 0
$$

split by the specialization map $s_{t^{-1}}^i : H^i(K,\mu_m^{\otimes j}) \to H^i(k,\mu_m^{\otimes j})$ associated with the local parameter t^{-1} at ∞.

Proof The exact sequence results from that of the theorem, and the statement about the splitting from Proposition 6.8.7 (after embedding K into the Laurent series field $k((t^{-1}))$). □

The split exact sequence of the corollary allows us to define maps

$$
\psi_P^i : H^{i-1}(\kappa(P),\mu_m^{\otimes(j-1)}) \to H^i(K,\mu_m^{\otimes j})
$$

satisfying $\partial_P^i \circ \psi_P^i = \mathrm{id}$ for each closed point P of $\mathbf{P}^1_k$, which we may call *coresidue maps*. We then get the following useful description of corestrictions.

Corollary 6.9.4 *The corestriction maps*

$$
\mathrm{Cor}_P : H^{i-1}(\kappa(P),\mu_m^{\otimes(j-1)}) \to H^{i-1}(k,\mu_m^{\otimes(j-1)})
$$

satisfy the formula

$$
\mathrm{Cor}_P = -\partial_\infty^i \circ \psi_P^i,
$$

where ∂_∞^i is the residue map associated with the point ∞.

Proof Let α be an element of $H^{i-1}(\kappa(P), \mu_m^{\otimes(j-1)})$. In the exact sequence of Theorem 6.9.1, consider the element of $\bigoplus H^{i-1}(\kappa(P), \mu_m^{\otimes(j-1)})$ given by α in the component indexed by P, $-\mathrm{Cor}_P(\alpha)$ in the component indexed by ∞, and 0 elsewhere. Since Cor_∞ is the identity map, this element maps to 0 in $H^{i-1}(k, \mu_m^{\otimes(j-1)})$ by the sum of corestriction maps, hence it is the residue of some element in $H^i(\kappa(P), \mu_m^{\otimes j})$, which is none but $\psi_P^i(\alpha)$. This proves the corollary. $\square$

Remarks 6.9.5

1. The results of this section generalize in a straightforward way to the case of an arbitrary $\mathbf{Z}/m\mathbf{Z}$-module A equipped with a $\mathrm{Gal}\,(k_s|k)$-action instead of $\mu_m^{\otimes j}$ (still assuming m prime to the characteristic). The role of the $\mathrm{Gal}\,(k_s|k)$-modules $\mu_m^{\otimes(j-1)}$ is then played by the groups $\mathrm{Hom}(\mu_m, A)$ equipped with the usual Galois action.

2. It is again possible to obtain the results of this section via methods of étale cohomology, namely using the localization theory and the so-called purity isomorphisms (see Milne [2]). But it is not obvious to check that the residue maps in the two theories are the same up to a sign.

Exercises

1. Let G be a profinite group of finite cohomological dimension, and let H be an open subgroup of G. Prove that $\mathrm{cd}_p(G) = \mathrm{cd}_p(H)$ for all primes p.
   [*Hint:* Show that the corestriction map $\mathrm{Cor} : H^n(H, A) \to H^n(G, A)$ is surjective for $n = \mathrm{cd}(G)$ and a torsion G-module A.]

2. Let G be a finite group. Show that $\mathrm{cd}_p(G) = \infty$ if p divides the order of G, and $\mathrm{cd}_p(G) = 0$ otherwise. [*Hint:* Use the previous exercise.]

3. (Kato, Kuzumaki) Let k be a perfect field such that the absolute Galois group $\Gamma = \mathrm{Gal}\,(k_s|k)$ has no nontrivial elements of finite order, and let X be a Severi–Brauer variety over k. Prove that for all primes p not dividing $\mathrm{char}(k)$ the product of restriction maps

$$H^1(k, \mathbf{Z}/p\mathbf{Z}) \to \prod_{P \in X_0} H^1(\kappa(P), \mathbf{Z}/p\mathbf{Z})$$

is injective, where the sum is over all closed points of X. [*Hint:* By the assumption on Γ, the subgroup topologically generated by an element $\sigma \in \Gamma$ is isomorphic to $\hat{\mathbf{Z}}$, and hence its fixed field has trivial Brauer group.]

[*Remark*: The condition on Γ is not very restrictive, for by Chapter 4, Exercise 6 fields of characteristic 0 having no ordered field structure enjoy this property.]

4. (a) Show that finite fields satisfy the C_1' property (Remark 6.2.2 (2)).
   (b) Same question for the function field of a curve over an algebraically closed field.

5. Let $L|K$ be a purely inseparable extension of fields of characteristic $p > 0$. Prove that the natural map $\mathrm{Br}\,(K) \to \mathrm{Br}\,(L)$ is surjective, and its kernel is a p-primary torsion group. [*Hint:* Take a separable closure K_s of K, exploit the exact sequence $1 \to K_s^\times \to (LK_s)^\times \to (LK_s)^\times/K_s^\times \to 1$ and use $\mathrm{cd}_p(K) \leq 1$.]

6. Let k be a field and p a prime invertible in k. Let $\chi \in H^1(k, \mathbf{Z}/p\mathbf{Z})$ be a surjective character, and $K|k$ the associated cyclic extension. Let $A|k$ be a central simple algebra. Over the rational function field $k(t)$ consider the $k(t)$-algebra $A_{k(t)} := A \otimes_k k(t)$ and the cyclic $k(t)$-algebra (χ, t). Finally, define $B := A_{k(t)} \otimes_{k(t)} (\chi, t)$ and $\widehat{B} := B \otimes_{k(t)} k((t))$.

   (a) Show that $\mathrm{ind}_{k((t))}(\widehat{B})$ divides $p\,\mathrm{ind}_K(A \otimes_k K)$. [*Hint:* use Corollary 4.5.4.]
   (b) Let $L|k$ be a field extension such that $L((t))|k((t))$ splits $\widehat{B}$. Show that L contains K. [*Hint:* Use Corollary 6.3.5.]
   (c) Show that $\mathrm{ind}_{k((t))}(\widehat{B}) = p\,\mathrm{ind}_K(A \otimes_k K)$ and $\mathrm{ind}_{k(t)}(B) = p\,\mathrm{ind}_K(A \otimes_k K)$.
   (d) Conclude that $A \otimes_k K$ is a division algebra if and only if B is a division algebra.

7. (suggested by Colliot-Thélène) Let $F = k(x, y)$ be the rational function field in the two indeterminates x, y, and let a, b be elements of $k^\times$.
   Prove that the biquaternion algebra $(a, x) \otimes_F (b, y)$ over F is a division algebra if and only if the images of the elements a and b in the $\mathbf{F}_2$-vector space $k^\times/k^{\times 2}$ are linearly independent. [*Hint:* Use the previous exercise.]

   [*Remark:* Generalizing the technique of this exercise one may construct algebras of period 2 and index 2^d over the purely transcendental extension $\mathbf{Q}(x_1, \ldots, x_d)$ for all $d > 1$.]

8. Let C be a smooth projective conic over a perfect field k with $C(k) = \emptyset$. Show that there is an exact sequence

   $$0 \to \mathbf{Z}/2\mathbf{Z} \to \mathrm{Br}\,(k) \to \mathrm{Br}\,(k(C)) \xrightarrow{\oplus r_P} \bigoplus_{P \in C_0} H^1(\kappa(P), \mathbf{Q}/\mathbf{Z}) \xrightarrow{\Sigma\,\mathrm{Cor}_P} H^1(k, \mathbf{Q}/\mathbf{Z}).$$

   What can you say about the cokernel of the last map?

9. Let K be the function field of a smooth projective curve over a perfect field k. Let (χ, b) be a cyclic algebra over K, where $b \in K$ and χ defines a degree m cyclic Galois extension of k with group G. We view χ as an element of $H^1(G, \mathbf{Z}/m\mathbf{Z})$.

   (a) For a closed point P of C, show that the residue map r_P is given by

   $$r_P((\chi, b)) = \mathrm{Res}_{G_P}^G(\chi) \cup v_P(b) \in H^1(G_P, \mathbf{Z}/m\mathbf{Z}),$$

   where G_P is the stabilizer of P in G, and $v_P(b)$ is viewed as an element of $H^0(G_P, \mathbf{Z})$.
   (b) Assuming moreover that k is finite, deduce a formula of Hasse:

   $$r_P((\chi, b)) = \frac{[\kappa(P) : k]\, v_P(b)}{m} \in \mathbf{Q}/\mathbf{Z}.$$

(c) Still assuming k finite, show that the Residue Theorem for (χ, b) is equivalent to the formula $\deg(\operatorname{div}(b)) = 0$.

[*Remark*: This exercise gives some hint about the origin of the name of the Residue Theorem, because the formula $\deg(\operatorname{div}(f)) = 0$ for an algebraic function f is equivalent (in characteristic 0) to the fact that the sum of the residues of the logarithmic differential form $f^{-1}df$ equals 0.]

10. Consider the affine surface X of equation $x^3 - x = y^2 + z^2$ over the field $\mathbf{R}$ of real numbers.

    (a) Show that X is unirational over $\mathbf{R}$. [*Hint*: Find an extension $\mathbf{R}(z)|\mathbf{R}(x)$ which splits the quaternion algebra $(-1, x^3 - x)$ over $\mathbf{R}(x)$, and use Proposition 1.3.2.]
    (b) Show that X is not rational by examining $\mathrm{Br}_{\mathrm{nr}}(k(X))$. [*Hint*: Consider the class of the quaternion algebra $(-1, x)$.]

11. ('No-name lemma' for finite groups) Let G be a finite group, and let V and W be vector spaces over a field k equipped with a faithful linear action of G, of dimensions n and m, respectively.

    (a) Prove that $k(V \oplus W)^G \cong k(V)^G(t_1, \ldots, t_m)$ for some independent variables t_i. [*Hint*: Apply Speiser's lemma to the extension $k(V)|k(V)^G$ and the vector space $W \otimes_k k(V)$.]
    (b) Conclude that $k(V)^G(t_1, \ldots, t_m) \cong k(W)^G(u_1, \ldots, u_n)$ for some independent variables t_i and u_j, and hence $\mathrm{Br}_{\mathrm{nr}}(k(V)^G) \cong \mathrm{Br}_{\mathrm{nr}}(k(W)^G)$.

12. This exercise gives another proof of the Steinberg relation for Galois cohomology by using Theorem 6.9.1. Let k be a field, m an integer invertible in k, and $k(t)$ the rational function field.

    (a) Verify the relation $(t) \cup (1 - t) = 0$ in $H^2(k(t), \mu_m^{\otimes 2})$ by calculating the residues of both sides and specializing at 0.
    (b) Given $a \in k^\times$, $a \neq 0, 1$, deduce by specialization that $(a) \cup (1 - a) = 0$ in $H^2(k, \mu_m^{\otimes 2})$.

7

Milnor K-theory

In this chapter we study the Milnor K-groups introduced in Chapter 4. There are two basic constructions in the theory: that of *tame symbols*, which are analogues of the residue maps in cohomology, and *norm maps* that generalize the field norm $N_{K|k} : K^\times \to k^\times$ for a finite extension $K|k$ to higher K-groups. Of these the first is relatively easy to construct, but showing the well-definedness of the second involves some rather intricate checking. This foreshadows that the chapter will be quite technical, but nevertheless it contains a number of interesting results. Among these, we mention Weil's reciprocity law for the tame symbol over the function field of a curve, a reciprocity law of Rosset and Tate, and considerations of Bloch and Tate about the Bloch–Kato conjecture.

Most of the material in this chapter stems from the three classic papers of Milnor [1], Bass–Tate [1] and Tate [4]. Kato's theorem on the well-definedness of the norm map appears in the second part of his treatise on the class field theory of higher dimensional local fields (Kato [1]), with a sketch of the proof.

7.1 The tame symbol

Recall that we have defined the Milnor K-groups $K_n^M(k)$ attached to a field k as the quotient of the n-th tensor power $(k^\times)^{\otimes n}$ of the multiplicative group of k by the subgroup generated by those elements $a_1 \otimes \cdots \otimes a_n$ for which $a_i + a_j = 1$ for some $1 \leq i < j \leq n$. Thus $K_0^M(k) = \mathbf{Z}$ and $K_1^M(k) = k^\times$. Elements of $K_n^M(k)$ are called *symbols*; we write $\{a_1, \ldots, a_n\}$ for the image of $a_1 \otimes \cdots \otimes a_n$ in $K_n^M(k)$. The relation $a_i + a_j = 1$ will be often referred to as the *Steinberg relation*.

Milnor K-groups are functorial with respect to field extensions: given an inclusion $\phi : k \subset K$, there is a natural map $i_{K|k} : K_n^M(k) \to K_n^M(K)$ induced by ϕ. Given $\alpha \in K_n^M(K)$, we shall often abbreviate $i_{K|k}(\alpha)$ by α_K.

There is also a natural product structure

$$K_n^M(k) \times K_m^M(k) \to K_{n+m}^M(k), \quad (\alpha, \beta) \mapsto \{\alpha, \beta\} \qquad (7.1)$$

coming from the tensor product pairing $(k^\times)^{\otimes n} \times (k^\times)^{\otimes m} \to (k^\times)^{\otimes n+m}$ which obviously preserves the Steinberg relation. This product operation equips the direct sum

$$K_*^M(k) = \bigoplus_{n \geq 0} K_n^M(k)$$

with the structure of a graded ring indexed by the nonnegative integers. The ring $K_*^M(k)$ is commutative in the graded sense:

Proposition 7.1.1 *The product operation (7.1) is graded-commutative, i.e. it satisfies*

$$\{\alpha, \beta\} = (-1)^{mn}\{\beta, \alpha\}$$

for $\alpha \in K_n^M(k)$, $\beta \in K_m^M(k)$.

For the proof we first establish an easy lemma:

Lemma 7.1.2 *The group $K_2^M(k)$ satisfies the relations*

$$\{x, -x\} = 0 \quad \text{and} \quad \{x, x\} = \{x, -1\}.$$

Proof For the first relation, we compute in $K_2^M(k)$

$$\{x, -x\} + \{x, -(1-x)x^{-1}\} = \{x, 1-x\} = 0,$$

and so

$$\{x, -x\} = -\{x, -(1-x)x^{-1}\} = -\{x, 1-x^{-1}\} = \{x^{-1}, 1-x^{-1}\} = 0.$$

The second one follows by bilinearity. $\square$

Proof of Proposition 7.1.1 By the previous lemma, in $K_2^M(k)$ we have the equalities

$$0 = \{xy, -xy\} = \{x, -x\} + \{x, y\} + \{y, x\} + \{y, -y\} = \{x, y\} + \{y, x\},$$

which takes care of the case $n = m = 1$. The proposition follows from this by a straightforward induction. $\square$

These basic facts are already sufficient for calculating the following example.

Example 7.1.3 For a finite field $\mathbf{F}$ the groups $K_n^M(\mathbf{F})$ are trivial for all $n > 1$.

To see this it is enough to treat the case $n = 2$. Writing ω for a generator of the cyclic group $\mathbf{F}^\times$, we see from bilinearity of symbols that it suffices to show $\{\omega, \omega\} = 0$. By Lemma 7.1.2 this element equals $\{\omega, -1\}$ and hence it has order at most 2. We show that it is also annihilated by an odd integer, which will prove the claim. If $\mathbf{F}$ has order 2^m for some m, we have $0 = \{1, \omega\} = \{\omega^{2^m - 1}, \omega\} = (2^m - 1)\{\omega, \omega\}$, and we are done. If $\mathbf{F}$ has odd order, then the same counting argument as in Example 1.3.6 shows that we may find elements $a, b \in \mathbf{F}^\times$ that are *not* squares in $\mathbf{F}$ which satisfy $a + b = 1$. But then $a = \omega^k$, $b = \omega^l$ for some odd integers k, l and hence $0 = \{a, b\} = kl\{\omega, \omega\}$, so we are done again.

As we have seen in Chapter 6, a fundamental tool for studying the Galois cohomology of discrete valuation fields is furnished by the residue maps. We now construct their analogue for Milnor K-theory; the construction will at the same time yield specialization maps for K-groups.

Let K be a field equipped with a discrete valuation $v : K^\times \to \mathbf{Z}$. Denote by A the associated discrete valuation ring and by κ its residue field. Once a local parameter π (i.e. an element with $v(\pi) = 1$) is fixed, each element $x \in K^\times$ can be uniquely written as a product $u\pi^i$ for some unit u of A and integer i. From this it follows by bilinearity and graded-commutativity of symbols that the groups $K_n^M(K)$ are generated by symbols of the form $\{\pi, u_2, \ldots, u_n\}$ and $\{u_1, \ldots, u_n\}$, where the u_i are units in A.

Proposition 7.1.4 *For each $n \geq 1$ there exists a unique homomorphism*

$$\partial^M : K_n^M(K) \to K_{n-1}^M(\kappa)$$

satisfying

$$\partial^M(\{\pi, u_2, \ldots, u_n\}) = \{\bar{u}_2, \cdots, \bar{u}_n\} \qquad (7.2)$$

for all local parameters π and all $(n-1)$-tuples $(u_2, \ldots, u_n)$ of units of A, where $\bar{u}_i$ denotes the image of u_i in κ.

Moreover, once a local parameter π is fixed, there is a unique homomorphism

$$s_\pi^M : K_n^M(K) \to K_n^M(\kappa)$$

with the property

$$s_\pi^M(\{\pi^{i_1}u_1,\cdots,\pi^{i_n}u_n\}) = \{\overline{u}_1\cdots,\overline{u}_n\} \tag{7.3}$$

for all n-tuples of integers $(i_1,\ldots,i_n)$ and units $(u_1,\ldots,u_n)$ of A.

The map ∂^M is called the *tame symbol* or the *residue map* for Milnor K-theory; the maps s_π^M are called *specialization maps*. We stress the fact that the s_π^M depend on the choice of π, whereas ∂^M does not, as seen from its definition.

Proof Unicity for s_π^M is obvious, and that of ∂^M follows from the above remark on generators of $K_n^M(K)$, in view of the fact that a symbol of the form $\{u_1,\ldots,u_n\}$ can be written as a difference $\{\pi u_1, u_2,\ldots,u_n\} - \{\pi, u_2,\ldots,u_n\}$ with local parameters π and πu_1, and hence it must be annihilated by ∂^M.

We prove existence simultaneously for ∂^M and the s_π^M via a construction due to Serre. Consider the free graded-commutative $K_*^M(\kappa)$-algebra $K_*^M(\kappa)[x]$ on one generator x of degree 1. By definition, its elements can be identified with polynomials with coefficients in $K_*^M(\kappa)$, but the multiplication is determined by $\alpha x = -x\alpha$ for $\alpha \in K_*^M(\kappa)$. Now take the quotient $K_*^M(\kappa)[\xi]$ of $K_*^M(\kappa)[x]$ by the ideal $(x^2 - \{-1\}x)$ where $\{-1\}$ is regarded as a symbol in $K_1(\kappa)$. The image ξ of x in the quotient satisfies $\xi^2 = \{-1\}\xi$. The ring $K_*^M(\kappa)[\xi]$ has a natural grading in which ξ has degree 1: one has

$$K_*^M(\kappa)[\xi] = \bigoplus_{n\geq 0} L_n,$$

where $L_n = K_n^M(\kappa) \oplus K_{n-1}^M(\kappa)\xi$ for $n > 0$ and $L_0 = K_0^M(\kappa) = \mathbf{Z}$.

Now fix a local parameter π and consider the group homomorphism

$$d_\pi : K^\times \to L_1 = \kappa^\times \oplus \xi\mathbf{Z}$$

given by $\pi^i u \mapsto (\overline{u}, \xi i)$. Taking tensor powers and using the product structure in $K_*^M(\kappa)[\xi]$, we get maps

$$d_\pi^{\otimes n} : (K^\times)^{\otimes n} \to L_n = K_n^M(\kappa) \oplus K_{n-1}^M(\kappa)\xi.$$

Denoting by $\pi_1 : L_n \to K_n^M(\kappa)$ and $\pi_2 : L_n \to K_{n-1}^M(\kappa)$ the natural projections, put

$$\partial^M := \pi_2 \circ d_\pi^{\otimes n} \quad \text{and} \quad s_\pi^M := \pi_1 \circ d_\pi^{\otimes n}.$$

One sees immediately that these maps satisfy the properties (7.2) and (7.3). Therefore the construction will be complete if we show that $d_\pi^{\otimes n}$ factors through $K_n^M(K)$, for then so do ∂^M and s_π^M.

Concerning our claim about $d_\pi^{\otimes n}$, it is enough to establish the Steinberg relation $d_\pi(x)d_\pi(1-x) = 0$ in L_2. To do so, note first that the multiplication map $L_1 \times L_1 \to L_2$ is given by

$$(x, \xi i)(y, \xi j) = (\{x, y\}, \{(-1)^{ij} x^j y^i\}\xi), \tag{7.4}$$

where apart from the definition of the L_i we have used the fact that the multiplication map $K_0(\kappa) \times K_1(\kappa) \to K_1(\kappa)$ is given by $(i, x) \mapsto x^i$.

Now take $x = \pi^i u$. If $i > 0$, the element $1 - x$ is a unit, hence $d_\pi(1-x) = 0$ and the Steinberg relation holds trivially. If $i < 0$, then $1 - x = (-u + \pi^{-i})\pi^i$ and $d_\pi(1-x) = (-\overline{u}, \xi i)$. It follows from (7.4) that

$$d_\pi(x)d_\pi(1-x) = (\overline{u}, \xi i)(-\overline{u}, \xi i) = (\{\overline{u}, -\overline{u}\}, \xi(\{(-1)^{-i^2} \overline{u}^i (-\overline{u})^{-i}\})),$$

which is 0 in L_2. It remains to treat the case $i = 0$. If $v(1-x) \neq 0$, then replacing x by $1 - x$ we arrive at one of the above cases. If $v(1-x) = 0$, i.e. x and $1 - x$ are both units, then $d_\pi(x)d_\pi(1-x) = (\{\overline{u}, 1-\overline{u}\}, 0 \cdot \xi) = 0$, and the proof is complete. □

Example 7.1.5 The tame symbol $\partial^M : K_1(K) \to K_0(\kappa)$ is none but the valuation map $v : K^\times \to \mathbf{Z}$. The tame symbol $\partial^M : K_2(K) \to K_1(\kappa)$ is given by the formula

$$\partial^M(\{a, b\}) = (-1)^{v(a)v(b)} \overline{a^{-v(b)} b^{v(a)}},$$

where the line denotes the image in κ as usual. One checks this using the definition of ∂^M and the second statement of Lemma 7.1.2.

This is the inverse of the classical formula for the tame symbol in number theory; it has its origin in the theory of the Hilbert symbol.

Remarks 7.1.6

1. The reader may have rightly suspected that tame symbols and specialization maps are not unrelated. In fact, for $\{a_1, \ldots, a_n\} \in K_n^M(K)$ one has the formula

$$s_\pi^M(\{a_1, \ldots, a_n\}) = \partial^M(\{-\pi, a_1, \ldots, a_n\})$$

for all local parameters π.
Indeed, if $a_1 = \pi^i u_1$ for some unit u_1 and integer i, one has

$$\{-\pi, a_1, \ldots, a_n\} = i\{-\pi, \pi, a_2, \ldots, a_n\} + \{-\pi, u_1, a_2, \ldots, a_n\},$$

where the first term on the right is trivial by the first statement in Lemma 7.1.2. Continuing this process, we may eventually assume that all the a_i are units, in which case the formula follows from the definitions.

2. The behaviour of tame symbols under field extensions can be described as follows. Let $L|K$ be a field extension, and v_L a discrete valuation of L extending v with residue field κ_L and ramification index e. Denoting the associated tame symbol by ∂_L^M, one has for all $\alpha \in K_n^M(K)$

$$\partial_L^M(\alpha_L) = e\,\partial^M(\alpha).$$

To see this, write a local parameter π for v as $\pi = \pi_L^e u_L$ for some local parameter π_L and unit u_L for v_L. Then for all $(n-1)$-tuples $(u_2, \ldots, u_n)$ of units for v one gets

$$\{\pi, u_2, \ldots, u_n\}_L = e\,\{\pi_L, u_2, \ldots, u_n\} + \{u_L, u_2, \ldots, u_n\},$$

where the second term is annihilated by ∂_L^M. The formula follows.

We close this section with the determination of the kernel and the cokernel of the tame symbol.

Proposition 7.1.7   *We have exact sequences*

$$0 \to U_n \to K_n^M(K) \xrightarrow{\partial^M} K_{n-1}^M(\kappa) \to 0$$

and

$$0 \to U_n^1 \to K_n^M(K) \xrightarrow{(s_\pi^M, \partial^M)} K_n^M(\kappa) \oplus K_{n-1}^M(\kappa) \to 0,$$

where U_n is the subgroup of $K_n^M(K)$ generated by those symbols $\{u_1, \ldots, u_n\}$ where all the u_i are units in A, and $U_n^1 \subset K_n^M(K)$ is the subgroup generated by symbols $\{x_1, \ldots, x_n\}$ with x_1 a unit in A satisfying $\bar{x}_1 = 1$.

The proof uses the following lemma, whose elegant proof is taken from Dennis–Stein [1].

Lemma 7.1.8   *With notations as in the proposition, the subgroup U_n^1 is contained in U_n.*

Proof   By writing elements of $K^\times$ as $x = u\pi^i$ with some unit u and prime element π one easily reduces the general case to the case $n = 2$ using bilinearity and the relation $\{\pi, -\pi\} = 0$. Then it suffices to show that symbols of the form $\{1 + a\pi, \pi\}$ with some $a \in A$ are contained in U_2.

Case 1: a is a unit in A. Then

$$\{1 + a\pi, \pi\} = \{1 + a\pi, -a\pi\} + \{1 + a\pi, -a^{-1}\} = \{1 + a\pi, -a^{-1}\}$$

by the Steinberg relation, and the last symbol lies in U_n.

Case 2: a lies in the maximal ideal of A. Then

$$\{1 + a\pi, \pi\} = \left\{1 + \frac{1+a}{1-\pi}\pi, \pi\right\} + \{1 - \pi, \pi\} = \left\{1 + \frac{1+a}{1-\pi}\pi, \pi\right\}.$$

Since here the element $(1 + a)(1 - \pi)^{-1}$ is a unit in A, we conclude by the first case. $\square$

Proof of Proposition 7.1.7 It follows from the definitions that ∂^M and s_π^M are surjective, and that the two sequences are complexes. By the lemma, for the exactness of the first sequence it is enough to check that each element in $\ker(\partial^M)$ is a sum of elements from U_n and U_n^1. Consider the map

$$\psi : K_{n-1}^M(\kappa) \to K_n^M(K)/U_n^1$$

defined by $\{\bar{u}_1, \cdots, \bar{u}_{n-1}\} \mapsto \{\pi, u_1, \cdots, u_{n-1}\} \bmod U_n^1$, where the u_i are arbitrary liftings of the $\bar{u}_i$. This is a well-defined map, because replacing some u_i by another lifting u_i' modifies $\{\pi, u_1, \cdots, u_{n-1}\}$ by an element in U_n^1. Now for $\alpha \in U_n$ we have $(\psi \circ \partial^M)(\alpha) = 0$, and for an element in $\ker(\partial^M)$ of the form $\beta = \{\pi, u_2, \ldots, u_n\}$ with the u_i units we have $0 = (\psi \circ \partial^M)(\beta) = \beta$ mod U_n^1, i.e. $\beta \in U_n^1$. Since the β of this form generate $\ker(\partial^M)$ together with U_n, we are done.

We now turn to the second sequence. Define a map $K_n^M(\kappa) \to U_n/U_n^1$ by sending $\{\bar{u}_1, \cdots, \bar{u}_n\}$ to $\{u_1, \ldots, u_n\} \bmod U_n^1$, again with some liftings u_i of the $\bar{u}_i$. We see as above that this map is well defined, and moreover it is an inverse to the map induced by the restriction of s_π^M to U_n (which is of course trivial on U_n^1). $\square$

Remark 7.1.9 It follows from the first sequence above (and was implicitly used in the second part of the proof) that the restriction of s_π^M to $\ker(\partial^M)$ is independent of the choice of π.

Corollary 7.1.10 *Assume moreover that K is complete with respect to v, and let $m > 0$ be an integer invertible in κ. Then the pair (s_π, ∂^M) induces an isomorphism*

$$K_n^M(K)/mK_n^m(K) \xrightarrow{\sim} K_n^M(\kappa)/mK_n^M(\kappa) \oplus K_{n-1}^M(\kappa)/mK_{n-1}^M(\kappa).$$

Proof By virtue of the second exact sequence of the proposition it is enough to see that in this case $mU_n^1 = U_n^1$, which in turn needs only to be checked for $n = 1$ by multilinearity of symbols. But since m is invertible in κ, for each unit $u \in U_1^1$ Hensel's lemma (cf. Appendix, Proposition A.5.5) applied to the polynomial $x^m - u$ shows that $u \in mU_1^1$. $\square$

7.2 Milnor's exact sequence and the Bass–Tate lemma

We now describe the Milnor K-theory of the rational function field $k(t)$ and establish an analogue of Faddeev's exact sequence due to Milnor.

Recall that the discrete valuations of $k(t)$ trivial on k correspond to the local rings of closed points P on the projective line $\mathbf{P}_k^1$. As before, we denote by $\kappa(P)$ their residue fields and by v_P the associated valuations. At each closed point $P \neq \infty$ a local parameter is furnished by a monic irreducible polynomial $\pi_P \in k[t]$; at $P = \infty$ one may take $\pi_P = t^{-1}$. The degree of the field extension $[\kappa(P) : k]$ is called the degree of the closed point P; it equals the degree of the polynomial π_P.

By the theory of the previous section we obtain tame symbols

$$\partial_P^M : K_n^M(k(t)) \to K_{n-1}^M(\kappa(P))$$

and specialization maps

$$s_{\pi_P}^M : K_n^M(k(t)) \to K_n^M(\kappa(P)).$$

Note that since each element in $k(t)^\times$ is a unit for all but finitely many valuations v_P, the image of the product map

$$\partial^M := (\partial_P^M) : K_n^M(k(t)) \to \prod_{P \in \mathbf{P}_0^1 \setminus \{\infty\}} K_{n-1}^M(\kappa(P))$$

lies in the direct sum.

Theorem 7.2.1 (Milnor) *The sequence*

$$0 \to K_n^M(k) \to K_n^M(k(t)) \xrightarrow{\partial^M} \bigoplus_{P \in \mathbf{P}_0^1 \setminus \{\infty\}} K_{n-1}^M(\kappa(P)) \to 0$$

is exact and split by the specialization map $s_{t^{-1}}^M$ at ∞.

Note that for $i = 1$ we get the sequence

$$1 \to k^\times \to k(t)^\times \xrightarrow{\partial^M} \bigoplus_\pi \mathbf{Z} \to 0,$$

which is equivalent to the decomposition of a rational function into a product of irreducible factors.

The proof exploits the filtration on $K_n^M(k(t))$

$$K_n^M(k) = L_0 \subset L_1 \subset \cdots \subset L_d \subset \ldots \tag{7.5}$$

where L_d is the subgroup of $K_n^M(k(t))$ generated by those symbols $\{f_1, ..., f_n\}$ where the f_i are polynomials in $k[t]$ of degree $\leq d$.

The key statement is the following.

Lemma 7.2.2 *For each $d > 0$ consider the homomorphism*

$$\partial_d^M : K_n^M(k(t)) \longrightarrow \bigoplus_{\deg(P)=d} K_{n-1}^M(\kappa(P))$$

defined as the direct sum of the maps ∂_P^M for all closed points P of degree d. Its restriction to L_d induces an isomorphism

$$\overline{\partial}_d^M : L_d/L_{d-1} \xrightarrow{\sim} \bigoplus_{\deg(P)=d} K_{n-1}^M(\kappa(P)).$$

Proof If P is a closed point of degree d, the maps ∂_P^M are trivial on the elements of L_{d-1}, hence the map $\overline{\partial}_d^M$ exists. To complete the proof we construct an inverse for $\overline{\partial}_d^M$.

Let P be a closed point of degree d. For each element $\overline{a} \in \kappa(P)$, there exists a unique polynomial $a \in k[t]$ of degree $\leq d-1$ whose image in $\kappa(P)$ is $\overline{a}$. Define maps

$$h_P : K_{n-1}^M(\kappa(P)) \to L_d/L_{d-1}$$

by the assignment

$$h_P(\{\overline{a}_2, \cdots, \overline{a}_n\}) = \{\pi_P, a_2, \cdots, a_n\} \mod L_{d-1}.$$

The maps h_P obviously satisfy the Steinberg relation, for $\overline{a}_i + \overline{a}_j = 1$ implies $a_i + a_j = 1$. So if we show that they are linear in each variable, we get that each h_P is a homomorphism, and then by construction the direct sum $\oplus\, h_P$ yields an inverse for $\overline{\partial}_d^M$.

We check linearity in the case $n = 2$, the general case being similar. For $\overline{a}_2 = \overline{b}_2\overline{c}_2$, we compare the polynomials a_2 and $b_2 c_2$. If they are equal, the claim is obvious. If not, we perform Euclidean division of $b_2 c_2$ by π_P to get $b_2 c_2 = a_2 - \pi_P f$ with some polynomial $f \in k[t]$ of degree $\leq d-1$ (note that the rest of the division must be a_2 by uniqueness). Therefore

$$\frac{\pi_P f}{a_2} = 1 - \frac{b_2 c_2}{a_2}, \tag{7.6}$$

and so in $K_2^M(k(t))$ we have the equalities

$$\{\pi_P, b_2 c_2\} - \{\pi_P, a_2\} = \left\{\pi_P, \frac{b_2 c_2}{a_2}\right\} = -\left\{\frac{f}{a_2}, \frac{b_2 c_2}{a_2}\right\} + \left\{\frac{\pi_P f}{a_2}, \frac{b_2 c_2}{a_2}\right\}$$

$$= -\left\{\frac{f}{a_2}, \frac{b_2 c_2}{a_2}\right\},$$

where we used the equality (7.6) in the last step. The last symbol lies in L_{d-1}, and the claim follows. $\square$

Proof of Theorem 7.2.1   Using induction on d, we derive from the previous lemma exact sequences

$$0 \to L_0 \to L_d \to \bigoplus_{\deg(P) \le d} K_{n-1}^M(\kappa(P)) \to 0$$

for each $d > 0$. These exact sequences form a natural direct system with respect to the inclusions coming from the filtration (7.5). As $L_0 = K_n^M(k)$ and $\bigcup L_d = K_n^M(k(t))$, we obtain the exact sequence of the theorem by passing to the limit. The statement about s_{t-1}^M is straightforward. $\square$

Note that Milnor's exact sequence bears a close resemblance to Faddeev's exact sequence in the form of Corollary 6.9.3. As in that chapter, the fact that the sequence splits allows us to define *coresidue* maps

$$\psi_P^M : K_{n-1}^M(\kappa(P)) \to K_n^M(k(t))$$

for all closed points $P \ne \infty$, enjoying the properties $\partial_P^M \circ \psi_P^M = \mathrm{id}_{\kappa(P)}$ and $\partial_P^M \circ \psi_Q^M = 0$ for $P \ne Q$. We thus obtain the following formula useful in calculations.

Corollary 7.2.3   *We have the equality*

$$\alpha = s_{t-1}(\alpha)_{k(t)} + \sum_{P \in \mathbf{A}_0^1} (\psi_P^M \circ \partial_P)(\alpha)$$

for all $n > 0$ and all $\alpha \in K_n^M(k(t))$.

For all $P \ne \infty$ we define *norm maps* $N_P : K_n^M((\kappa(P)) \to K_n^M(k)$ by the formula

$$N_P := -\partial_\infty^M \circ \psi_P^M$$

for all $n \ge 0$. For $P = \infty$ we define N_P to be the identity map of $K_n^M(k)$.

With the above notations, Milnor's exact sequence implies

Corollary 7.2.4   (Weil reciprocity law)   *For all $\alpha \in K_n^M(k(t))$ we have*

$$\sum_{P \in \mathbf{P}_0^1} (N_P \circ \partial_P^M)(\alpha) = 0.$$

Proof For $P \neq \infty$ we have from the defining property of the maps ψ_P

$$\partial_P^M \left(\alpha - \sum_{P \neq \infty} (\psi_P^M \circ \partial_P^M)(\alpha) \right) = \partial_P^M(\alpha) - \partial_P^M(\alpha) = 0,$$

so by Milnor's exact sequence

$$\alpha - \sum_{P \neq \infty} (\psi_P^M \circ \partial_P^M)(\alpha) = \beta$$

for some β coming from $K_n^M(k)$. We have $\partial_\infty^M(\beta) = 0$, so the corollary follows by applying $-\partial_\infty^M$ to both sides. $\qquad\square$

Weil's original reciprocity law concerned the case $n = 2$ and had the form

$$\sum_{P \in \mathbf{P}_0^1} (N_{\kappa(P)|k} \circ \partial_P^M)(\alpha) = 0.$$

Note that in this case the tame symbols ∂_P^M have an explicit description by Example 7.1.5. To relate this form to the previous corollary, it suffices to use the second statement of the following proposition.

Proposition 7.2.5 *For $n = 0$ the map $N_P : K_0^M(\kappa(P)) \to K_0^M(k)$ is given by multiplication with $[\kappa(P) : k]$, and for $n = 1$ it coincides with the field norm $N_{\kappa(P)|k} : \kappa(P)^\times \to k^\times$.*

The proof relies on the following behaviour of the norm map under extensions of the base field.

Lemma 7.2.6 *Let $K|k$ be a field extension, and P a closed point of $\mathbf{P}_k^1$. Then the diagram*

$$
\begin{array}{ccc}
K_n^M(\kappa(P)) & \xrightarrow{\ N_P\ } & K_n^M(k) \\
{\scriptstyle \oplus i_{\kappa(Q)|\kappa(P)}} \downarrow & & \downarrow {\scriptstyle i_{K|k}} \\
\displaystyle\bigoplus_{Q \mapsto P} K_n^M(\kappa(Q)) & \xrightarrow{\ \Sigma e_Q N_Q\ } & K_n^M(K)
\end{array}
$$

commutes, where the notation $Q \mapsto P$ stands for the closed points of $\mathbf{P}_K^1$ lying above P, and e_Q is the ramification index of the valuation v_Q extending v_P to $K(t)$.

Proof   According to Remark 7.1.6 (2), the diagram

$$
\begin{array}{ccc}
K_{n+1}^M(k(t)) & \xrightarrow{\partial_P^M} & K_n^M(\kappa(P)) \\
\Big\downarrow{\scriptstyle i_{K(t)|k(t)}} & & \Big\downarrow{\scriptstyle \oplus_Q i_{\kappa(Q)|\kappa(P)}} \\
K_{n+1}^M(K(t)) & \xrightarrow{\oplus \partial_Q^M} & \displaystyle\bigoplus_{Q \mapsto P} K_n^M(\kappa(Q))
\end{array}
$$

commutes. Hence so does the diagram

$$
\begin{array}{ccc}
K_{n+1}^M(k(t)) & \xleftarrow{\psi_P^M} & K_n^M(\kappa(P)) \\
\Big\downarrow{\scriptstyle i_{K(t)|k(t)}} & & \Big\downarrow{\scriptstyle \oplus i_{\kappa(Q)|\kappa(P)}} \\
K_{n+1}^M(K(t)) & \xleftarrow{\Sigma e_Q \psi_Q^M} & \displaystyle\bigoplus_{Q \mapsto P} K_n^M(\kappa(Q))
\end{array}
$$

whence the compatibility of the lemma in view of the definition of the norm
maps N_P. $\square$

Proof of Proposition 7.2.5   Apply the above lemma with K an algebraic closure of k. In this case the points Q have degree 1 over K, so the maps N_Q are
identity maps. Moreover, the vertical maps are injective for $n = 0, 1$. The statement for $n = 0$ then follows from the formula $\sum e_Q = [\kappa(P) : k]$ (a particular
case of Proposition A.6.7 of the Appendix), and for $n = 1$ from the definition
of the field norm $N_{\kappa(P)|k}(\alpha)$ as the product of the roots in K (considered with
multiplicity) of the minimal polynomial of α. $\square$

Remark 7.2.7   For later use, let us note that the norm maps N_P satisfy the
projection formula: for $\alpha \in K_n^M(k)$ and $\beta \in K_m^M(\kappa(P))$ one has

$$
N_P(\{\alpha_{\kappa(P)}, \beta\}) = \{\alpha, N_P(\beta)\}.
$$

This is an immediate consequence of the definitions.

   We conclude this section by a very useful technical statement which is not a
consequence of Milnor's exact sequence itself, but is proven in a similar vein.
Observe that if $K|k$ is a field extension, the graded ring $K_*^M(K)$ becomes a
(left) $K_*^M(k)$-module via the change-of-fields map $K_*^M(k) \to K_*^M(K)$ and
the product structure.

Proposition 7.2.8 (Bass–Tate Lemma)   *Let $K = k(a)$ be a field extension
obtained by adjoining a single element a of degree d to k. Then $K_*^M(K)$ is
generated as a left $K_*^M(k)$-module by elements of the form*

$$\{\pi_1(a), \pi_2(a), \cdots, \pi_m(a)\},$$

where the π_i are monic irreducible polynomials in $k[t]$ satisfying $\deg(\pi_1) < \deg(\pi_2) < \cdots < \deg(\pi_m) \leq d - 1$.

The proof is based on the following property of the subgroups L_d introduced in Lemma 7.2.2.

Lemma 7.2.9 *The subgroup $L_d \subset K_n^M(k(t))$ is generated by symbols of the shape*

$$\{a_1, \cdots, a_m, \pi_{m+1}, \pi_{m+2}, \cdots, \pi_n\}, \tag{7.7}$$

where the a_i belong to $k^\times$ and the π_i are monic irreducible polynomials in $k[t]$ satisfying $\deg(\pi_{m+1}) < \deg(\pi_{m+2}) < \cdots < \deg(\pi_n) \leq d$.

Proof By factoring polynomials into irreducible terms and using bilinearity and graded-commutativity of symbols, we obtain generators for the group L_d of the shape (7.7), except that the π_i a priori only satisfy $\deg(\pi_{m+1}) \leq \cdots \leq \deg(\pi_n) \leq d$. The point is to show that the inequalities may be chosen to be strict, which we do in the case $n = 2$ for polynomials π_1, π_2 of the same degree, the general case being similar. We use induction on d starting from the case $d = 0$ where we get constants $\pi_1 = a_1$, $\pi_2 = a_2$. So assume $d > 0$. If $\deg(\pi_1) = \deg(\pi_2) < d$, we are done by induction. It remains the case $\deg(\pi_1) = \deg(\pi_2) = d$, where we perform Euclidean division to get $\pi_2 = \pi_1 + f$ with some f of degree $\leq d - 1$. So $1 = \pi_1/\pi_2 + f/\pi_2$ and therefore $\{\pi_1/\pi_2, f/\pi_2\} = 0$ in $K_2^M(k(t))$. Using Lemma 7.1.2 we may write

$$\{\pi_1, \pi_2\} = \{\pi_1/\pi_2, \pi_2\} + \{\pi_2, -1\} = -\{\pi_1/\pi_2, f/\pi_2\}$$
$$+ \{\pi_1/\pi_2, f\} + \{\pi_2, -1\},$$

which equals $-(\{f, \pi_1\} + \{-f, \pi_2\})$ by bilinearity and graded-commutativity. We conclude by decomposing the polynomial f into irreducible factors. $\square$

Proof of Proposition 7.2.8 Let π_P be the minimal polynomial of a over k; it defines a closed point P of degree d on $\mathbf{P}_k^1$. It follows from Lemma 7.2.2 that the tame symbol ∂_P^M induces a surjection of L_d onto $K_n^M(\kappa(P))$. Applying the previous lemma, we conclude that $K_n^M(\kappa(P))$ is generated by symbols of the form

$$\partial_P^M\{a_1, \cdots, a_m, \pi_{m+1}, \pi_{m+2}, \cdots, \pi_n\},$$

where the a_i belong to $k^\times$ and the π_i are monic irreducible polynomials satisfying $\deg(\pi_1) < \deg(\pi_2) < \cdots < \deg(\pi_n) \leq d$. If $\pi_n \neq \pi_P$, all the π_i

satisfy $v_\pi(\pi_i) = 0$ and the above symbols are zero. For $\pi_n = \pi_P$, they equal $\{a_1, \cdots, a_m, \pi_{m+1}(a), \pi_{m+2}(a), \cdots, \pi_{n-1}(a)\}$ up to a sign by the defining property of ∂_P^M, and the proposition follows. $\square$

In what follows we shall use the Bass–Tate lemma several times via the following corollary.

Corollary 7.2.10   *Let $K|k$ be a finite field extension. Assume one of the following holds:*

- $K|k$ *is a quadratic extension,*
- $K|k$ *is of prime degree p and k has no nontrivial finite extensions of degree prime to p.*

Then $K_*^M(K)$ is generated as a left $K_*^M(k)$-module by $K_1^M(K) = K^\times$. In other words, the product maps $K_{n-1}^M(k) \otimes K^\times \to K_n^M(K)$ are surjective.

Proof   In both cases, K is obtained by adjoining a single element a to k, and the only monic irreducible polynomials in $k[t]$ of degree strictly smaller than $[K : k]$ are the linear polynomials $x - a$. We conclude by applying the proposition. $\square$

Remark 7.2.11   A typical case when the second condition of the corollary is satisfied is when k is a *maximal prime-to-p extension* of some field $k_0 \subset k$. This is an algebraic extension $k|k_0$ such that all finite subextensions have degree prime to p and which is maximal with respect to this property. If k_0 is perfect or has characteristic p, we can construct such an extension k by taking the subfield of a separable closure k_s of k_0 fixed by a pro-p Sylow subgroup of $\mathrm{Gal}\,(k_s|k_0)$. If k_0 is none of the above, we may take k to be a maximal prime to p extension of a perfect closure of k_0.

7.3 The norm map

Let $K|k$ be a finite field extension. In this section we construct norm maps $N_{K|k} : K_n^M(K) \to K_n^M(k)$ for all $n \geq 0$ satisfying the following properties:

1. The map $N_{K|k} : K_0^M(K) \to K_0^M(k)$ is multiplication by $[K : k]$.
2. The map $N_{K|k} : K_1^M(K) \to K_1^M(k)$ is the field norm $N_{K|k} : K^\times \to k^\times$.
3. (Projection formula) Given $\alpha \in K_n^M(k)$ and $\beta \in K_m^M(K)$, one has

$$N_{K|k}(\{\alpha_K, \beta\}) = \{\alpha, N_{K|k}(\beta)\}.$$

4. (Composition) Given a tower of field extensions $K'|K|k$, one has

$$N_{K'|k} = N_{K|k} \circ N_{K'|K}.$$

Furthermore, a reasonable norm map should be compatible (for finite separable extensions) with the corestriction maps on cohomology via the Galois symbol. This issue will be discussed in the next section.

Remark 7.3.1 For any norm map satisfying the above properties (1)–(3) the composite maps $N_{K|k} \circ i_{K|k} : K_n^M(k) \to K_n^M(k)$ are given by multiplication with the degree $[K : k]$ for all n. This is obvious for $n = 0, 1$, and the case $n > 1$ follows from the case $n = 1$ by an easy induction using the projection formula.

In the case when $K = k(a)$ is a simple field extension, the minimal polynomial of a defines a closed point P on $\mathbf{P}_k^1$ for which $K \cong \kappa(P)$. The norm map N_P of the previous section satisfies properties (1) and (2) by virtue of Proposition 7.2.5, as well as property (3) by Remark 7.2.7, so it is a natural candidate for $N_{K|k}$. But even in this case one has to check that the definition depends only on K and not on the choice of P.

Changing the notation slightly, for a simple finite field extension $K = k(a)$ define $N_{a|k} : K_n^M(k(a)) \to K_n^M(k)$ by $N_{a|k} := N_P$, where P is the closed point of $\mathbf{P}_k^1$ considered above. Given an arbitrary finite field extension $K|k$, write $K = k(a_1, \ldots, a_r)$ for some generators $a_1, \ldots, a_r$ and consider the chain of subfields

$$k \subset k(a_1) \subset k(a_1, a_2) \subset \cdots \subset k(a_1, \ldots, a_r) = K.$$

Now put

$$N_{a_1, \ldots, a_r|k} := N_{a_r|k(a_1, \ldots, a_{r-1})} \circ \cdots \circ N_{a_2|k(a_1)} \circ N_{a_1|k}.$$

Note that by the preceding discussion the maps $N_{a_1, \ldots, a_r|k}$ satisfy properties (1)–(4) above, and also the formula $N_{a_1, \ldots, a_r|k} \circ i_{K|k} = [K : k]$, by virtue of Remark 7.3.1.

Theorem 7.3.2 (Kato) *The maps $N_{a_1, \ldots, a_r|k} : K_n^M(K) \to K_n^M(k)$ do not depend on the choice of the generating system $(a_1, \ldots, a_r)$.*

The theorem allows us to define without ambiguity

$$N_{K|k} := N_{a_1, \ldots a_r|k} : K_n^M(K) \to K_n^M(k)$$

for all $n \geq 0$. We have the following immediate corollary:

Corollary 7.3.3 *For a k-automorphism $\sigma : K \to K$ one has $N_{K|k} \circ \sigma = N_{K|k}$.*

Proof Indeed, according to the theorem $N_{a_1,\dots,a_r|k} = N_{\sigma(a_1),\dots,\sigma(a_r)|k}$ for every system of generators $(a_1, \dots, a_r)$. $\square$

The rest of this section will be devoted to the proof of Kato's theorem. A major step in the proof is the following reduction statement, essentially due to Bass and Tate.

Proposition 7.3.4 *Assume that Theorem 7.3.2 holds for all fields k that have no nontrivial finite extension of degree prime to p for some prime number p. Then the theorem holds for arbitrary k.*

For the proof we need some auxiliary statements.

Lemma 7.3.5 *For an algebraic extension $K|k$ the kernel of the change-of-fields map $i_{K|k} : K_n^M(k) \to K_n^M(K)$ is a torsion group. It is annihilated by the degree $[K : k]$ in the case of a finite extension.*

Proof Considering $K_n^M(K)$ as the direct limit of the groups $K_n^M(K_i)$ for all finite subextensions $k \subset K_i \subset K$ we see that it suffices to prove the second statement. Write $K = k(a_1, \dots, a_r)$ for some generators a_i. As noted above, the norm map $N_{a_1,\dots,a_r|k}$ satisfies the formula $N_{a_1,\dots,a_r|k} \circ i_{K|k} = [K : k]$, whence the claim. $\square$

Before stating the next lemma, recall the following well-known facts from algebra (see e.g. Atiyah–Macdonald [1], Chapter 8). Given a finite field extension $K|k$ and an arbitrary field extension $L|k$, the tensor product $K \otimes_k L$ is a finite-dimensional (hence Artinian) L-algebra, and as such decomposes as a finite direct sum of local L-algebras R_j in which the maximal ideal M_j is nilpotent. Let e_j be the smallest positive integer with $M_j^{e_j} = 0$. In the case when $K = k(a)$ is a simple field extension, the e_j correspond to the multiplicities of the irreducible factors in the decomposition of the minimal polynomial $f \in k[t]$ of a over L. In particular, for $K|k$ separable all the e_j are equal to 1.

Lemma 7.3.6 *In the above situation, write $K = k(a_1, \dots, a_r)$ with suitable $a_i \in K$. Denote by L_j the residue field R_j/M_j and by $p_j : L \otimes_k K \to L_j$ the natural projections. Then the diagram*

$$K_n^M(K) \xrightarrow{N_{a_1,\ldots,a_r|k}} K_n^M(k)$$

$$\oplus i_{L_j|K} \downarrow \qquad\qquad \downarrow i_{L|k}$$

$$\bigoplus_{j=1}^{m} K_n^M(L_j) \xrightarrow{\Sigma\, e_j N_{p_j(a_1),\ldots,p_j(a_r)|L}} K_n^M(L)$$

commutes.

Proof By the discussion above, for $r = 1$ we are in the situation of Lemma 7.2.6 and thus the statement has been already proven (modulo a straightforward identification of the e_i with the ramification indices of the corresponding valuations on $k(t)$). We prove the general case by induction on r. Write $k(a_1) \otimes_k L \cong \oplus R_j$ for some local L-algebras R_j, and decompose the finite dimensional L-algebra $K \otimes_{k(a_1)} R_j$ as $K \otimes_{k(a_1)} R_j = \oplus R_{ij}$ for some R_{ij}. Note that $K \otimes_k L \cong \oplus_{i,j} R_{ij}$. Write L_j (resp. L_{ij}) for the residue fields of the L-algebras R_j (resp. R_{ij}), and similarly e_j and e_{ij} for the corresponding nilpotence indices. In the diagram

$$K_n^M(K) \xrightarrow{N_{a_2,\ldots,a_r|k(a_1)}} K_n^M(k(a_1)) \xrightarrow{N_{a_1|k}} K_n^M(k)$$

$$\oplus i_{L_{ij}|K} \downarrow \qquad\qquad \oplus i_{L_j|k(a_1)} \downarrow \qquad\qquad \downarrow i_{L|k}$$

$$\bigoplus_{i,j} K_n^M(L_{ij}) \xrightarrow{\underset{j}{\oplus} \underset{i}{\Sigma}(e_{ij}e_j^{-1})N_{p_{ij}(a_2),\ldots,p_{ij}(a_r)|L_j}} \bigoplus_{j} K_n^M(L_j) \xrightarrow{\Sigma\, e_j N_{p_j(a_1)|L}} K_n^M(L)$$

both squares commute by the inductive hypothesis. The lemma follows. □

Proof of Proposition 7.3.4 Write $K = k(a_1,\ldots,a_r) = k(b_1,\ldots,b_s)$ in two different ways. Let $\Delta \subset K_n^M(K)$ be the subgroup generated by elements of the form $N_{a_1,\ldots,a_r|k}(\alpha) - N_{b_1,\ldots,b_s|k}(\alpha)$ for some $\alpha \in K_n^M(K)$. Our job is to prove $\Delta = 0$. Consider the diagram of the previous lemma with $L = \bar{k}$, an algebraic closure of k. Then $L_j \cong L$ for all j and in the bottom row we have a sum of identity maps. Considering the similar diagram for $N_{b_s,\ldots,b_s|k}$ we get an equality $i_{\bar{k}|k} \circ N_{a_1,\ldots,a_r|k} = i_{\bar{k}|k} \circ N_{b_1,\ldots,b_s|k}$, whence $\Delta \subset \ker(i_{\bar{k}|k})$. We thus conclude from Lemma 7.3.5 that Δ is a torsion group. Denoting by Δ_p its p-primary component it is therefore enough to show that $\Delta_p = 0$ for all prime numbers p. Fix a prime p, and let L be a maximal prime to p extension of k (cf. Remark 7.2.11). As all finite subextensions of $L|k$ have degree prime to p, an application of Lemma 7.3.5 shows that the restriction of $i_{L|k}$ to Δ_p is injective. On the other hand, the assumption of the proposition applies to L and hence the map $\Sigma\, e_j N_{p_j(a_1),\ldots,p_j(a_r)|L}$ of Lemma 7.3.6 does not depend on the a_i. Therefore $i_{L|K}(\Delta_p) = 0$, which concludes the proof. □

For the rest of this section p will be a fixed prime number, and k will always denote a field having no nontrivial finite extensions of degree prime to p.

Concerning such fields, the following easy lemma will be helpful.

Lemma 7.3.7   *Let $K|k$ be a finite extension.*

1. *The field K inherits the property of having no nontrivial finite extension of degree prime to p.*
2. *If $K \neq k$, there exists a subfield $k \subset K_1 \subset K$ such that $K_1|k$ is a normal extension of degree p.*

Proof   For the first statement let $L|K$ be a finite extension of degree prime to p. If $L|k$ is separable, take a Galois closure $\widetilde{L}$. By our assumption on k, the fixed field of a p-Sylow subgroup in $\mathrm{Gal}\,(\widetilde{L}|k)$ must equal k, so that $L = K$. If $K|k$ is purely inseparable, then $L|K$ must be separable, so $L|k$ has a subfield $L_0 \neq k$ separable over k unless $L = K$. Finally, if $K|k$ is separable but $L|K$ is not, we may assume the latter to be purely inseparable. Taking a normal closure $\widetilde{L}$, the fixed field of $\mathrm{Aut}_k(\widetilde{L})$ defines a nontrivial prime to p extension of k unless $L = K$.

The second statement is straightforward in the case when the extension $K|k$ is purely inseparable, so by replacing K with the maximal separable subextension of $K|k$ we may assume that $K|k$ is a separable extension. Consider the Galois closure $\widetilde{K}$ of K. The first statement implies that the Galois group $G := \mathrm{Gal}\,(\widetilde{K}|k)$ is a p-group. Now let H be a maximal subgroup of G containing $\mathrm{Gal}\,(\widetilde{K}|K)$. By the theory of finite p-groups (see e.g. Suzuki [1], Corollary of Theorem 1.6), it is a normal subgroup of index p in G, so we may take K_1 to be its fixed field.   □

We now start the proof of Theorem 7.3.2 with the case of a degree p extension, still due to Bass and Tate.

Proposition 7.3.8   *Assume that $[K : k] = p$, and write $K = k(a)$ for some $a \in K$. The norm maps $N_{a|k} : K_n^M(k(a)) \to K_n^M(k)$ do not depend on the choice of a.*

Proof   Let P be the closed point of $\mathbf{P}_k^1$ defined by the minimal polynomial of a. According to Corollary 7.2.10, the group $K_n^M(K)$ is generated by symbols of the form $\{\alpha_K, b\}$, with $\alpha \in K_{n-1}^M(k)$ and $b \in K^\times$. We compute using the projection formula for N_P (Remark 7.2.7) and Proposition 7.2.5:

$$N_{a|k}(\{\alpha_K, b\}) = N_P(\{\alpha_K, b\}) = \{\alpha, N_P(b)\} = \{\alpha, N_{K|k}(b)\}.$$

Here the right-hand side does not depend on a, as was to be shown.   □

Henceforth the notation $N_{L|K} : K_n^M(L) \to K_n^M(K)$ will be legitimately used for extensions of degree p (and for those of degree 1).

Next we need the following compatibility statement with the tame symbol (which does not concern k, so there is no assumption on the fields involved). For a generalization, see Proposition 7.4.1 in the next section.

Proposition 7.3.9 *Let K be a field complete with respect to a discrete valuation v with residue field κ, and $K'|K$ a normal extension of degree p. Denote by κ' the residue field of the unique extension v' of v to K'. Then for all $n > 0$ the diagram*

$$
\begin{array}{ccc}
K_n^M(K') & \xrightarrow{\ \partial_{K'}^M\ } & K_{n-1}^M(\kappa') \\[4pt]
{\scriptstyle N_{K'|K}}\Big\downarrow & & \Big\downarrow{\scriptstyle N_{\kappa'|\kappa}} \\[4pt]
K_n^M(K) & \xrightarrow{\ \partial_K^M\ } & K_{n-1}^M(\kappa)
\end{array}
$$

commutes.

The notes of Sridharan [1] have been helpful to us in writing up the following proof. We begin with a special case.

Lemma 7.3.10 *The compatibility of the proposition holds for symbols of the form $\alpha = \{a', a_2, \ldots, a_n\} \in K_n^M(K')$, with $a' \in K'^\times$ and $a_i \in K^\times$.*

Proof Using Lemma 7.1.2, multilinearity and graded-commutativity we may assume that $v(a_i) = 0$ for $i > 2$ and $0 \le v(a'), v(a_2) \le 1$. Setting $f := [\kappa' : \kappa]$ and denoting by e the ramification index of $v'|v$ we have the formula

$$f \cdot v' = v \circ N_{K'|K} \tag{7.8}$$

(see Appendix, Proposition A.6.8 (2)). Now there are four cases to consider.

Case 1: $v'(a') = v(a_2) = 0$. Then $v(N_{K'|K}(a')) = 0$, so using the projection formula we obtain $\partial_K^M(N_{K'|K}(\alpha)) = 0$, and likewise $\partial_{K'}^M(\alpha) = 0$.

Case 2: $v'(a') = 1, v(a_2) = 0$. In this case Remark 7.3.1 implies that with the usual notations $N_{\kappa'|\kappa}(\partial_{K'}^M(\alpha)) = f\{\bar{a}_2, \ldots, \bar{a}_n\}$. On the other hand, from (7.8) we infer that $N_{K'|K}(a') = u\pi^f$ for some unit u and local parameter π for v. So using the projection formula and the multilinearity of symbols we get $N_{K'|K}(\alpha) = f\{\pi, a_2, \ldots, a_n\} + \{u, a_2, \ldots, a_n\}$. This element has residue $f\{\bar{a}_2, \ldots, \bar{a}_n\}$ as well.

Case 3: $v'(a') = 0, v(a_2) = 1$. Then $a_2 = u'\pi'^e$ for some unit u' and local parameter π' for v', so using graded-commutativity and multilinearity of symbols we obtain $\partial_{K'}^M(\alpha) = -e\{\bar{a}', \bar{a}_3, \ldots, \bar{a}_n\}$. This element has norm

$-e\{N_{\kappa'|\kappa}(\bar{a}'), \bar{a}_3, \ldots, \bar{a}_n\}$ by the projection formula. On the other hand, $\partial_K^M(N_{K'|K}(\alpha)) = -\{\overline{N_{K'|K}(a')}, \bar{a}_3, \ldots, \bar{a}_n\}$. The claim now follows from the equality $\overline{N_{K'|K}(a')} = N_{\kappa'|\kappa}(\bar{a}')^e$, which is easily verified in both the unramified and the totally ramified case.

Case 4: $v'(a') = v(a_2) = 1$. Write $a' = \pi'$, $a_2 = \pi$, $\pi = u'\pi'^e$, $N_{K'|K}(\pi') = u\pi^f$ as above. Then using multilinearity and Lemma 7.1.2 we get

$$\partial_{K'}^M(\alpha) = \partial_{K'}^M(\{\pi', u', a_3, \ldots, a_n\} + e\{\pi', -1, a_3, \ldots, a_n\})$$
$$= \{(-1)^e \bar{u}', \bar{a}_3, \ldots, \bar{a}_n\},$$

which has norm $\{(-1)^{ef} N_{\kappa'|\kappa}(\bar{u}'), \bar{a}_3, \ldots, \bar{a}_n\}$. On the other hand, using the projection formula we obtain as above

$$\partial_K^M(N_{K'|K}(\alpha)) = \partial_K^M(\{u\pi^f, \pi, a_3, \ldots, a_n\}) = \partial_K^M(-\{\pi, u, a_3, \ldots, a_n\}$$
$$+ f\{\pi, -1, a_3, \ldots, a_n\}) = \{(-1)^f \bar{u}^{-1}, \bar{a}_3, \ldots, \bar{a}_n\}.$$

So it is enough to see $(-1)^{ef} N_{\kappa'|\kappa}(\bar{u}') = (-1)^f \bar{u}^{-1}$. Notice that in the above computations we are free to modify π and π' by units. In particular, in the case when $e = 1$ and $f = p$ we may take $\pi = \pi'$, so that $u' = u = 1$ and the equality is obvious. In the case $e = p$, $f = 1$ the element π' is a root of an Eisenstein polynomial $x^p + a_{p-1}x^{p-1} + \cdots + a_0$ and we may take $\pi = a_0$. Then $u = (-1)^p$ and $\bar{u}' = -1$, so we are done again. $\square$

Proof of Proposition 7.3.9 Let α be an element of $K_n^M(K')$, and set $\delta := \partial_K^M(N_{K'|K}(\alpha)) - N_{\kappa'|\kappa}(\partial_{K'}^M(\alpha))$. We prove $\delta = 0$ by showing that δ is annihilated both by some power of p and by some integer prime to p.

By Corollary 7.2.10, if $K^{(p)}$ denotes a maximal prime to p extension of K, the image of α in $K_n^M(K'K^{(p)})$ is a sum of symbols of the shape as in Lemma 7.3.10 above (for the extension $KK^{(p)}|K^{(p)}$). These symbols are all defined at a finite level, so the lemma enables us to find some extension $L|K$ of degree prime to p so that

$$\delta^L := \partial_L^M(N_{LK'|L}(i_{LK'|K'}(\alpha))) - N_{\kappa'_L|\kappa_L}(\partial_{LK'}^M(i_{LK'|K'}(\alpha))) = 0.$$

Now since $K'|K$ has degree p, we have $LK' \cong L \otimes_K K'$. This implies that the valuations $v'|v$ and their unique extensions $v'_L|v_L$ have the same ramification index e, and hence by Remark 7.1.6 (2) the tame symbol $\partial_{LK'}^M$ is the e-th multiple of ∂_L^M on symbols coming from $K_n^M(L)$, just like the tame symbol $\partial_{K'}^M$ is the e-th multiple of ∂_K^M on $i_{K'|K}(K_n^M(K))$. On the other hand, by Lemma 7.3.6 the norm map $N_{LK'|L}$ is the base change of $N_{K'|K}$ to $K_n^M(LK')$. It follows from these remarks that we have $i_{L|K}(\delta) = \delta^L$, and hence $i_{L|K}(\delta) = 0$. Thus $[L : K]\delta = 0$ by Lemma 7.3.5.

To see that δ is annihilated by some power of p, we look at the base change $K' \otimes_K K'$. Assume first that $K'|K$ is separable. Then it is Galois by assumption, so $K' \otimes_K K'$ splits as a product of p copies of K'. Therefore it is obvious that the required compatibility holds for α after base change to K'. But now there is a difference between the unramified and the ramified case. In the unramified case the residue fields in the copies of K' all equal κ, so the compatibilities of Remark 7.1.6 (2) and Lemma 7.3.6 apply with all ramification indices equal to 1, and we conclude as above that $i_{K'|K}(\delta) = 0$, hence $p\delta = 0$. In the ramified case the said compatibilities apply with ramification indices equal to p on the level of residue fields, so we conclude $p i_{K'|K}(\delta) = 0$ and $p^2\delta = 0$. Finally, in the case when $K'|K$ is purely inseparable, the tensor product $K' \otimes_K K'$ is a local ring with residue field κ' and nilpotent maximal ideal of length p. After base change to K' we therefore arrive at a diagram where both norm maps are identity maps, so the required compatibility is a tautology. Remark 7.1.6 (2) and Lemma 7.3.6 again apply with ramification indices equal to p, so we conclude as in the previous case that $p^2\delta = 0$. $\square$

Corollary 7.3.11 *Let $L|k$ be a normal extension of degree p, and let P be a closed point of the projective line $\mathbf{P}_k^1$. Then the diagram*

$$
\begin{array}{ccc}
K_n^m(L(t)) & \xrightarrow{\oplus \partial_Q^M} & \bigoplus_{Q \mapsto P} K_{n-1}^M(\kappa(Q)) \\
{\scriptstyle N_{L(t)|k(t)}} \downarrow & & \downarrow {\scriptstyle \Sigma N_{\kappa(Q)|\kappa(P)}} \\
K_n^m(k(t)) & \xrightarrow{\partial_P^M} & K_{n-1}^M(\kappa(P))
\end{array}
$$

commutes for all $n > 0$.

Proof Denote by $\widehat{K}_P$ (resp. $\widehat{L}_Q$) the completions of $k(t)$ (resp. $L(t)$) with respect to the valuations defined by P and Q. In the diagram

$$
\begin{array}{ccccc}
K_n^M(L(t)) & \longrightarrow & \bigoplus_{Q \mapsto P} K_n^M(\widehat{L}_Q) & \xrightarrow{\oplus \partial_Q^M} & \bigoplus_{Q \mapsto P} K_{n-1}^M(\kappa(Q)) \\
{\scriptstyle N_{L(t)|k(t)}} \downarrow & & \downarrow {\scriptstyle \Sigma N_{\widehat{L}_Q|\widehat{K}_P}} & & \downarrow {\scriptstyle \Sigma N_{\kappa(Q)|\kappa(P)}} \quad (7.9) \\
K_n^M(k(t)) & \longrightarrow & K_n^M(\widehat{K}_P) & \xrightarrow{\partial_P^M} & K_{n-1}^M(\kappa(P))
\end{array}
$$

the right square commutes by the above proposition. Commutativity of the left square follows from Lemma 7.2.6 (or Lemma 7.3.6), noting that $L(t) \otimes_{k(t)} \widehat{K}_P$ is a direct product of fields according to Proposition A.6.4 (1) of the Appendix (and the remark following it). The corollary follows. $\square$

Now comes the crucial step in the proof of Theorem 7.3.2.

Lemma 7.3.12 *Let $L|k$ be a normal extension of degree p, and let $k(a)|k$ be a simple finite field extension. Assume that L and $k(a)$ are both subfields of some algebraic extension of k, and denote by $L(a)$ their composite. Then for all $n \geq 0$ the diagram*

$$
\begin{array}{ccc}
K_n^M(L(a)) & \xrightarrow{\ N_{a|L}\ } & K_n^M(L) \\
{\scriptstyle N_{L(a)|k(a)}}\Big\downarrow & & \Big\downarrow{\scriptstyle N_{L|k}} \\
K_n^M(k(a)) & \xrightarrow{\ N_{a|k}\ } & K_n^M(k)
\end{array}
$$

commutes.

Proof   Let P (resp. Q_0) be the closed point of $\mathbf{P}_k^1$ (resp. $\mathbf{P}_L^1$) defined by the minimal polynomial of a over k (resp. L). Given $\alpha \in K_n^M(L(a))$, we have $N_{a|L}(\alpha) = -\partial_\infty^M(\beta)$ for some $\beta \in K_{n+1}^M(L(t))$ satisfying $\partial_{Q_0}^M(\beta) = \alpha$ and $\partial_Q^M(\beta) = 0$ for $Q \neq Q_0$. Corollary 7.3.11 yields

$$
\partial_P^M\big(N_{L(t)|k(t)}(\beta)\big) = \sum_{Q \mapsto P} N_{\kappa(Q)|\kappa(P)}(\partial_Q^M(\beta)) = N_{\kappa(Q_0)|\kappa(P)}(\alpha),
$$

and, by a similar argument, $\partial_{P'}^M\big(N_{L(t)|k(t)}(\beta)\big) = 0$ for $P \neq P'$. Hence by definition of $N_{a|k}$ we get

$$
N_{a|k}\big(N_{L(a)|k(a)}(\alpha)\big) = -\partial_\infty^M\big(N_{L(t)|k(t)}(\beta)\big).
$$

On the other hand, since the only point of $\mathbf{P}_L^1$ above ∞ is ∞, another application of Corollary 7.3.11 gives

$$
\partial_\infty^M\big(N_{L(t)|k(t)}(\beta)\big) = N_{L|k}(\partial_\infty^M(\beta)).
$$

Hence finally

$$
N_{a|k}\big(N_{L(a)|k(a)}(\alpha)\big) = -N_{L|k}(\partial_\infty^M(\beta)) = N_{L|k}\big(N_{a|L}(\alpha)\big).
$$

$\square$

At last, we come to:

Proof of Theorem 7.3.2   As noted before, it is enough to treat the case when k has no nontrivial extension of degree prime to p for a fixed prime p. Let p^m be the degree of the extension $K|k$. We use induction on m, the case $m = 1$ being Proposition 7.3.8. Write $K = k(a_1, \ldots a_r) = k(b_1, \ldots, b_s)$ in two different ways. By Lemma 7.3.7 (2) the extension $k(a_1)|k$ contains a normal subfield $k(\bar{a}_1)$ of degree p over k. Applying Lemma 7.3.12 with $a = a_1$ and $L = k(\bar{a}_1)$

yields $N_{a_1|k} = N_{\bar{a}_1|k} \circ N_{a_1|k(\bar{a}_1)}$. So by inserting $\bar{a}_1$ in the system of the a_i and reindexing we may assume that $[k(a_1) : k] = p$, and similarly $[k(b_1) : k] = p$. Write K_0 for the composite of $k(a_1)$ and $k(b_1)$ in K, and choose elements c_i with $K = K_0(c_1, \ldots, c_t)$. Note that by Lemma 7.3.7 (1) the fields $k(a_1)$ and $k(b_1)$ have no nontrivial prime to p extensions, so we may apply induction to conclude that

$$N_{a_2,\ldots,a_r|k(a_1)} = N_{K_0|k(a_1)} \circ N_{c_1,\ldots c_t|K_0} \quad \text{and}$$

$$N_{b_2,\ldots,b_s|k(b_1)} = N_{K_0|k(b_1)} \circ N_{c_1,\ldots c_t|K_0}.$$

On the other hand, Lemma 7.3.12 for $a = a_1$ and $L = k(b_1)$ implies

$$N_{a_1|k} \circ N_{K_0|k(a_1)} = N_{b_1|k} \circ N_{K_0|k(b_1)}.$$

The above equalities imply $N_{a_1,\ldots,a_r|k} = N_{b_1,\ldots,b_s|k}$, as desired. $\qquad\square$

7.4 Reciprocity laws

As an application of the existence of norm maps, we now prove two theorems which both go under the name 'reciprocity law', though they are quite different. The first one will be the general form of the Weil reciprocity law. For its proof we need a compatibility between the tame symbol and the norm map (generalizing Proposition 7.3.9), which we explain first.

Proposition 7.4.1 *Let K be a field complete with respect to a discrete valuation v with residue field κ. Let $K'|K$ be a finite extension, and denote by κ' the residue field of the unique extension v' of v to K'. Then for all $n > 0$ the diagram*

$$
\begin{array}{ccc}
K_n^M(K') & \xrightarrow{\ \partial_{K'}^M\ } & K_{n-1}^M(\kappa') \\[2pt]
\scriptstyle N_{K'|K} \downarrow & & \downarrow \scriptstyle N_{\kappa'|\kappa} \\[2pt]
K_n^M(K) & \xrightarrow{\ \partial_K^M\ } & K_{n-1}^M(\kappa)
\end{array}
$$

commutes.

Proof We may split up $K'|K$ into a separable and a purely inseparable extension. The latter can be written as the union of a tower of radical extensions of degree equal to the characteristic of K. By applying Proposition 7.3.9 to each of these extensions we reduce to the case when $K'|K$ is a separable extension.

We next fix a prime number p, and let $K^{(p)}$ denote a maximal prime to p extension of K. Then $K^{(p)} \otimes_K K'$ splits up into a product of finite separable

238   *Milnor K-theory*

extensions $K_i | K^{(p)}$ with $[K_i : K^{(p)}]$ a power of p. Using Lemma 7.3.7 inductively, we may write $K_i | K^{(p)}$ as the union of a tower of normal extensions of degree p. A repeated application of Proposition 7.3.9 therefore implies the claim for $K_i | K^{(p)}$. Arguing as in the proof of that proposition, we obtain that for each $\alpha \in K_n^M(K')$ the element $\delta = \partial_K^M(N_{K'|K}(\alpha)) - N_{\kappa'|\kappa}(\partial_{K'}^M(\alpha))$ is annihilated by some integer prime to p. As p was arbitrary here, the proof is complete. $\qquad\square$

Corollary 7.4.2   *Assume moreover that there exist local parameters π and π' for v and v', respectively, satisfying $(-\pi')^e = -\pi$, where e is the ramification index. Then for all $n > 0$ the diagram*

$$
\begin{array}{ccc}
K_n^M(K') & \xrightarrow{\;s_{\pi'}^M\;} & K_n^M(\kappa') \\
{\scriptstyle N_{K'|K}}\downarrow & & \downarrow{\scriptstyle e\,N_{\kappa'|\kappa}} \\
K_n^M(K) & \xrightarrow{\;s_{\pi}^M\;} & K_n^M(\kappa)
\end{array}
$$

commutes.

Note that the assumption of the corollary is satisfied in the cases when the ramification is tame (see Appendix, Proposition A.6.8 (4)) or the extension is purely inseparable.

Proof   By Remark 7.1.6 (1) and the projection formula we have

$$ s_\pi^M(N_{K'|K}(\alpha)) = \partial_K^M(\{-\pi, N_{K'|K}(\alpha)\}) = \partial_K^M(N_{K'|K}(\{-\pi, \alpha\})) $$

for all $\alpha \in K_n^M(K')$. By our assumption on π and Proposition 7.4.1, the last term here equals $e\,N_{\kappa'|\kappa}(\partial_{K'}^M(\{-\pi', \alpha\})) = e\,N_{\kappa'|\kappa}(s_{\pi'}^M(\alpha))$, as desired. $\quad\square$

The proposition has the following globalization.

Corollary 7.4.3   *Let K be a field equipped with a discrete valuation v with residue field $\kappa(v)$, and let $K'|K$ be a finite extension. Assume that the integral closure of the valuation ring A of v in K' is a finite A-module, and for an extension w of v to K' denote by $\kappa(w)$ the corresponding residue field. Then for all $n > 0$ the diagram*

$$
\begin{array}{ccc}
K_n^M(K') & \xrightarrow{\;\oplus\partial_w^M\;} & \bigoplus_{w|v} K_{n-1}^M(\kappa(w)) \\
{\scriptstyle N_{K'|K}}\downarrow & & \downarrow{\scriptstyle \Sigma N_{\kappa(w)|\kappa(v)}} \\
K_n^M(K) & \xrightarrow{\;\partial_K^M\;} & K_{n-1}^M(\kappa(v))
\end{array}
$$

commutes, where the sum is over the finitely many extensions w of v.

Proof This is proven by exactly the same argument as Corollary 7.3.11: one has a diagram analogous to diagram (7.9) considered there, whose right square commutes by Proposition 7.4.1, and the left square by Proposition 7.3.6. □

We may now extend the Weil reciprocity law to the case of curves.

Proposition 7.4.4 (Weil reciprocity law for a curve) *Let C be a smooth projective curve over k. For a closed point P let $\partial_P^M : K_n^M(k(C)) \to K_{n-1}^M(\kappa(P))$ be the tame symbol coming from the valuation on $k(C)$ defined by P. Then for all $\alpha \in K_n^M(k(C))$ we have*

$$\sum_{P \in C_0} (N_{\kappa(P)|k} \circ \partial_P^M)(\alpha) = 0.$$

Proof Take a finite morphism $\phi : C \to \mathbf{P}^1$ defined over k. It induces a diagram

$$
\begin{array}{ccccc}
K_n^M(k(C)) & \xrightarrow{\oplus \partial_Q^M} & \displaystyle\bigoplus_{Q \in C_0} K_{n-1}^M(\kappa(Q)) & \xrightarrow{\Sigma N_{\kappa(Q)|k}} & K_{n-1}^M(k) \\
{\scriptstyle N_{k(C)|k(t)}} \downarrow & & \downarrow {\scriptstyle \oplus_P \Sigma_{Q \mapsto P} N_{\kappa(Q)|\kappa(P)}} & & \downarrow {\scriptstyle \mathrm{id}} \\
K_n^M(k(t)) & \xrightarrow{\oplus \partial_P^M} & \displaystyle\bigoplus_{P \in \mathbf{P}_0^1} K_{n-1}^M(\kappa(P)) & \xrightarrow{\Sigma N_{\kappa(P)|k}} & K_{n-1}^M(k),
\end{array}
$$

where commutativity of the left square follows from Corollary 7.4.3 (applicable in view of Remark A.6.5 of the Appendix), and that of the right square from property (4) of the norm map. According to Corollary 7.2.4, the lower row is a complex, hence so is the upper row by commutativity of the diagram. □

Remark 7.4.5 A special case of Weil's reciprocity law often occurs in the following form. For a divisor $D = \sum n_P P \in \mathrm{Div}(C)$ and a rational function $f \in k(C)$ such that $n_P = 0$ at all poles of f put

$$f(D) := \prod_P N_{\kappa(P)|k}(f(P))^{n_P},$$

where $f(P)$ is defined as the image of f in $\kappa(P)$.

Now suppose f and g are rational functions on C such that $\mathrm{div}(f)$ and $\mathrm{div}(g)$ have disjoint support, i.e. no closed point of C has a nonzero coefficient in both $\mathrm{div}(f)$ and $\mathrm{div}(g)$. Then by applying the case $n = 2$ of the Weil reciprocity law to the symbol $\{f, g\}$ and using the explicit description of Example 7.1.5 one obtains the simple formula

$$f(\mathrm{div}(g)) = g(\mathrm{div}(f)).$$

The second reciprocity law we discuss in this section is due to Rosset and Tate, and only concerns K_2. Let $f, g \in k[t]$ be nonzero relatively prime polynomials. Define the *Rosset–Tate symbol* $(f|g) \in K_2^M(k)$ by

$$(f|g) := s_t^M \Big(\sum_{\{P:\, g(P)=0\}} (\psi_P^M \circ \partial_P^M)(\{g, f\}) \Big), \qquad (7.10)$$

where s_t^M is a specialization map at 0 and ψ_P^M is the coresidue map introduced before Corollary 7.2.3. For g constant we set $(f|g) := 0$.

The symbol is additive in both variables, in the sense that $(f|g_1 g_2) = (f|g_1)+(f|g_2)$ and $(f_1 f_2|g) = (f_1|g)+(f_2|g)$. The following lemma describes it explicitly.

Lemma 7.4.6 *The Rosset–Tate symbol has the following properties.*

1. *If g is constant or $g = t$, then $(f|g) = 0$.*
2. *If g is a nonconstant irreducible polynomial different from t and a is a root of g in some algebraic closure of k, then*

$$(f|g) = N_{k(a)|k}(\{-a, f(a)\}).$$

Proof In statement (1) we only have to treat the case $g = t$. In this case the sum in (7.10) defining $(f|g)$ has only one term coming from $P = 0$. Applying Corollary 7.2.3 with $\alpha = \{t, f\}$ yields $\{t, f\} = (\psi_0^M \circ \partial_0^M)(\{t, f\})$, so that $(t|f) = s_t^M(\{t, f(0)\}) = 0$.

For (2), the only closed point P of $\mathbf{A}_k^1$ contributing to the sum is that defined by the polynomial g. Let a be the image of g in $\kappa(P)$, so that $\kappa(P) = k(a)$. Since $\{-a, f(a)\} = s_t^M(\{t - a, f(a)\})$ (where the specialization takes place in $\kappa(P)(t)$), applying Corollary 7.4.2 to the unramified extension $\kappa(P)((t))|k((t))$ yields

$$N_{\kappa(P)|k}(\{-a, f(a)\}) = s_t^M(N_{\kappa(P)(t)|k(t)}(\{t - a, f(a)\})).$$

Therefore the claim is a consequence of the equality

$$N_{\kappa(P)(t)|k(t)}(\{t - a, f(a)\}) = (\psi_P^M \circ \partial_P^M)(\{g, f\}).$$

This equivalently means $N_{\kappa(P)(t)|k(t)}(\{t - a, f(a)\}) = \psi_P^M(f(a))$, which in turn follows from Corollary 7.2.3 applied to $\alpha = N_{\kappa(P)|k(t)}(\{t - a, f(a)\})$, noting that $\partial_P^M(N_{\kappa(P)(t)|k(t)}(\{t - a, f(a)\})) = N_{\kappa(P)|\kappa(P)}(f(a)) = f(a)$ according to Corollary 7.4.3 applied with $K = k(t)$, $K' = \kappa(P)(t)$ and v the valuation defined by P, and moreover $s_{t-1}(N_{\kappa(P)(t)|k(t)}(\{t-a, f(a)\})) = 0$ by Corollary 7.4.2. □

The second statement of the lemma implies:

Corollary 7.4.7 *The symbol $(f|g)$ depends only on the image of f in the quotient ring $k[t]/(g)$.*

Remark 7.4.8 The properties (1) and (2) of the lemma together with the additivity property characterize the symbol. In fact, Rosset and Tate [1] defined their symbol in such an explicit way, with a slight difference: according to their definition, the right-hand side of the formula in property (2) is $N_{k(a)|k}(\{a, f(a)\})$.

To state the main theorem on the Rosset–Tate symbol, introduce the following notation for polynomials $f \in k[t]$: if $f = a_n t^n + a_{n-1} t^{n-1} + \cdots + a_m t^m$ with $a_n a_m \neq 0$, put $\ell(f) := a_n$ (the leading coefficient) and $c(f) := a_m$ (the last nonzero coefficient). They depend multiplicatively on f.

Theorem 7.4.9 (Rosset–Tate reciprocity law) *Let $f, g \in k[t]$ be nonzero relatively prime polynomials. Then*

$$(f|g) + \{c(f), c(g)\} = (g|f) + \{\ell(f), \ell(g)\}.$$

Proof By Corollary 7.2.3 we have

$$\{f, g\} = s_{t-1}(\{f, g\})_{k(t)} + \sum_{\{P:\, f(P)=0\}} (\psi_P^M \circ \partial_P^M)(\{f, g\}) + \sum_{\{P:\, g(P)=0\}} (\psi_P^M \circ \partial_P^M)(\{f, g\}),$$

so that

$$\sum_{\{P:\, g(P)=0\}} (\psi_P^M \circ \partial_P^M)(\{g, f\}) + \{f, g\} = s_{t-1}(\{f, g\})_{k(t)}$$

$$+ \sum_{\{P:\, f(P)=0\}} (\psi_P^M \circ \partial_P^M)(\{f, g\}).$$

By applying the specialization map s_t^M at 0, we obtain

$$(f|g) + s_t(\{f, g\}) = s_{t-1}(\{f, g\}) + (g|f).$$

Finally, writing $f = c(f)t^m \tilde{f}$ and $g = c(g)t^l \tilde{g}$ with $\tilde{f}(0) = \tilde{g}(0) = 1$ we get $s_t^M(\{f, g\}) = \{c(f), c(g)\}$ by definition of s_t^M (Proposition 7.1.4). A similar computation shows $s_{t-1}(\{f, g\}) = \{\ell(f), \ell(g)\}$, and the theorem follows. □

As a corollary, we obtain a bound on the length of the symbol $(f|g)$.

Corollary 7.4.10 *Let f and g be relatively prime polynomials. Then the symbol $(f|g) \in K_2^M(k)$ is of length at most $\deg(g)$, i.e. it is a sum of at most $\deg(g)$ terms of the form $\{a_i, b_i\}$.*

Proof The proof goes by induction on the degree of g. The degree zero case means $(f|g) = 0$, which holds by Lemma 7.4.6 (1). The same statement and additivity of the symbol allows one to assume in the higher degree case that g is monic. Corollary 7.4.7 allows us to assume $\deg(f) < \deg(g)$, after performing Euclidean division of g by f. Theorem 7.4.9 then shows $(f|g) + \{c(f), c(g)\} = (g|f)$. By the inductive hypothesis the symbol $(g|f)$ has length at most $\deg(f)$, so that $(f|g)$ has length at most $\deg(f) + 1 \leq \deg(g)$. $\qquad\qquad\square$

Corollary 7.4.11 *Let $K|k$ be a finite field extension, and let a, b be elements of $K^\times$. Then the symbol $N_{K|k}(\{a, b\}) \in K_2^M(k)$ has length at most $[k(a) : k]$.*

Proof By the projection formula, we have

$$N_{K|k}(\{a, b\}) = N_{k(a)|k}(N_{K|k(a)}(\{a, b\})) = N_{k(a)|k}(\{a, N_{K|k(a)}(b)\}).$$

Let g be the minimal polynomial of $-a$ over k. Then $N_{K|k(a)}(b) = f(-a)$ for some polynomial $f \in k[t]$ and $N_{k(a)|k}(\{a, N_{K|k(a)}(b)\}) = (f|g)$ by Lemma 7.4.6 (2), so the previous corollary applies. $\qquad\qquad\square$

Remark 7.4.12 The Euclidean division process by which we have proven Corollary 7.4.10 also provides an explicit algorithm for computing the symbol $N_{K|k}(\{a, b\})$.

The main motivation for Rosset and Tate to prove their reciprocity law was the following application to central simple algebras.

Proposition 7.4.13 *Let p be a prime number, and let k be a field of characteristic prime to p containing a primitive p-th root of unity ω. Every central simple k-algebra A of degree p is Brauer equivalent to a tensor product of at most $(p - 1)!$ cyclic k-algebras of degree p.*

Note that the proposition proves a special case of the (surjectivity part of the) Merkurjev–Suslin theorem, and moreover it yields a bound on the length of a symbol of order p.

The proof will use the fact that norm maps on K-theory and corestrictions in Galois cohomology are compatible via the Galois symbol. Let's admit this for the moment; a proof will be given in the next section (Proposition 7.5.4).

Proof If A is split, the statement is trivial. If A is nonsplit, then it is a division algebra split by a degree p extension $K|k$. As k has characteristic prime to p by assumption, the extension $K|k$ is separable. Denote by $\widetilde{K}|k$ a Galois closure. Note that $\mathrm{Gal}\,(\widetilde{K}|k)$ is a subgroup of the degree p symmetric group S_p, so it has order dividing $p!$. In particular, each p-Sylow subgroup $P \subset \mathrm{Gal}\,(\widetilde{K}|K)$ has order p, and its fixed field $L := \widetilde{K}^P$ has degree at most $(p-1)!$ over k. Choose an integer $m > 1$ with $[L : k]m \equiv 1 \bmod p$. Since A has period p, the algebra $B := A^{\otimes m}$ satisfies $[L : k][B] = [A]$ in $\mathrm{Br}\,(k)$. Moreover, since $A \otimes_k L$ is split by the extension $\widetilde{K}|L$, so does $B \otimes_k L$. Hence by Corollary 4.7.8 there exist $a, b \in L^\times$ with $B \otimes_k L \cong (a, b)_\omega$. We have

$$[A] = [L : k][B] = \mathrm{Cor}_k^L([B \otimes_k L]) = \mathrm{Cor}_k^L(h_{L,p}^2(\{a, b\}))$$
$$= h_{k,p}^2(N_{L|k}(\{a, b\}))$$

using Propositions 4.7.1 and 7.5.4. By the previous corollary, the symbol $N_{L|k}(\{a, b\})$ has length at most $[L : k] \le (p-1)!$, whence the proposition. □

Remark 7.4.14 The case $p = 2$ gives back Corollary 1.2.1. In this case, the bound $(p-1)!$ is trivially optimal. However, for $p > 2$ it may be improved to $(p-1)!/2$ in the presence of a p-th root of unity (see Exercise 10). We know little about the optimality of the latter bound. In fact, the following famous question is attributed to Albert: *Is every degree p division algebra isomorphic to a cyclic algebra?* A positive answer for $p = 3$ follows from the above bound (the result is originally due to Wedderburn [3]; see Exercise 9), but the question is open for $p > 3$. Albert proposed conjectural counterexamples for $p = 5$, which were shown to be actually cyclic in the paper Rowen [3], where new putative counterexamples are put forward. A positive answer to the question in characteristic 0 would imply the same in positive characteristic (see Chapter 9, Exercise 4). Positive answers are known in important special cases, such as arithmetic fields (see Corollary 6.3.8 as well as Remarks 6.5.4 and 6.5.5).

7.5 Applications to the Galois symbol

In this section we collect some useful elementary remarks about the Galois symbol and the Bloch–Kato conjecture. To begin with, we examine compatibility properties for the Galois symbol.

Proposition 7.5.1   *Let K be a field equipped with a discrete valuation v with residue field κ. Let m be an integer invertible in K. Then for all $n > 0$ the diagram*

$$
\begin{array}{ccc}
K_n^M(K) & \xrightarrow{\ \partial^M\ } & K_{n-1}^M(\kappa) \\[4pt]
{\scriptstyle h_{K,m}^n}\downarrow & & \downarrow{\scriptstyle h_{\kappa,m}^{n-1}} \\[4pt]
H^n(K,\mu_m^{\otimes n}) & \xrightarrow{\ \partial_v^n\ } & H^{n-1}(\kappa,\mu_m^{\otimes(n-1)})
\end{array}
$$

commutes, where ∂_v^n is the residue map introduced in Chapter 6, Section 6.8.

Proof   Without loss of generality we may assume K is complete with respect to v. The case $n = 1$ follows immediately from the construction of the maps concerned. In the general case it suffices, as usual, to consider symbols of the shape $\{a, u_2, \ldots, u_n\} \in K_n^M(K)$, where the u_i are units for v. Corollary 7.1.10 yields a well-defined section $\lambda_m : K_{n-1}^M(\kappa)/m \to K_{n-1}^M(K)/m$ to any specialization map modulo m, sending $\{\bar{u}_2, \ldots, \bar{u}_n\} \in K_{n-1}^M(\kappa)$ to $\{u_2, \ldots, u_n\} \in K_{n-1}^M(K)$, where the u_i are arbitrary liftings u_i of the $\bar{u}_i$ to units in K. Moreover, the diagram

$$
\begin{array}{ccc}
K_{n-1}^M(\kappa)/m & \xrightarrow{\ \lambda_m\ } & K_{n-1}^M(K)/m \\[4pt]
{\scriptstyle h_{\kappa,m}^{n-1}}\downarrow & & \downarrow{\scriptstyle h_{K,m}^{n-1}} \\[4pt]
H^{n-1}(\kappa,\mu_m^{\otimes(n-1)}) & \xrightarrow{\ \mathrm{Inf}\ } & H^{n-1}(K,\mu_m^{\otimes(n-1)})
\end{array}
$$

commutes, as one verifies using the explicit description of the Kummer maps $h_{K,m}^1$ and $h_{\kappa,m}^1$ in terms of cocycles (see e.g. Remark 3.2.4). The proposition now follows by induction from the case $n = 1$ via Lemma 6.8.4 (applied with $p = 1, q = n - 1$, G the absolute Galois group of K, H the inertia group of v, $A = \mu_m$ and $B = \mu_m^{\otimes(n-1)}$).   □

As a corollary, we get the compatibility between specialization maps.

Corollary 7.5.2   *Assume moreover that t is a local parameter for v. Then the diagram of specialization maps*

$$
\begin{array}{ccc}
K_n^M(K) & \xrightarrow{\ s_t^M\ } & K_n^M(\kappa) \\[4pt]
{\scriptstyle h_{K,m}^n}\downarrow & & \downarrow{\scriptstyle h_{\kappa,m}^n} \\[4pt]
H^n(K,\mu_m^{\otimes n}) & \xrightarrow{\ s_t^n\ } & H^n(\kappa,\mu_m^{\otimes(n-1)})
\end{array}
$$

commutes, where s_t^n is the specialization map introduced in Chapter 6, Section 6.8.

Proof This follows from the proposition in view of Remark 7.1.6 (1) and the construction of s_t^n in Construction 6.8.6. □

Another immediate corollary is the compatibility between Milnor's and Faddeev's exact sequences.

Corollary 7.5.3 *The diagram with exact rows*

$$0 \to \quad K_n^M(k) \quad \to \quad K_n^M(k(t)) \quad \xrightarrow{\oplus \partial_P^M} \quad \bigoplus_{P \in \mathbf{P}_0^1 \setminus \{\infty\}} K_{n-1}^M(\kappa(P)) \quad \to 0$$

$$h_{k,m}^n \downarrow \qquad\qquad h_{k(t),m}^n \downarrow \qquad\qquad\qquad\qquad \downarrow \oplus h_{\kappa(P),m}^{n-1}$$

$$0 \to H^n(k, \mu_m^{\otimes n}) \to H^n(k(t), \mu_m^{\otimes n}) \quad \xrightarrow{\oplus \partial_P^n} \quad \bigoplus_{P \in \mathbf{P}_0^1 \setminus \{\infty\}} H^{n-1}(\kappa(P), \mu_m^{\otimes(n-1)}) \to 0$$

commutes, where the upper row is the sequence of Theorem 7.2.1, and the lower row that of Corollary 6.9.3.

Finally, we give the already announced compatibility between norm maps in K-theory and corestrictions in cohomology.

Proposition 7.5.4 *Let $K|k$ be a finite separable extension, and m an integer invertible in k. Then for all $n \geq 0$ the diagram*

$$\begin{array}{ccc} K_n^M(K) & \xrightarrow{\ N_{K|k}\ } & K_n^M(k) \\[4pt] {\scriptstyle h_{K,m}^n} \downarrow & & {\scriptstyle h_{k,m}^n} \downarrow \\[4pt] H^n(K, \mu_m^{\otimes n}) & \xrightarrow{\ \mathrm{Cor}\ } & H^n(k, \mu_m^{\otimes n}) \end{array}$$

commutes.

Proof By property (4) of the norm map and the similar property of corestrictions (which follows easily from their construction), we reduce to the case when $K = k(a)$ is a simple field extension. In this case $N_{K|k} = -\partial_\infty^M \circ \psi_P$, where P is the closed point $\mathbf{P}_k^1$ defined by the minimal polynomial of a. By Corollary 6.9.4, a similar formula holds for the corestriction map. The two are compatible via the Galois symbol by virtue of Corollary 7.5.3. □

We now turn to applications to the Bloch–Kato conjecture. The first one is an immediate consequence of Corollary 7.5.3.

Proposition 7.5.5 (Bloch) *Let $m, n > 0$ be integers, with m invertible in k.*

1. The Galois symbol

$$h_{k(t),m}^n : K_n^M(k(t))/m \to H^n(k(t), \mu_m^{\otimes n})$$

is injective (resp. surjective or bijective) if and only if the Galois symbols

$$h_{k,m}^n : K_n^M(k)/m \to H^n(k, \mu_m^{\otimes n}) \text{ and}$$
$$h_{L,m}^{n-1} : K_{n-1}^M(L)/m \to H^{n-1}(L, \mu_m^{\otimes(n-1)})$$

have the same property for all finite simple extensions $L|k$.

2. *Assume that $h_{L,m}^{n-1} : K_{n-1}^M(L)/mK_{n-1}^M(L) \to H^{n-1}(L, \mu_m^{\otimes(n-1)})$ is bijective for all finite simple extensions $L|k$. Then*

$$\ker(h_{k,m}^n) \cong \ker(h_{k(t),m}^n) \quad \text{and} \quad \operatorname{coker}(h_{k,m}^n) \cong \operatorname{coker}(h_{k(t),m}^n).$$

This gives a means for proving bijectivity of the Galois symbol for $k(t)$ if the bijectivity is already known for fields of smaller transcendence degree.

Here is another (unpublished) criterion of Bloch for the surjectivity of the Galois symbol.

Proposition 7.5.6 (Bloch)   *Let $m, n > 0$ be integers, with m invertible in k. Assume that*

- *the Galois symbol $h_{L,m}^{n-1} : K_{n-1}^M(L)/mK_{n-1}^M(L) \to H^{n-1}(L, \mu_m^{\otimes n-1})$ is an isomorphism for all finitely generated extensions $L|k$;*
- *the Galois symbol $h_{k,m}^n : K_n^M(k)/mK_n^M(k) \to H^n(k, \mu_m^{\otimes n})$ is surjective.*

Then the following statements are equivalent:

1. *The Galois symbol $h_{K,m}^n$ is surjective for all fields K containing k.*
2. *For all field extensions $K|k$ equipped with a discrete valuation v the restriction map $H^n(K, \mu_m^{\otimes n}) \to H^n(\widehat{K}_v, \mu_m^{\otimes n})$ is surjective, where $\widehat{K}_v$ stands for the completion of K with respect to v.*

In particular, the two statements are equivalent in the case when $n = 2$ and $\operatorname{cd}(k) \leq 1$.

Proof   For (1) $\Rightarrow$ (2) it is enough to establish the surjectivity of the map $K_n^M(K)/mK_n^M(K) \to K_n^M(\widehat{K}_v)/mK_n^M(\widehat{K}_v)$ induced by $i_{K_v|K}$, which readily follows by combining the second sequence in Proposition 7.1.7 for K with Corollary 7.1.10 for K_v.

For (2) $\Rightarrow$ (1), note first that the first assumption and part (2) of the previous corollary yield an isomorphism $\operatorname{coker}(h_{k,m}^n) \cong \operatorname{coker}(h_{k(t),m}^n)$. Applying statement (2) to the completion $\widehat{K}_P$ of $k(t)$ with respect to the discrete valuation defined by a closed point P gives the surjectivity of the map $H^n(k(t), \mu_m^{\otimes n}) \to H^n(\widehat{K}_P, \mu_m^{\otimes n})$. By Proposition 6.8.7, the latter group surjects onto $H^n(\kappa(P), \mu_m^{\otimes n})$ via every specialization map. Taking Corollary 7.5.2 into account, we thus get a surjection $\operatorname{coker}(h_{k,m}^n) \to \operatorname{coker}(h_{\kappa(P),m}^n)$.

Proceeding by induction using this statement and part (2) of the previous corollary, we get surjective maps $\mathrm{coker}\,(h^n_{k,m}) \to \mathrm{coker}\,(h^n_{K,m})$ for all finitely generated extensions $K|k$. Finally, one may write an arbitrary extension $K|k$ as a direct limit of finitely generated fields $K_i|k$, and obtain $\mathrm{coker}\,(h^n_{K,m}) \cong \varinjlim \mathrm{coker}\,(h^n_{K_i,m})$. Thus the surjectivity of $h^n_{k,m}$ implies that of $h^n_{K,m}$ for all extensions $K|k$.

The last statement of the proposition is obvious, since $h^1_{K,m}$ is an isomorphism for all fields K by Kummer theory, and $H^2(k, \mu^{\otimes 2}_m)$ vanishes for fields of cohomological dimension ≤ 1. $\qquad\qquad\square$

Remark 7.5.7 For fields containing a field of cohomological dimension 1 (in particular, for fields of positive characteristic) we thus get a purely cohomological reformulation of the surjectivity part of the Merkurjev–Suslin theorem, which in the case of fields containing a primitive m-th root of unity reduces to an even more suggestive surjectivity statement about the map $_m\mathrm{Br}\,(K) \to {}_m\mathrm{Br}\,(K_v)$. Bloch found this argument in the 1970s well before the Merkurjev–Suslin theorem was proven. By a result of Tate, however, the theorem was already known for number fields (see the next section), so in fact Bloch's result rephrased the surjectivity of $h^2_{K,m}$ for all fields K. For higher n it gives an inductive strategy for proving the Bloch–Kato conjecture.

We close this section by an important reduction statement due to Tate, which reduces the proof of the Bloch–Kato conjecture to the case of p-torsion coefficients. It will be used in the proof of the Merkurjev–Suslin theorem.

Proposition 7.5.8 (Tate) *Let $m, n > 0$ be integers, with m invertible in k. Assume that the Galois symbol $h^{n-1}_{k,m}$ is surjective, and that $h^n_{k,p}$ is bijective for all prime divisors p of m. Then the Galois symbol $h^n_{k,m}$ is bijective.*

For the proof we need the following lemma.

Lemma 7.5.9 *Assume k contains a primitive p-th root of unity ω, where p is a prime invertible in k. Then for all $r > 0$ we have a commutative diagram*

$$
\begin{array}{ccc}
\mu_p \otimes K^M_{n-1}(k) & \xrightarrow{\{\,,\,\}} & K^M_n(k)/p^r K^M_n(k) \\
{\scriptstyle [\omega]\cup h^{n-1}_{k,p}} \downarrow & & \downarrow {\scriptstyle h^n_{k,p^r}} \\
H^{n-1}(k, \mu^{\otimes n}_p) & \xrightarrow{\;\delta^n\;} & H^n(k, \mu^{\otimes n}_{p^r}),
\end{array}
$$

where $[\omega]$ denotes the class of ω in $H^0(k, \mu_p)$, the upper horizontal map associates with a pair (ω, a) the symbol $\{\omega, a\}$ modulo p^r, and δ^n is a so-called

Bockstein homomorphism, i.e. a boundary map coming from the long exact sequence associated with the sequence

$$1 \to \mu_{p^r}^{\otimes n} \to \mu_{p^{r+1}}^{\otimes n} \xrightarrow{p^r} \mu_p^{\otimes n} \to 1 \qquad (7.11)$$

of Galois modules.

Proof   First a word about exact sequence (7.11). For $n = 0$ it is none but the natural exact sequence

$$0 \to \mathbf{Z}/p^r\mathbf{Z} \to \mathbf{Z}/p^{r+1}\mathbf{Z} \xrightarrow{p^r} \mathbf{Z}/p\mathbf{Z} \to 0,$$

which can be regarded as an exact sequence of $\mathbf{Z}/p^{r+1}\mathbf{Z}$-modules via the natural maps $\mathbf{Z}/p^{r+1}\mathbf{Z} \to \mathbf{Z}/p^r\mathbf{Z}$ and $\mathbf{Z}/p^{r+1}\mathbf{Z} \to \mathbf{Z}/p\mathbf{Z}$ given by multiplication by p and p^r, respectively. The general sequence is obtained by tensoring this sequence by $\mu_{p^{r+1}}^{\otimes n}$ over $\mathbf{Z}/p^{r+1}\mathbf{Z}$. Given a symbol $\alpha \in K_{n-1}^M(k)$, the element $y := h_{k,p}^{n-1}(\alpha)$ comes from the element $y_{r+1} := h_{k,p^{r+1}}^{n-1}(\alpha)$ via the map $H^{n-1}(k, \mu_{p^{r+1}}^{\otimes(n-1)}) \to H^{n-1}(k, \mu_p^{\otimes(n-1)})$ induced by raising the coefficients to the p^r-th power. Similarly, the element $y_r := h_{k,p^r}^{n-1}(\alpha)$ is the image of y_{r+1} via the map that raises coefficients to the p-th power. Now using Proposition 3.4.8 and the preceding discussion, we have

$$\delta_n([\omega] \cup h_{k,p}^{n-1}(\alpha)) = \delta^n([\omega] \cup y) = \delta^n([\omega] \cup y_{r+1}) = \delta^1([\omega]) \cup y_{r+1}$$
$$= \delta^1([\omega]) \cup y_r.$$

It is immediately seen by examining the Kummer sequence that $\delta^1([\omega])$ is none but $h_{k,p^r}^1([\omega])$. Hence the right-hand side is $h_{k,p^r}^n(\{\omega, \alpha\})$ by definition of the Galois symbol, and the proof is complete. $\square$

Proof of Proposition 7.5.8   By decompositing m into a product of prime powers we see that it is enough to consider the case $m = p^r$. Moreover, we may and do assume that k contains a primitive p-th root of unity ω. Indeed, if not, consider the commutative diagrams

$$
\begin{array}{ccc}
K_n^M(k)/p^r & \xrightarrow{h_{k,p^r}^n} & H^n(k, \mu_{p^r}^{\otimes n}) \\
\scriptstyle i_{k(\omega)|k} \downarrow & & \downarrow \scriptstyle \mathrm{Res} \\
K_n^M(k(\omega))/p^r & \xrightarrow{h_{k(\omega),p^r}^n} & H^n(k(\omega), \mu_{p^r}^{\otimes n})
\end{array}
\quad \text{and} \quad
\begin{array}{ccc}
K_n^M(k)/p^r & \xrightarrow{h_{k,p^r}^n} & H^n(k, \mu_{p^r}^{\otimes n}) \\
\scriptstyle N_{k(\omega)|k} \uparrow & & \uparrow \scriptstyle \mathrm{Cor} \\
K_n^M(k(\omega))/p^r & \xrightarrow{h_{k(\omega),p^r}^n} & H^n(k(\omega), \mu_{p^r}^{\otimes n}),
\end{array}
$$

where the second diagram commutes by Proposition 7.5.4. The composite maps Cor $\circ$ Res and $N_{k(\omega)|k} \circ i_{k(\omega)|k}$ are both multiplication by the degree $[k(\omega) : k]$ which is prime to p. As the groups involved are p-primary torsion groups, these composite maps are isomorphisms, which implies that the vertical maps are injective in the first diagram and surjective in the second. It follows that the bijectivity of $h^n_{k(\omega),p^r}$ implies that of h^n_{k,p^r}.

For $m = p^r$ the proof goes by induction on r using the exact sequence (7.11). It induces the bottom row in the exact commutative diagram

$$\begin{array}{ccccccc}
K^M_n(k)/p^r K^M_n(k) & \xrightarrow{\ p\ } & K^M_n(k)/p^{r+1}K^M_n(k) & \longrightarrow & K^M_n(k)/pK^M_n(k) & \longrightarrow & 0 \\
\ \downarrow{\scriptstyle h^n_{k,p^r}} \wr & & \ \downarrow{\scriptstyle h^n_{k,p^{r+1}}} & & \ \downarrow{\scriptstyle h^n_{k,p}} \wr & & \\
H^n(k,\mu^{\otimes n}_{p^r}) & \longrightarrow & H^n(k,\mu^{\otimes n}_{p^{r+1}}) & \longrightarrow & H^n(k,\mu^{\otimes n}_{p}). & &
\end{array}$$

By the inductive hypothesis the left and the right vertical maps are isomorphisms. A diagram chase then shows that $h^n_{k,p^{r+1}}$ is surjective. For injectivity, we complete the left-hand side of the diagram as

$$\begin{array}{ccccc}
\mu_p \otimes K^M_{n-1}(k) & \xrightarrow{\ \{\ ,\ \}\ } & K^M_n(k)/p^r K^M_n(k) & \xrightarrow{\ p\ } & K^M_n(k)/p^{r+1}K^M_n(k) \\
\ \downarrow{\scriptstyle \omega \cup h^{n-1}_{k,p}} & & \ \downarrow{\scriptstyle h^n_{k,p^r}} \wr & & \ \downarrow{\scriptstyle h_{k,p^{r+1}}} \\
H^{n-1}(k,\mu^{\otimes n}_p) & \xrightarrow{\ \delta\ } & H^n(k,\mu^{\otimes n}_{p^r}) & \longrightarrow & H^n(k,\mu^{\otimes n}_{p^{r+1}})
\end{array}$$

using the lemma above, where the upper row is not necessarily exact but is a complex since $p\{\omega, a\} = 0$ for all $a \in k^\times$. If $\alpha \in K^M_n(k)$ is such that $h^n_{k,p^{r+1}}(p\alpha) = 0$ in $H^2(k, \mu^{\otimes n}_{p^{r+1}})$, the diagram shows that $h^n_{k,p^r}(\alpha)$ is in the image of δ. Now the left vertical map is surjective, as so is $h^{n-1}_{k,p}$ by assumption, and tensor product by μ_p is the identity map by our assumption that $\omega \in k$. Thus we may modify α by a symbol of the form $\{\omega, a\}$ to get $h^n_{k,p^r}(\alpha) = 0$ without changing $p\alpha$. Hence $\alpha \in p^r K^M_n(k)$ by injectivity of h^2_{k,p^r}, so $p\alpha \in p^{r+1}K^M_n(k)$, i.e. $\ker(h^n_{k,p^{r+1}}) = 0$. $\qquad\qquad\square$

7.6 The Galois symbol over number fields

In this section we establish the following basic theorem which was the first substantial result in the direction of the Merkurjev–Suslin theorem.

Theorem 7.6.1 (Tate) *If k is a number field, then the Galois symbol $h^2_{k,m}$ is bijective for all positive integers m.*

Remarks 7.6.2

1. It is known that a number field k has p-cohomological dimension 2 if $p > 2$, or if $p = 2$ and k is totally imaginary (see Serre [4], II.4.4), so the theorem answers (but historically predates) the full Bloch–Kato conjecture for odd m or totally imaginary k.

2. Surjectivity of the Galois symbol is a consequence of the fact that all central simple algebras are cyclic over k (Remark 6.5.5). Recall that this difficult result uses the main theorems of class field theory.

In view of the last remark, we only prove injectivity of the Galois symbol here. This will also use facts from class field theory, but there are purely algebraic ideas involved as well, which are interesting in their own right. We begin by explaining these. From now on, we only consider the case when $m = p$ is a prime (which is allowed by Proposition 7.5.8).

The starting point is the following easy observation.

Lemma 7.6.3 *Let k be a field containing the p-th roots of unity for some prime p invertible in k, and let a, b be elements in $k^\times$. If $h^2_{k,p}(\{a, b\}) = 0$, then $\{a, b\} \in pK^M_2(k)$.*

Proof By Proposition 4.7.1 and Corollary 4.7.5 we find $c \in k(\sqrt[p]{a})$ with $b = N_{k(\sqrt[p]{a})|k}(c)$. Using the projection formula we compute

$$\{a, b\} = \{a, N_{k(\sqrt[p]{a})|k}(c)\} = N_{k(\sqrt[p]{a})|k}(\{a, c\}) = pN_{k(\sqrt[p]{a})|k}(\{\sqrt[p]{a}, c\}),$$

whence the lemma. $\square$

Now given $a, b, x \in k^\times$ with $h^2_{k,p}(\{a, b\}) = h^2_{k,p}(\{b, x\})$, an application of the lemma to $\{a, b\} - \{b, x\} = \{a, b\} + \{x, b\} = \{ax, b\}$ shows that $\{a, b\} = \{b, x\}$ modulo $pK^M_2(k)$. We can then continue this procedure with some $y \in k^\times$ satisfying $h^2_{k,p}(\{b, x\}) = h^2_{k,p}(\{x, y\})$, and so on. If every other pair $(c, d) \in k^\times$ can be reached by a chain of this type, then injectivity of $h^2_{k,p}$ follows, at least for symbols of length 1. The following definition formalizes this idea.

Definition 7.6.4 Let k be a field, and p a prime number invertible in k. We say that the *chain lemma holds for k and p* if for any two pairs (a, b) and $(c, d) \in k^\times \times k^\times$ satisfying $h^2_{k,p}(\{a, b\}) = h^2_{k,p}(\{c, d\})$ there exist an integer $n \geq 0$ and elements $x_{-1} = a, x_0 = b, x_1, ..., x_{n-1} = c, x_n = d$ in $k^\times$ such that

$$h^2_{k,p}(\{x_i, x_{i+1}\}) = h^2_{k,p}(\{x_{i+1}, x_{i+2}\})$$

holds for all $i = -1, 0, \ldots, n - 2$. We say that the *chain lemma holds with length N* if for all pairs (a, b) and $(c, d) \in k^\times \times k^\times$ we may choose a chain as above with $n \leq N$.

Remarks 7.6.5

1. It is conjectured that the chain lemma holds for all fields k and all primes p. For $p = 2$ we shall prove this in a moment; for $p = 3$ see Rost [2]. Rost (unpublished) has also proven that the chain lemma always holds for a prime p and a field k having no nontrivial finite extensions of degree prime to p.
2. Variants of the chain lemma occur in quadratic form theory (see Elman and Lam [1]), and in a more general context in Rost [4].

Note that the argument after Lemma 7.6.3 yields:

Corollary 7.6.6 *Assume that the chain lemma holds for k and p. Then the identity $h_{k,p}^2(\{a, b\}) = h_{k,p}^2(\{c, d\})$ implies $\{a, b\} = \{c, d\} \mod p K_2^M(k)$ for all $a, b, c, d \in k^\times$.*

We can now formalize the strategy for proving Theorem 7.6.1.

Proposition 7.6.7 *Let k be a field containing the p-th roots of unity and satisfying the following two conditions:*

- *the chain lemma holds for k;*
- *for each finite set of elements $a_1, b_1, a_2, b_2, \ldots, a_r, b_r$ in $k^\times$ we may find a degree p cyclic extension $K|k$ so that $\mathrm{Res}_k^K(h_{k,p}^2(\{a_i, b_i\})) = 0$ for $1 \leq i \leq r$.*

Then the Galois symbol $h_{k,p}^2$ is injective.

Proof Let $\alpha = \sum_{i=1}^r \{a_i, b_i\}$ be a symbol in the kernel of $h_{k,p}^2$. Take an extension $K|k$ as in the second condition above, and write it as $K = k(\sqrt[p]{c})$ for some $c \in k^\times$ using Kummer theory. By Proposition 4.7.1 and Corollary 4.7.7 we find elements $d_i \in k^\times$ with $h_{k,p}^2(\{a_i, b_i\}) = h_{k,p}^2(\{c, d_i\})$ for $1 \leq i \leq r$. Corollary 7.6.6 shows that under the first assumption $\{a_i, b_i\} = \{c, d_i\} \mod pK_2^M(k)$ for all i, so setting $d = d_1 d_2 \cdots d_r$ yields $\alpha = \{c, d\} \mod pK_2^M(k)$. The proposition then follows from Lemma 7.6.3. $\square$

We next verify that the chain lemma holds for all primes and all number fields. The first step in this direction is:

Lemma 7.6.8 *If $p = 2$, the chain lemma holds with length 3 for all fields k.*

In Chapter 1 we gave a sketch of a proof by Tate in an exercise. We now give another proof due to Rost.

Proof   The condition $h_{k,2}^2(\{a,b\}) = h_{k,2}^2(\{c,d\})$ means that the quaternion algebras (a,b) and (c,d) are isomorphic over k. We may assume they are non-split (otherwise use the isomorphisms $(a,b) \cong (1,b)$ and $(c,d) \cong (c,1)$). Choose $X, Y \in (a,b) \setminus k$ such that $X^2 = b$ and $Y^2 = c$, and define $Z = XY - YX$. Consider the reduced characteristic polynomial $N(t - Z)$ of Z, where N is the quaternion norm. It has degree 2, and the coefficient of t is the quaternion trace $T(Z) = Z + \overline{Z}$ which is 0. Therefore $N(t - Z) = t^2 - z$ with $z := Z^2 \in k$. Notice that $XZ + ZX = X(XY - YX) + (XY - YX)X = 0$ and similarly $YZ + ZY = 0$. So if $Z \neq 0$, then (X, Z) and (Y, Z) are both quaternion bases of (a,b), and hence $(a, b) \cong (b, z) \cong (z, c) \cong (c, d)$ is a suitable chain. If $Z = 0$, then Y lies in the 2-dimensional commutative subalgebra $k[X]$. Since moreover its minimal polynomial over k is $t^2 - c$, we must have $Y = \lambda X$ for suitable $\lambda \in k$ and hence $c = \lambda^2 b$. Thus we have a length 3 chain $(a, b) \cong (b, a) \cong (a, c) \cong (c, d)$ in this case as well.   □

Next we have the following lemma of Tate.

Lemma 7.6.9   *If k contains the p-th roots of unity and the p-torsion subgroup $_p\mathrm{Br}(k)$ is cyclic, the chain lemma holds for k and p with length 4.*

Proof   Lemma 7.6.8 allows us to assume that p is odd. Assume given a, b, c, d such that $h_{k,p}^2(\{a,b\}) = h_{k,p}^2(\{c,d\})$. As in the above proof, we may assume that both sides are nonzero, and therefore yield a generator α of $_p\mathrm{Br}(k) \cong \mathbf{F}_p$. Via this last isomorphism $h_{k,p}^2$ may be identified with a bilinear map $\phi : k^\times/k^{\times p} \times k^\times/k^{\times p} \to \mathbf{F}_p$ of $\mathbf{F}_p$-vector spaces, which is moreover anticommutative by Proposition 7.1.1. Our task is to find $x, y \in k^\times$ satisfying $\phi(b, x) = \phi(x, y) = \phi(y, c) = \alpha$. The linear forms $t \mapsto \phi(b, t)$ and $t \mapsto \phi(t, c)$ are nonzero and hence surjective. If these forms are either linearly independent or equal, then we can take $y = c$ and find an x such that $\phi(b, x) = \alpha$ and $\phi(x, y) = \phi(x, c) = \alpha$. Suppose now that these two linear forms are dependent but not equal. The forms $t \mapsto \phi(t, c)$ and $t \mapsto \phi(d, t) = \phi(t, d^{-1})$ are independent of each other because $\phi(d, c) = -\phi(c, d) \neq 0$. Thus by assumption for $y = cd^{-1}$ the linear forms $t \mapsto \phi(b, t)$ and $t \mapsto \phi(t, y)$ must be linearly independent. As above, we find x satisfying $\phi(b, x) = \phi(x, y) = \alpha$. Finally, note that since p is odd, the anticommutative form ϕ is actually alternating, so that $\phi(c, c) = 0$ and therefore $\phi(y, c) = \phi(cd^{-1}, c) = \phi(d^{-1}, c) = \phi(c, d) = \alpha$, which yields the end of the chain.   □

These were the purely algebraic statements involved in the proof of Theorem 7.6.1. To proceed further, we need some facts from class field theory.

Facts 7.6.10 Let k be a number field. Denote by Ω the set of all places of k, and for each $v \in \Omega$ denote by k_v the completion of k at v. For v finite k_v is a finite extension of $\mathbf{Q}_p$ for some prime p, and for v infinite k_v is isomorphic to $\mathbf{R}$ or $\mathbf{C}$.

1. For each finite place v there is an isomorphism $\mathrm{inv}_{k_v} : \mathrm{Br}\,(k_v) \xrightarrow{\sim} \mathbf{Q}/\mathbf{Z}$, and for a finite extension $L_w|k_v$ one has $\mathrm{inv}_{L_w} \circ \mathrm{Res}^{L_w}_{k_v} = [L_w : k_v]\,\mathrm{inv}_{k_v}$. See Serre [2], Chapter XIII, Propositions 6 and 7.
2. The restriction maps $\mathrm{Br}\,(k) \to \mathrm{Br}\,(k_v)$ are trivial for all but finitely many $v \in \Omega$, and the map $\mathrm{Br}\,(k) \to \oplus_{v \in \Omega}\mathrm{Br}\,(k_v)$ is injective. These statements are contained in the Albert–Brauer–Hasse–Noether theorem already recalled in Remark 6.5.5.
3. If p is an odd prime, then given a finite set S of places of k and characters $\chi_v \in H^1(k_v, \mathbf{Z}/p\mathbf{Z})$ for all $v \in S$, there exists a global character $\chi \in H^1(k, \mathbf{Z}/p\mathbf{Z})$ inducing the χ_v by restriction to k_v. This is a particular case of the Grunwald–Wang theorem (Artin–Tate [1], Chapter X).
4. Let $\alpha_1, \ldots, \alpha_r$ be a finite set of elements in ${}_p\mathrm{Br}\,(k)$ and $a_1, \ldots, a_r \in k^\times$. Assume given for each place v of k a character $\chi_v \in H^1(k_v, \mathbf{Z}/p\mathbf{Z})$ such that $\chi_v \cup h^1_{k_v,p}(a_i) = \mathrm{Res}^{k_v}_k(\alpha_i)$ for $1 \leq i \leq r$. Then there exists a character $\chi \in H^1(k, \mathbf{Z}/p\mathbf{Z})$ such that $\chi \cup h^1_{k,p}(a_i) = \alpha_i$ for $1 \leq i \leq r$. This follows from global class field theory: almost the same statement is proven in Cassels–Fröhlich [1], Ex. 2.16, p. 355 (note that condition (i) there follows from the global reciprocity law and that one may choose $\chi_v = 0$ for all but finitely many v). One may also consult Lemma 5.2 of Tate [4].

We can now prove the chain lemma for number fields.

Lemma 7.6.11 *If k is a number field containing the p-th roots of unity, then the chain lemma holds with length 4 for k and p.*

Proof Assume given $a, b, c, d \in k^\times$ with $h^2_{k,p}(\{a,b\}) = h^2_{k,p}(\{c,d\})$. We have to find $x, y \in k^\times$ such that

$$h^2_{k,p}(\{b,x\}) = h^2_{k,p}(\{x,y\}) = h^2_{k,p}(\{y,c\}) = h^2_{k,p}(\{c,d\}). \qquad (7.12)$$

Let S be the set of places such that $\mathrm{Res}^{k_v}_k(h^2_{k,p}(\{c,d\})) \neq 0$. For each $v \in S$ the group ${}_p\mathrm{Br}\,(k_v)$ is cyclic (by Fact 7.6.10 (1) and by $\mathrm{Br}\,(\mathbf{R}) \cong \mathbf{Z}/2\mathbf{Z}$), so we may apply Lemma 7.6.9 to find elements $x_v, y_v \in k_v^\times$ for each $v \in S$ such that

$$h_{k_v,p}^2(\{b, x_v\}) = h_{k_v,p}^2(\{x_v, y_v\}) = h_{k_v,p}^2(\{y_v, c\}) = h_{k_v,p}^2(\{c, d\}). \quad (7.13)$$

The last of these equalities implies that $h_{k_v,p}^2(\{dy_v, c\}) = 0$, and therefore $dy_v = N_{k_v(\sqrt[p]{c})|k_v}(t_v)$ for some $t_v \in k_v(\sqrt[p]{c})$ according to Proposition 4.7.1 and Corollary 4.7.5. Fact 7.6.10 (3) together with the Kummer isomorphism $k(\sqrt[p]{c})^\times / k(\sqrt[p]{c})^{\times p} \cong H^1(k(\sqrt[p]{c}), \mathbf{Z}/p\mathbf{Z})$ enable us to find $t \in k(\sqrt[p]{c})$ such that $t_v^{-1}t \in k_v(\sqrt[p]{c})^{\times p}$ for all $v \in S$. Put $y = d^{-1}N_{k(\sqrt[p]{c})|k}(t)$. Then for all $v \in S$ we have $h_{k,p}^2(\{y, c\}) = h_{k,p}^2(\{d^{-1}, c\}) = h_{k,p}^2(\{c, d\})$ by Proposition 4.7.1, and moreover

$$h_{k_v,p}^2(\{y, c\}) = h_{k_v,p}^2(\{y_v, c\}), \quad (7.14)$$

since $N_{k_v(\sqrt[p]{c})|k_v}(t_v^{-1}t) \in k_v^{\times p}$ by our choice of t. Fixing this y, it remains to find $x \in k^\times$ satisfying the first two equalities in (7.12). According to (7.13) and (7.14), the equations $h_{k,p}^2(\{b, x\}) = h_{k,p}^2(\{y, c\})$ and $h_{k,p}^2(\{x, y\}) = h_{k,p}^2(\{y, c\})$ have simultaneous solutions x_v over k_v for each $v \in S$, and for $v \notin S$ they have the trivial solution by the choice of S. We conclude by Fact 7.6.10 (4), applied with $r = 2$, $a_1 = b^{-1}$, $a_2 = y$, $\alpha_1 = \alpha_2 = h_{k,p}^2(\{y, c\})$ and $\chi_v = h_{k_v,p}^1(x_v)$.   □

We finally come to:

Proof of Theorem 7.6.1 By Proposition 7.5.8 it is enough to treat the case $m = p$. As in the proof of that proposition, we may also assume that k contains the p-th roots of unity. It then suffices to check the conditions of Proposition 7.6.7. The first one is the previous lemma. To check the second, it is enough to find for a given finite set $\alpha_1, \dots, \alpha_r$ of classes in $_p\mathrm{Br}\,(k)$ a cyclic extension $L|k$ of degree p so that $\mathrm{Res}_k^L(\alpha_i) = 0$ for all i. By Fact 7.6.10 (2) we find a finite set S of places so that $\mathrm{Res}_k^{k_v}(\alpha_i) = 0$ for all i and all $v \notin S$. Choose an element $b \in k^\times$ which does not lie in $k_v^{\times p}$ for any $v \in S$. For instance, one may take $b = u\pi_1 \dots \pi_s$, where u is a unit and the π_i are prime elements for the finite places in S. For p odd this is already sufficient; for $p = 2$ one uses Dirichlet's Unit Theorem (Neukirch [1], Chapter I, Theorem 7.4) to choose u so that b becomes negative in the completions for the real places in S. The extension $L = k(\sqrt[p]{b})|k$ is then cyclic of degree p, and so are the extensions $Lk_v|k_v$ for $v \in S$. Using Fact 7.6.10 (1) and the vanishing of $\mathrm{Br}\,(\mathbf{C})$ we therefore see that $\mathrm{Res}_k^{Lk_v}(\alpha_i) = 0$ for all i and all v in S. For the other places we already have $\mathrm{Res}_k^{k_v}(\alpha_i) = 0$ by assumption, so that Fact 7.6.10 (2) implies $\mathrm{Res}_k^L(\alpha_i) = 0$ for all i, as required.   □

EXERCISES

1. (Bass, Tate) This exercise studies the K-groups of an algebraically closed field k.

   (a) Let A, B be two divisible abelian groups. Show that $A \otimes_{\mathbf{Z}} B$ is uniquely divisible, i.e. a $\mathbf{Q}$-vector space.
   (b) Show that $K_2^M(k)$ is uniquely divisible. [*Hint:* Use the presentation $0 \to R \to k^\times \otimes k^\times \to K_2^M(k) \to 0$.]
   (c) Let $K|k$ be a field extension. Show that the map $K_2^M(k) \to K_2^M(K)$ is injective.

2. (a) Given a field extension $K|k$, show that the natural maps $K_n^M(k) \otimes_{\mathbf{Z}} \mathbf{Q} \to K_n^M(K) \otimes_{\mathbf{Z}} \mathbf{Q}$ are injective for all $n \geq 1$. [*Hint:* First consider the cases $K|k$ finite and $K = k(t)$.]
   (b) If k is an uncountable field, show that $K_n^M(k) \otimes_{\mathbf{Z}} \mathbf{Q}$ is uncountable for all $n \geq 1$.

3. Establish isomorphisms $K_n^M(\mathbf{R})/2K_n^M(\mathbf{R}) \cong \mathbf{Z}/2\mathbf{Z}$ for all $n \leq 0$.

4. This exercise gives a simpler proof of Theorem 7.3.2 for $n = 2$ and fields of characteristic 0. Let k be such a field, and let $K|k$ be a finite field extension. Let $N_1 = N_{a_1, \ldots, a_r | k}$ and $N_2 = N_{b_1, \ldots, b_s | k}$ be two candidates for the norm map $K_2^M(K) \to K_2^M(k)$. Denote by δ_k the difference $N_1 - N_2$.

   (a) Observe that Im (δ_k) is a torsion group.
   (b) Show that $\delta_{k(t)} : K_2^M(K(t)) \to K_2^M(k(t))$ takes values in $K_2^M(k)$, where $k(t)|k$ is a rational function field.
   (c) Given $a, b \in k^\times$, show that $\delta_{k(t)}(\{a, (1-t) + tb\}) = 0$. [*Hint:* Use the fact that the evaluation map $k[[t]]^\times \to k^\times$ has divisible kernel.]
   (d) Conclude by specialization that $\delta_k(\{a, b\}) = 0$.

5. (Tate) Let m be an integer invertible in k, and assume that k contains a primitive m-th root of unity ω. Denote by A the subgroup of $_m K_2^M(k)$ consisting of elements of the form $\{\omega, a\}$ with $a \in k^\times$.

   (a) Show that the equality $A = {}_m K_2^M(k)$ is equivalent to the existence of a homomorphism $f : mK_2^M(k) \to K_2^M(k)/A$ such that $f(m\alpha) = \alpha \bmod A$ for all $\alpha \in K_2^M(k)$.
   (b) If such an f exists, show that it is unique.
   (c) Given $a, b \in k^\times$, show that $\{a, b\} \in mK_2^M(k)$ if and only if there exists a finite extension $K|k$ and elements $\alpha, \beta \in K$ such that $\alpha^m = a$ and $N_{K|k}(\beta) = b$.
   (d) Verify that for $\{a, b\} \in mK_2^M(k)$ the image of $N_{K|k}(\{\alpha, \beta\})$ in the quotient $K_2^M(K)/A$ depends only on the pair a, b.
   (e) Assume that $\operatorname{cd}(k) \leq 1$. Conclude from (c) that $K_2^M(k)$ is m-divisible, and use (a) and (d) to show that $A = {}_m K_2^M(k)$. [*Hint:* Use that Br $(L|k)$ is trivial for all finite cyclic extensions $L|k$ of degree m.]

6. Let k be a field, $n > 0$ an integer and α an element of $K_n^M(k(t))$. For each closed point P of the affine line $\mathbf{A}_k^1$, write $\kappa(P) = k(a_P)$ with suitable a_P.

   (a) Check that the norm $N_{\kappa(P)(t)|k(t)}(\{t - a_P, \partial_P^M(\alpha)\})$ is independent of the choice of a_P.

(b) Establish the following more explicit variant of Corollary 7.2.3:

$$\alpha = s_{t-1}(\alpha)_{k(t)} + \sum_{P \in \mathbf{A}_0^1} N_{k(P)(t)|k(t)}(\{t - a_P, \partial_P^M(\alpha)\}).$$

[*Remark:* The analogous formula for Galois cohomology may be found in Garibaldi–Merkurjev–Serre [1], Exercise 9.23.]

7. (Optimality of the Rosset–Tate bound) Let p be a prime number, and k a field containing a primitive p-th root of unity ω. Consider the purely transcendental extension $E = k(x_1, y_1, x_2, y_2, ..., x_p, y_p)$ in $2p$ indeterminates. Make the cyclic group $\mathbf{Z}/p\mathbf{Z} = \langle \sigma \rangle$ act on E by $\sigma(x_i) = x_{i+p}$ and $\sigma(y_i) = y_{i+p}$ (where $i + p$ is taken mod p). Let $F \subset E$ be the fixed field under this action. Prove that $N_{E|F}(\{x_1, y_1\})$ cannot be represented in $K_2^M(F)$ by a symbol of length $p - 1$. [*Hint:* Use the Galois symbol and Exercise 6 of Chapter 6.]

8. Let $n > 1$ be an odd integer, and k a field containing a primitive n-th root of unity ω. Let $K|k$ be a finite Galois extension whose Galois group is the dihedral group D_n, i.e. it has a presentation of the form

$$\langle \sigma, \tau \mid \sigma^n = 1, \ \tau^2 = 1, \ \sigma\tau\sigma = \tau \rangle.$$

Let L be the fixed field of σ in K. This exercise shows that a central simple k-algebra A of degree n split by K is isomorphic to a cyclic algebra.

(a) Show that there exists $a \in L^\times$ with $K = L(\sqrt[n]{a})$ and $N_{L|K}(a) \in (k^\times)^n$. [*Hint:* if $K = L(\sqrt[n]{c})$, take $a = c^n$.]

(b) Show that $A \otimes_k L \cong (a, b)_\omega$ for some $b \in L^\times$.

(c) Conclude that $[A] = \mathrm{Cor}_k^L([(a, b)_\omega])$ in $\mathrm{Br}(k)$. If $b \in k^\times$, conclude moreover that A is isomorphic to a cyclic algebra.

(d) Assume that $b \in L^\times \setminus k^\times$. Show that there exist $a', b' \in k^\times$ such that $aa' + bb' = 0$ or 1, and prove that the relation

$$[(a, b)_\omega] + [(a, b')_\omega] + [(a', bb')_\omega] = 0$$

holds in $\mathrm{Br}(L)$. Conclude that A is isomorphic to a cyclic algebra in this case as well.

(e) Adapt the preceding arguments to show that the conclusion also holds in the case when k is of characteristic $p \geq 3$ and $n = p$. [*Hint:* Show that $K = L(c)$ for some c such that $a := c^p - c$ lies in L and $\mathrm{Tr}_{L|k}(a) = u^n - u$ for some $u \in k$.]

[*Remark:* The theorem of this exercise is due to Rowen–Saltman [1]. The above proof is that of Tignol [1].]

9. (Wedderburn) Show that every central simple algebra of degree 3 over a field k is isomorphic to a cyclic algebra. [*Hint:* Use the previous exercise.]

10. Let $p > 2$ be a prime number, and k a field containing a primitive p-th root of unity. Improve the bound of Proposition 7.4.13 by showing that every central simple k-algebra of degree p is Brauer equivalent to the tensor product of at most $(p-1)!/2$ cyclic k-algebras of degree p. [*Hint:* Use Exercise 8 and the fact that the symmetric group S_p contains D_p as a subgroup.]

11. Let k be a field, and $m > 0$ an integer invertible in k.

(a) Show that injectivity (resp. surjectivity) of the Galois symbols

$$h_{k,m}^{n-1} : K_{n-1}^M(k)/m \to H^{n-1}(k, \mu_m^{\otimes(n-1)}) \text{ and}$$
$$h_{k,m}^n : K_n^M(k)/m \to H^n(k, \mu_m^{\otimes n})$$

imply the corresponding property for the Galois symbol

$$h_{k((t)),m}^n : K_n^M(k((t)))/m \to H^n(k((t)), \mu_m^{\otimes n}).$$

(b) Verify the Bloch–Kato conjecture for the Laurent series field $\mathbf{F}_q((t))$ and m prime to q.

8

The Merkurjev–Suslin theorem

The bulk of the present chapter is devoted to the central result of this book, the celebrated theorem of Merkurjev and Suslin on the bijectivity of the Galois symbol $h^2_{k,m} : K^M_2(k)/mK^M_2(k) \to H^2(k, \mu_m^{\otimes 2})$ for all fields k and all integers m invertible in k. Following a method of Merkurjev, we shall deduce the theorem by a specialization argument from the partial results obtained at the end of the last chapter, using a powerful tool which is interesting in its own right, the K_2-analogue of Hilbert's Theorem 90. Apart from the case when m is a power of 2, no elementary proof of this theorem is known. To establish it, we first develop the foundations of the theory of Gersten complexes in Milnor K-theory. This material requires some familiarity with the language of schemes. Next comes an even deeper input, a technical statement about the K-cohomology of Severi–Brauer varieties. Its proof involves techniques outside the scope of the present book, so at this point our discussion will not be self-contained. The rest of the argument is then much more elementary and requires only the tools developed earlier in this book, so some readers might wish to take the results of the first three sections on faith and begin with Section 8.4. In the final sections we shall apply the techniques used in the proof of the Merkurjev–Suslin theorem to the study of reduced norm maps for cyclic division algebras of squarefree degree. As a consequence, we shall obtain an elegant characterization of fields of cohomological dimension 2 due to Suslin.

The main theorem was first proven in Merkurjev [1] in the case when m is a power of 2, relying on a computation by Suslin of the Quillen K-theory of a conic. Later several elementary proofs of this case were found which use no algebraic K-theory at all, at the price of rather involved calculations. Merkurjev himself gave three such proofs (see Wadsworth [1], Merkurjev [6] and Elman–Karpenko–Merkurjev [1]); another one by Rost is contained in the book of Kersten [1]. For a proof in the language of quadratic forms, see Arason [2]. The general theorem first appeared in the seminal paper of Merkurjev and Suslin [1]. Its proof was later improved and simplified in Suslin [1], [2] and Merkurjev [2]. Several ideas involved in these proofs, most notably generalizations of Hilbert's Theorem 90 to higher K-groups, play a prominent role in Voevodsky's proof of the general Bloch–Kato conjecture.

8.1 Gersten complexes in Milnor K-theory

Let X be a variety over a field k. We regard X as a k-scheme, and denote by X_i the set of its points of dimension i (i.e. those scheme-theoretic points whose Zariski closure in X has dimension i). Following Kato, we construct in this section for each integer n complexes of abelian groups

$$S_n(X) : \bigoplus_{P \in X_d} K^M_{n+d}(\kappa(P)) \xrightarrow{\partial} \bigoplus_{P \in X_{d-1}} K^M_{n+d-1}(\kappa(P)) \xrightarrow{\partial} \cdots \xrightarrow{\partial} \bigoplus_{P \in X_0} K^M_n(\kappa(P))$$

called *Gersten complexes in Milnor K-theory*. The degree i term in such a complex will be the one indexed by the points in X_i. By convention, we put $K^M_n(\kappa(P)) := 0$ for $n < 0$. Therefore the complex $S_n(X)$ will be trivial for $n < -d$, and concentrated between the terms of degree d and $-n$ for $-d \le n < 0$.

When investigating properties of these complexes, we shall be sometimes forced to work at the level of local rings. These are not finitely generated k-algebras any more, so in the remainder of this section *we shall also allow X to be the affine scheme defined by a local ring of a variety over a field, or by the completion of such a local ring*. These all give rise to excellent schemes according to Grothendieck [4], (7.8.3), and moreover the dimension of the scheme-theoretic closure of a point defines a well-behaved dimension function on them. We shall denote by d the dimension of X.

Construction 8.1.1 We construct the maps ∂ in the sequence $S_n(X)$ as follows. Take a point $P \in X_{i+1}$, and let Z_P be its Zariski closure in X. Each point Q of codimension 1 on Z_P corresponds to a point in X_i. On the normalization $\tilde{Z}_P$ of Z_P there are finitely many points $Q_1, \ldots, Q_r$ lying above Q. The local ring of each Q_j on $\tilde{Z}_P$ is a discrete valuation ring, hence it defines a discrete valuation on the function field $\kappa(P)$ of $\tilde{Z}_P$. Denoting by $\partial^M_{Q_j}$ the associated tame symbol, we may define maps $\partial^P_Q : K^M_{n+i+1}(\kappa(P)) \to K^M_{n+i}(\kappa(Q))$ by setting

$$\partial_Q^P := \sum_{j=1}^{r} N_{\kappa(Q_j)|\kappa(Q)} \circ \partial_{Q_j}^M.$$

Since each function $f \in \kappa(P)$ has only finitely many zeroes and poles on $\tilde{Z}_P$, the valuations $v_{Q_j}(f)$ associated with the codimension 1 points on X are trivial for all but finitely many Q_j (here Q_j runs over the set of *all* codimension 1 points). *A fortiori*, for fixed $\alpha \in K_{n+i+1}^M(\kappa(P))$ the tame symbols $\partial_{Q_j}^M$ are trivial for all but finitely many Q_j. It therefore makes sense to consider the sum

$$\partial_P := \sum_{Q \in Z_P} \partial_Q^P,$$

and finally, we may define the map ∂ as the direct sum of the maps ∂_P for all $P \in X_{i+1}$.

Theorem 8.1.2 (Kato)   *The sequence $S_n(X)$ is a complex for all $n \geq -d$.*

Proof   The proof is in several steps.

Step 1: Reduction to the local case. Let P_0 be a point of dimension i, and α an element of $K_{n+i}(\kappa(P_0))$. We have to prove that $(\partial \circ \partial)(\alpha)$ is the zero element in $\bigoplus K_{n+i-2}(\kappa(P))$. This sum is indexed by points of dimension $i - 2$. We may assume $i \geq 2$ (otherwise there is nothing to prove) and reason for each direct summand separately. The construction of ∂ shows that in doing so we may replace X by the normalization of the closure of P_0 in X; in particular, we may assume that X is normal. Let P_2 be a point of codimension 2 in X, and denote by A its local ring. The points of codimension 1 involved in the construction of the component of $\partial(\alpha)$ indexed by P_2 all correspond to prime ideals of height 1 in A. This shows that for the proof of $(\partial \circ \partial)(\alpha) = 0$ we may replace $S_n(X)$ by $S_n(\mathrm{Spec}\,(A))$, and assume that X is the spectrum of an integrally closed local ring of dimension 2.

Step 2: Reduction to the complete case. Next we show that we may replace A by its completion $\widehat{A}$. To see this, write $\widehat{X}$ for the spectrum of $\widehat{A}$, K for the fraction field of A and $\widehat{K}$ for that of $\widehat{A}$. Since A is excellent, here $\widehat{A}$ is integrally closed as well according to Theorem A.5.2 of the Appendix. Furthermore, the first statement of the same theorem applied to A/P shows that $\widehat{A}/P\widehat{A}$ has no nilpotents. Thus each prime ideal P of height 1 in X decomposes as a finite product $P = P_1 \ldots P_r$ of distinct height 1 prime ideals in $\widehat{A}$ (note that since A and $\widehat{A}$ are integrally closed, these are actually principal ideals). In terms of discrete valuations, the valuation defined by each P_i continues that of P with ramification index 1, which shows the commutativity of the first square in the diagram

$$S_n(X): \qquad K^M_{n+2}(K) \xrightarrow{\ \partial\ } \bigoplus_{P \in X_1} K^M_{n+1}(\kappa(P)) \xrightarrow{\ \partial\ } K^M_n(\kappa)$$

$$\downarrow i_{\widehat{K}|K} \qquad\qquad \downarrow \oplus_{Q \to P} i_{\kappa(Q)|\kappa(P)} \qquad\qquad \downarrow \mathrm{id}$$

$$S_n(\widehat{X}): \qquad K^M_{n+2}(\widehat{K}) \xrightarrow{\ \partial\ } \bigoplus_{Q \in \widehat{X}_1} K^M_{n+1}(\kappa(Q)) \xrightarrow{\ \partial\ } K^M_n(\kappa)$$

in view of Remark 7.1.6 (2). To check the commutativity of the second square, we consider the quotient ring A/P. Its completion with respect to the maximal ideal M of A is none but the direct sum $\oplus \widehat{A}/P_i$, as seen using Proposition A.5.1 of the Appendix and the Chinese Remainder Theorem. But the integral closure of $\oplus \widehat{A}/P_i$ in the direct sum of the fraction fields of the A/P_i is none but the completion of the integral closure of A/P, again by Theorem A.5.2 of the Appendix. The maximal ideals in these rings are all induced by M, so there is no ramification and the required commutativity again follows from Remark 7.1.6 (2).

Step 3: Reduction to the case of a power series ring. We have arrived at the case when X is the spectrum of a complete local ring A of dimension 2; denote by κ its residue field. According to a version of the Cohen structure theorem (see Appendix, Theorem A.5.4 (1)), such an A can be written as a finitely generated module over a subring of the form $B[[t]]$, where B is a complete discrete valuation ring with the same residue field κ. Denote by Y the spectrum of $B[[t]]$, by K the fraction field of A and by K_0 that of B. The left square of the diagram

$$S_n(X): \qquad K^M_{n+2}(K) \xrightarrow{\ \partial\ } \bigoplus_{Q \in X_1} K^M_{n+1}(\kappa(Q)) \xrightarrow{\ \partial\ } K^M_n(\kappa)$$

$$\downarrow N_{K|K_0} \qquad\qquad \downarrow \oplus_P \Sigma_{Q \to P} N_{\kappa(Q)|\kappa(P)} \qquad \mathrm{id} \downarrow$$

$$S_n(Y): \qquad K^M_{n+2}(K_0) \xrightarrow{\ \partial\ } \bigoplus_{P \in Y_1} K^M_{n+1}(\kappa(P)) \xrightarrow{\ \partial\ } K^M_n(\kappa)$$

commutes because of Corollary 7.4.3, and the right square because of Proposition 7.4.1. An inspection of the diagram reveals that if $S_n(Y)$ is a complex, then so is $S_n(X)$, so we may assume $A = B[[t]]$.

Step 4: Conclusion. The ring $B[[t]]$ is a regular local ring, hence a unique factorization domain. Its prime ideals of height 1 are all principal, generated by either a local parameter π of B or a so-called Weierstrass polynomial, i.e. a monic irreducible polynomial in $B[t]$ whose coefficients, except for the leading one, are divisible by π (see e.g. Lang [3], Chapter IV, Theorem 9.3). Thus the multiplicative group of the fraction field K of $B[[t]]$ is generated by the units of B, by π and by Weierstrass polynomials. In order to verify $(\partial \circ \partial)(\alpha) = 0$

for $\alpha \in K_{n+2}^M(K)$, we may reduce, by construction of the tame symbol, to the case $n = 0$ and moreover using bilinearity of symbols we may assume α is a symbol of the form $\{a, b\}$, with a and b chosen among the generators of $K^\times$ described above.

The cases when a or b are units of B are straightforward. Next consider the case when $a = \pi$ and b is a Weierstrass polynomial of degree N. Note that the residue field $\kappa(P)$ of the prime ideal $P = (b)$ is a degree N finite extension of the fraction field F of B, and hence the discrete valuation v of B extends uniquely to a valuation v_P of $\kappa(P)$, with some ramification index e_P. Its residue field is a finite extension of κ; write f_P for the degree of this extension. Our assumption that we are dealing with excellent rings implies that $e_P f_P = N$. On the other hand, the image of b in the residue field $\kappa((t))$ of (π) is t^N, whose t-adic valuation is N. Therefore, by definition of the tame symbol we get $(\partial \circ \partial)(\{\pi, P\}) = N - e_P f_P = 0$.

We still have to deal with the cases where a and b are Weierstrass polynomials. These are units for the valuation associated with π, and hence the corresponding tame symbol is trivial. The other discrete valuations to be considered are those coming from Weierstrass polynomials, and these in turn define closed points of the projective line $\mathbf{P}_F^1$. The associated tame symbols on K and $F(t)$ are given by the same formula. Viewing $\alpha = \{a, b\}$ as an element of $K_2(F(t))$, we now show that

$$(\partial \circ \partial)(\{a, b\}) = \sum_{P \in \mathbf{P}_{F,0}^1} f_P \, v_P(\partial_P^M(\{a, b\})), \qquad (8.1)$$

i.e. that the terms coming from points on $\mathbf{P}_F^1$ other than those defined by Weierstrass polynomials do not contribute to the sum. For the points coming from irreducible polynomials in $F[t]$ which are not Weierstrass polynomials this is straightforward, because the associated tame symbols are trivial on α. There is still one point of $\mathbf{P}_F^1$ to consider, namely the one at infinity, where t^{-1} is a local parameter. To handle it, write $a = t^N a_1, b = t^M b_1$, with $a_1, b_1 \in B[t^{-1}]$ satisfying $a_1(0) = b_1(0) = 1$. Using Lemma 7.1.2 and bilinearity of symbols we get

$$\{a, b\} = \{a_1, b_1\} - N\{t^{-1}, b_1\} - M\{a_1, t^{-1}\} + MN\{t^{-1}, -1\}.$$

We see using the condition $a_1(0) = b_1(0) = 1$ that the tame symbol associated with t^{-1} annihilates the first three terms, and the fourth gets mapped to $(-1)^{MN}$ in F. But $(-1)^{MN}$ is a unit for the valuation of F, and we are done.

Now we may rewrite the right-hand side of (8.1) as

$$\sum_{P \in \mathbf{P}^1_{F,0}} f_P \, v_P(\partial_P^M(\{a,b\})) = v\Big(\sum_{P \in \mathbf{P}^1_{F,0}} N_{\kappa(P)|F}(\partial_P^M(\{a,b\})) \Big).$$

Indeed, this follows from the equality $f_P \, v_P = v \circ N_{\kappa(P)|F}$, which is a very special case of Proposition 7.4.1, because here v_P is none but the tame symbol on $\kappa(P)^\times$ equipped with its canonical valuation, and multiplication by f_P is the norm map on K_0 of the residue field of $\kappa(P)$. To conclude the proof it remains to observe that the sum in parentheses is trivial, by Weil's reciprocity law (Corollary 7.2.4). $\qquad\square$

8.2 Properties of Gersten complexes

In this section X will be a scheme of the type considered in the previous one, but the reader may safely assume it is a variety over a field k.

Notice that given an open subscheme $U \subset X$, there are natural restriction maps

$$\bigoplus_{P \in X_i} K^M_{n+i}(\kappa(P)) \to \bigoplus_{P \in U_i} K^M_{n+i}(\kappa(P))$$

for all $n \in \mathbf{Z}$ and $0 \le i \le d$, induced by the inclusion map $U \subset X$. These manifestly commute with the boundary maps ∂ in the complex $S_n(X)$, whence a map of complexes $j^* : S_n(X) \to S_n(U)$. One defines similarly a pushforward map $i_* : S_n(Z) \to S_n(X)$ induced by the inclusion $i : Z \to X$ of a closed subscheme.

Proposition 8.2.1 *Let $U \subset X$ be an open subscheme with complement Z.*

1. (Localization) For all $n \in \mathbf{Z}$ the natural sequence of complexes

$$0 \to S_n(Z) \xrightarrow{i_*} S_n(X) \xrightarrow{j^*} S_n(U) \to 0$$

is exact.

2. (Mayer–Vietoris) Let $V \subset X$ be a second open subscheme satisfying $U \cup V = X$. Then the sequence of complexes

$$0 \to S_n(X) \xrightarrow{j_U^* \oplus j_V^*} S_n(U) \oplus S_n(V) \xrightarrow{j_{U \cap V}^* - j_{U \cap V}^*} S_n(U \cap V) \to 0$$

is exact.

3. *(Mayer–Vietoris for closed subsets) Assume there exists a closed subscheme $T \subset X$ such that $Z \cup T = X$. Then the sequence of complexes*

$$0 \to S_n(Z \cap T) \xrightarrow{i_{(Z \cap T)*} \oplus i_{(Z \cap T)*}} S_n(Z) \oplus S_n(T) \xrightarrow{i_{Z*} - i_{T*}} S_n(X) \to 0$$

is exact.

Proof In all three cases the required exactness is readily checked at the level of each term $\bigoplus_{P \in X_i} K_{n+i}^M(\kappa(P))$ of the complex $S_n(X)$. □

Definition 8.2.2 For $0 \leq i \leq d$ denote the i-th homology group of the complex $S_n(X)$ (i.e. the homology at the term indexed by the points in X_i) by $A_i(X, K_n^M)$. It is the *i-th homology group of X with values in K_n^M*.

Example 8.2.3 The case $n = -i$ is especially important for $0 \leq i \leq d$. Here we obtain the group

$$A_i(X, K_{-i}^M) = \operatorname{coker}\Big(\bigoplus_{P \in X_{i+1}} (\kappa(P))^\times \to \bigoplus_{P \in X_i} \mathbf{Z} \Big),$$

the *Chow group of dimension i cycles on X*. This observation is the starting point for the application of K-theoretic methods to the study of algebraic cycles. See Colliot-Thélène [3] and Murre [1] for informative surveys on this research area.

With the above notations, Proposition 8.2.1 together with Proposition 3.1.1 yields:

Corollary 8.2.4 *Under the assumptions of Proposition 8.2.1 one has natural long exact sequences*

$$\cdots \to A_i(Z, K_n^M) \xrightarrow{i_*} A_i(X, K_n^M) \xrightarrow{j^*} A_i(U, K_n^M) \longrightarrow A_{i-1}(Z, K_n^M) \to \cdots,$$

$$\cdots \to A_i(X, K_n^M) \xrightarrow{j_U^* \oplus j_V^*} A_i(U, K_n^M) \oplus A_i(V, K_n^M) \to$$

$$\xrightarrow{j_{U \cap V}^* - j_{U \cap V}^*} A_i(U \cap V, K_n^M) \to A_{i-1}(X, K_n^M) \to \cdots$$

and

$$\cdots \longrightarrow A_i(Z \cap T, K_n^M) \xrightarrow{i_{(Z \cap T)*} \oplus i_{(Z \cap T)*}} A_i(Z, K_n^M) \oplus A_i(T, K_n^M) \to$$

$$\xrightarrow{i_{Z*} - i_{T*}} A_i(X, K_n^M) \to A_{i-1}(Z \cap T, K_n^M) \to \cdots.$$

Next we turn to the homotopy invariance property of K_n^M-homology groups. First some notation: for an integer $j \in \mathbf{Z}$, we define the *shifted complex* $S_n(X)[j]$ as the one whose degree i term is the degree $i - j$ term in $S_n(X)$.

Let Y be another k-scheme satisfying the same assumptions as X (e.g. a k-variety), and let j be its dimension. Given a point in X_i, its closure in X is an integral closed subscheme Z. Then $Z \times_k Y$ is an integral closed subscheme in $X \times_k Y$ of dimension $i + j$, and thus corresponds to a point in $(X \times_k Y)_{i+j}$. This construction defines for all $n \in \mathbf{Z}$ natural maps

$$\bigoplus_{P \in X_i} K_{n+i+j}^M(\kappa(P)) \to \bigoplus_{P \in (X \times Y)_{i+j}} K_{n+i+j}^M(\kappa(P)),$$

commuting with the differentials in the complexes $S_{n+j}(X)[j]$ and $S_n(X \times_k Y)$. Therefore we obtain maps

$$S_{n+j}(X)[j] \to S_n(X \times_k Y) \tag{8.2}$$

for all $n \in \mathbf{Z}$.

Proposition 8.2.5 *For all $n \geq 0$ the natural map*

$$S_{n+1}(X)[1] \to S_n(X \times_k \mathbf{A}_k^1)$$

defined above induces isomorphisms on homology, i.e. the induced maps

$$A_{i-1}(X, K_{n+1}^M) \to A_i(X \times_k \mathbf{A}_k^1, K_n^M)$$

are isomorphisms for all i.

The proof is an adaptation of an argument by Quillen proving the homotopy invariance of Quillen K-theory. It is based on a suggestion of Joël Riou.

Proof Without loss of generality we may assume X is reduced. By the noetherian assumption, we may decompose it into a finite union of irreducible closed subschemes, and then a finite number of applications of the Mayer–Vietoris sequence for closed subsets (and induction on dimension) shows that the irreducible case implies the reducible case.

We can therefore assume X is reduced and irreducible, and use induction on dimension. If $\dim(X) = 0$, then $X = \mathrm{Spec}(F)$ for some field extension $F \supset k$, i.e. a point defined over F. The Gersten complex $S_n(\mathbf{A}_F^1)$ takes the shape

$$K_{n+1}^M(F(t)) \to \bigoplus_{P \in \mathbf{A}_0^1} K_n^M(\kappa(P)).$$

By Milnor's exact sequence (Theorem 7.2.1) we have $A_0(\mathbf{A}_F^1, K_n^M) = 0$ and $A_1(\mathbf{A}_F^1, K_n^M) = K_n^M(F)$ for all $n \geq 0$, which proves the statement for

Spec (F). Now take a general X and assume that the statement holds for all reduced closed subschemes $Z \subset X$ properly contained in X. Consider the commutative diagram

$$\cdots \to A_{i-1}(Z, K_{n+1}^M) \to A_{i-1}(X, K_{n+1}^M) \to A_{i-1}(X \setminus Z, K_{n+1}^M) \to \cdots$$

$$\downarrow \alpha_Z^i \qquad\qquad \downarrow \alpha_X^i \qquad\qquad \downarrow \alpha_{X \setminus Z}^i$$

$$\cdots \to A_i(Z \times_k \mathbf{A}_k^1, K_n^M) \to A_i(X \times_k \mathbf{A}_k^1, K_n^M) \to A_i((X \setminus Z) \times_k \mathbf{A}_k^1, K_n^M) \to \cdots$$

whose exact rows come from the first sequence in Corollary 8.2.4. The map α_Z^i is an isomorphism by the inductive assumption.

Now consider the system of (possibly reducible) reduced closed subschemes Z properly contained in X. The natural inclusion maps make it into a directed partially ordered set. With respect to this directed index set the complexes $S_n(Z)$ together with the pushforward maps i_{Z*} form a direct system, hence so do their homology groups. Similarly, the complexes $S_n(X \setminus Z)$ together with the pullback maps $j_{X \setminus Z}^*$ also form a direct system. The direct limit of this system is $S_{n+d}(\mathrm{Spec}\,(k(X))[d]$, because the only point of X contained in all of the $X \setminus Z$ is the generic point (the shift in degree comes from the fact that the closure of the generic point in X has dimension d, whereas $\mathrm{Spec}\,(k(X))$ has dimension 0). We get a similar statement for the homology groups, so by the exactness property of the direct limit (Lemma 4.3.2) we obtain a commutative diagram

$$\cdots \to \varinjlim A_{i-1}(Z, K_{n+1}^M) \to A_{i-1}(X, K_{n+1}^M) \to A_{i-d-1}(\mathrm{Spec}\,(k(X)), K_{n+d+1}^M) \to \cdots$$

$$\downarrow \varinjlim \alpha_Z^i \qquad\qquad \downarrow \alpha_X^i \qquad\qquad \downarrow \alpha_{k(X)}^{i-d}[d]$$

$$\cdots \to \varinjlim A_i(Z \times_k \mathbf{A}_k^1, K_n^M) \to A_i(X \times_k \mathbf{A}_k^1, K_n^M) \to A_{i-d}(\mathrm{Spec}\,(k(X))) \times_k \mathbf{A}_k^1, K_{n+d}^M) \to \cdots$$

Here the first vertical map is an isomorphism as a direct limit of the isomorphisms α_Z^i, and the third one is an isomorphism by the zero-dimensional case. The diagram then implies that the map α_X^i in the middle is an isomorphism as well. $\qquad\qquad\qquad\qquad\qquad\qquad\qquad\qquad\qquad\qquad\qquad\square$

As an application of the homotopy invariance property, we now compute the K_n^M-homology groups of projective spaces.

Proposition 8.2.6 *For all integers $n, d \geq 0$ we have*

$$A_i(\mathbf{P}^d, K_n^M) \cong \begin{cases} K_{n+i}^M(k) & \text{if } 0 \leq i \leq d; \\ 0 & \text{otherwise.} \end{cases}$$

Proof The proof goes by induction on d, the case $d = 0$ being obvious. Consider $\mathbf{P}^{d-1}$ embedded in $\mathbf{P}^d$ as a hyperplane; the complement is naturally isomorphic to affine d-space $\mathbf{A}^d$. In this situation we may combine the localization sequence of Corollary 8.2.4 with an iterated application of the isomorphism of Proposition 8.2.5 in the commutative diagram

$$\cdots \to A_i(\mathbf{P}^{d-1}, K_n^M) \xrightarrow{i^*} A_i(\mathbf{P}^d, K_n^M) \xrightarrow{j^*} A_i(\mathbf{A}_k^d, K_n^M) \xrightarrow{\partial} A_{i-1}(\mathbf{P}^{d-1}, K_n^M) \to \cdots$$

$$A_{i-d}(k, K_{n+d}^M) = A_{i-d}(k, K_{n+d}^M),$$

where the vertical maps are morphisms of the type (8.2). The diagram provides a splitting of the maps j^*, so we get decompositions

$$A_i(\mathbf{P}^d, K_n^M) \cong A_{i-d}(k, K_{n+d}^M) \oplus A_i(\mathbf{P}^{d-1}, K_n^M)$$

for all i and n. Here we have

$$A_{i-d}(k, K_{n+d}^M) \cong \begin{cases} K_{n+d}^M(k) & \text{if } i = d; \\ 0 & \text{otherwise.} \end{cases}$$

Moreover, for $i > d$ the groups $A_i(\mathbf{P}^d, K_n^M)$ obviously vanish. We therefore obtain the result by induction on d. □

Example 8.2.7 For us the most important case will be that of $n = 2 - d$. In this case we get:

$$A_d(\mathbf{P}^d, K_{2-d}^M) \cong K_2^M(k), \quad A_{d-1}(\mathbf{P}^d, K_{2-d}^M) \cong k^\times, \quad A_{d-2}(\mathbf{P}^d, K_{2-d}^M) \cong \mathbf{Z},$$

and the other groups are 0.

Remark 8.2.8 The *Gersten conjecture* for Milnor K-theory states that if X is the spectrum of an excellent regular local ring of finite dimension d, the complexes $S_n(X)$ are acyclic for all n in all degrees smaller than d. In the case of local rings of a smooth variety over a perfect field, this has been proven by Gabber; see Colliot-Thélène/Hoobler/Kahn [1] or Rost [1] for the proof (both papers work in a more general axiomatic setup).

This theorem has the following remarkable consequence. Given a smooth variety X over a perfect field, the rule $U \mapsto S_n(U)$ for all open subsets of X together with the restriction maps j^* introduced at the beginning of this section define a complex $S_{n,X}$ of presheaves for the Zariski topology on X. Moreover, one checks easily that this is actually a complex of *flabby* sheaves (i.e. the restriction maps $S_n(U) \to S_n(V)$ are surjective for $V \subset U$). Hence if we denote by $\mathcal{K}_{n+d}^M$ the sheaf associated with the presheaf $U \mapsto A_d(U, K_{n+d}^M)$, Gabber's acyclicity theorem implies that $S_{n,X}$ furnishes a *flabby resolution* of the sheaf $\mathcal{K}^M_{n+d}$. Thus it can be used to calculate the cohomology groups of the Zariski sheaf $\mathcal{K}^M_n$, and we get isomorphisms

$$H^i_{\mathrm{Zar}}(X, \mathcal{K}^M_{n+d}) \cong A_{d-i}(X, K^M_n).$$

Kerz [1] proved that $\mathcal{K}^M_{n+d}$ equals the Zariski sheaf associated with the presheaf $U \mapsto K^M_{n+d}(\mathcal{O}_X(U))$, where $K^M_{n+d}(A)$ for a commutative ring A is defined as the quotient of the tensor power $(A^\times)^{\otimes n+d}$ of the group of units $A^\times$ by the subgroup generated by elements $a_1 \otimes \cdots \otimes a_{n+d}$ where $a_i + a_j = 1$ for some $i \neq j$.

8.3 A property of Severi–Brauer varieties

We now begin the proof of the Merkurjev–Suslin theorem by establishing a crucial technical ingredient needed for the proof of Hilbert's Theorem 90 for K_2, to be discussed in the next section.

Theorem 8.3.1 *Let k be a field, p a prime invertible in k, and X a Severi–Brauer variety of dimension $d = p - 1$ over k. If $K|k$ is a finite extension of degree p which splits X, the natural maps*

$$A_{d-i}(X, K^M_{i+1-d}) \to A_{d-i}(X_K, K^M_{i+1-d})$$

are injective for all $0 \leq i \leq p - 1$.

Remarks 8.3.2

1. Via the isomorphism of Remark 8.2.8, the statement of the theorem becomes equivalent to the injectivity of the maps $H^i_{\mathrm{Zar}}(X, \mathcal{K}^M_{i+1}) \to H^i_{\mathrm{Zar}}(X_K, \mathcal{K}^M_{i+1})$, but we shall not need this interpretation. In fact, this injectivity holds for X of any dimension (see Kahn [2]).
2. The theorem also serves in the proof of the general Bloch–Kato conjecture, during the verification of the 'multiplication principle' for splitting varieties (see Suslin–Joukhovitski [1]).

Oddly enough, the only currently known proof of the theorem is a somewhat mysterious argument relying on Quillen's calculation of the algebraic K-theory of X. As a result we cannot give a self-contained exposition of the argument here. Still, we shall explain the method, referring to the literature for some facts from algebraic K-theory. The best short introduction to Quillen K-theory is Swan [3]; the original paper Quillen [1] still makes valuable reading, and the book of Srinivas [1] is a useful account. We also have to assume familiarity with spectral sequences, for which we refer to Weibel [1].

We begin by a very succinct review of the construction of Quillen's K-groups. Given a scheme X, denote by $\mathcal{M}(X)$ the category of coherent sheaves on X, and by $\mathcal{P}(X) \subset \mathcal{M}(X)$ the full subcategory of vector bundles (by which we mean locally free sheaves of finite rank). In the fundamental paper of Quillen [1] a purely categorical construction is introduced, producing new categories $Q\mathcal{M}(X)$ and $Q\mathcal{P}(X)$ out of $\mathcal{M}(X)$ and $\mathcal{P}(X)$. Taking the geometric realizations of the nerves of these categories yields topological spaces $|BQ\mathcal{M}(X)|$ and $|BQ\mathcal{P}(X)|$. One then defines for all $n \geq 0$ the groups $G_n(X)$ and $K_n(X)$ as the homotopy groups $\pi_{n+1}(|BQ\mathcal{M}(X)|)$ and $\pi_{n+1}(|BQ\mathcal{P}(X)|)$, respectively. When X is regular (e.g. a smooth variety over a field), each object in $\mathcal{M}(X)$ has a finite resolution by objects of $\mathcal{P}(X)$. From this one infers via the so-called resolution theorem of Quillen that in this case $G_n(X) = K_n(X)$ for all n.

For a commutative ring A we define the groups $G_n(A)$ and $K_n(A)$ to be $G_n(\mathrm{Spec}\,(A))$ and $K_n(\mathrm{Spec}\,(A))$, respectively. As $\mathcal{M}(\mathrm{Spec}\,A)$ is equivalent to the category $\mathcal{M}(A)$ of finitely generated A-modules and $\mathcal{P}(\mathrm{Spec}\,A)$ to the category $\mathcal{P}(A)$ of finitely generated projective A-modules, one may also define $G_n(A)$ and $K_n(A)$ as homotopy groups of the spaces $|BQ\mathcal{M}(A)|$ and $|BQ\mathcal{P}(A)|$, respectively. This latter construction immediately generalizes to not necessarily commutative A.

Facts 8.3.3 For a field F there exists a natural map from the n-th Milnor K-group $K_n^M(F)$ to the n-th Quillen K-group $K_n(F)$ which is an isomorphism for $n \leq 2$, but the two groups differ in general for $n > 2$. The comparison result for $n \leq 2$ is rather difficult, especially in the case $n = 2$, where it is a famous theorem of Matsumoto (see Milnor [2]).

Moreover, for $n = 1$ and A a not necessarily commutative ring the construction gives back the group $K_1(A)$ defined in Chapter 2.

Let $\mathcal{M}^i$ be the full subcategory of $\mathcal{M}(X)$ consisting of coherent sheaves whose support is of codimension $\geq i$ in X. They define a decreasing filtration of the category $\mathcal{M}(X)$, whence decreasing filtrations

$$G_n(X) = G_n(\mathcal{M}^0) \supset G_n(\mathcal{M}^1) \supset G_n(\mathcal{M}^2) \supset \ldots \qquad (8.3)$$

of the groups $G_n(X)$. Quillen has shown that there are long exact sequences

$$\cdots \to G_n(\mathcal{M}^{i+1}) \to G_n(\mathcal{M}^i) \to G_n(\mathcal{M}^i/\mathcal{M}^{i+1}) \to G_{n-1}(\mathcal{M}^{i+1}) \to \cdots$$
$$(8.4)$$

and isomorphisms

$$G_n(\mathcal{M}^i/\mathcal{M}^{i+1}) \cong \bigoplus_{P \in X^i} K_n(\kappa(P)), \qquad (8.5)$$

for all $i \geq 0$, where X^i stands for the set of scheme-theoretic points whose closure in X has codimension i.

By the general theory of spectral sequences, the filtration (8.3) gives rise to an exact couple via the exact sequence (8.4), and hence to a spectral sequence which converges to $G_n(X)$ if the filtration is finite, i.e. if X has finite dimension. The isomorphism (8.5) allows one to identify its E_1-term, thus one obtains a spectral sequence of the shape

$$E_1^{r,s}(X) = \bigoplus_{P \in X^r} K_{-r-s}(\kappa(P)) \Rightarrow G_{-n}(X)$$

called the *spectral sequence of Brown–Gersten–Quillen*. It is a fourth quadrant spectral sequence (i.e. $r \geq 0$ and $s \leq 0$), so the E_1-terms are zero for $|s| < |r|$. The filtration induced on $G_n(X)$ is precisely (8.3); the fact that it is a descending filtration accounts for the negative indices. We denote its i-th term $G_n(\mathcal{M}^i)$ by $F^i(G_n(X))$ and its i-th graded piece $G_n(\mathcal{M}^i/\mathcal{M}^{i+1})$ by $\mathrm{gr}^i(G_n(X))$.

By definition, the E_2-terms of the spectral sequence are obtained as the homology groups of a complex

$$\bigoplus_{P \in X^0} K_n(\kappa(P)) \to \bigoplus_{P \in X^1} K_{n-1}(\kappa(P)) \to \cdots \to \bigoplus_{P \in X^n} K_0(\kappa(P)), \quad (8.6)$$

the *Gersten complex in Quillen K-theory*. Up to reindexing, this complex is of a similar shape as the Gersten–Milnor complex constructed in Section 8.1. Moreover, the last three terms of the two complexes are isomorphic by Fact 8.3.3. Quillen has checked that the last coboundary map $\oplus K_1(\kappa(P)) \to \oplus K_0(\kappa(P))$ in (8.6) is induced by the valuation map, and Suslin has checked (Suslin [1], Proposition 6.8) that the penultimate coboundary $\oplus K_2(\kappa(P)) \to \oplus K_1(\kappa(P))$ is induced by a map which equals the tame symbol up to a character with values in $\{-1, 1\}$. These facts imply:

Lemma 8.3.4 *Let X be a smooth variety of dimension d over a field. Then there are natural isomorphisms*

$$E_2^{i,-i} \cong CH^i(X) := CH_{d-i}(X) \quad \text{and} \quad E_2^{i,-i-1} \cong A_{d-i}(X, K_{i+1-d}^M)$$

for all $0 \leq i \leq d$.

From now on we assume that X is a smooth variety over a field, so that $K_n(X) = G_n(X)$ for all n. By the lemma and the Brown–Gersten–Quillen spectral sequence we obtain maps $\rho_i : CH^i(X) \to \mathrm{gr}^i(K_0(X))$. In fact, they are induced by sending a closed subvariety Z of codimension i to the class of $\mathcal{O}_Z$ in $G_0(X) \cong K_0(X)$; in particular they are surjective. The general theory of Chern classes introduced by Grothendieck provides a rational splitting of these maps.

Fact 8.3.5 For each $i \geq 0$ there exists a canonical group homomorphism $c_i : K_0(X) \to CH^i(X)$ called the *i-th Chern class map* which is trivial on $K_0(\mathcal{M}^{i+1})$, and moreover the composite map $c_i \circ \rho_i : CH^i(X) \to \mathrm{gr}^i(K_0(X)) \to CH^i(X)$ equals multiplication by $(-1)^{i-1}(i-1)!$. Consequently, the collection of the Chern class maps induces a direct sum decomposition

$$K_0(X) \otimes \mathbf{Q} \cong \bigoplus_{i=0}^{d} CH^i(X) \otimes \mathbf{Q}.$$

The claim about the map $c_i \circ \rho_i$ boils down to the formula $[c_i(\mathcal{O}_Z)] = (-1)^{i-1}(i-1)! \, [Z]$ in $CH^i(X)$ for a closed subvariety $Z \subset X$ of codimension i. When Z is smooth, this follows from the Riemann–Roch formula without denominators (Fulton [1], Example 15.3.1). The general case reduces easily to the smooth case (Suslin [1], Proposition 9.3).

We can now easily prove:

Lemma 8.3.6 *Let p be a prime number. For a Severi–Brauer variety X of dimension $d = p - 1$ over a field k, the natural maps $CH^i(X) \to \mathrm{gr}^i(K_0(X))$ are isomorphisms for all $0 \leq i \leq p - 1$.*

Proof We have already remarked that the maps in question are surjective. For injectivity it will be enough to show in view of Fact 8.3.5 that the groups $CH^i(X)$ have no torsion elements of order dividing $(i-1)!$ for $0 \leq i \leq p-1$. In fact, they have no torsion prime to p. Indeed, by Proposition 7.4.1 we have a commutative diagram

$$
\begin{array}{ccc}
\bigoplus\limits_{Q \in X_K^{i-1}} K_1(\kappa(Q)) & \longrightarrow & \bigoplus\limits_{Q \in X_K^i} K_0(\kappa(Q)) \\
{\scriptstyle \oplus N_{\kappa(Q)|\kappa(P)}} \Big\downarrow & & \Big\downarrow {\scriptstyle \oplus N_{\kappa(Q)|\kappa(P)}} \\
\bigoplus\limits_{P \in X^{i-1}} K_1(\kappa(P)) & \longrightarrow & \bigoplus\limits_{P \in X^i} K_0(\kappa(P))
\end{array}
$$

where the horizontal maps are induced by valuation maps. According to Example 8.2.3 the cokernels of these maps are respectively $CH^i(X_K)$ and $CH^i(X)$, so the diagram defines a norm map $CH^i(X_K) \to CH^i(X)$. By a basic property of norm maps, the composite of this norm with the natural flat pullback map $CH^i(X) \to CH^i(X_K)$ induced by the maps $\iota_{\kappa(Q)|\kappa(P)}$ is multiplication by p. On the other hand, applying Proposition 8.2.6 with $n = -i$ shows that $CH^i(X_K) \cong \mathbf{Z}$ for all i. In particular, these groups are torsion free, so the groups $CH^i(X)$ can only have torsion elements of order dividing p. $\qquad\square$

Corollary 8.3.7 *With notations as in the above lemma, the natural maps $E_2^{i,-i-1} \to \mathrm{gr}^i(K_1(X))$ coming from the Brown–Gersten–Quillen spectral sequence are isomorphisms for all $0 \le i \le p - 1$.*

Proof All differentials whose target or source is a term $E_m^{i,-i-1}$ come from or land in terms of the shape $E_m^{j,-j}$. According to the first isomorphism in Lemma 8.3.4 and the statement of the lemma above these terms map isomorphically onto $\mathrm{gr}^i(K_0(X))$ for $m = 2$, and hence for all $m \ge 2$. But then all differentials in question must be zero, whence the corollary. $\qquad\square$

Consider now the pullback map $\pi^* : K_1(X) \to K_1(X_K)$ induced by pulling back vector bundles along the projection $\pi : X_K \to X$. Combining the corollary with the second isomorphism of Lemma 8.3.4 we see that the statement of Theorem 8.3.1 is equivalent to the injectivity of the maps $\mathrm{gr}^i(K_1(X)) \to \mathrm{gr}^i(K_1(X_K))$ induced by π^*. Or in other words:

Proposition 8.3.8 *Let X be a Severi–Brauer variety of dimension $d = p - 1$ over a field k split by an extension $K|k$ of degree p. The filtration by codimension of support on $K_1(X_K)$ induces that on $K_1(X)$, i.e.*

$$F^i K_1(X_K) \cap \pi^* K_1(X) = \pi^* F^i K_1(X)$$

for all i.

The proof relies on Quillen's calculation of the K-theory of Severi–Brauer varieties.

Fact 8.3.9 Let X be a Severi–Brauer variety of dimension $d-1$ split by $K|k$, and let A be a corresponding central simple algebra. Then for all $n \ge 0$ there is a decomposition

$$K_n(X) \cong \bigoplus_{j=0}^{d} K_n(A^{\otimes j}). \tag{8.7}$$

To construct the decomposition, Quillen shows that there is a rank d vector bundle $\mathcal{J}$ on X equipped with a left action by A which becomes isomorphic to $\mathcal{O}(-1)^{\oplus d}$ after pullback to X_K. The decomposition is then induced (from right to left) by the map $(u_0, \ldots, u_d)$, where $u_j : \mathcal{P}(A^{\otimes j}) \to \mathcal{P}(X)$ maps a projective $A^{\otimes j}$-module M to $\mathcal{J}^{\otimes j} \otimes_{A^{\otimes j}} M$. This decomposition is best explained in §12 of Swan [3]; see also the other references cited above.

In the split case $X \cong \mathbf{P}_k^{d-1}$ the decomposition reduces to

$$K_n(X) \cong \bigoplus_{j=0}^{d} K_n(k), \tag{8.8}$$

the isomorphism being induced (from right to left) by the map $(v_0, \ldots, v_d)$, where $v_j : \mathcal{P}(k) \to \mathcal{P}(X)$ maps a k-vector space V to $\mathcal{O}(-j) \otimes V$. To see that this is a special case of the previous construction one uses an easy case of *Morita equivalence* (see e.g. Rowen [2]) which shows that the module categories of k and $M_d(k)$ are equivalent.

Corollary 8.3.10 *Assume A is a division algebra of prime degree p. For $n = 1$ the isomorphism (8.7) becomes $K_1(X) \cong \mathrm{Nrd}(A^\times)^{\oplus d} \oplus k^\times$, where $\mathrm{Nrd} : A^\times \to k^\times$ is the reduced norm map.*

Proof By the last fact in 8.3.3, on the right-hand side of (8.7) we are dealing with the groups of Chapter 2 in the case $n = 1$. The term $j = 0$ contributes the summand $k^\times$. For each $j > 0$ we may find i with $ij \equiv 1 \bmod p$, so that $A^{\otimes ij}$ is Brauer equivalent to A, and hence $K_1(A) \cong K_1(A^{\otimes ij})$ by Lemma 2.10.5. There are injective base change maps $A \to A^{\otimes j} \to A^{\otimes ij}$ whose composite induces the isomorphism, so $K_1(A) \cong K_1(A^{\otimes j})$. Finally, the reduced norm map gives rise to an isomorphism $K_1(A) \xrightarrow{\sim} \mathrm{Nrd}(A^\times)$ by Wang's theorem (Theorem 2.10.12). $\square$

To attack the proof of Proposition 8.3.8, we first rewrite the isomorphism (8.8) for X_K in another way. Consider first the case $n = 0$, and denote by γ the class of $\mathcal{O}(-1)$ in $K_0(X_K)$. The decomposition for $K_0(X_K)$ is then just the direct sum of the $\mathbf{Z}\gamma^j$. But according to the description of (8.8), the case of general n reduces to $n = 0$, for it is induced by a product $K_0(X_K) \otimes_{\mathbf{Z}} K_n(K) \to K_n(X_K)$. Thus we may write

$$K_n(X_K) \cong \bigoplus_{j=0}^{d} K_n(K)\gamma^j.$$

Lemma 8.3.11   *The filtration by codimension of support on $K_n(X_K)$ is described by*

$$F^i K_n(X_K) \cong \bigoplus_{j=i}^{d} K_n(K)(\gamma - 1)^j.$$

Proof   Let H be a hyperplane in $X_K \cong \mathbf{P}_K^{d-1}$. The exact sequence

$$0 \to \mathcal{O}(-1) \to \mathcal{O} \to \mathcal{O}_H \to 0$$

shows that the class of $\mathcal{O}_H$ in $K_0(X_K)$ is precisely $\gamma - 1$. It generates $\mathrm{gr}^1 K_0(X_K) \cong \mathbf{Z}$, and its powers $[\mathcal{O}_H]^j$ generate $\mathrm{gr}^j K_0(X) \cong \mathbf{Z}$ (see the proof of Lemma 8.3.6), whence the case $n = 0$. To treat the general case, we exploit the isomorphism $K_0(X_K) \otimes_{\mathbf{Z}} K_n(K) \xrightarrow{\sim} K_n(X_K)$. In fact, it is induced by a product map on K-groups coming from tensoring vector bundles on X with trivial bundles coming from $\mathrm{Spec}\,(K)$ (Swan [3], §8 or Suslin [1], §6). As such, it preserves filtration by codimension of support, and the result follows from the case $n = 0$. $\qquad\qquad\square$

For the proof of Proposition 8.3.8 we need one last fact from Quillen K-theory: any proper morphism of finite-dimensional noetherian schemes $\phi : S \to T$ gives rise to pushforward maps $\phi_* : G_n(S) \to G_n(T)$ for all n, induced by taking higher direct images of coherent sheaves (Suslin [1], §6). It yields the norm map when $n = 1$ and $S = \mathrm{Spec}\,(E), T = \mathrm{Spec}\,(F)$ for a finite extension $E|F$ of fields, and satisfies the projection formula $\phi_*(x \cdot \phi^*(y)) = \phi_*(x)y$ with respect to the product encountered in the last proof.

Proof of Proposition 8.3.8   We may assume A is a division algebra, for otherwise it is split and the claim is obvious. The pullback map $\pi^* : K_1(X) \to K_1(X_K)$ respects the decompositions (8.8) and (8.7), and moreover using Morita equivalence and Wang's theorem may be identified on the components with the inclusion $\mathrm{Nrd}(A^\times) \hookrightarrow K^\times$. Hence it is injective, and its image may be described as the subgroup

$$\bigoplus_{j=0}^{d} \mathrm{Nrd}(A^\times)\gamma^j \subset \bigoplus_{j=0}^{d} K^\times \gamma^j.$$

By virtue of the previous lemma, the intersection of this subgroup with $F^i K_1(X_K)$ is given by

$$\left(\bigoplus_{j=0}^{d} \mathrm{Nrd}(A^\times)\gamma^j \right) \cap \left(\bigoplus_{j=i}^{d} K^\times (\gamma - 1)^j \right) = \bigoplus_{j=i}^{d} \mathrm{Nrd}(A^\times)(\gamma - 1)^j.$$

To see the equality here, notice that if

$$\sum_{j=i}^{d} a_j (\gamma - 1)^j = \sum_{j=i}^{d} b_j \gamma^j$$

with $a_j \in K^\times$, $b_j \in \mathrm{Nrd}(A^\times)$, then we must have $a_j \in \mathrm{Nrd}(A^\times)$ by descending induction on j.

So to conclude the proof it remains to show that $\mathrm{Nrd}(A^\times)(\gamma - 1)^j \subset \pi^*(F^i K_1(X))$ for $j \geq i$. Pick $\alpha \in \mathrm{Nrd}(A^\times)$ and $x \in A$ satisfying $\mathrm{Nrd}(x) = \alpha$. Proposition 2.6.8 shows that $\alpha = N_{L|k}(x)$ for a degree p subfield $L \subset A$ which is moreover a splitting field for A. Apply the push-forward map $\phi_* : K_1(X_L) \to K_1(X)$ coming from $\phi : X_L \to X$ to $x(\gamma - 1)^j$, and consider the decomposition (8.8) for X_L. Since the class γ is the pullback of the class of the vector bundle $\mathcal{J}$ on X by the construction of Fact 8.3.9, the projection formula for the pushforward map implies $\alpha(\gamma - 1)^j = \phi_*(x(\gamma - 1)^j)$. The previous lemma applied with L in place of K shows that $x(\gamma - 1)^j \in F^i K_1(X_L)$ for all $j \geq i$. But by its construction ϕ_* preserves filtration by codimension of support, whence $\alpha(\gamma - 1)^j \in F^i K_1(X)$ for $j \geq i$, which proves our claim in view of the description of the map $\pi^* : K_1(X) \to K_1(X_K)$ given above. This proves the proposition, and thereby Theorem 8.3.1. $\square$

Remark 8.3.12 The proof above shows that the injections

$$A_{d-i}(X, K_{i+1-d}^M) \to A_{d-i}(X_K, K_{i+1-d}^M)$$

can be identified via Proposition 8.3.4 and Corollary 8.3.7 with the inclusion maps $\mathrm{Nrd}(A^\times)(\gamma - 1)^i \subset K^\times(\gamma - 1)^i$, or in other words with the inclusion $\mathrm{Nrd}(A^\times) \subset K^\times$.

8.4 Hilbert's Theorem 90 for K_2

Recall from Example 2.3.4 that the classical form of Hilbert's Theorem 90 is the following statement: *In a cyclic Galois extension $K|k$ of fields each element of norm 1 can be written in the form $\sigma(a)a^{-1}$, where $a \in K^\times$ and σ is a fixed generator of* $\mathrm{Gal}\,(K|k)$. In other words, the complex

$$K^\times \xrightarrow{\sigma - 1} K^\times \xrightarrow{N_{K|k}} k^\times$$

is exact. In this section we prove an analogue of this statement for the group K_2^M, which is the crucial ingredient in the proof of the Merkurjev–Suslin theorem. Let $K|k$ be a cyclic extension as above, and consider the sequence of maps

$$K_2^M(K) \xrightarrow{\sigma-1} K_2^M(K) \xrightarrow{N_{K|k}} K_2^M(k).$$

Here σ acts on $K^\times \otimes_{\mathbf{Z}} K^\times$ by $\sigma(a \otimes b) = \sigma(a) \otimes \sigma(b)$, and $K_2^M(K)$ carries the induced action. The sequence is a complex, because the norm map satisfies $N_{K|k}(\sigma(\alpha)) = N_{K|k}(\alpha)$ for all $\alpha \in K_2^M(K)$ (Corollary 7.3.3).

Now the promised result is:

Theorem 8.4.1 (Hilbert's Theorem 90 for K_2) *Let $K|k$ be a cyclic Galois extension of prime degree p, and let σ be a generator of $\mathrm{Gal}\,(K|k)$. Then the complex*

$$K_2^M(K) \xrightarrow{\sigma-1} K_2^M(K) \xrightarrow{N_{K|k}} K_2^M(k) \qquad (8.9)$$

is exact.

Remarks 8.4.2

1. The theorem in fact holds for arbitrary cyclic extensions (see Exercise 5), but the prime degree case is the crucial one.
2. Thanks to Voevodsky's work on the Bloch–Kato conjecture, we now know that the sequences

$$K_n^M(K) \xrightarrow{\sigma-1} K_n^M(K) \xrightarrow{N_{K|k}} K_n^M(k)$$

   are exact for all $n > 0$ and $K|k$ as in the theorem. In the case where k has characteristic $p = [K : k]$ this was proven earlier by Izhboldin [1]; see Theorem 9.8.8 below.

First we establish the theorem in an important special case.

Proposition 8.4.3 *Let $K|k$ be a cyclic Galois extension of degree p as above. Assume that*

- *k has no nontrivial finite extensions of degree prime to p;*
- *the norm map $N_{K|k} : K^\times \to k^\times$ is surjective.*

Then Theorem 8.4.1 holds for the extension $K|k$.

Before starting the proof, we introduce some notation. Let $K|k$ be a cyclic extension of degree p as above, and let $F \supset k$ be a field. If $K \otimes_k F$ is not a field, then it splits into a direct sum $K \otimes_k F \cong F^{\oplus p}$ of copies of F. In this last case we shall use the notation $K_2^M(K \otimes_k F)$ for $K_2^M(F)^{\oplus p}$ and denote by $N_{K \otimes_k F|F} : K_2^M(K \otimes_k F) \to K_2^M(F)$ the map $K_2^M(F)^{\oplus p} \to K_2^M(F)$ given by the sum of the identity maps.

Proof Consider the map

$$\overline{N}_{K|k} : K_2^M(K)/(\sigma - 1)K_2^M(K) \to K_2^M(k)$$

induced by the norm $N_{K|k}$. The idea of the proof is to construct an inverse

$$\psi : K_2^M(k) \to K_2^M(K)/(\sigma - 1)K_2^M(K)$$

for the map $\overline{N}_{K|k}$. To do so, we first define a map

$$\widetilde{\psi} : k^\times \times k^\times \to K_2^M(K)/(\sigma - 1)K_2^M(K)$$

as follows. Let $a, b \in k^\times$. By assumption $a = N_{K|k}(c)$ for some $c \in K^\times$, and we set

$$\widetilde{\psi}(a, b) := \{c, b_K\} \in K_2^M(K)/(\sigma - 1)K_2^M(K),$$

where b_K means b viewed as an element of K. To see that $\widetilde{\psi}$ is well defined, take another element $c' \in K^\times$ with $N_{K|k}(c') = a$. Then $c'c^{-1}$ has norm 1, so it is of the form $\sigma(e)e^{-1}$ by the original Theorem 90 of Hilbert recalled above. As $\sigma(b_K) = b_K$, we have

$$\{c', b_K\} - \{c, b_K\} = \{\sigma(e)e^{-1}, b_K\} = (\sigma - 1)\{e, b_K\} \in (\sigma - 1)K_2^M(K).$$

The map $\widetilde{\psi}$ is manifestly bilinear, so it extends to a map

$$\widetilde{\psi} : k^\times \otimes_{\mathbf{Z}} k^\times \to K_2^M(K)/(\sigma - 1)K_2^M(K).$$

We next check that $\widetilde{\psi}$ respects the Steinberg relation, i.e. $\widetilde{\psi}(a \otimes 1 - a) = 0$ for $a \neq 0, 1$. Set $L = k(\alpha)$ for some $\alpha \in \bar{k}$ satisfying $\alpha^p = a$ if $a \notin k^{\times p}$, and set $L = K$ otherwise. The tensor product $M = K \otimes_k L$ is a field for $L \neq K$, and a direct sum of p copies of K for $L = K$. We have $H^1(G, M^\times) = 0$ for $G = \mathrm{Gal}\,(K|k)$, in the first case by Hilbert's Theorem 90, and in the second one because the group of invertible elements $M^\times$ is a co-induced G-module. Observe that

$$1 - a_K = N_{M|K}(1 - \alpha_M). \tag{8.10}$$

Indeed, if the extension $L|k$ is purely inseparable, so is $M|K$, and we have $N_{M|K}(1 - \alpha_M) = (1 - \alpha_M)^p = 1 - a_K$. Otherwise, $L|k$ is a cyclic Galois extension generated by some automorphism τ (as by our first assumption k contains the p-th roots of unity), so in L we have a product decomposition

$$x^p - a = \prod_{i=0}^{p-1} (x - \tau^i(\alpha)).$$

Setting $x = 1$ we get $1 - a = \prod(1 - \tau^i(\alpha)) = N_{L|k}(1 - \alpha)$. Since $M|K$ is either a degree p cyclic Galois extension, or a direct sum of p copies of K, this implies (8.10). We now compute using the projection formula

$$\{c, 1 - a\} = N_{M|K}(\{c_M, 1 - \alpha_M\})$$
$$= N_{M|K}(\{c_M \alpha_M^{-1}, 1 - \alpha_M\} + \{\alpha_M, 1 - \alpha_M\}).$$

Here $N_{M|L}(c_M \alpha_M^{-1}) = a(a^p)^{-1} = 1$, because $\sigma(\alpha_M) = \alpha_M$. By the vanishing of $H^1(G, M^\times)$ we may thus write $c_M \alpha_M^{-1} = \sigma(d)d^{-1}$ for some $d \in M$ using Example 3.2.9 (this is just the classical Hilbert 90 if $L \neq K$), and conclude that

$$\{c, 1 - a\} = N_{M|K}(\{\sigma(d)d^{-1}, 1 - \alpha_M\}) = (\sigma - 1)N_{M|K}(\{d, 1 - \alpha_M\}),$$

noting that $N_{M|K}$ commutes with the action of σ and $\sigma(\alpha_M) = \alpha_M$. This is an element lying in $(\sigma - 1)K_2^M(K)$, which finishes the verification.

We have now shown that the map $\tilde{\psi}$ induces a map

$$\psi : K_2^M(k) \to K_2^M(K)/(\sigma - 1)K_2^M(K).$$

Moreover, for all $a, b \in k^\times$ one has

$$N_{K|k}(\psi(\{a, b\}) = N_{K|k}(\{c, b\}) = \{N_{K|k}(c), b\} = \{a, b\}$$

by the projection formula, which implies in particular that ψ is injective. For surjectivity, note that thanks to the first assumption of the proposition we may apply the corollary to the Bass–Tate lemma (Corollary 7.2.10) according to which the symbol map $K^\times \otimes_\mathbf{Z} k^\times \to K_2^M(K)$ is surjective. Hence so is ψ, and the proof is finally finished. $\square$

The idea of the proof of Theorem 8.4.1 is then to embed k into some (very large) field extension $F_\infty \supset k$ satisfying the conditions of the proposition above, so that moreover the induced map between the homologies of the complex (8.9) and the similar one associated with the extension $F_\infty K|F_\infty$ is injective. The proposition will then enable us to conclude. The required injectivity property will be assured by the two propositions below.

Denote by $V(F)$ the homology of the complex

$$K_2^M(K \otimes_k F) \xrightarrow{\sigma - 1} K_2^M(K \otimes_k F) \xrightarrow{N_{K \otimes_k F|F}} K_2^M(F),$$

where we keep the notation introduced before the previous proof, the automorphism σ acts on $K \otimes_k F$ via the first factor, and $K_2^M(K \otimes_k F)$ is equipped with the induced action. For a tower of field extensions $E|F|k$, we have natural morphisms $V(k) \to V(F) \to V(E)$. We can now state:

Proposition 8.4.4 *Let $k'|k$ be an algebraic extension of degree prime to p. Then the map $V(k) \to V(k')$ is injective.*

Here in the case of an infinite algebraic extension we mean an extension whose finite subextensions all have degree prime to p.

Proof Let $L|k$ be a finite extension. The diagram

$$
\begin{array}{ccccc}
K_2^M(K \otimes_k L) & \xrightarrow{\ \sigma-1\ } & K_2^M(K \otimes_k L) & \xrightarrow{\ N_{K \otimes_k L|L}\ } & K_2^M(L) \\
{\scriptstyle N_{K \otimes_k L|K}}\downarrow & & {\scriptstyle N_{K \otimes_k L|K}}\downarrow & & {\scriptstyle N_{L|k}}\downarrow \\
K_2^M(K) & \xrightarrow{\ \sigma-1\ } & K_2^M(K) & \xrightarrow{\ N_{K|k}\ } & K_2^M(k)
\end{array}
$$

gives rise to a norm map $N_{L|k} : V(L) \to V(k)$. It follows from the results of Chapter 7, Section 7.3 that the composite $V(k) \to V(L) \xrightarrow{N_{L|k}} V(k)$ is multiplication by the degree $[L : k]$. In particular, for $L = K$ we get that the composite $V(k) \to V(K) \xrightarrow{N_{K|k}} V(k)$ is multiplication by p. On the other hand, notice that $V(K) = 0$. Indeed, by Galois theory $K \otimes_k K$ splits as a direct product of p copies of K, and σ acts *trivially* on each component, because we defined its action on $K \otimes_k K$ via the first factor. From this the vanishing of $V(K)$ is immediate.

Putting the above together we get that $p V(k) = 0$. On the other hand, if $L = k'$ for a finite extension $k'|k$ of degree prime to p, the composite map $V(k) \to V(k') \to V(k)$ is multiplication by $[k' : k]$, and therefore the map $V(k) \to V(k')$ must be injective. To pass from here to the general case of an algebraic extension $k'|k$ of degree prime to p, we write k' as a direct limit of its finite subextensions. The claim than follows from the fact that the functor $L \to V(L)$ commutes with direct limits (because so do tensor products and exact sequences). $\square$

The second injectivity property, which is the main step in the proof of Theorem 8.4.1, is given by the following proposition.

Proposition 8.4.5 *Let X be a Severi–Brauer variety split by the degree p cyclic extension $K|k$. Then the map $V(k) \to V(k(X))$ is injective.*

Proof An element of $\ker\big(V(k) \to V(k(X))\big)$ is given by some $\alpha \in K_2^M(K)$ of trivial norm such that there exists $\beta \in K_2^M(K(X))$ satisfying $\alpha_{K(X)} = (\sigma - 1)\beta$. We would like to prove that β may be chosen in $K_2^M(K)$. Consider the commutative diagram

$$
\begin{array}{ccccc}
K_2^M(K) & \xrightarrow{\sigma-1} & K_2^M(K) & \xrightarrow{N_{K|k}} & K_2^M(k) \\
\downarrow & & \downarrow & & \downarrow \\
\beta \in K_2^M(K(X)) & \xrightarrow{\sigma-1} & K_2^M(K(X)) & \xrightarrow{N_{K|k}} & K_2^M(k(X)), \\
\partial^M \downarrow & & \partial^M \downarrow & & \partial^M \downarrow \\
\bigoplus_{P \in X_K^1} \kappa(P)^\times & \xrightarrow{\sigma-1} & \bigoplus_{P \in X_K^1} \kappa(P)^\times & \xrightarrow{N_{K|k}} & \bigoplus_{P \in X^1} \kappa(P)^\times
\end{array}
$$

whose rows and columns are complexes. The lower right square commutes by Proposition 7.4.1; commutativity of the other squares is straightforward. The lower left square shows that $(\sigma - 1)\partial^M(\beta) = 0$, so $\partial^M(\beta)$ is fixed by the action of $\mathrm{Gal}(K|k)$, i.e. it comes from an element of $\bigoplus_{P \in X^1} \kappa(P)^\times$. Moreover, the valuations of this element coming from points of X^2 must be trivial, as so are those of $\partial^M(\beta)$. This means that there is an element γ_0 in

$$
Z(X) := \ker\Big(\bigoplus_{P \in X^1} \kappa(P)^\times \longrightarrow \bigoplus_{P \in X^2} \mathbf{Z} \Big)
$$

such that $\partial^M(\beta) = \gamma_{0,K}$. Now look at the commutative diagram

$$
\begin{array}{ccccccc}
 & & 0 & & 0 & & \\
 & & \downarrow & & \downarrow & & \\
K_2^M(k(X))/K_2^M(k) & \xrightarrow{\partial^M} & Z(X) & \longrightarrow & A_{d-1}(X, K_{2-d}^M) & \longrightarrow & 0 \\
\downarrow & & \downarrow & & \downarrow & & \\
K_2^M(K(X))/K_2^M(K) & \xrightarrow{\partial^M} & Z(X_K) & \longrightarrow & A_{d-1}(X_K, K_{2-d}^M) & \longrightarrow & 0.
\end{array}
$$

The exact rows of this diagram are given by the definition of the group $A_{d-1}(X, K_{2-d}^M)$. Injectivity of the second vertical map is obvious, and that of the third comes from the case $i = 1$ of Theorem 8.3.1. It follows from the diagram that there exists $\beta_0 \in K_2^M(k(X))$ such that $\partial^M(\beta_0) = \gamma_0$. We may therefore replace β by $\beta - \beta_{0,K}$ without affecting the relation $\alpha = (\sigma - 1)\beta$, so that $\partial^M(\beta) = 0$. This means that β lies in $A_d(X_K, K_{2-d}^M) \subset K_2^M(K(X))$. As X_K is a projective space over K, we get from Example 8.2.7 that β lies in $K_2^M(K)$, which is what we wanted to show. □

We finally come to the

Proof of Theorem 8.4.1 Define a tower of fields

$$
k = F_0 \subset F_1 \subset F_2 \subset F_3 \subset \cdots \subset F_\infty = \bigcup_n F_n
$$

inductively as follows:

1. The field F_{2n+1} is a maximal prime to p extension of F_{2n};

2. the field F_{2n+2} is the compositum of all function fields of Severi–Brauer varieties associated with cyclic algebras of the form (χ, b), where χ is a fixed character of $\mathrm{Gal}\,(F_{2n+1}K|F_{2n+1}) \cong \mathrm{Gal}\,(K|k)$ and $b \in F_{2n+1}^{\times}$.

Here some explanations are in order. Concerning the maximal prime to p extension, see Remark 7.2.11. By the compositum of all function fields of a family $\{X_i : i \in I\}$ of varieties we mean the direct limit of the direct system of the function fields of all $X_{i_1} \times \ldots X_{i_r}$ for finite subsets $\{i_1, \ldots, i_r\} \subset I$, partially ordered by the natural inclusions. Finally, the extensions $KF_j|F_j$ are all cyclic of degree p. Indeed, this property is preserved when passing to an extension of degree prime to p, and also when taking the function field of a product of Severi–Brauer varieties, as these varieties are geometrically integral, and hence the base field is algebraically closed in their function field.

Now Proposition 8.4.4 shows that $V(F_{2n})$ injects into $V(F_{2n+1})$ for $n \geq 0$, and an iterated application of Proposition 8.4.5 implies that $V(F_{2n+1})$ injects into $V(F_{2n+2})$. Therefore $V(k) = V(F_0)$ injects into $V(F_\infty)$. We now show that the field F_∞ satisfies the conditions in Proposition 8.4.3, which will conclude the proof. Indeed, it has no algebraic extension of degree prime to p by construction. To verify the surjectivity of the norm map $N_{KF_\infty|F_\infty}$, use Corollary 4.7.3 which states that under the isomorphism $F_\infty^{\times}/N_{KF_\infty|F_\infty}((KF_\infty)^{\times}) \cong \mathrm{Br}\,(KF_\infty|F_\infty)$ the classes of all elements $b \in F_\infty^{\times}$ are mapped to the classes of the algebras (χ, b). But each b comes from some F_{2m+1}, and hence the class of the algebra (χ, b) in $\mathrm{Br}\,(KF_{2m+2}|F_{2m+2})$ is trivial by Châtelet's theorem. As $\mathrm{Br}\,(KF_\infty|F_\infty)$ is the direct limit of the $\mathrm{Br}\,(KF_j|F_j)$ by Lemma 4.3.3, this concludes the verification of the assumptions of Proposition 8.4.3, and therefore the proof of the theorem. $\square$

Remark 8.4.6 The technique of building towers of fields like the one in the proof above has turned out to be useful in other situations as well. Merkurjev [3] used such a technique for constructing fields of cohomological dimension 2 over which there exist division algebras of period 2 and index 2^d for d arbitrarily large (this is to be compared with the discussion of Remark 2.8.12). In the same paper, Merkurjev gave a counterexample to a conjecture of Kaplansky's in the theory of quadratic forms. Colliot-Thélène and Madore [1] (see also Colliot-Thélène [5]) constructed by means of the above technique a field k of cohomological dimension 1 and characteristic 0 over which there exist smooth projective varieties that are birational to projective space over $\bar{k}$ but have no rational point (note that Severi–Brauer varieties always do, by Theorem 6.1.8).

The theorem has the following important applications.

Theorem 8.4.7 *Let $m > 1$ be an integer invertible in k, and assume that k contains a primitive m-th root of unity ω. Then the m-torsion subgroup $_mK_2^M(k)$ consists of elements of the form $\{\omega, b\}$, with $b \in k^\times$.*

Proof We begin with the crucial case when $m = p$ is a prime number. Let K be the Laurent series field $k((t))$, and consider the cyclic Galois extension $K'|K$ given by $K' = k((t'))$, where $t = t'^p$. Let σ be the generator of $\mathrm{Gal}(K'|K)$ satisfying $\sigma(t') = \omega t'$, and let $\alpha = \Sigma\{a_i, b_i\}$ be an element of $_pK_2^M(k)$. We compute

$$N_{K'|K}(\alpha_{K'}) = N_{K'|K}(\sum\{a_i, b_i\}) = \sum\{N_{K'|K}(a_i), b_i\}$$
$$= \sum\{a_i^p, b_i\} = p\alpha = 0$$

using the projection formula. Hilbert's Theorem 90 for K_2^M then implies that there exists $\beta \in K_2^M(K')$ such that $\alpha_{K'} = (\sigma - 1)\beta$. We denote by $\partial' : K_2^M(K') \to k^\times$ the tame symbol associated with the canonical valuation of K' and set $\gamma := \partial'(\beta)$. The element $\tilde{\beta} := \beta - \{t', \gamma\}$ then satisfies $\partial'(\tilde{\beta}) = 0$. Since $(\sigma - 1)\{t', \gamma\} = \{\sigma(t')t'^{-1}, \gamma\} = \{\omega, \gamma\}$, replacing β by $\tilde{\beta}$ and α by $\alpha - \{\omega, \gamma\}$ we may assume $\partial'(\beta) = 0$. By definition we have $\alpha = s_{t'}(\alpha_{K'}) = s_{t'}((\sigma - 1)\beta)$, where $s_{t'}$ is the specialization map associated with t'. But $\partial'(\beta) = 0$ implies, by the first part of Proposition 7.1.7, that β is a sum of symbols of the form $\{u_i, v_i\}$ with some units u_i, v_i in $k[[t']]^\times$. As the extension $K'|K$ is totally ramified, u_i and $\sigma(u_i)$ have the same image modulo (t'), and similarly for the v_i. Thus $s_{t'}((\sigma - 1)\{u_i, v_i\}) = 0$, so that $\alpha = s_{t'}((\sigma - 1)\beta) = 0$, which concludes the proof in the case $m = p$.

Next assume that $m = p^r$ is a prime power with $r > 1$, and that the statement is already known for all $m = p^j$ with $0 < j < r$. Given $\alpha \in {_{p^r}}K_2^M(k)$, we have $p\alpha = \{\omega^p, b\} = p\{\omega, b\}$ for some $b \in k^\times$ by induction, as ω^p is a primitive p^{r-1}-st root of unity. Therefore $\alpha - \{\omega, b\} = \{\omega^{p^{r-1}}, c\}$ for some $c \in k^\times$ by the case $m = p$, which proves the theorem in this case. Finally, the general case follows by decomposing each m-torsion element in $K_2^M(k)$ into a sum of p_i-primary torsion elements for the prime divisors p_i of m. $\quad\square$

Theorem 8.4.8 *Assume that $\mathrm{char}(k) = p > 0$. Then $_pK_2^M(k) = 0$.*

Proof We now exploit the Artin–Schreier extension K' of $K = k((t))$ given by $t'^p - t' = t^{-1}$, and denote by σ the generator of $\mathrm{Gal}(K'|K)$ mapping t' to $t' + 1$. Let $\alpha \in {_p}K_2^M(k)$. By the same argument as in the previous proof, we find $\beta \in K_2^M(K')$ satisfying $\partial'(\beta) = 0$ and $\gamma \in k^\times$ such that

$$\alpha_{K'} = (\sigma - 1)(\beta + \{t', \gamma\}) = (\sigma - 1)\beta + \{t' + 1, \gamma\} - \{t', \gamma\}.$$

As above, we use the specialization map $s_{t'}$ and note that $\partial'(\beta) = 0$ implies $s_{t'}((\sigma-1)\beta) = 0$. Hence $\alpha = s_{t'}(\alpha_{K'}) = s_{t'}((\sigma-1)\beta)+\{1,r\}-\{1,r\} = 0$, as was to be shown. $\qquad\qquad\square$

The above theorem is due to Suslin. It will be generalized in Section 9.8 to Milnor K-groups of arbitrary degree.

8.5 The Merkurjev–Suslin theorem: a special case

In this section we prove the injectivity part of the Merkurjev–Suslin theorem for some special fields. Namely, fix a prime number p, and let k_0 be a finite extension of $\mathbf{Q}$, or else an algebraically closed field or a finite field of characteristic prime to p. The fields k we shall consider in this section will be extensions of k_0 of the following type: there exists a subfield $k_0 \subset k_p \subset k$ which is a finitely generated purely transcendental extension of k_0, and moreover $k|k_p$ is a finite Galois extension whose degree is a power of p.

Theorem 8.5.1 *Let k be a field of the above type. Then the Galois symbol*

$$h_{k,p}^2 : K_2^M(k)/pK_2^M(k) \to H^2(k,\mu_p^{\otimes 2})$$

is injective.

The proof will be by induction on the transcendence degree, so the case of a base field has to be considered first. The case of a finite extension of $\mathbf{Q}$ results from Theorem 7.6.1. For k algebraically closed, the statement is obvious, as $H^2(k,\mu_p^{\otimes 2})$ is then trivial and $K_2^M(k)$ divisible by p (since $k^{\times p} = k^{\times}$). For k finite, both $K_2^M(k)$ and $H^2(k,\mu_p^{\otimes 2})$ are trivial, the first one by Example 7.1.3, and the second because $\mathrm{cd}(k) \leq 1$ in this case.

Thus we can turn to extensions of the base field. The core of the argument is the following proposition due to Merkurjev.

Proposition 8.5.2 *Let k be an arbitrary field containing a primitive p-th root of unity ω, and let $K = k(\sqrt[p]{a})$ be a cyclic extension of degree p. Assume that the Galois symbol $h_{k,p}^2 : K_2^M(k)/pK_2^M(k) \to H^2(k,\mu_p^{\otimes 2})$ is injective. Then the complex*

$$K_2^M(k)/pK_2^M(k) \xrightarrow{\,i_{K|k}\oplus h_{k,p}^2\,} K_2^M(K)/pK_2^M(K) \oplus H^2(k,\mu_p^{\otimes 2})$$

$$\xrightarrow{\,h_{K,p}^2 - \mathrm{Res}_k^K\,} H^2(K,\mu_p^{\otimes 2})$$

is exact.

For the proof we need some preliminary lemmas. The first is a well-known one from group theory.

Lemma 8.5.3 *Let G be a group and p a prime number. In the group algebra $\mathbf{F}_p[G]$ we have an equality*

$$(\sigma - 1)^{p-1} = \sigma^{p-1} + \cdots + \sigma + 1$$

for all $\sigma \in G$, where 1 is the unit element of G.

Proof   By Newton's binomial formula

$$(\sigma - 1)^{p-1} = \sum_{i=0}^{p-1} \binom{p-1}{i} \sigma^i (-1)^{p-1-i},$$

where

$$\binom{p-1}{i} = \frac{(p-1)\dots(p-i)}{1\cdots i} \equiv (-1)^i \bmod p,$$

and it remains to use the congruence $(-1)^{p-1} \equiv 1 \bmod p$.   □

Lemma 8.5.4 *For k and K as in the proposition, the natural map*

$$\ker(\iota_{K|k}) \to \ker(\mathrm{Res}_k^K)$$

induced by the commutative diagram

$$
\begin{array}{ccc}
K_2^M(k)/pK_2^M(k) & \xrightarrow{\;h_{k,p}^2\;} & H^2(k,\mu_p^{\otimes 2}) \\[4pt]
{\scriptstyle \iota_{K|k}}\downarrow & & \downarrow{\scriptstyle \mathrm{Res}_k^K} \\[4pt]
K_2^M(K)/pK_2^M(K) & \xrightarrow{\;h_{K,p}^2\;} & H^2(K,\mu_p^{\otimes 2})
\end{array}
$$

is surjective.

Proof   Let c be a class in the kernel of Res_k^K. Since $\mu_p \subset k$, we may identify c with an element in the kernel $\mathrm{Br}\,(K|k)$ of the map $\mathrm{Br}\,(k) \to \mathrm{Br}\,(K)$, and hence with the class of a cyclic algebra $(a,b)_\omega$ for appropriate $b \in k^\times$, by Corollaries 4.4.6 and 4.7.4. By Proposition 4.7.1 we therefore obtain $c = h_{k,p}^2(\{a,b\})$. But $\{a,b\}$ goes to 0 in $K_2^M(K)/pK_2^M(K)$ as a is a p-th power in K. The lemma follows.   □

Note that in the above proof we did not use the assumed injectivity of $h_{k,p}^2$. Under this hypothesis, the lemma (and its proof) yields:

Corollary 8.5.5   *Assume moreover that $h_{k,p}^2$ is injective. Then the natural map*

$$\ker(\iota_{K|k}) \to \ker(\mathrm{Res}_k^K)$$

is an isomorphism, and each element of $\ker(\iota_{K|k})$ *can be represented by a symbol of the form* $\{a, b\}$ *with some* $b \in k^{\times}$.

Proof of Proposition 8.5.2   Take $\beta \in K_2^M(K)/pK_2^M(K)$ and $c \in H^2(k, \mu_p^{\otimes 2})$ such that the pair (β, c) lies in the kernel of $h_{K,p}^2 - \mathrm{Res}_k^K$. It will be enough to find $\alpha \in K_2^M(k)/pK_2^M(k)$ for which $\iota_{K|k}(\alpha) = \beta$. Indeed, for such an α the commutative diagram of the previous lemma shows that $h_{k,p}^2(\alpha)$ differs from c by an element of $\ker(\mathrm{Res}_k^K)$. But by the lemma we may modify α by an element of $\ker(\iota_{K|k})$ so that $h_{k,p}^2(\alpha) = c$ holds.

Next, denote by σ the generator of $G = \mathrm{Gal}\,(K|k)$ mapping $\sqrt[p]{a}$ to $\omega \sqrt[p]{a}$. The idea of the proof is to construct inductively for each $i = 1, \dots, p-1$ elements $\alpha_i \in K_2^M(k)/pK_2^M(k)$ and $\beta_i \in K_2^M(K)/pK_2^M(K)$ satisfying

$$\beta = \iota_{K|k}(\alpha_i) + (\sigma - 1)^i(\beta_i). \tag{8.11}$$

This will prove the proposition, because for $i = p - 1$ we may write using Lemma 8.5.3

$$(\sigma - 1)^{p-1}(\beta_{p-1}) = \sum_{j=0}^{p-1} \sigma^i(\beta_{p-1}) = i_{K|k}(N_{K|k}(\beta_{p-1}))$$

in $K_2^M(K)/pK_2^M(K)$, and hence $\alpha = \alpha_{p-1} + N_{K|k}(\beta_{p-1})$ will do the job.

We begin the construction of the elements α_i and β_i by the case $i = 1$. Using the compatibility of the Galois symbol with norm maps (Proposition 7.5.4), we may write

$$0 = pc = \mathrm{Cor}_k^K\left(\mathrm{Res}_k^K(c)\right) = \mathrm{Cor}_k^K\left(h_{K,p}^2(\beta)\right) = h_{k,p}^2\left(N_{K|k}(\beta)\right).$$

By the assumed injectivity of $h_{k,p}^2$, we find $\alpha_1 \in K_2^M(k)$ with $N_{K|k}(\beta) = p\alpha_1$. Since $N_{K|k}(\alpha_{1,K}) = p\alpha_1$ by Remark 7.3.1 (here as always the subscript K means image by $i_{K|k}$), an application of Hilbert's Theorem 90 for K_2 shows that

$$\beta - \alpha_{1,K} = (\sigma - 1)\beta_1$$

for some $\beta_1 \in K_2^M(K)$, whence the case $i = 1$ of the required identity. In particular, in the case $p = 2$ this concludes the proof, so in what follows we assume $p > 2$.

Now assume that the elements α_j and β_j have been constructed for all $j \leq i$. Making the element $(\sigma - 1)^{p-1-i} \in \mathbf{Z}[G]$ act on both sides of the identity (8.11) we obtain

$$(\sigma-1)^{p-1-i}(\beta) = (\sigma-1)^{p-1-i}(\iota_{K|k}(\alpha_i)) + (\sigma-1)^{p-1}(\beta_i) = (\sigma-1)^{p-1}(\beta_i),$$

because σ acts trivially on $\alpha_{i,K}$.

Consider now the group $H^2(K, \mu_p^{\otimes 2})$ equipped with its natural G-action and apply $h_{K,p}^2$ to both sides. Using the compatibility of the Galois symbol with the action of G (an immediate consequence of Lemma 3.3.15 and Proposition 3.4.10 (4)), we get

$$(\sigma - 1)^{p-1-i}\mathrm{Res}_k^K(c) = (\sigma - 1)^{p-1-i}h_{K,p}^2(\beta) = h_{K,p}^2\big((\sigma - 1)^{p-1}(\beta_i)\big).$$

Here the left-hand side is 0, because the image of the map Res_k^K is fixed by G (this follows from the constructions of the conjugation action and the restriction map in Chapter 3), and $p - 1 - i \geq 1$. Using Lemma 8.5.3 we get from the above

$$0 = h_{K,p}^2\big(N_{K|k}(\beta_i)_K\big) = \mathrm{Res}_k^K\big(h_{k,p}^2\big(N_{K|k}(\beta_i)\big)\big).$$

Applying Corollary 8.5.5 we infer

$$h_{k,p}^2\big(N_{K|k}(\beta_i)\big) = h_{k,p}^2(\{a, b_i\})$$

for some $b_i \in k^\times$, and from the injectivity of $h_{k,p}^2$ we conclude that

$$N_{K|k}(\beta_i) - \{a, b_i\} = p\delta_i$$

for some $\delta_i \in K_2^M(k)$. Using the equality $N_{K|k}(\sqrt[p]{a}) = a$, we may rewrite this as

$$N_{K|k}\big(\beta_i - \{\sqrt[p]{a}, b_i\}\big) = p\delta_i$$

using the projection formula. Therefore, taking the identity $N_{K|k}(\delta_{i,K}) = p\delta_i$ into account, we get that $\beta_i - \{\sqrt[p]{a}, b_i\} - \delta_{i,K}$ is an element of trivial norm. Hence Hilbert's Theorem 90 for K_2^M implies

$$\beta_i - \{\sqrt[p]{a}, b_i\} - \delta_{i,K} = (\sigma - 1)\beta_{i+1}$$

for some $\beta_{i+1} \in K_2^M(K)$. Applying $(\sigma - 1)^i$ yields

$$(\sigma - 1)^i(\beta_i) = (\sigma - 1)^i(\{\sqrt[p]{a}, b_i\}) + (\sigma - 1)^{i+1}\beta_{i+1}.$$

Since $b_i \in k$, here for $i = 1$ we have $(\sigma - 1)(\{\sqrt[p]{a}, b_1\}) = \{\omega, b_1\}_K$ by definition of σ, and therefore $(\sigma - 1)^i(\{\sqrt[p]{a}, b_i\}) = 0$ for $i > 1$. All in all, putting $\alpha_2 := \alpha_1 + \{\omega, b_1\}$ and $\alpha_{i+1} := \alpha_i$ for $i > 1$ completes the inductive step. $\square$

Notice that the commutative diagram of Lemma 8.5.4 induces a natural map $\operatorname{coker}(h_{k,p}^2) \to \operatorname{coker}(h_{K,p}^2)$.

Corollary 8.5.6 *Under the assumptions of the previous proposition, the Galois symbol $h_{K,p}^2$ is injective, and so is the map $\operatorname{coker}(h_{k,p}^2) \to \operatorname{coker}(h_{K,p}^2)$ just defined.*

Proof For the first statement, take a class $\beta \in \ker(h_{K,p}^2)$ and a class c in the kernel of the restriction map $H^2(k, \mu_p^{\otimes 2}) \to H^2(K, \mu_p^{\otimes 2})$. By the proposition above, there is a class $\alpha \in K_2^M(k)/pK_2^M(k)$ with $\iota_{K|k}(\alpha) = \beta$ and $h_{k,p}^2(\alpha) = c$. Moreover, such an α is unique by injectivity of $h_{k,p}^2$. On the other hand, by Lemma 8.5.4 our c comes from an element in the kernel of $\iota_{K|k}$, so the above unicity yields $\alpha \in \ker(\iota_{K|k})$ and $\beta = 0$, as desired. The second statement is an immediate consequence of the proposition. $\square$

We can now proceed to:

Proof of Theorem 8.5.1 To begin with, a restriction-corestriction argument as in the proof of Proposition 7.5.8 shows that for a finite extension $L|K$ of fields which has degree prime to p the natural maps $\ker(h_{K,p}^2) \to \ker(h_{L,p}^2)$ and $\operatorname{coker}(h_{K,p}^2) \to \operatorname{coker}(h_{L,p}^2)$ are injective. By virtue of this fact we may enlarge k_0 so that it contains a primitive p-th root of unity.

We then proceed by induction on the transcendence degree. As noted after the statement of the theorem, the statement holds for k_0. Next choose a tower of field extensions $k_0 \subset k_p \subset k$ as at the beginning of this section. As the Galois symbol $h_{F,p}^1$ is bijective for all fields by virtue of Kummer theory (Theorem 4.3.6), an iterated application of Proposition 7.5.5 (1) shows that the injectivity of $h_{k_0,p}^2$ implies that of $h_{k_p,p}^2$. Finally, we write the Galois extension $k|k_p$ as a tower of Galois extensions of degree p, and obtain injectivity of $h_{k,p}^2$ by an iterated application of the first statement in Corollary 8.5.6. $\square$

To conclude this section, we note that the above arguments also yield the following interesting consequence.

Proposition 8.5.7 *Assume that k is a field having the property that the Galois symbol $h_{L,p}^2$ is injective for all finite extensions $L|k$. Then $h_{k,p}^2$ is surjective.*

Proof As in the previous proof, we may assume that k contains a primitive p-th root of unity. Take an element $c \in H^2(k, \mu_p^{\otimes 2})$. There exists a finite Galois extension $L|k$ such that c restricts to 0 in $H^2(L, \mu_p^{\otimes 2})$. Let P be a p-Sylow subgroup of $\operatorname{Gal}(L|k)$ and L^P its fixed field. The extension $L^P|k$ has prime to

p degree, so the remark at the beginning of the previous proof shows that the natural map coker $(h_{k,p}) \to$ coker $(h_{L^P,p})$ is injective. We may then replace k by L^P, and assume that the degree $[L : k]$ is a power of p. Thus, just like in the previous proof, an iterated application of the second statement of Corollary 8.5.6 shows that the restriction map coker $(h^2_{k,p}) \to$ coker $(h^2_{L,p})$ is injective, which implies that $c \in \mathrm{Im}\,(h^2_{k,p})$. $\qquad\qquad\qquad\qquad\qquad\qquad\qquad\qquad\square$

Thus in order to complete the proof of the Merkurjev–Suslin theorem, it remains to establish the injectivity of the Galois symbol for arbitrary fields. This is the content of the next section.

8.6 The Merkurjev–Suslin theorem: the general case

Following Merkurjev, we shall now use the results of the previous section to prove the following crucial fact. The general case of the Merkurjev–Suslin theorem will then be an easy consequence.

Theorem 8.6.1 *Let k be a field containing a primitive p-th root of unity ω, and let $K = k(\sqrt[p]{a})$ be a Galois extension of degree p. The kernel of the natural map*

$$K_2^M(k)/pK_2^M(k) \to K_2^M(K)/pK_2^M(K)$$

consists of images of symbols of the form $\{a, b\}$ with some $b \in k^\times$.

Note that this result was established in Corollary 8.5.5 under the assumption that $h^2_{k,p}$ is injective. Thus it can be also viewed as a consequence of the injectivity part of the Merkurjev–Suslin theorem. In particular, it holds for fields of the type considered in Theorem 8.5.1.

The idea of the proof is to reduce the statement to the case of the particular fields considered in Theorem 8.5.1 via a specialization argument. Recall that in Chapter 7 we constructed for a field K equipped with a discrete valuation v with residue field κ specialization maps $s_t^M : K_n^M(K) \to K_n^M(\kappa)$ for all $n \geq 0$, depending on the choice of a local parameter t for v. We shall use these maps here for $n = 1$ or 2, employing the more precise notations s_t^1 and s_t^2, respectively. Note that they satisfy the formula $s_t^2\{x, y\} = \{s_t^1(x), s_t^1(y)\}$ for all $x, y \in k$. The strategy of the proof is then summarized by the following proposition, which immediately implies Theorem 8.6.1.

Proposition 8.6.2 *Under the assumptions of the theorem, let*

$$\alpha = \{a_1, b_1\} + \cdots + \{a_n, b_n\} \in K_2^M(k)$$

be a symbol satisfying $\iota_{K|k}(\alpha) \in pK_2^M(K)$. *Then there exist a subfield* $k_0 \subset k$ *containing the elements* $\omega, a, a_1, \ldots, a_n, b_2, \ldots, b_n$, *an integer* $d > 0$ *and an iterated Laurent series field* $k_d := k_0((t_1)) \ldots ((t_d))$, *so that for suitable* $B \in k_d^\times$ *the element*

$$b := (s_{t_1}^1 \circ s_{t_2}^1 \circ \cdots \circ s_{t_d}^1)(B) \in k_0 \subset k$$

satisfies $\{a, b\} = \alpha$ *modulo* $pK_2^M(k)$, *where*

$$s_{t_i} : k_0((t_1)) \ldots ((t_i)) \to k_0((t_1)) \ldots ((t_{i-1}))$$

is the specialization map associated with t_i.

To ensure the existence of a suitable $B \in k_d$, the idea is to find elements $A_i, B_i \in k_d$ specializing to a_i and b_i, respectively, so that moreover the elements $A_1, \ldots, A_n, B_1, \ldots, B_n, a$ and ω all lie in a subfield of k_d which is of the type considered in Theorem 8.5.1, and the symbol $\sum \{A_i, B_i\}$ becomes divisible by p after we adjoin the p-th root $\sqrt[p]{a}$. Once this is done, Theorem 8.5.1 and Corollary 8.5.5 will guarantee the existence of the required B.

This argument prompts the necessity of finding a general criterion for elements $a_1, \ldots, a_n, b_n, \ldots, b_n$ in a field F which forces the p-divisibility of the symbol $\sum \{a_i, b_i\}$ in $K_2^M(F)$. The next proposition, due to Merkurjev, gives such a criterion.

First some notation: given an integer $N > 0$, denote by $\mathcal{A}_N$ the set of nonzero functions $\alpha : \{1, 2, \ldots, N\} \to \{0, 1, \ldots, p-1\}$. For a field F containing a primitive p-th root of unity and elements $a_1, \ldots, a_N \in F^\times$ set

$$a_\alpha := \prod_{i=1}^{N} a_i^{\alpha(i)} \quad \text{and} \quad F_\alpha := F(\sqrt[p]{a_\alpha}),$$

and denote by N_α the norm map $N_{F_\alpha | F} : F_\alpha^\times \to F^\times$.

Proposition 8.6.3 *Let* F *be a field containing a primitive* p-*th root of unity, and let* $a_1, \ldots, a_n$ *be elements in* $F^\times$ *whose images are linearly independent in the* $\mathbf{F}_p$-*vector space* $F^\times / F^{\times p}$. *Then for all* $b_1, \ldots, b_n \in F^\times$ *the symbol*

$$\sum_{i=1}^{n} \{a_i, b_i\} \in K_2^M(F)$$

lies in $pK_2^M(F)$ *if and only if there exist an integer* $N \geq n$ *and elements* $a_{n+1}, \ldots, a_N, c_1, \ldots c_N \in F^\times$ *and* $w_\alpha \in F_\alpha^\times$ *for each* $\alpha \in \mathcal{A}_N$ *satisfying*

$$c_i^p b_i = \prod_{\alpha \in \mathcal{A}_N} N_\alpha(w_\alpha)^{\alpha(i)} \qquad (8.12)$$

for all $1 \leq i \leq N$, *where we set* $b_i = 1$ *for* $n < i \leq N$.

Note that although formula (8.12) does not involve the a_i, the w_α depend on them. For the proof we need:

Lemma 8.6.4   *Let $\mathcal{N}$ be the subgroup of $F^\times \otimes_\mathbf{Z} F^\times$ generated by those elements $a \otimes b$ for which b is a norm from the extension $F(\sqrt[p]{a})|F$. The natural map $F^\times \otimes_\mathbf{Z} F^\times \to K_2^M(F)$ induces an isomorphism*

$$(F^\times \otimes_\mathbf{Z} F^\times/\mathcal{N}) \otimes \mathbf{Z}/p\mathbf{Z} \xrightarrow{\sim} K_2^M(F)/pK_2^M(F).$$

Proof   Denote by F_a the extension $F(\sqrt[p]{a})$. If $a \otimes b \in \mathcal{N}$, then by definition $b = N_{F_a|F}(b')$ for some $b' \in F_a$, so that in $K_2^M(F)$ we may write using the projection formula

$$\{a, b\} = \{a, N_{F_a|F}(b')\} = N_{F_a|F}(\{a, b'\}) = N_{F_a|F}(\{(\sqrt[p]{a})^p, b'\})$$
$$= pN_{F_a|F}(\{\sqrt[p]{a}, b'\}).$$

This shows that the natural surjection $F^\times \otimes_\mathbf{Z} F^\times \to K_2^M(F)/pK_2^M(F)$ factors through $(F^\times \otimes_\mathbf{Z} F^\times/\mathcal{N}) \otimes \mathbf{Z}/p\mathbf{Z}$. Finally, if some $a \otimes b$ maps to 0 in $K_2^M(k)/pK_2^M(k)$, then it is congruent modulo p to a sum of elements of the form $c \otimes (1 - c)$. But the equality $N_{F_c|F}(1 - \sqrt[p]{c}) = 1 - c$ (obtained as in the proof of Proposition 8.4.3) shows that these elements lie in $\mathcal{N}$, and the proof is complete. $\square$

Proof of Proposition 8.6.3   We first prove sufficiency, which does not require the independence assumption on the a_i. Equation (8.12) yields an equality $b_i = \prod_{\alpha \in \mathcal{A}_N} N_\alpha(w_\alpha)^{\alpha(i)}$ in $F^\times/F^{\times p}$, so that using $b_i = 1$ for $i > n$ we get

$$\sum_{i=1}^n \{a_i, b_i\} = \sum_{i=1}^N \{a_i, b_i\} = \sum_{i=1}^N \sum_{\alpha \in \mathcal{A}_N} \{a_i, N_\alpha(w_\alpha)^{\alpha(i)}\} \text{ in } K_2^M(F)/pK_2^M(F).$$

But here

$$\sum_{i=1}^N \{a_i, N_\alpha(w_\alpha)^{\alpha(i)}\} = \{a_\alpha, N_\alpha(w_\alpha)\}$$

by bilinearity of symbols, and $\{a_\alpha, N_\alpha(w_\alpha)\} \in pK_2^M(F)$ by Lemma 8.6.4.

For the converse, assume that $\sum\{a_i, b_i\}$ lies in $pK_2^M(F)$. By Lemma 8.6.4, we may then find elements $e_1, \ldots, e_r \in F^\times$ and $w_j \in F(\sqrt[p]{e_j})$ for $i \le j \le r$ so that we have an equality

$$\sum_{i=1}^n a_i \otimes b_i = \sum_{j=1}^r e_j \otimes N_{F(\sqrt[p]{e_j})|F}(w_j) \quad \text{in } F^\times \otimes_\mathbf{Z} \otimes F^\times \otimes_\mathbf{Z} \mathbf{Z}/p\mathbf{Z}.$$

We may assume the e_i map to independent elements in $F^\times/F^{\times p}$. Let V be the $\mathbf{F}_p$-subspace of $F^\times/F^{\times p}$ generated by the images of $a_1, \ldots, a_n, e_1, \ldots, e_r$, and choose elements $a_{n+1}, \ldots a_N \in F^\times$ so that $a_1, \ldots, a_N$ modulo $F^{\times p}$ yield a basis of V. For each j we may then find $\alpha_j \in \mathcal{A}_N$ so that the images of a_{α_j} and e_j in V are the same. For each $\alpha \in \mathcal{A}_N$ write $w_\alpha := w_j$ if $\alpha = \alpha_j$ for one of the α_j's just defined, and $w_\alpha := 1$ otherwise. In $F^\times \otimes_{\mathbf{Z}} F^\times \otimes_{\mathbf{Z}} \mathbf{Z}/p\mathbf{Z}$ we may then write

$$\sum_{i=1}^n a_i \otimes b_i = \sum_{j=1}^r e_j \otimes N_{F(\sqrt[p]{e_j})|F}(w_j) = \sum_{\alpha \in \mathcal{A}_N} a_\alpha \otimes N_\alpha(w_\alpha)$$

$$= \sum_{\alpha \in \mathcal{A}_N} \left(\prod_{i=1}^N a_i^{\alpha(i)} \right) \otimes N_\alpha(w_\alpha).$$

Introducing $b_i = 1$ for $i > n$ and using bilinearity, we rewrite the above as

$$\sum_{i=1}^N a_i \otimes b_i = \sum_{i=1}^N a_i \otimes \left(\prod_{\alpha \in \mathcal{A}_N} N_\alpha(w_\alpha)^{\alpha(i)} \right) \quad \text{in } F^\times \otimes_{\mathbf{Z}} F^\times \otimes_{\mathbf{Z}} \mathbf{Z}/p\mathbf{Z}.$$

As the images of the $a_i \bmod F^{\times p}$ are linearly independent, this is only possible if $b_i = \prod_{\alpha \in \mathcal{A}_N} N_\alpha(w_\alpha)^{\alpha(i)} \bmod F^{\times p}$, whence the existence of the required c_i. $\qquad\square$

We now turn to the proof of Proposition 8.6.2. Our proof is a variant of that of Klingen [1], itself a simplification of Merkurjev's original argument.

Proof of Proposition 8.6.2 Consider the element $\alpha = \sum\{a_i, b_i\}$ of the proposition. Without loss of generality we may assume that the images of $a, a_1, \ldots, a_n$ are linearly independent in $k^\times/k^{\times p}$, and therefore the images of $a_1, \ldots, a_n$ are linearly independent in $K^\times/K^{\times p}$. We may then apply Proposition 8.6.3, of which we keep the notations. We find elements $a_{n+1}, \ldots, a_N, c_1, \ldots, c_N \in K^\times$ as well as $w_\alpha \in K_\alpha^\times$ satisfying (8.12). Introduce the notations $u := \sqrt[p]{a}$ and $u_\alpha := \sqrt[p]{a_\alpha}$ for all $\alpha \in \mathcal{A}_N$. The u^j for $0 \le j \le p-1$ form a k-basis of K, and the $u^i u_\alpha^j$ for $0 \le i, j \le p-1$ form a k-basis of K_α for each α. Thus we find elements $a_{ij} \in k$ for $0 \le j \le p-1$ and $n+1 \le i \le N$, as well as $w_{ij\alpha} \in k$ with $0 \le i, j \le p-1$ and $\alpha \in \mathcal{A}_N$ satisfying

$$a_i = \sum_{j=0}^{p-1} a_{ij} u^j \quad \left(\text{resp. } w_\alpha = \sum_{i,j=0}^{p-1} w_{ij\alpha} u^i u_\alpha^j \right)$$

for all $n+1 \le i \le N$ (resp. $\alpha \in \mathcal{A}_N$).

Now let f_0 be the subfield of k generated over the prime field by a and the p-th root of unity ω. Introduce independent variables

$$A_i, B_i \; (1 \le i \le n), \quad A_{ij} \; (n+1 \le i \le N, \; 0 \le j \le p-1)$$

and

$$W_{ij\alpha} \; (0 \le i, j \le p-1, \; \alpha \in \mathcal{A}_N),$$

and take the purely transcendental extension

$$F_0 := f_0(A_i, B_i, A_{ij}, W_{ij\alpha}).$$

We first construct a field extension $L_0|F_0$ containing an element B for which $\sum \{A_i, B_i\} = \{a, B\}$ holds modulo an element in $pK_2^M(L_0)$. To do so, for each $1 \le i \le N$ introduce elements $D_i \in F_0(u)$ via the formula

$$D_i := B_i^{-1} \prod_{\alpha \in \mathcal{A}_N} N_\alpha(W_\alpha),$$

where the W_α and the norm maps N_α are defined by setting

$$A_\alpha := \prod_{i=1}^N A_i^{\alpha(i)}, \; U_\alpha := \sqrt[p]{A_\alpha}, \; W_\alpha := \sum_{i,j=0}^{p-1} W_{ij\alpha} u^i U_\alpha^j \text{ and}$$

$$N_\alpha := N_{F_0(u,U_\alpha)|F_0(u)},$$

with the conventions $A_i := \sum A_{ij} u^j$ and $B_i := 1$ for $i > n$. Denoting by σ the generator of the Galois group $\mathrm{Gal}\,(F_0(u)|F_0) \cong \mathbf{Z}/p\mathbf{Z}$ sending u to ωu, pick elements $C_{i,l}$ in some fixed algebraic closure of $F_0(u)$ satisfying $C_{i,l}^p = \sigma^l(D_i)$ for $1 \le i \le N$ and $0 \le l \le p-1$.

Let L be the finite extension of $F_0(u)$ obtained by adjunction of all the $C_{i,l}$. Observe that the choice of the $C_{i,l}$ among the roots of the polynomials $x^p - \sigma^l(C_i)$ defines an extension of σ to an automorphism in $\mathrm{Gal}\,(L|F_0)$; we call it again σ, and denote by $L_0 := L^\sigma \subset L$ its fixed field. Note that $L = L_0(u)$. The construction together with Proposition 8.6.3 show that the symbol $\sum \{A_i, B_i\}$ lies in $K_2^M(L_0)$ and its image in $K_2^M(L)$ is divisible by p. On the other hand, the field L_0 is a p-power degree Galois extension of the purely transcendental field F_0, hence by Theorem 8.5.1 the Galois symbol $h_{L_0,p}^2$ is injective. An application of Corollary 8.5.5 then allows one to find $B \in L_0$ with $\sum \{A_i, B_i\} = \{a, B\}$ modulo $pK_2^M(L_0)$.

To complete the proof, we embed L_0 into a field $k_d = k_0((t_1)) \ldots ((t_d))$ so that k_0 is a subfield of k containing the elements a_i, b_i, which are moreover exactly the images of the A_i and B_i in k_0. Define an f_0-algebra homomorphism $f_0[A_i, B_i, A_{ij}, W_{ij\alpha}] \to k$ by sending $A_i \mapsto a_i, B_i \mapsto b_i, A_{ij} \mapsto a_{ij}$,

$W_{ij\alpha} \mapsto w_{ij\alpha}$. Its kernel is a prime ideal P in the polynomial ring $f_0[A_i, B_i, A_{ij}, W_{ij\alpha}]$; denote by $\mathcal{O}_P$ the associated localization. The local ring $\mathcal{O}_P$ is regular (being the local ring of a scheme-theoretic point on affine space), hence by the Cohen structure theorem (see Appendix, Theorem A.5.4 (2)) its completion $\widehat{\mathcal{O}}_P$ is isomorphic to a formal power series ring of the form $\kappa_0[[v_1, \ldots, v_d]]$, where $\kappa_0 \subset k$ is the residue field of $\mathcal{O}_P$. Its fraction field naturally embeds into the iterated Laurent series field $\kappa_d := \kappa_0((v_1)) \ldots ((v_d))$. In particular, we have a natural embedding $F_0 \hookrightarrow \kappa_d$, since F_0 is the fraction field of $\mathcal{O}_P$.

Write K_d for the composite $L\kappa_d$, where $L|F_0$ is the extension constructed above. The automorphism $\sigma \in \mathrm{Gal}\,(L|F_0)$ induces an element in $\mathrm{Gal}\,(K_d|\kappa_d)$ which we again denote by σ. Let $k_d := K_d^\sigma$ be its fixed field; note that $L_0 \subset k_d$ and $K_d = k_d(u)$. Extending the discrete valuation defined by v_d to the finite extension $k_d|\kappa_d$ we again get a Laurent series field of the form $k_{d-1}((t_d))$, where k_{d-1} is a finite extension of $\kappa_0((v_1)) \ldots ((v_{d-1}))$. This in turn becomes equipped with the unique extension of the valuation defined by v_{d-1}, and so is a Laurent series field $k_{d-2}((t_{d-1}))$ for suitable t_{d-1}. Continuing this process, we may finally write $k_d = k_0((t_1)) \ldots ((t_d))$ with some finite extension $k_0|\kappa_0$.

It remains to show that the field k_0 may be embedded into k. For this observe first that the elements $d_i := b_i^{-1} \prod_{\alpha \in \mathcal{A}_N} N_\alpha(w_\alpha)$ of K considered at the beginning of the proof all lie in $\kappa_0(u)$ by construction, and they are precisely the images of the elements D_i in $\kappa_0(u)$ (which are units for all valuations concerned). The elements $\sigma^l(D_i)$ map to $\bar{\sigma}^l(d_i)$ in $\kappa_0(u)$, where $\bar{\sigma}$ is the generator of $\mathrm{Gal}\,(\kappa_0(u)|\kappa_0)$ sending u to ωu. Extending $\bar{\sigma}$ to the automorphism in $\mathrm{Gal}\,(K|k)$ with the same property, we see that denoting by $\bar{C}_{i,l}$ the image of $C_{i,l}$ in the residue field $k_0(u)$ of K_d the map $\bar{C}_{i,l} \mapsto \bar{\sigma}^l(c_i)$ induces an embedding $k_0(u) \hookrightarrow K$ compatible with the action of $\bar{\sigma}$. We conclude by taking invariants under $\bar{\sigma}$. □

We have finally arrived at the great moment when we can prove in full generality:

Theorem 8.6.5 (Merkurjev–Suslin) *Let k be a field, and $m > 0$ an integer invertible in k. Then the Galois symbol*

$$h_{k,m}^2 : K_2^M(k)/mK_2^M(k) \to H^2(k, \mu_m^{\otimes 2})$$

is an isomorphism.

Proof By virtue of Proposition 7.5.8 it is enough to treat the case when $m = p$ is a prime number, and in view of Proposition 8.5.7 it suffices to prove injectivity. As in the proof of that proposition, we may assume that k contains a primitive p-th root of unity. Take a symbol $\alpha = \{a_1, b_1\} + \cdots + \{a_n, b_n\}$ in $K_2^M(k)$ whose mod p image lies in the kernel of $h_{k,p}^2$. We have to prove $\alpha \in pK_2^M(k)$. We proceed by induction on n, the case $n = 1$ being Lemma 7.6.3. If $a_n \in k^{\times p}$, we are done by the case of symbols of length $n - 1$. Otherwise, set $K = k(\sqrt[p]{a_n})$. Then $i_{K|k}(\alpha) = i_{K|k}(\{a_1, b_1\} + \cdots + \{a_{n-1}, b_{n-1}\})$ in $K_2^M(K)/pK_2^M(K)$, and it lies in the kernel of $h_{K,p}^2$ by compatibility of the Galois symbol with restriction maps. Hence by induction we may assume $i_{K|k}(\alpha) \in pK_2^M(K)$. But then by Theorem 8.6.1 the image of α in $K_2^M(k)/pK_2^M(k)$ equals that of a symbol of the form $\{a_n, b\}$ for some $b \in k^{\times}$. We conclude by the case $n = 1$. □

Remarks 8.6.6

1. The above proof of the theorem relies on Theorem 8.5.1, and hence in characteristic 0 on Tate's result for number fields, which has a nontrivial input from class field theory. Note however that this arithmetic input is not necessary in positive characteristic, or for fields containing an algebraically closed subfield. The original proof of injectivity in Merkurjev–Suslin [1] did not use a specialization argument, but a deep fact from the K-theory of central simple algebras which amounts to generalizing the isomorphism $A_d(\mathbf{P}^d, K_{2-d}^M) \cong K_2^M(k)$ of Example 8.2.7 to arbitrary Severi–Brauer varieties of squarefree degree. They however used a specialization technique for proving surjectivity of the Galois symbol, which could later be eliminated by the method of Proposition 8.5.2. All in all, there exists a proof of the theorem which uses no arithmetic at all, at the price of hard inputs from K-theory.

2. One may ask whether the Merkurjev–Suslin theorem holds in a stronger form, namely whether in the presence of a primitive m-th root of unity every central simple algebra of period m is actually *isomorphic* to a tensor product of cyclic algebras. There are counterexamples to this statement related to Amitsur's examples cited in Remark 2.2.13 (2). But these involve fields of cohomological dimension ≥ 3, so the question remains open for fields of cohomological dimension ≤ 2. The case $m = 2$ was answered positively by Sivatsky in his thesis using quadratic form techniques (see Kahn [3] for an exposition, where the result is attributed to Merkurjev).

The Merkurjev–Suslin theorem implies in particular that in the presence of a primitive m-th root of unity the m-torsion of the Brauer group can be generated by classes of algebras of degree m. It is not known whether this property holds over arbitrary fields. In fact, Albert asked a more precise question at the end of his paper Albert [3]: given a prime p invertible in a field k and an integer $r > 1$, is the p^r-torsion part of $\mathrm{Br}\,(k)$ generated by classes of division algebras D with $\mathrm{per}(D) = \mathrm{ind}(D) \le p^r$?

The following theorem, taken from Merkurjev [6], gives a positive answer to this question for $r = 1$.

Theorem 8.6.7 (Merkurjev) *Let k be a field, and p a prime number invertible in k. The p-torsion of $\mathrm{Br}\,(k)$ is generated by classes of division algebras of degree p.*

We begin by the following lemma.

Lemma 8.6.8 *In the situation of the theorem let moreover $K|k$ be a finite Galois extension with group G. If the order of G is prime to p, then every degree p division algebra over K whose class in $\mathrm{Br}\,(K)$ is invariant by G comes by base change from a degree p division algebra over k.*

Proof Since the order n of G is prime to p, the abelian group μ_p is uniquely n-divisible. Therefore the profinite version of Proposition 3.3.20 (Corollary 4.3.5) applies and shows that the natural map $H^2(k, \mu_p) \to H^2(K, \mu_p)^G$ is an isomorphism. In particular, a degree p division algebra D over K whose class in $\mathrm{Br}\,(K)$ is invariant by G is Brauer equivalent to $\widetilde{D} \otimes_k K$ for a unique division algebra $\widetilde{D}$ over k of period p. By Theorem 2.8.7 the index of D is a power of p and hence prime to n. Corollary 4.5.4 (2) then shows that $\widetilde{D} \otimes_k K$ is a division algebra Brauer equivalent to D, hence $\widetilde{D} \otimes_k K \cong D$ and $\widetilde{D}$ has degree p. $\qquad\square$

Remark 8.6.9 The lemma and its proof work more generally with any positive integer invertible in k and prime to the order of G in place of p. The proof moreover shows that the k-algebra $\widetilde{D}$ is unique up to isomorphism.

We also need a lemma from linear algebra.

Lemma 8.6.10 *Let p be a prime, and V an $\mathbf{F}_p$-vector space equipped with a linear action of a subgroup G of the multiplicative group $\mathbf{F}_p^\times$. There is a direct sum decomposition in $\mathbf{F}_p^\times$-invariant subspaces*

$$V = \bigoplus_\theta V_\theta$$

where the sum is over the character group of G and G acts on V_θ by $\sigma(v) = \theta(\sigma)v$.

Proof   Over an algebraic closure $\overline{\mathbf{F}}_p$ the elements of G are simultaneously diagonalizable, so there is a direct sum decomposition in eigenspaces as above. But the characters involved all take values in $\mu_{p-1} \subset \mathbf{F}_p$, so the decomposition exists already over $\mathbf{F}_p$.   □

Proof of Theorem 8.6.7   In the case when $\mu_p \subset K$ the statement is an immediate consequence of the Merkurjev–Suslin theorem. Otherwise consider the extension $K = k(\mu_p)$ where the Merkurjev–Suslin theorem may be applied. After choosing an isomorphism $\mu_p \cong \mathbf{Z}/p\mathbf{Z}$ it shows that each class in $H^2(K, \mu_p)$ is a sum of classes of the form $\chi_i \cup c_i$ with $\chi_i \in H^1(K, \mathbf{Z}/p\mathbf{Z})$ and $c_i \in H^1(K, \mu_p)$. Viewing $\chi_i \cup c_i$ as a p-torsion element in $\mathrm{Br}\,(K)$, it is represented by a degree p cyclic division algebra by Proposition 4.7.1, so in view of Lemma 8.6.8 it suffices to show that each $\chi_i \cup c_i$ can be chosen to be invariant by the action of $G = \mathrm{Gal}\,(K|k)$ on $H^2(K, \mu_p)$.

To do so, apply the direct sum decomposition of the previous lemma to the $\mathbf{F}_p$-vector spaces $H^1(K, \mathbf{Z}/p\mathbf{Z})$, $H^1(K, \mu_p)$ and $H^2(K, \mu_p)$ equipped with the action of $G \subset \mathbf{F}_p^\times$. After decomposing the χ_i and the c_i in their θ-components we may assume that for all i we have $\chi_i \in H^1(K, \mathbf{Z}/p\mathbf{Z})_{\theta_i}$ and $c_i \in H^1(K, \mu_p)_{\theta_i'}$ with some characters θ_i, θ_i'. But then $\chi_i \cup c_i \in H^2(K, \mu_p)_{\theta_i \theta_i'}$, so the sum of the $\chi_i \cup c_i$ can only be invariant by G if each $\theta_i \theta_i'$ is the trivial character. This concludes the proof.   □

Remark 8.6.11   If every division algebra of degree p were cyclic (as asked in Remark 7.4.14), then the above theorem would imply that the p-torsion of the Brauer group is generated by classes of cyclic algebras for an arbitrary field k of characteristic different from p. This latter fact is currently known only for $p = 3$ (by the result of Wedderburn [3] already cited) and for $p = 5$ by work of Matzri [1]. There are also positive results over fields k such that $[k(\mu_p) : k] \leq 3$ in Merkurjev [6].

8.7   Reduced norms and K_2-symbols

In the remainder of this chapter we apply the techniques and results of the previous sections to the study of reduced norm maps for cyclic division algebras of squarefree degree. All theorems come from the classic papers Merkurjev–Suslin [1] and Suslin [1].

This section is devoted to the proof of the following theorem. Our method of proof differs from that of the original sources.

Theorem 8.7.1 *Let $K|k$ be a cyclic Galois extension of fields whose degree m is squarefree and invertible in k. Fix an isomorphism $\chi : \mathrm{Gal}\,(K|k) \xrightarrow{\sim} \mathbf{Z}/m\mathbf{Z}$, and for $b \in k^\times$ consider the cyclic algebra $A = (\chi, b)$. An element $c \in k^\times$ is in the image of the reduced norm map $\mathrm{Nrd} : A^\times \to k^\times$ if and only if the symbol $\{b, c\} \in K_2^M(k)$ lies in the image of the norm map $N_{K|k} : K_2^M(K) \to K_2^M(k)$.*

We start the proof of the theorem with some auxiliary statements. The first is a general property of Gersten complexes. Suppose X is a smooth variety of dimension d over k, and $K|k$ is a finite Galois extension. The group $G := \mathrm{Gal}\,(K|k)$ acts naturally on the terms of the Gersten complex

$$K_2^M(K(X)) \xrightarrow{\partial^1} \bigoplus_{P \in X_K^1} \kappa(P)^\times \xrightarrow{\partial^2} \bigoplus_{P \in X_K^2} \mathbf{Z} \qquad (8.13)$$

associated with the base change X_K, and hence also on its homology groups $A_{d-i}(X_K, K_{2-d}^M)$. Concerning these homology groups, we have:

Lemma 8.7.2 *In the above situation there is an exact sequence*

$$A_{d-1}(X, K_{2-d}^M) \to A_{d-1}(X_K, K_{2-d}^M)^G \to H^1\big(G, K_2^M(K(X))/A_d(X_K, K_{2-d}^M)\big)$$

of abelian groups. Moreover, the natural map

$$K_2^M(k(X))/A_d(X, K_{2-d}^M) \to (K_2^M(K(X))/A_d(X_K, K_{2-d}^M))^G$$

is injective.

The exact sequence of the lemma can in fact be continued further; see Exercise 1.

Proof As in the proof of Proposition 8.4.5, denote by $Z(X_K)$ the kernel of the map ∂^2 in the Gersten complex (8.13). By definition of the homology groups, we then have an exact sequence of G-modules

$$0 \to K_2^M(K(X))/A_d(X_K, K_{2-d}^M) \to Z(X_K) \to A_{d-1}(X_K, K_{2-d}^M) \to 0. \qquad (8.14)$$

The beginning of its G-cohomology sequence reads

$$0 \to (K_2^M(K(X))/A_d(X_K, K_{2-d}^M))^G \to Z(X_K)^G \to A_{d-1}(X_K, K_{2-d}^M)^G$$
$$\to H^1\big(G, K_2^M(K(X))/A_d(X_K, K_{2-d}^M)\big).$$

We may insert part of this sequence in the commutative diagram

$$0 \longrightarrow (K_2^M(K(X))/A_d(X_K, K_{2-d}^M))^G \longrightarrow Z(X_K)^G \longrightarrow A_{d-1}(X_K, K_{2-d}^M)^G$$

$$0 \longrightarrow K_2^M(k(X))/A_d(X, K_{2-d}^M) \longrightarrow Z(X) \longrightarrow A_{d-1}(X, K_{2-d}^M) \longrightarrow 0$$

where the bottom row is obtained from the Gersten complex of X in an analogous manner.

We contend that here the middle vertical map $Z(X) \to Z(X_K)^G$ is an isomorphism, from which both statements of the lemma follow by an immediate diagram chase. To justify this, notice that by a similar argument as in the proof of Lemma 5.4.5 we have isomorphisms of G-modules

$$\bigoplus_{Q \in X_K^1} \kappa(Q)^\times \cong \bigoplus_{P \in X^1} \left(\bigoplus_{Q \mapsto P} \kappa(Q)^\times \right) \cong \bigoplus_{P \in X^1} M_{G_P}^G(\kappa(P)^\times),$$

where the notation $Q \mapsto P$ stands for the points of X_K lying above a fixed point $P \in X^1$, and G_P is the stabilizer of P in G. By taking G-invariants, we obtain

$$\left(\bigoplus_{Q \in X_K^1} \kappa(Q)^\times \right)^G \cong \bigoplus_{P \in X^1} \kappa(P)^\times.$$

Similarly, we have

$$\left(\bigoplus_{Q \in X_K^2} \mathbf{Z} \right)^G \cong \bigoplus_{P \in X^2} \mathbf{Z},$$

whence the claim. $\square$

Consider now the particular case where G is cyclic and X is a Severi–Brauer variety split by $K|k$. By Example 8.2.7 we then have G-equivariant isomorphisms

$$A_d(X_K, K_{2-d}^M) \cong K_2^M(K), \quad A_{d-1}(X_K, K_{2-d}^M) \cong K^\times,$$

where the first isomorphism is induced by the inclusion $A_d(X_K, K_{2-d}^M) \to K_2^M(K)$. Therefore the second map in the exact sequence of Lemma 8.7.2 has target $H^1(G, K_2^M(K(X))/K_2^M(K))$. Moreover, using the exact sequence of G-modules

$$0 \to K_2^M(K) \to K_2^M(K(X)) \to K_2^M(K(X))/K_2^M(K) \to 0$$

we obtain a coboundary map

$$H^1(G, K_2^M(K(X))/K_2^M(K)) \to H^2(G, K_2^M(K)).$$

Since the extension $K|k$ is cyclic, we can calculate G-cohomology as in Example 3.2.9, and write

$$H^2(G, K_2^M(K)) \cong K_2^M(K)^G / N(K_2^M(K)),$$

where $N : K_2^M(K) \to K_2^M(K)$ is given by $\alpha \mapsto \sum_i \sigma^i \alpha$ for a generator σ of G. Notice that we have

$$N = i_{K|k} \circ N_{K|k}$$

with the inclusion and norm maps in Milnor K-theory on the right-hand side, and therefore the inclusion map $K_2^M(k) \to K_2^M(K)$ induces a map $N_{K|k}(K_2^M(K)) \to N(K_2^M(K))$.

All in all, we have a map

$$\delta : H^1\big(G, K_2^M(K(X))/K_2^M(K)\big) \to K_2^M(K)^G / N(K_2^M(K)),$$

and moreover the following holds.

Lemma 8.7.3 *In the situation above, the diagram*

$$
\begin{array}{ccc}
k^\times \cong A_{d-1}(X_K, K_{2-d}^M)^G & \xrightarrow{\ \gamma\ } & H^1\big(G, K_2^M(K(X))/K_2^M(K)\big) \\
\downarrow & & \downarrow{\scriptstyle\delta} \\
K_2^M(k)/N_{K|k}(K_2^M(K)) & \longrightarrow & K_2^M(K)^G/N(K_2^M(K))
\end{array}
$$

commutes. Here the map γ comes from Lemma 8.7.2, and the left vertical map is induced by $c \mapsto \{b, c\}$, where $b \in k^\times$ defines the cyclic algebra $A = (\chi, b)$.

Proof During the proof of Proposition 5.4.4 we constructed a sequence of maps

$$(\operatorname{Pic} X_K)^G \to H^1(G, K(X)^\times / K^\times) \to H^2(G, K^\times).$$

Fixing an element $c \in k^\times$, we can insert it in a commutative diagram

$$
\begin{array}{ccccc}
(\operatorname{Pic} X_K)^G & \longrightarrow & H^1(G, K(X)^\times/K^\times) & \longrightarrow & H^2(G, K^\times) \\
\downarrow{\scriptstyle c} & & \downarrow{\scriptstyle c} & & \downarrow{\scriptstyle c} \\
A_{d-1}(X_K, K_{2-d}^M)^G & \xrightarrow{\ \gamma\ } & H^1(G, K_2^M(K(X))/K_2^M(K)) & \longrightarrow & H^2(G, K_2^M(K)).
\end{array}
$$
$$(8.15)$$

Here the second and third vertical maps are induced by cup-product with the class of $c \in H^0(G, K^\times)$ using the product structure in Milnor K-theory. The first map is induced by the commutative diagram with exact rows

$$K(X)^\times \longrightarrow \mathrm{Div}X_K \longrightarrow \mathrm{Pic}\,X_K \longrightarrow 0$$

$$\downarrow \qquad\qquad \downarrow \qquad\qquad \downarrow$$

$$K_2^M(K(X)) \longrightarrow Z(X_K) \longrightarrow A_{d-1}(X_K, K_{2-d}^M) \longrightarrow 0$$

where the first two vertical maps are again induced by the product in Milnor K-theory with $c \in K^\times = K_1^M(K)$.

Since X_K is a projective space, we have $(\mathrm{Pic}\,X_K)^G \cong \mathbf{Z}$, and the composite of the maps in the upper row of diagram (8.15) sends the class of 1 to the class of A by Theorem 5.4.11. Furthermore, we have a commutative diagram

$$H^2(G, K^\times) \xrightarrow{\;\cong\;} k^\times/N_{K|k}(k^\times)$$

$$\downarrow \qquad\qquad\qquad\qquad \downarrow \qquad\qquad (8.16)$$

$$H^2(G, K_2^M(K)) \xrightarrow{\;\cong\;} K_2^M(K)^G/N(K_2^M(K))$$

where the left vertical map comes from diagram (8.15) and the right one is induced by sending $x \in k^\times$ to the class of $\{x, c\}$ in $K_2^M(k)/N_{K|k}(K_2^M(K))$, and then composing with the map

$$K_2^M(k)/N_{K|k}(K_2^M(K)) \to K_2^M(K)^G/N(K_2^M(K)).$$

Using Corollary 4.7.4, we see that the class of $A = (\chi, b)$ in $H^2(G, K^\times)$ is mapped to that of the symbol $\{b, c\}$ by the composite of the upper and right maps of the diagram. This proves the lemma, since by construction $\delta \circ \gamma$ is the composite of the lower horizontal maps in diagrams (8.15) and (8.16). $\qquad\qquad\qquad\qquad\qquad\qquad\qquad\qquad\qquad\qquad\qquad\qquad\square$

Lemma 8.7.4 *Assume moreover that $K|k$ is of prime degree p. Then the map* $\delta: H^1\big(G, K_2^M(K(X))/K_2^M(K)\big) \to K_2^M(K)^G/N(K_2^M(K))$ *is injective.*

Proof In order to ease notation, we apply the abbreviations $A_d(X) := A_d(X, K_{2-d}^M)$ and $A_d(X_K) := A_d(X_K, K_{2-d}^M)$ throughout the proof. In the construction of δ we used the identification $A_d(X_K) = K_2^M(K)$ but the argument to follow will be more transparent if we stick to $A_d(X_K)$.

By applying the calculation of the cohomology of cyclic groups to the source of δ as well, we may identify δ with a map

$$_N(K_2^M(K(X))/A_d(X_K))/(\sigma - 1)(K_2^M(K(X))/A_d(X_K)) \to A_d(X_K)^G/N(A_d(X_K))$$

where the subscript $_N$ denotes the kernel of N. From the explicit resolution used in Example 3.2.9 we obtain the following description of this map. Given $\alpha \in {}_N(K_2^M(K(X))/A_d(X_K))$, we may lift it to $\tilde\alpha \in K_2^M(K(X))$ satisfying

$N(\widetilde{\alpha}) \in A_d(X_K)$. Since $(\sigma - 1) \circ N = 0$, we have $N(\widetilde{\alpha}) \in A_d(X_K)^G$ and $\delta(\alpha) = N(\widetilde{\alpha}) \bmod N(A_d(X_K))$. This description does not depend on the choice of $\widetilde{\alpha}$ as $N \circ (\sigma - 1) = 0$.

As already recalled, the map $N : K_2^M(K(X)) \to K_2^M(K(X))$ factors as a composite $N = \iota_{K|k} \circ N_{K|k}$ of the inclusion and norm maps in Milnor K-theory. By the compatibility between residue and norm maps (Corollary 7.4.3), the map $N_{K|k}$ sends the subgroup $A_d(X_K) \subset K_2^M(K(X))$ to $A_d(X)$. Therefore we may factor $N : K_2^M(K(X)/A_d(X_K)) \to K_2^M(K(X))/A_d(X_K)$ as

$$K_2^M(K(X))/A_d(X_K) \overset{N_{K|k}}{\to} K_2^M(k(X))/A_d(X) \overset{\iota_{K|k}}{\to} K_2^M(K(X))/A_d(X_K).$$

Here the second map is injective by the second statement of Lemma 8.7.2. Therefore when we lift $\alpha \in {}_N(K_2^M(K(X))/A_d(X_K))$ to $\widetilde{\alpha} \in K_2^M(K(X))$ in the above description, we actually have $N_{K|k}(\widetilde{\alpha}) \in A_d(X)$. It follows that δ factors as the composite of the map

$$\widetilde{\delta} : {}_N(K_2^M(K(X))/A_d(X_K))/(\sigma - 1)(K_2^M(K(X))/A_d(X_K))$$
$$\to A_d(X)/\mathrm{Im}(N_{K|k}(A_d(X_K)))$$

defined by

$$\widetilde{\delta}(\alpha) := N_{K|k}(\widetilde{\alpha}) \bmod \mathrm{Im}(N_{K|k}(A_d(X_K)))$$

followed by the map

$$A_d(X)/\mathrm{Im}(N_{K|k}(A_d(X_K))) \to A_d(X_K)^G/N(A_d(X_K))$$

induced by the inclusion maps $\iota_{K|k}$. Since this latter map is injective by construction, we are reduced to showing the injectivity of $\widetilde{\delta}$. If $\widetilde{\delta}(\alpha) = 0$, then by modifying $\widetilde{\alpha}$ by an element of $A_d(X_K)$ we may assume that $N_{K|k}(\widetilde{\alpha}) = 0$. But then $\widetilde{\alpha} \in (\sigma - 1)K_2^M(K(X))$ by Hilbert's Theorem 90 for K_2 (Theorem 8.4.1), and therefore $\alpha = 0$ as required. □

Proof of Theorem 8.7.1 Assume first that the cyclic extension $K|k$ is of prime degree p. In this case, suppose that $\{b, c\} \in K_2^M(k)$ is a norm from $K_2^M(K)$. As explained before Lemma 8.7.3, the element $b \in k^\times$ gives rise to an element of the group $A_{d-1}(X_K, K_{2-d}^M)^G$. This element lies in the kernel of the map $\gamma : A_{d-1}(X_K, K_{2-d}^M)^G \to H^1(G, K_2^M(K(X))/K_2^M(K))$ by Lemmas 8.7.3 and 8.7.4, and therefore comes from $A_{d-1}(X, K_{2-d}^M)$ by Lemma 8.7.2. But the map $A_{d-1}(X, K_{2-d}^M) \to A_{d-1}(X_K, K_{2-d}^M)^G$ is identified with the inclusion $\mathrm{Nrd}(A^\times) \hookrightarrow k^\times$ by Remark 8.3.12.

The proof of the converse is much more elementary. Indeed, suppose that $c \in \mathrm{Nrd}(A^\times)$. If $A = (\chi, b)$ is split, then b is a norm from the extension $K|k$

by Corollary 4.7.5, hence $\{b, c\}$ is also a norm from $K_2^M(K)$ by the projection formula. If A is nonsplit, then by Proposition 2.6.8 it contains a subfield L of degree p over k such that $A \otimes_k L$ is split and $c = N_{L|k}(d)$ for some $d \in L$. If $L = K$, we are again done by the projection formula. Otherwise $LK|L$ is a cyclic extension of degree p obtained via base change from $K|k$, and therefore, by the same argument as above, we find $b' \in KL$ with $N_{KL|L}(b') = b$ since $A \otimes_k L$ is split. Applying the projection formula twice, we compute

$$\{b, c\} = \{b, N_{L|k}(d)\} = N_{L|k}(\{b, d\}) = N_{L|k}(\{N_{LK|L}(b'), d\})$$
$$= N_{LK|k}(\{b', d\})$$

and therefore $\{b, c\} \in N_{K|k}(K_2^M(K))$ since $N_{LK|k} = N_{K|k} \circ N_{LK|K}$. This concludes the proof of the case $m = p$.

Finally, assume that $m = p_1 \cdots p_r$ with distinct primes p_i. In this case we find for each i an intermediate field $k \subset K_i \subset K$ such that $K|K_i$ and $K_i|k$ are both cyclic Galois extensions and $[K_i : k] = p_i$. Since we know that the theorem holds in the prime degree case, we may assume it holds for the algebras $A_{K_i} := A \otimes_k K_i$ for each i using induction on r. Now if $\{b, c\} \in N_{K|K_i}(K_2^M(K))$ for each i, then the formulae $N_{K|k} = N_{K_i|k} \circ N_{K|K_i}$ and $N_{K_i|k} \circ \iota_{K_i|k} = p_i$ imply $p_i\{b, c\} \in N_{K|k}(K_2^M(K))$ for each i. Hence by choosing integers n_i such that $\sum n_i p_i = 1$ we see that $\{b, c\} \in N_{K|k}(K_2^M(K))$. In a similar vein, if $c \in \mathrm{Nrd}(A_{K_i}^{\times})$ for all i, then we conclude $c \in \mathrm{Nrd}(A^{\times})$ using Proposition 2.6.6. This completes the inductive step.   □

8.8 A useful exact sequence

We now prove an exact sequence in Galois cohomology that will be needed for the cohomological characterization of reduced norms in the next section. The proof is rather technical, so the reader willing to accept the statement on faith may skip it. Note, however, that a generalization of this exact sequence is also used in Voevodsky's proof of the Bloch–Kato conjecture.

Theorem 8.8.1 *Let $K|k$ be a cyclic Galois extension of squarefree degree m invertible in k. The following sequence of cohomology groups is exact:*

$$H^2(k, \mu_m^{\otimes 2}) \oplus H^2(K, \mu_m^{\otimes 2}) \xrightarrow{(\mathrm{Res}, \sigma - 1)} H^2(K, \mu_m^{\otimes 2}) \to$$
$$\xrightarrow{\mathrm{Cor}} H^2(k, \mu_m^{\otimes 2}) \xrightarrow{\cup [\chi]} H^3(k, \mu_m^{\otimes 2}) \xrightarrow{\mathrm{Res}} H^3(K, \mu_m^{\otimes 2}).$$

Here, as usual, σ is a generator of $\mathrm{Gal}\,(K|k)$ acting by conjugation on $H^2(K, \mu_m^{\otimes 2})$, and $[\chi] \in H^1(k, \mathbf{Z}/m\mathbf{Z})$ is the class of the character of $\mathrm{Gal}\,(k_s|k)$ that lifts the isomorphism $\chi : \mathrm{Gal}\,(K|k) \xrightarrow{\sim} \mathbf{Z}/m\mathbf{Z}$ sending σ to 1.

We begin the proof of the theorem with some easy observations.

Lemma 8.8.2 *The sequence of Theorem 8.8.1 is a complex.*

Proof At the term $H^2(K, \mu_m^{\otimes 2})$ the composition of the maps is zero because of the formula $\mathrm{Cor} \circ \mathrm{Res} = m$ (Proposition 3.3.7) and the definition of corestriction maps, noting that all groups are annihilated by m. At the next term we use the projection formula $\mathrm{Cor}(\alpha) \cup [\chi] = \mathrm{Cor}(\alpha \cup \mathrm{Res}([\chi]))$ (Proposition 3.4.10 (3)) together with the fact that the restriction of χ to $\mathrm{Gal}\,(k_s|K)$ is trivial by definition. Finally, at the term $H^3(k, \mu_m^{\otimes 2})$ we apply the formula $\mathrm{Res}(\alpha \cup [\chi]) = \mathrm{Res}(\alpha) \cup \mathrm{Res}([\chi])$ (Proposition 3.4.10 (1)) and the triviality of $\mathrm{Res}([\chi])$. □

Now comes the deep input in the proof of the theorem.

Lemma 8.8.3 *With notations as in the theorem, the sequence*

$$H^2(k, \mu_m^{\otimes 2}) \oplus H^2(K, \mu_m^{\otimes 2}) \xrightarrow{(\mathrm{Res},\sigma-1)} H^2(K, \mu_m^{\otimes 2}) \xrightarrow{\mathrm{Cor}} H^2(k, \mu_m^{\otimes 2})$$

is exact.

Proof Since m is squarefree, after decomposing the groups in their p-primary components we reduce to the case where m is a prime number. By the Merkurjev–Suslin theorem and Proposition 7.5.4 the sequence of the lemma is isomorphic to the sequence of Milnor K-groups

$$K_2^M(k)/mK_2^M(k) \oplus K_2^M(K)/mK_2^M(K) \xrightarrow{(\iota_{K|k},\sigma-1)} K_2^M(K)/mK_2^M(K)$$
$$\xrightarrow{N_{K|k}} K_2^M(k)/mK_2^M(k).$$

But the latter sequence is exact by Hilbert's Theorem 90 for K_2 (Theorem 8.4.1). □

The rest of the proof of Theorem 8.8.1 will only use arguments in group cohomology. Consider the following situation: G is a finite group, and $H \subset G$ a normal subgroup such that $G/H \cong \mathbf{Z}/p\mathbf{Z}$ for a prime number p. Denote by σ a generator of G/H and by $\chi : G \to \mathbf{Z}/p\mathbf{Z}$ the character lifting the isomorphism $G/H \cong \mathbf{Z}/p\mathbf{Z}$ that sends σ to 1. Let $[\chi]$ be the class of χ in the cohomology group $H^1(G, \mathbf{Z}/p\mathbf{Z})$ and $\delta([\chi])$ its image by the boundary map $\delta : H^1(G, \mathbf{Z}/p\mathbf{Z}) \to H^2(G, \mathbf{Z})$ in the long exact sequence of

$$0 \to \mathbf{Z} \to \mathbf{Z} \to \mathbf{Z}/p\mathbf{Z} \to 0.$$

Fix also a G-module A that will serve as a coefficient module.

Proposition 8.8.4 *Assume the G-module A satisfies $pA = 0$. For an integer $i \geq 0$ assume moreover that the sequence*

$$H^i(G, A) \oplus H^i(H, A) \xrightarrow{(\mathrm{Res}, \sigma - 1)} H^i(H, A) \xrightarrow{\mathrm{Cor}} H^i(G, A)$$

is exact. Then the sequence

$$H^{i-1}(G, A) \oplus H^i(G, A) \xrightarrow{(\cup \delta([\chi]), \cup[\chi])} H^{i+1}(G, A) \xrightarrow{\mathrm{Res}} H^{i+1}(H, A)$$

is exact as well.

For the proof of the proposition consider the map given by multiplication by $\sigma - 1$ on the group algebra $\mathbf{Z}[G/H]$. Its kernel is the free $\mathbf{Z}$-submodule generated by the element $1 + \sigma + \cdots + \sigma^{p-1}$. Identifying this submodule with $\mathbf{Z}$ shows that the free resolution of Example 3.2.9 computing the cohomology of G/H breaks up in short exact sequences of the form

$$0 \to \mathbf{Z} \to \mathbf{Z}[G/H] \xrightarrow{\sigma - 1} \mathbf{Z}[G/H] \to \mathbf{Z} \to 0.$$

We consider this as a sequence of G-modules. Tensoring by A over $\mathbf{Z}$ and splitting the resulting exact sequence in two yields short exact sequences of G-modules

$$0 \to A \to A \otimes_{\mathbf{Z}} \mathbf{Z}[G/H] \to C \to 0, \tag{8.17}$$

$$0 \to C \to A \otimes_{\mathbf{Z}} \mathbf{Z}[G/H] \to A \to 0. \tag{8.18}$$

Lemma 8.8.5 *Let A be an arbitrary G-module.*

1. *The maps $H^i(G, A) \to H^i(G, A \otimes_{\mathbf{Z}} \mathbf{Z}[G/H])$ in the long exact cohomology sequence associated with (8.17) identify with restriction maps $\mathrm{Res} : H^i(G, A) \to H^i(H, A)$.*
2. *Similarly, the maps $H^i(G, A \otimes_{\mathbf{Z}} \mathbf{Z}[G/H]) \to H^i(G, A)$ in the long exact cohomology sequence associated with (8.18) identify with corestriction maps $\mathrm{Cor} : H^i(H, A) \to H^i(G, A)$.*
3. *The map $H^{i-1}(G, A) \to H^{i+1}(G, A)$ given by cup-product with $\delta([\chi])$ equals the composite of the boundary maps $H^{i-1}(G, A) \to H^i(G, C)$ and $H^i(G, C) \to H^{i+1}(G, A)$ coming from (8.18) and (8.17), respectively.*

Proof The first two statements follow from Remark 3.3.10. To prove the third one, we may reduce to the case $i = 1$. Indeed, each class $\alpha \in H^{i-1}(G, A)$ is the image of $1 \in H^0(G, \mathbf{Z})$ via the map $H^0(G, \mathbf{Z}) \to H^{i-1}(G, A)$ given by cup-product with α. But cup-product by α is compatible with cup-product by $\delta([\chi])$, and so are boundary maps in long exact sequences by virtue of Proposition 3.4.8. So we may assume $A = \mathbf{Z}$ with trivial action and $i = 1$. But χ comes by inflation from a character of G/H, and $1 \in H^0(G, \mathbf{Z})$ is trivially the inflation of $1 \in H^0(G/H, \mathbf{Z})$. Therefore we reduce to the case $H = \{1\}$ using the compatibility of Proposition 3.4.10 (2), in which case the statement is a consequence of Proposition 3.4.11 (3). □

The previous lemma enables us to do the following construction. Given an element α in the kernel of the corestriction map $H^i(H, A) \to H^i(G, A)$, we may lift it to an element $\beta \in H^i(G, C)$ by Lemma 8.8.5 (1); such a lifting is well defined up to an element in the image of the boundary map $H^{i-1}(G, A) \to H^i(G, C)$ coming from exact sequence (8.18). Taking the image of β by the boundary map $H^i(G, C) \to H^{i+1}(G, A)$ coming from exact sequence (8.17) induces a well-defined map

$$\Psi : \ker(H^i(H, A) \xrightarrow{\text{Cor}} H^i(G, A)) \to H^{i+1}(G, A)/\text{Im}\,(\cup\delta([\chi]))$$

where $\cup\delta([\chi]) : H^{i-1}(G, A) \to H^{i+1}(G, A)$ is the cup-product map.

Lemma 8.8.6 *Assume A satisfies $pA = 0$. The diagram*

$$
\begin{array}{ccc}
H^i(G, A) & \xrightarrow{\text{Res}} & \ker(H^i(H, A) \xrightarrow{\text{Cor}} H^i(G, A)) \\
{\scriptstyle \cup[\chi]}\big\downarrow & & \big\downarrow{\scriptstyle \Psi} \\
H^{i+1}(G, A) & \longrightarrow & H^{i+1}(G, A)/\text{Im}\,(\cup\delta([\chi]))
\end{array}
$$

commutes, where the bottom horizontal map is the natural projection and the top horizontal map exists because $(\text{Cor} \circ \text{Res})H^i(G, A) = pH^i(G, A) = 0$.

Proof As in the proof of the previous lemma, we reduce to the case $i = 0$ and $A = \mathbf{F}_p$. Consider the exact sequences (8.17) and (8.18) for $A = \mathbf{F}_p$. The element $1 \in H^0(G, \mathbf{F}_p)$ induces the element $\nu := 1 + \sigma + \cdots + \sigma^{p-1}$ in $H^0(G, C \otimes_{\mathbf{Z}} \mathbf{F}_p) = \ker(H^0(G, \mathbf{F}_p[G/H])) \to H^0(G, \mathbf{F}_p)$. When constructing the image of ν by the boundary map $H^0(G, C \otimes_{\mathbf{Z}} \mathbf{F}_p) \to H^1(G, \mathbf{F}_p)$ coming from exact sequence (8.17), we first lift ν to an element μ in $H^0(G, \mathbf{F}_p[G/H])$ satisfying $(\sigma - 1)\mu = \nu$, and then take the 1-cocycle $a_\tau : \tau \mapsto (\tau - 1)\mu$ for $\tau \in G$. Since the elements of H act trivially on $\mathbf{F}_p[G/H]$, the 1-cocycle a_τ is the inflation of the 1-cocycle $G/H \to \mathbf{F}_p$ given

by $\sigma \mapsto (\sigma - 1)\mu = \nu$. Having identified the kernel of $(\sigma - 1) : \mathbf{F}_p[G/H] \to$
$\mathbf{F}_p[G/H]$ with $\mathbf{F}_p$ by sending ν to 1, we thus obtain $a_\tau = \chi$, and the lemma
is proven. $\square$

Proof of Proposition 8.8.4   First of all, the second sequence in the proposition
is a complex since the restriction of χ (and hence of $\delta([\chi])$) to H is trivial and
the cup-product is compatible with restriction maps by Proposition 3.4.10 (1).
     Now pick a class γ in the kernel of the restriction map $H^{i+1}(G, A) \to$
$H^{i+1}(H, A)$. By Lemma 8.8.5 (2) it comes from a class $\beta \in H^i(G, C)$
which we may send to an element α in the kernel of the corestriction map
$H^i(H, A) \to H^i(G, A)$ by Lemma 8.8.5 (1). The element β is well defined
up to an element of $H^i(H, A)$ and the composite of the maps $H^i(H, A) \to$
$H^i(G, C) \to H^i(H, A)$ induced by exact sequences (8.17) and (8.18), respectively, is $\sigma - 1$ by construction. Thus by the assumption of the proposition
up to modifying β by an element of $H^i(H, A)$ we may assume that α comes
by restriction from an element of $H^i(G, A)$. Now comparison with the construction of the map Ψ before Lemma 8.8.6 shows that the image of γ
in $H^{i+1}(G, A)/\mathrm{Im}\,(\cup\delta([\chi]))$ is exactly $\Psi(\alpha)$, and therefore Lemma 8.8.6
finishes the proof. $\square$

Proposition 8.8.7   *Let A be a G-module satisfying $pA = 0$. For an integer
$i \geq 0$ assume that the sequence*

$$H^i(G, A) \xrightarrow{\cup[\chi]} H^{i+1}(G, A) \xrightarrow{\mathrm{Res}} H^{i+1}(H, A)$$

is exact. Then the sequence

$$H^i(H, A) \xrightarrow{\mathrm{Cor}} H^i(G, A) \xrightarrow{\cup[\chi]} H^{i+1}(G, A)$$

is also exact.

     The proof of the proposition uses an auxiliary filtration on the G/H-module
$M := \mathbf{F}_p[G/H]$ defined by

$$M_r := \mathrm{Im}\big(M \xrightarrow{(\sigma-1)^{(p-r)}} M\big)$$

for $1 \leq r \leq p$.
     The M_r form an increasing filtration $M_1 \subset \cdots \subset M_p = M$ of the $\mathbf{F}_p$-
vector space M with $\dim_{\mathbf{F}_p} M_r = r$; in particular, there is an isomorphism
$M_1 \cong \mathbf{F}_p$. To see this, note that sending x to σ induces an isomorphism of
$\mathbf{F}_p$-vector spaces $\mathbf{F}_p[x]/(x-1)^p \xrightarrow{\sim} M$, whence isomorphisms

$$M_r \cong \mathrm{Im}(\mathbf{F}_p[x]/(x-1)^p \xrightarrow{(x-1)^{(p-r)}} \mathbf{F}_p[x]/(x-1)^p) \xrightarrow{\sim} \mathbf{F}_p[x]/(x-1)^r.$$

The above isomorphisms also show that for each pair (r, s) of positive integers satisfying $r + s \leq p$ we have an exact sequence

$$0 \to M_r \to M_{r+s} \xrightarrow{(\sigma-1)^s} M_s \to 0.$$

In particular, for the pairs $(r, 1)$ and $(1, s)$ we have the exact sequences of G-modules

$$0 \to M_r \to M_{r+1} \xrightarrow{v_r} \mathbf{F}_p \to 0 \tag{8.19}$$

and

$$0 \to \mathbf{F}_p \xrightarrow{u_s} M_{s+1} \to M_s \to 0. \tag{8.20}$$

We shall need another compatibility result.

Lemma 8.8.8 *The boundary map $H^i(G, A) \to H^{i+1}(G, A)$ coming from exact sequence (8.20) for $s = 1$ equals cup-product with the class of χ in $H^1(G, \mathbf{Z}/p\mathbf{Z})$.*

Proof As in the proof of Lemma 8.8.6, we reduce to the case $i = 0$ and $A = \mathbf{F}_p$. In view of the exact commutative diagram

$$
\begin{array}{ccccccccc}
0 & \longrightarrow & \mathbf{F}_p & \longrightarrow & M_p & \longrightarrow & M_{p-1} & \longrightarrow & 0 \\
& & \downarrow{\scriptstyle \mathrm{id}} & & \downarrow{\scriptstyle (\sigma-1)^{p-2}} & & \downarrow{\scriptstyle (\sigma-1)^{p-2}} & & \\
0 & \longrightarrow & \mathbf{F}_p & \longrightarrow & M_2 & \longrightarrow & \mathbf{F}_p & \longrightarrow & 0
\end{array}
$$

the proof reduces to that of Lemma 8.8.6, noting that $M_p = \mathbf{F}_p[G]$ and $M_{p-1} = C \otimes_{\mathbf{Z}} \mathbf{F}_p$ with the notations there. $\qquad\square$

Proof of Proposition 8.8.7 To handle the case $p = 2$, it suffices to take the long exact sequence associated with (8.20) for $s = 1$ tensored by A. By Lemma 8.8.8 it takes the shape

$$\cdots \to H^i(G, A) \xrightarrow{\mathrm{Res}} H^i(H, A) \xrightarrow{\mathrm{Cor}} H^i(G, A) \xrightarrow{\cup[\chi]} H^{i+1}(G, A) \to \cdots$$

whence the claim. We henceforth assume $p \geq 3$, and consider the exact commutative diagram

$$
\begin{array}{ccc}
0 & & 0 \\
\downarrow & & \downarrow \\
\mathbf{F}_p & = & \mathbf{F}_p \\
\scriptstyle u_r \downarrow & & \scriptstyle u_{r+1} \downarrow \\
\end{array}
$$

$$
\begin{array}{ccccccccc}
0 & \longrightarrow & M_r & \longrightarrow & M_{r+1} & \xrightarrow{v_{r+1}} & \mathbf{F}_p & \longrightarrow & 0 \\
 & & \scriptstyle \sigma-1 \downarrow & & \scriptstyle \sigma-1 \downarrow & & \scriptstyle \mathrm{id} \downarrow & & \\
0 & \longrightarrow & M_{r-1} & \longrightarrow & M_r & \xrightarrow{v_r} & \mathbf{F}_p & \longrightarrow & 0 \\
 & & \downarrow & & \downarrow & & & & \\
 & & 0 & & 0 & & & &
\end{array}
$$

coming from exact sequences (8.19) and (8.20) for $2 \le r \le p - 1$. After tensoring with A and passing to cohomology we obtain the exact commutative diagram:

$$
\begin{array}{ccccccc}
H^i(G, A \otimes M_r) & \longrightarrow & H^i(G, A \otimes M_{r+1}) & \xrightarrow{v_{r+1,*}} H^i(G, A) & \longrightarrow & H^{i+1}(G, A \otimes M_r) \\
\downarrow & & \downarrow & \quad\downarrow \scriptstyle \mathrm{id} & & \downarrow \\
H^i(G, A \otimes M_{r-1}) & \longrightarrow & H^i(G, A \otimes M_r) & \xrightarrow{v_{r,*}} H^i(G, A) & \longrightarrow & H^{i+1}(G, A \otimes M_{r-1}) \\
\downarrow & & \downarrow & & & \\
H^{i+1}(G, A) & \xrightarrow{\mathrm{id}} & H^{i+1}(G, A) & & & \\
\downarrow & & \downarrow & & & \\
H^{i+1}(G, A \otimes M_r) & \longrightarrow & H^{i+1}(G, A \otimes M_{r+1}). & & &
\end{array}
$$

$$(8.21)$$

The upper right square of the diagram induces an inclusion

$$
\ker(H^i(G, A) \to H^{i+1}(G, A \otimes M_r)) \subset \ker(H^i(G, A) \to H^{i+1}(G, A \otimes M_{r-1})).
$$
$$(8.22)$$

We shall prove that this inclusion is in fact an equality. This will imply the proposition, since by applying it inductively for r descending from $p - 1$ to 2 one obtains an equality

$$
\ker(H^i(G, A) \to H^{i+1}(G, A \otimes M_{p-1})) = \ker(H^i(G, A) \to H^{i+1}(G, A \otimes M_1))
$$

which can be rewritten as

$$
\mathrm{Im}\,(H^i(H, A) \xrightarrow{\mathrm{Cor}} H^i(G, A)) = \ker(H^i(G, A) \xrightarrow{\cup [\chi]} H^{i+1}(G, A))
$$

according to Lemma 8.8.5 (2) and Lemma 8.8.8.

To show the reverse inclusion in (8.22), we pick an element α in $\ker(H^i(G, A) \to H^{i+1}(G, A \otimes M_{r-1}))$ and lift it to $\beta \in H^i(G, A \otimes M_r)$. We shall show that up to modifying β by an element of $H^i(G, A \otimes M_{r-1})$ we may choose it so that it lies in the image of the map $H^i(G, A \otimes M_{r+1}) \to H^i(G, A \otimes M_r)$, which will imply the claim by diagram (8.21).

To do so, consider the second exact column of diagram (8.21). The image γ of β in $H^{i+1}(G, A)$ maps to zero in $H^{i+1}(G, A \otimes M_{r+1})$, hence also in $H^{i+1}(G, A \otimes M) \cong H^{i+1}(H, A)$. The assumption of the proposition then implies that we may write $\gamma = \chi \cup \gamma_0$ for some $\gamma_0 \in H^i(G, A)$. We now consider the commutative diagram

$$
\begin{array}{ccccccccc}
0 & \longrightarrow & \mathbf{F}_p & \xrightarrow{u_p} & M_2 & \longrightarrow & \mathbf{F}_p & \longrightarrow & 0 \\
& & \downarrow{\scriptstyle \mathrm{id}} & & \downarrow & & \downarrow & & \\
0 & \longrightarrow & \mathbf{F}_p & \xrightarrow{u_{r-1}} & M_r & \longrightarrow & M_{r-1} & \longrightarrow & 0
\end{array}
$$

coming from exact sequence (8.20) for $s = 1$ and $s = r - 1$. It induces a commutative diagram

$$
\begin{array}{ccccc}
H^i(G, A \otimes M_2) & \longrightarrow & H^i(G, A) & \xrightarrow{\cup[\chi]} & H^{i+1}(G, A) \\
\downarrow & & \downarrow & & \downarrow{\scriptstyle \mathrm{id}} \\
H^i(G, A \otimes M_r) & \longrightarrow & H^i(G, M_{r-1}) & \longrightarrow & H^{i+1}(G, A)
\end{array}
$$

on cohomology, where the upper right map was computed in Lemma 8.8.8. Chasing $\gamma_0 \in H^i(G, A)$ through the right square shows that $\gamma = \chi \cup \gamma_0 \in H^{i+1}(G, A)$ belongs to the image of the map $H^i(G, M_{r-1}) \to H^{i+1}(G, A)$. By the middle square in the left columns of diagram (8.21) this map factors through $H^i(G, M_r)$, and hence we may modify β as required. $\qquad\square$

Proof of Theorem 8.8.1 By Lemmas 8.8.2 and 8.8.3 the sequence is a complex that is exact at the second term. As in the proof of Lemma 8.8.3 we reduce to the case where $m = p$ is a prime number. Next, observe that Propositions 8.8.4 and 8.8.7 immediately generalize to the case where G is profinite and H is an open normal subgroup of index p, with the same proofs. Applying the profinite version of Proposition 8.8.4 with $G = \mathrm{Gal}(k_s|k)$, $H = \mathrm{Gal}(k_s|K)$ and $A = \mu_p^{\otimes 2}$ shows that Lemma 8.8.3 implies the exactness of the sequence

$$
H^1(k, \mu_p^{\otimes 2}) \oplus H^2(k, \mu_p^{\otimes 2}) \xrightarrow{(\cup\delta([\chi]), \cup[\chi])} H^3(k, \mu_p^{\otimes 2}) \xrightarrow{\mathrm{Res}} H^3(K, \mu_p^{\otimes 2}).
$$

We claim that the image of the map $\cup[\chi] : H^2(k, \mu_p^{\otimes 2}) \to H^3(k, \mu_p^{\otimes 2})$ contains that of $\cup\delta[\chi] : H^1(k, \mu_p^{\otimes 2}) \to H^3(k, \mu_p^{\otimes 2})$. For this we may replace k

by $k(\mu_p)$ by a restriction-corestriction argument using that $[k(\mu_p) : k]$ is prime to p, and choose a primitive p-th root of unity $\omega \in H^0(k, \mu_p) \subset H^0(k, k_s^\times)$. Cup-product with ω then induces an isomorphism $H^1(k, \mu_p) \xrightarrow{\sim} H^1(k, \mu_p^{\otimes 2})$. But as in the proof of Proposition 4.7.1, we have $\omega \cup \delta([\chi]) = \partial(\omega) \cup [\chi]$, where $\partial : H^0(k, k_s^\times) \to H^1(k, \mu_p)$ is the Kummer boundary map. This implies the claim.

We thus conclude that the sequence

$$H^2(k, \mu_p^{\otimes 2}) \xrightarrow{\cup [\chi]} H^3(k, \mu_p^{\otimes 2}) \xrightarrow{\mathrm{Res}} H^3(K, \mu_p^{\otimes 2})$$

is exact. The remaining exactness in the sequence of the theorem follows from the profinite version of Proposition 8.8.7.   □

8.9 Reduced norms and cohomology

Having Theorems 8.7.1 and 8.8.1 at our disposal, we can now prove:

Theorem 8.9.1 *Let A be a central simple algebra of squarefree degree m invertible in k. Denote by ∂ the Kummer coboundary $k^\times \to H^1(k, \mu_m)$ and by $[A]$ the class of A in $H^2(k, \mu_m)$.*

An element $c \in k^\times$ is in the image of the reduced norm map $\mathrm{Nrd}: A^\times \to k^\times$ if and only if $\partial(c) \cup [A] = 0$ in $H^3(k, \mu_m^{\otimes 2})$.

Note an important special case: if $K|k$ is a degree m cyclic Galois extension and $A = (\chi, b)$ the cyclic k-algebra associated with a character $\chi :$ $\mathrm{Gal}(K|k) \xrightarrow{\sim} \mathbf{Z}/m\mathbf{Z}$, then

$$\partial(c) \cup [A] = \partial(c) \cup \partial(b) \cup [\chi] = -h_{k,m}^2(\{b, c\}) \cup [\chi] \qquad (8.23)$$

by Corollary 4.7.4 (and its proof), where $[\chi]$ is the class of χ in $H^1(k, \mathbf{Z}/m\mathbf{Z})$. Thus in case of a cyclic algebra of squarefree degree the criterion of the theorem is equivalent to the vanishing of $h_{k,m}^2(\{b, c\}) \cup [\chi]$.

Proof We may assume A is nonsplit, as the split case is almost a tautology. The 'only if' part in fact holds without any assumption on the degree of A. Indeed, assume $c \in \mathrm{Nrd}(A^\times)$. By Proposition 2.6.8 we find a finite separable field extension $K|k$ splitting A and an element $d \in K^\times$ such that $N_{K|k}(d) = c$. By Lemma 4.6.2 and Proposition 3.4.10 (3) we have

$$\partial(c) \cup [A] = \partial(N_{K|k}(d)) \cup [A] = \mathrm{Cor}(\partial_K(d)) \cup [A] = \mathrm{Cor}(\partial_K(d) \cup [A \otimes_k K]).$$

But $A \otimes_k K$ is split, whence the vanishing of $\partial(c) \cup [A]$.

We prove the converse first in the case where $A = (\chi, b)$ is a cyclic algebra of degree p split by a degree p cyclic extension $K|k$. By formula (8.23) our assumption is then equivalent to the vanishing of $h_{k,p}^2(\{b, c\}) \cup [\chi]$. Since the Galois symbol $h_{k,p}^2$ is compatible with norm maps (Proposition 7.5.4), the Merkurjev–Suslin theorem implies that $h_{k,p}^2(\{b, c\})$ is in the image of the corestriction map $H^2(K, \mu_p^{\otimes 2}) \to H^2(k, \mu_p^{\otimes 2})$ if and only if $\{b, c\} = N_{K|k}(\alpha) + p\beta$ for some $\alpha \in K_2^M(K)$ and $\beta \in K_2^M(k)$. But $N_{K|k}(\alpha) + p\beta = N_{K|k}(\alpha + i_{K|k}(\beta))$, and therefore $h_{k,p}^2(\{b, c\})$ is in the image of the corestriction map if and only if $\{b, c\}$ is a norm from $K_2^M(K)$. We can now conclude by applying Theorem 8.7.1 together with the exactness of the sequence of Theorem 8.8.1 at the third term.

Next, assume A is an arbitrary degree p division algebra over k. We then find a finite separable field extension $L|k$ of degree prime to p such that $A \otimes_k L$ is cyclic of degree p. This is proven as in the proof of Proposition 7.4.13. Namely, start with a (necessarily separable) degree p field extension $K|k$ splitting A, and define L as the fixed field of a p-Sylow subgroup in the Galois group $\mathrm{Gal}\,(\widetilde{K}|k)$ for a Galois closure $\widetilde{K}|k$ of $K|k$. Since $\mathrm{Gal}\,(\widetilde{K}|L)$ is cyclic of degree p (being a subgroup of S_p), we have found a degree p cyclic extension splitting $A \otimes_k L$, and hence $A \otimes_k L$ is indeed a cyclic algebra by Corollary 4.7.7. By the cyclic case treated above, we then find $d \in (A \otimes_k L)^\times$ with $\mathrm{Nrd}(d) = c$. But then $c^{[L:k]} = N_{L|k}(c) \in \mathrm{Nrd}(A^\times)$ by Proposition 2.6.6. Choosing r with $r[L : k] \equiv 1 \bmod p$, we obtain $c^{r[L:k]} = c \bmod k^{\times p}$. Since we also have $k^{\times p} \subset \mathrm{Nrd}(A^\times)$ (by definition of the reduced norm or by Proposition 2.6.8) and $\mathrm{Nrd}(A^\times) \subset k^\times$ is a subgroup, this shows $c \in \mathrm{Nrd}(A^\times)$.

Finally, the case where m is squarefree reduces to the prime degree case by the same argument as at the end of the proof of Theorem 8.7.1. $\qquad\square$

Remark 8.9.2 In the first paragraph of the previous proof we have in fact constructed a map $k^\times/\mathrm{Nrd}(A^\times) \to H^3(k, \mu_m^{\otimes 2})$ induced by cup-product with the class of an arbitrary central simple algebra A of degree m. Given a finite separable splitting field K of A, composition with the isomorphism $H^1(\mathrm{Gal}\,(K|k), \mathrm{SL}_1(A \otimes_k K)) \leftrightarrow k^\times/\mathrm{Nrd}(A^\times)$ of Proposition 2.7.3 thus yields a map of pointed sets

$$H^1(k, \mathrm{SL}_1(A \otimes_k k_s)) \to H^3(k, \mu_m^{\otimes 2}).$$

(Here, in order to replace $H^1(\mathrm{Gal}\,(K|k), \mathrm{SL}_1(A \otimes_k K))$ we also need to know that $H^1(K, \mathrm{SL}_1(A \otimes_k k_s)) \cong H^1(K, \mathrm{SL}_m(k_s))$ is trivial, which was an easy exercise in Chapter 2.) The theorem asserts that this map has trivial kernel if A has squarefree degree.

The above map, often called the Suslin invariant, is an example of a *cohomological invariant* for a linear algebraic group. By definition, given a linear algebraic group G and a torsion Galois module C over a field k, a cohomological invariant is a collection of maps of pointed sets

$$a_F : H^1(F, G(F_s)) \to H^i(F, C)$$

for some i, where F runs over all field extensions $F \supset k$. These are subject to the compatibility relations given by the commutativity of

$$
\begin{array}{ccc}
H^1(F, G(F_s)) & \xrightarrow{\ a_F\ } & H^i(F, C) \\
\downarrow & & \downarrow \\
H^1(F', G(F_s')) & \xrightarrow{\ a_{F'}\ } & H^i(F', C)
\end{array}
$$

for each field extension $F'|F$, where the vertical restriction maps are induced by choosing compatible separable closures $F_s' \supset F_s \supset k_s$. In the special case considered above, we obtain such a compatible system of maps for $G = SL_1(A)$ by performing the construction for the base change algebras $A \otimes_k F$.

Cohomological invariants are basic tools in the study of linear algebraic groups over general fields. The most useful to date is the Rost invariant, defined for arbitrary semi-simple simply connected G using $C = \mu_m^{\otimes 2}$ for suitable m and $i = 3$. In the case of $SL_1(A)$ it gives back the Suslin invariant. For details, see the book of Garibaldi–Merkurjev–Serre [1].

We are now ready to prove the following beautiful theorem of Suslin.

Theorem 8.9.3 *Let k be a field, and p a prime number invertible in k. The following properties are equivalent:*

1. *The p-cohomological dimension of k is at most 2.*
2. *For every central simple algebra A of degree p defined over a finite field extension $\tilde{k}$ of k the reduced norm map $\mathrm{Nrd} : A^\times \to \tilde{k}^\times$ is surjective.*

The theorem may be viewed as a noncommutative analogue of Theorem 6.1.8 for fields of cohomological dimension ≤ 2.

Proof To show (1) $\Rightarrow$ (2), it will suffice to treat the case $\tilde{k} = k$ by virtue of Lemma 6.1.4. We then conclude by applying Theorem 8.9.1, as $H^3(k, \mu_p^{\otimes 2}) = 0$ by assumption (1).

To prove the implication (2) $\Rightarrow$ (1), note first that thanks to Lemma 6.1.4 condition (1) holds for k if and only if it holds for its maximal prime-to-p extension. On the other hand, the criterion of Theorem 8.9.1 for $m = p$ and the usual restriction-corestriction argument imply that condition (2) is also insensitive to passing to a field extension of degree prime to p (of course, this may also be checked directly). Thus we may assume that k has no nontrivial extension of degree prime to p; in particular, we have $\mu_p \subset k$.

Assume now that (1) fails. Since $\mu_p \subset k$, Proposition 6.1.5 then implies $H^3(k, \mu_p^{\otimes 2}) \cong H^3(k, \mathbf{Z}/p\mathbf{Z}) \neq 0$. Let α be a nontrivial class in this group. Proposition 4.2.11 implies that there exists a finite separable field extension $K|k$ such that the restriction of α to $H^3(K, \mu_p^{\otimes 2})$ is trivial. We may assume that $K|k$ is Galois and is of minimal degree among the Galois extensions having this property. Since $\mathrm{Gal}\,(K|k)$ is a p-group by assumption, its centre contains an element σ of order p. The fixed field K_0 of σ has the properties that $K|K_0$ is a degree p cyclic Galois extension and moreover α has nontrivial image α_0 in $H^3(K_0, \mu_p^{\otimes 2})$, by minimality of $[K : k]$. Therefore by Theorem 8.8.1 applied to the extension $K|K_0$ we find $\beta \in H^2(K_0, \mu_p^{\otimes 2})$ mapping to α_0 by the map $H^2(K_0, \mu_p^{\otimes 2}) \rightarrow H^3(K_0, \mu_p^{\otimes 2})$ given by cup-product with $[\chi]$. By the Merkurjev–Suslin theorem β is a finite sum of elements of the form $h^2_{K_0,p}(\{b, c\})$ for $b, c \in \widetilde{k}^\times$, so we find b, c such that $h^2_{K_0,p}(\{b, c\})$ is not annihilated by cup-product with $[\chi]$. But then c cannot be a reduced norm from the degree p cyclic algebra (χ, b) by Theorem 8.9.1 and formula (8.23), which contradicts (2). □

Before stating the next corollary, recall from Remark 6.2.2 that a field k is called a C_r-field for an integer $r > 0$ if every homogeneous polynomial $f \in k[x_1, \ldots, x_n]$ of degree d with $d^r < n$ has a nontrivial zero in k^n. We need the following nonobvious generalization of Lemma 6.2.4.

Fact 8.9.4 If k is a C_r-field, then so is every finite field extension of k. The proof of this fact is based on the same idea as that of the case $r = 1$ but is more involved. See Chapter 3 of Greenberg [2] or the original paper Lang [1].

We can now prove the following analogue of Proposition 6.2.3.

Corollary 8.9.5 If k is a C_2-field, then $\mathrm{cd}(k) \leq 2$.

Proof Taking Proposition 6.1.10 into account, it will suffice to show $\mathrm{cd}_p(k) \leq 2$ for p invertible in k. This we do by applying the criterion of Theorem 8.9.3. Thanks to Fact 8.9.4 we may assume $\widetilde{k} = k$. Since nonsplit central simple algebras of prime degree are division algebras and the split case is straightforward, we conclude by observing that over a C_2-field the reduced norm map is surjective for an arbitrary central division algebra A of degree n. This is proven by an argument akin to that in Proposition 6.2.3: by fixing a

k-basis $v_1, \ldots, v_{n^2}$ of A and adding an extra variable t, we consider for each $c \in k^\times$ the polynomial

$$P_c := \mathrm{Nrd} \left(\sum_i x_i v_i \right) - ct^n.$$

It is homogeneous of degree n in the $n^2 + 1$ variables $x_1, \ldots, x_{n^2}, t$, and as such has a nontrivial zero $(a_1, \ldots, a_{n^2}, b)$ over a C_2-field. We cannot have $b = 0$ here as it would force all the a_i to be 0 by Proposition 2.6.2. Therefore we may assume $b = 1$, which shows that $c \in \mathrm{Nrd}(A)$. $\qquad\square$

Remark 8.9.6   The Milnor conjecture (now a theorem of Voevodsky [1]) implies that a C_r-field has 2-cohomological dimension $\leq r$. See Serre [4], Chapter II, §4.5.

EXERCISES

1. (Colliot-Thélène, Raskind) Let X be a smooth variety of dimension d over a field k, and let $K|k$ be a finite Galois extension with group G. Establish an exact sequence

$$A_{d-1}(X, K_{2-d}^M) \to A_{d-1}(X_K, K_{2-d}^M)^G \xrightarrow{\delta} H^1\big(G, K_2^M(K(X))/A_d(X_K, K_{2-d}^M)\big)$$
$$\to \ker\big(CH^2(X) \to CH^2(X_K)\big) \to H^1(G, A_{d-1}(X_K, K_{2-d}^M))$$

where X_K denotes the base change of X to K.

2. Let Γ be the Galois group $\mathrm{Gal}\,(\mathbf{C}|\mathbf{R})$.

   (a) Show that $_2 K_2^M(\mathbf{R})$ is a cyclic group of order 2 generated by $\{-1, -1\}$.
   (b) Show that $K_2^M(\mathbf{R})/N_{\mathbf{C}|\mathbf{R}}(K_2^M(\mathbf{C}))$ is also the cyclic group of order 2 generated by $\{-1, -1\}$.
   (c) Let X be the projective conic of equation $x_0^2 + x_1^2 + x_2^2 = 0$ in $\mathbf{P}_{\mathbf{R}}^2$. Show that $H^1(\Gamma, K_2^M(\mathbf{C}(X))/K_2^M(\mathbf{C})) \cong \mathbf{Z}/2\mathbf{Z}$. [*Hint:* Use the previous exercise.]
   (d) Conclude that $H^1(\Gamma, K_2^M(\mathbf{C}(X))) \cong \mathbf{Z}/2\mathbf{Z}$, and try to find an explicit generator.

3. Assume that k is a field having no nontrivial finite extension of degree prime to p, and that $K_2^M(k)/pK_2^M(k) = 0$. Let $K|k$ be a cyclic extension of degree p.

   (a) Show that the norm map $N_{K|k} : K^\times \to k^\times$ is surjective.
   (b) Conclude that $K_2^M(K)/pK_2^M(K) = 0$. [*Hint:* Use Proposition 8.4.3 and Lemma 8.5.3.]
   (c) Show that $\mathrm{cd}(k) \leq 1$.

4. Let p be a prime number invertible in k, and let $K|k$ be a cyclic Galois extension of degree p^r for some $r \geq 1$. Denote by σ a generator of $\mathrm{Gal}\,(K|k)$. Define a tower of fields

$$k = F_0 \subset F_1 \subset F_2 \subset F_3 \subset \cdots \subset F_\infty = \bigcup_n F_n$$

inductively as follows:

- the field F_{2n+1} is a maximal prime to p extension of F_{2n};
- the field F_{2n+2} is the compositum of all function fields of Severi–Brauer varieties associated with cyclic algebras of the form (a, b), where $a, b \in F_{2n+1}^\times$.

(a) Show that F_∞ has no nontrivial prime to p extension and that

$$K_2^M(F_\infty)/pK_2^M(F_\infty) = 0.$$

(b) Show that the sequence

$$K_2^M(KF_\infty) \xrightarrow{\sigma-1} K_2^M(KF_\infty) \xrightarrow{N_{KF_\infty|F_\infty}} K_2^M(F_\infty)$$

is exact.

[*Hint:* Argue as in the proof of Proposition 8.4.3 using step (b) of the previous exercise.]

5. (Hilbert's Theorem 90 for K_2 in the general case) Let m be an integer invertible in k, and let $K|k$ be a cyclic Galois extension of degree m. Denote by σ a generator of $\mathrm{Gal}(K|k)$. Show that the sequence

$$K_2^M(K) \xrightarrow{\sigma-1} K_2^M(K) \xrightarrow{N_{K|k}} K_2^M(k)$$

is exact. [*Hint:* First use a restriction-corestriction argument as in the proof of Proposition 8.4.4 to reduce to the case when $m = p^r$ is a prime power. Then mimic the proof of Theorem 8.4.1 using the previous exercise.]

6. Let p be a prime number, and k a field of characteristic prime to p whose p-cohomological dimension is ≤ 2. Show that for every central simple algebra A of p-power degree defined over a finite field extension $\tilde{k}|k$ the reduced norm map $\mathrm{Nrd} : A^\times \to \tilde{k}^\times$ is surjective. [*Hint:* Use induction on the index of A starting from Theorem 8.9.3.]

7. Let k be a field of characteristic 0. Show that $\mathrm{cd}(k) \leq 2$ if and only if for every tower of finite field extensions $L|K|k$ the norm map $N_{L|K} : K_2^M(L) \to K_2^M(K)$ is surjective. [*Hint:* Use arguments from the proof of Theorem 8.9.3.]

9

Symbols in positive characteristic

In the preceding chapters, when working with Galois cohomology groups or K-groups modulo some prime, a standing assumption was that the groups under study were torsion groups prime to the characteristic of the base field. We now remove this restriction. In the first part of the chapter the central result is Teichmüller's theorem, according to which the p-primary torsion subgroup in the Brauer group of a field of characteristic $p > 0$ is generated by classes of cyclic algebras – a characteristic p ancestor of the Merkurjev–Suslin theorem. We shall give two proofs of this statement: a more classical one due to Hochschild which uses central simple algebras, and a totally different one based on a presentation of the p-torsion in Br (k) via logarithmic differential forms. The key tool here is a famous theorem of Jacobson–Cartier characterizing logarithmic forms. The latter approach leads us to the second main topic of the chapter, namely the study of the differential symbol. This is a p-analogue of the Galois symbol which relates the Milnor K-groups modulo p to a certain group defined using differential forms. We shall prove the Bloch–Gabber–Kato theorem establishing its bijectivity, and obtain as an application the absence of p-torsion in Milnor K-groups of fields of characteristic p, a statement due to Izhboldin.

Teichmüller's result first appeared in the ill-famed journal *Deutsche Mathematik* (Teichmüller [1]); see also Jacobson [3] for an account of the original proof. The role of derivations and differentials in the theory of central simple algebras was noticed well before the Second World War; today the most important work seems to be that of Jacobson [1]. This line of thought was further pursued in papers by Hochschild [1], [2], and, above all, in the thesis of Cartier [1], which opened the way to a wide range of geometric developments. The original references for the differential symbol are the papers of Kato [2] and Bloch–Kato [1]; they have applied the theory to questions concerning higher-dimensional local fields and p-adic Hodge theory.

9.1 The theorems of Teichmüller and Albert

In what follows k will denote a field of characteristic $p > 0$, and k_s will be a fixed separable closure of k. According to the Merkurjev–Suslin theorem, the m-torsion subgroup of $\mathrm{Br}\,(k)$ is generated by classes of cyclic algebras for all m prime to p, provided that k contains the m-th roots of unity. For m a power of p, the statement is still valid (without, of course, the assumption on roots of unity); it was proven by Teichmüller as early as 1936. But around the same time Albert obtained an even stronger result: each class of p-power order in the Brauer group can actually be represented by a cyclic algebra. In this section we prove these classical theorems.

First recall some facts that will be used several times below. For all integers $r > 0$, classes in $H^1(k, \mathbf{Z}/p^r\mathbf{Z})$ correspond to characters $\tilde{\chi}$ of the absolute Galois group $\mathrm{Gal}\,(k_s|k)$ of order dividing p^r. We shall always denote by χ the character induced by $\tilde{\chi}$ on the finite quotient of $\mathrm{Gal}\,(k_s|k)$ defined by the kernel of $\tilde{\chi}$. We have a natural pairing

$$j_r : H^1(k, \mathbf{Z}/p^r\mathbf{Z}) \times H^0(k, k_s^\times) \to {}_{p^r}\mathrm{Br}\,(k)$$

sending a pair $(\tilde{\chi}, b)$ to $\delta(\tilde{\chi}) \cup b$, where $\delta : H^1(k, \mathbf{Z}/p^r\mathbf{Z}) \to H^2(k, \mathbf{Z})$ is the coboundary map coming from the exact sequence

$$0 \to \mathbf{Z} \to \mathbf{Z} \to \mathbf{Z}/p^r\mathbf{Z} \to 0.$$

According to Proposition 4.7.3, the element $j_r(\tilde{\chi}, b)$ equals the class of the cyclic algebra (χ, b) in $\mathrm{Br}\,(k)$. As a consequence, bilinearity of the cup-product implies that

$$[(\chi, pb)] = [(p\chi, b)] \quad \text{in} \quad \mathrm{Br}\,(k). \tag{9.1}$$

Also, recall from Chapter 2 that in the important case when χ defines a degree p Galois extension of k, the algebra (χ, b) has a presentation of the form

$$(\chi, b) = [a, b] = \langle x, y | x^p - x = a, y^p = b, y^{-1}xy = x + 1 \rangle, \tag{9.2}$$

for some $a \in k$, and conversely a k-algebra with such a presentation is cyclic.

Another frequently used fact will be the following. Given positive integers $r, s > 0$, consider the short exact sequence

$$0 \to \mathbf{Z}/p^r\mathbf{Z} \longrightarrow \mathbf{Z}/p^{r+s}\mathbf{Z} \xrightarrow{p^r} \mathbf{Z}/p^s\mathbf{Z} \to 0.$$

Since $\mathrm{cd}_p(k) \le 1$ (Proposition 6.1.10), the associated long exact sequence ends like this:

$$H^1(k, \mathbf{Z}/p^r\mathbf{Z}) \to H^1(k, \mathbf{Z}/p^{r+s}\mathbf{Z}) \xrightarrow{p^r} H^1(k, \mathbf{Z}/p^s\mathbf{Z}) \to 0. \tag{9.3}$$

Armed with these facts, we now begin the proof of Teichmüller's theorem using a method of Hochschild. The key statement is the following.

Theorem 9.1.1 (Albert, Hochschild)   Let $K = k(\sqrt[p^{r_1}]{b_1}, ..., \sqrt[p^{r_n}]{b_n})$ be a purely inseparable extension. For each class $\alpha \in \mathrm{Br}(K|k)$ we may find characters $\tilde{\chi}_i \in H^1(k, \mathbf{Z}/p^{r_i}\mathbf{Z})$ so that

$$\alpha = \sum_{i=1}^n [(\chi_i, b_i)] \quad \text{in} \quad \mathrm{Br}(k),$$

where χ_i is the injective character induced by $\tilde{\chi}_i$ on a finite quotient of $\mathrm{Gal}(k_s|k)$.

We start the proof by extending Proposition 4.5.5 to the case of purely inseparable extensions.

Lemma 9.1.2   *Let $K|k$ be a purely inseparable field extension of degree $n = p^r$. Then the boundary map $\delta_n : H^1(k, \mathrm{PGL}_n(k_s)) \to \mathrm{Br}(k)$ induces a bijection*

$$\ker(H^1(k, \mathrm{PGL}_n(k_s)) \to H^1(K, \mathrm{PGL}_n(Kk_s))) \cong \mathrm{Br}(K|k).$$

Moreover, if $A|k$ is a central simple algebra of degree n split by K, then K embeds as a commutative k-subalgebra into A.

Proof   We have already shown in the proof of Theorem 4.4.1 the injectivity of δ_n (even of δ_∞), so it suffices to see surjectivity. Denoting by G the Galois group $\mathrm{Gal}(k_s|k)$, consider the short exact sequence of G-modules

$$1 \to k_s^\times \to (Kk_s)^\times \to (Kk_s)^\times/k_s^\times \to 1,$$

where G naturally identifies to $\mathrm{Gal}(Kk_s|K)$, because the composite Kk_s is a separable closure of K. As $H^1(G, (Kk_s)^\times) = 0$ by Hilbert's Theorem 90, we get isomorphisms

$$H^1(G, (Kk_s)^\times/k_s^\times) \cong \ker(H^2(G, k_s^\times) \to H^2(G, (Kk_s)^\times)) \cong \mathrm{Br}(K|k).$$

On the other hand, the choice of a k-basis of K provides an embedding $K \hookrightarrow M_n(k)$, whence a G-equivariant map $\rho : Kk_s \cong K \otimes_k k_s \to M_n(k_s)$, and finally a map $\bar{\rho} : (Kk_s)^\times/k_s^\times \to \mathrm{PGL}_n(k_s)$. Arguing as in the proof of Theorem 4.4.1, we obtain a commutative diagram:

$$
\begin{array}{ccc}
H^1(G, (Kk_s)^\times/k_s^\times) & \xrightarrow{\phi} & H^2(G, k_s^\times) \\
\downarrow & & \downarrow{\scriptstyle\mathrm{id}} \\
H^1(G, \mathrm{PGL}_n(k_s)) & \xrightarrow{\delta_n} & H^2(G, k_s^\times).
\end{array}
$$

Since $\mathrm{Im}\,(\phi) = \mathrm{Br}\,(K|k)$ by the above, the diagram tells us that each element α in $\mathrm{Br}\,(K|k) \subset H^2(G, k_s^\times)$ comes from some element β in $H^1(G, \mathrm{PGL}_n(k_s))$. But α restricts to 0 in $H^2(G, (Kk_s)^\times)$, so the commutative diagram

$$
\begin{array}{ccc}
H^1(G, \mathrm{PGL}_n(k_s)) & \xrightarrow{\ \delta_{n,k}\ } & H^2(G, k_s^\times) \\
\downarrow & & \downarrow \\
H^1(G, \mathrm{PGL}_n(Kk_s)) & \xrightarrow{\ \delta_{n,K}\ } & H^2(G, (Kk_s)^\times)
\end{array}
$$

and the injectivity of $\delta_{n,K}$ imply that β maps to 1 in $H^1(G, \mathrm{PGL}_n(Kk_s))$, as required.

For the last statement, assume that the class $\alpha \in \mathrm{Br}\,(K|k)$ considered above is the class of a degree n algebra A. Since β comes from an element of $H^1(G, (Kk_s)^\times/k_s^\times)$, we get that A is isomorphic to the twisted form of M_n by a 1-cocycle z with values in the subgroup $\bar\rho\big((Kk_s)^\times/k_s^\times\big)$ of $\mathrm{PGL}_n(k_s)$ (see Chapter 2, Section 2.3). Therefore the twisted algebra $A \cong (_zM_n)^G$ contains $(_z\rho(Kk_s))^G$. But $(_z\rho(Kk_s))^G = (\rho(Kk_s))^G \cong K$, because the conjugation action of $(Kk_s)^\times$ on Kk_s is trivial, and ρ is G-equivariant. $\square$

We shall also need the following easy lemma.

Lemma 9.1.3   *Let k be a field of characteristic $p > 0$, and let A be a not necessarily commutative k-algebra. For $y \in A$ consider the k-vector space endomorphism $D_y : A \to A$ defined by $v \mapsto vy - yv$, and let $D_y^{[p]}$ be its p-th iterate. Then $D_y^{[p]} = D_{y^p}$.*

Proof   Consider the maps $L_y : v \mapsto yv$ and $R_y : v \mapsto vy$, and write $D_y = R_y - L_y$. As L_y and R_y commute in the endomorphism ring of the k-vector space A, the binomial formula implies $D_y^{[p]} = (R_y - L_y)^{[p]} = R_y^{[p]} + (-1)^p L_y^{[p]} = D_{y^p}$, as p divides the binomial coefficients $\binom{p}{i}$ for all $0 < i < p$. $\square$

Proof of Theorem 9.1.1   We start with the case of degree p, i.e. $K = k(\sqrt[p]{b})$. Let α be a nonzero class in $\mathrm{Br}\,(K|k)$. Lemma 9.1.2 shows that there exists a central simple k-algebra A of degree p containing K with $[A] = \alpha$; it is a division algebra as $\alpha \neq 0$. As $K \subset A$, we find $y \in A$ with $y^p = b$. Consider the k-endomorphism $D_y : A \to A$ of the lemma above. As b is in the centre of A, we get $D_y^{[p]} = D_{y^p} = D_b = 0$ using the lemma. Since y is not central in A, we find $w \in A$ with $z := D_y(w) \neq 0$ but $D_y(z) = 0$, i.e. $yz = zy$. Setting $x = z^{-1}yw$ we obtain $xy - yx = z^{-1}y(wy - yw) = z^{-1}yz = y$, and hence $y^{-1}xy = x + 1$. As A is a division algebra, the k-subalgebra $k(x)$ generated by x is a commutative subfield nontrivially containing k, so $[k(x) : k] = p$ by dimension reasons. Moreover, the formula $y^{-1}xy = x + 1$ implies that conjugation by y equips the extension $k(x)|k$ with a nontrivial k-automorphism of order p, so $k(x)|k$ is a cyclic Galois extension of degree p, and $x^p - x = x(x+1)\ldots(x+p-1)$ lies in k. Setting $a := x^p - x$ we see that A contains a cyclic subalgebra (χ, b) with presentation (9.2), and this inclusion must be an isomorphism for dimension reasons. This settles the degree p case.

To treat the general case, we use induction on the degree $[K : k]$. Denote by $E \subset K$ the subfield $k(\sqrt[p^{r_1}]{b_1}, \ldots, \sqrt[p^{r_n-1}]{b_n})$. Then K is a degree p purely inseparable extension of E generated by $\sqrt[p]{u_n}$, where $u_n := \sqrt[p^{r_n-1}]{b_n}$. Given $\alpha \in \mathrm{Br}\,(K|k)$, we have $\mathrm{Res}_k^E(\alpha) \in \mathrm{Br}\,(K|E)$, so by the degree p case we find $\tilde{\chi} \in H^1(k, \mathbf{Z}/p\mathbf{Z}) \cong H^1(E, \mathbf{Z}/p\mathbf{Z})$ with $\mathrm{Res}_k^E(\alpha) = [(\chi, u_n)]$. By exact sequence (9.3) we find a character $\tilde{\chi}_n \in H^1(k, \mathbf{Z}/p^{r_n}\mathbf{Z})$ with $p^{r_n-1}\tilde{\chi}_n = \tilde{\chi}$. In $\mathrm{Br}\,(E)$ we have

$$[(\chi, u_n)] = \mathrm{Res}_k^E(\delta(\tilde{\chi})) \cup u_n = \left(p^{r_n-1}\mathrm{Res}_k^E(\delta(\tilde{\chi}_n))\right) \cup u_n$$

$$= \mathrm{Res}_k^E\left(\delta(\tilde{\chi}_n)\right) \cup b_n = [(\chi_n, b_n)]$$

by bilinearity of the cup-product. Hence $\beta := \alpha - (\delta(\tilde{\chi}_n) \cup b_n) = \alpha - [(\chi_n, b_n)]$ lies in $\mathrm{Br}\,(E|k)$. By induction we may write β as a sum of classes of the form $[(\chi_i, b_i)]$, and the proof is complete. $\qquad\square$

We now come to

Theorem 9.1.4 (Teichmüller)   *The map*

$$j_r : H^1(k, \mathbf{Z}/p^r\mathbf{Z}) \otimes k^\times \longrightarrow {}_{p^r}\mathrm{Br}\,(k)$$

is surjective for all $r > 0$. In other words, every central simple k-algebra of p-power degree is Brauer equivalent to a tensor product of cyclic algebras.

Remark 9.1.5   Teichmüller's result holds up to Brauer equivalence, but not up to isomorphism. Indeed, McKinnie [1] gave examples of central simple k-algebras of period p not isomorphic to a tensor product of cyclic algebras of degree p. This is a characteristic p analogue of the counterexample of Amitsur–Rowen–Tignol cited at the end of Chapter 1. Non-cyclic division algebras of p-power degree were known before (see Amitsur–Saltman [1]).

Before starting the proof of Theorem 9.1.4, recall from field theory that a purely inseparable extension $K|k$ is said to be *of height $\leq r$* for some integer $r \geq 0$ if $K^{p^r} \subset k$. The least such r (if exists) is called the height of $K|k$. In particular, $K|k$ is of height 1 if $K \neq k$ and $K^p \subset k$, or equivalently if K can be generated by p-th roots of elements of k. We shall need the following easy facts.

Facts 9.1.6   The *maximal* height 1 purely inseparable extension $\tilde{k}$ of k is obtained by adjoining all p-th roots of elements of k. The composite $\tilde{k}k_s$ is none but the separable closure $\tilde{k}_s$ of $\tilde{k}$. It is also the maximal height 1 purely inseparable extension of k_s: indeed, if $\alpha^p = a$ for some $a \in k_s$, then $f(\alpha^p) = 0$ for a separable polynomial $f \in k[x]$, but then extracting p-th roots from the coefficients of f we get a separable polynomial $g \in \tilde{k}[x]$ with $g(\alpha)^p = g(\alpha) = 0$, so that $\alpha \in \tilde{k}_s$.

Lemma 9.1.7   *Every central simple k-algebra of period p is split by a finite extension $K|k$ of height 1.*

Proof   Consider the maximal height 1 purely inseparable extension $\tilde{k}$ of k described above. It will be enough to show that every central simple k-algebra of period p is split by $\tilde{k}$, for then it is also split by some finite subextension. As $\tilde{k}_s|k_s$ is the maximal purely inseparable extension of height 1, raising elements to the p-th power induces an isomorphism $\tilde{k}_s^\times \xrightarrow{\sim} k_s^\times$. On the other hand, the composite $k_s^\times \to \tilde{k}_s^\times \xrightarrow{p} k_s^\times$ is just the multiplication by p map on $k_s^\times$. Taking Galois cohomology over k (noting that $\tilde{k}_s$ is a $\mathrm{Gal}\,(k_s|k)$-module via the isomorphism $\mathrm{Gal}\,(k_s|k) \cong \mathrm{Gal}\,(\tilde{k}_s|\tilde{k})$), it follows that the multiplication by p map on $H^2(k, k_s^\times)$ coincides with the composite $H^2(k, k_s^\times) \to H^2(k, \tilde{k}_s^\times) \xrightarrow{p} H^2(k, k_s^\times)$. As the last map here is an isomorphism by the above, it follows that all p-torsion elements in $\mathrm{Br}\,(k) \cong H^2(k, k_s^\times)$ must map to 0 in $H^2(k, \tilde{k}_s^\times) \cong H^2(\tilde{k}, \tilde{k}_s^\times) \cong \mathrm{Br}\,(\tilde{k})$, as was to be shown. $\square$

Proof of Theorem 9.1.4   We prove surjectivity of j_r by induction on r. The case $r = 1$ follows from Lemma 9.1.7 and Theorem 9.1.1. Now assume that the statement is known for all integers $1 \le i \le r$, and consider the commutative diagram

$$H^1(k, \mathbf{Z}/p^r\mathbf{Z}) \otimes k^\times \xrightarrow{\mathrm{id} \otimes \mathrm{id}} H^1(k, \mathbf{Z}/p^{r+1}\mathbf{Z}) \otimes k^\times \xrightarrow{p^r \otimes \mathrm{id}} H^1(k, \mathbf{Z}/p\mathbf{Z}) \otimes k^\times \to 0$$

$$\downarrow j_r \qquad\qquad\qquad \downarrow j_{r+1} \qquad\qquad\qquad \downarrow j_1$$

$$_{p^r}\mathrm{Br}\,(k) \xrightarrow{\;\mathrm{id}\;} \;_{p^{r+1}}\mathrm{Br}\,(k) \xrightarrow{\;p^r\;} \;_p\mathrm{Br}\,(k)$$

whose exact upper row comes from (9.3). In view of the diagram, the surjectivity of j_{r+1} follows from that of j_1 and j_r, which we know from the inductive assumption.   □

We now come to the most powerful result of this section.

Theorem 9.1.8 (Albert)   *Every central simple k-algebra of p-power degree is Brauer equivalent to a cyclic algebra.*

The proof is based on the following proposition which is interesting in its own right.

Proposition 9.1.9   *Let A_1, A_2 be two cyclic k-algebras of degrees p^{r_1} and p^{r_2}, respectively. Then there exists an integer $r \leq r_1 + r_2$ and an element $b \in k^\times$ so that the extension $k(\sqrt[p^r]{b})$ splits both A_1 and A_2.*

Combined with Theorem 9.1.1, the proposition immediately yields:

Corollary 9.1.10   *For A_1 and A_2 as in the proposition, the tensor product $A_1 \otimes_k A_2$ is Brauer equivalent to a cyclic algebra of the form (χ, b) for some character χ of order dividing p^r.*

Once we have Corollary 9.1.10, we can easily *prove Albert's theorem* by exploiting what we already know. Indeed, by induction we get that tensor products of cyclic algebras of p-power degree are Brauer equivalent to a cyclic algebra, and so Albert's theorem follows from that of Teichmüller.

For the proof of Proposition 9.1.9 we need the following lemma from field theory.

Lemma 9.1.11   *Let $K = k(\sqrt[p^r]{b})|k$ be a purely inseparable extension of degree p^r, and let $L|k$ be a finite separable extension. Then there exists an element $v \in LK$ whose norm $N_{LK|K}(v)$ generates the extension $K|k$.*

Proof   We may assume $[K : k] > 1$. Setting $u = \sqrt[p^r]{b}$ we have $[K : k(u^p)] = p$, so it will be enough to find $v \in LK$ with $N_{LK|K}(v) \notin k(u^p)$. Using the theorem of the primitive element, we write $L = k(w)$ for appropriate $w \in L$. Let $f = t^m + \alpha_1 t^{m-1} + \cdots + \alpha_m$ be the minimal polynomial of w over k. Grouping exponents into residue classes mod p, we write

$$f = \sum_{i=0}^{p-1} f_i(t^p)t^i.$$

Since f is a separable polynomial, we find $j \neq 0$ such that $f_j \neq 0$. Then $f_j((ut)^p) \in K[t]$ is a nonzero polynomial, and since k is an infinite field (otherwise it would have no nontrivial inseparable extension), there exists $\alpha \in k^\times$ such that $f_j((\alpha u)^p) \neq 0$. Now put $v := w - \alpha u$. The minimal polynomial of v over K is $f(t + \alpha u) = t^m + \cdots + f(\alpha u)$, so $N_{LK|K}(v) = (-1)^m f(\alpha u)$ and

$$f(\alpha u) = \sum_{i=0}^{p-1} f_i((\alpha u)^p)\alpha^i u^i.$$

This is an expression for $f(\alpha u)$ as a linear combination of the basis elements $1, u, \ldots, u^{p-1}$ of the $k(u^p)$-vector space K. Since the coefficient $f_j((\alpha u)^p)\alpha^j$ is nonzero, we have $f(\alpha u) \notin k(u^p)$ and hence $N_{LK|K}(v) \notin k(u^p)$, as desired. $\qquad\square$

Proof of Proposition 9.1.9 For $i = 1, 2$ write $A_i = (\chi_i, b_i)$ with characters χ_i of order p^{r_i} and elements $b_i \in k^\times$. If $b_i = c_i^p$ for some $c_i \in k^\times$, then formula (9.1) shows that A_i is Brauer equivalent to the cyclic algebra $(p\chi_i, c_i)$. So up to replacing A_i by a Brauer equivalent algebra we may assume that $[k(\sqrt[p^{r_i}]{b_i}) : k] = p^{r_i}$ for $i = 1, 2$. Denote by k_2 the cyclic extension of k defined by the kernel of $\tilde\chi_2$. Lemma 9.1.11 provides $v \in k_2(\sqrt[p^{r_1}]{b_1})$ such that $z := N_{k_2(\sqrt[p^{r_1}]{b_1})|k(\sqrt[p^{r_1}]{b_1})}(v)$ generates the extension $k(\sqrt[p^{r_1}]{b_1})|k$. Consider now the purely inseparable extension $E = k(\sqrt[p^{r_1}]{b_1})(\sqrt[p^{r_2}]{zb_2})$ of k. Since z generates $k(\sqrt[p^{r_1}]{b_1})$ over k, the element $y := \sqrt[p^{r_2}]{zb_2}$ generates E over k, and thus we have $[E : k] = p^{r_1+r_2}$. As b_1 is a p^{r_1}-th power in E, the algebra $(\chi_1, b_1) \otimes_k E$ is split. To see that E also splits (χ_2, b_2), we write

$$(\chi_2, b_2) \otimes_k E \cong (\chi_2, y^{p^{r_2}}z^{-1}) \otimes_k E \cong (\chi_2, y^{p^{r_2}}) \otimes_{k(\sqrt[p^{r_1}]{b_1})} (\chi_2, z^{-1}) \otimes_k E.$$

Since χ_2 has order dividing p^{r_2}, the algebra $(\chi_2, y^{p^{r_2}}) \otimes_k E$ splits. On the other hand, the algebra (χ_2, z^{-1}) splits over $k(\sqrt[p^{r_1}]{b_1})$ according to Corollary 4.7.5, because z is a norm from the extension $k_2(\sqrt[p^{r_1}]{b_1})|k(\sqrt[p^{r_1}]{b_1})$, and therefore it also splits over E. Hence $(\chi_2, b_2) \otimes_k E$ splits, as desired. $\qquad\square$

To conclude this section, we discuss a positive characteristic analogue of Proposition 7.4.13 due to M. Florence [1].

Theorem 9.1.12 (Florence)   *A division algebra A of degree p^n and period p^r over k is Brauer equivalent to a tensor product of at most $p^n - 1$ cyclic algebras of degree p^r.*

In the case $r = n = 1$, the theorem thus says that a degree p division algebra over k is Brauer equivalent to a tensor product of at most $p - 1$ cyclic algebras of degree p. This is a much stronger bound than that of Proposition 7.4.13 in the prime-to-p case.

Remark 9.1.13   If we moreover assume that A is a *cyclic* division algebra (or, more generally, becomes cyclic over a finite separable extension of k), a better bound is known (Mammone and Merkurjev [1]): A is Brauer equivalent to the tensor product of at most p^{n-r} cyclic algebras of degree p^r. Note that by Theorem 9.1.8 a general A is always Brauer equivalent to a cyclic algebra, but the latter is not necessarily a division algebra.

Proof of Theorem 9.1.12   For ease of notation let us write $d := p^n$ for the degree of A. In view of Theorem 9.1.1, it will suffice to produce a splitting field for A that is of the form $k(\sqrt[p^{r_1}]{b_1}, \ldots, \sqrt[p^{r_m}]{b_m})$ with some $m \le d - 1$, $b_i \in \overline{k}$ and $r_i \le r$.

Let $K|k$ be a Galois splitting field of A, and denote by G its Galois group. Given a K-vector space V, endow it with another K-vector space structure where addition is defined in the same way and multiplication is given by $a \cdot v := a^{p^r} v$. This construction is functorial in V, so applying it to morphisms $K^d \to K^d$ we obtain a group homomorphism $F_r : \mathrm{GL}_d(K) \to \mathrm{GL}_d(K)$. It fits in the commutative exact diagram

$$
\begin{array}{ccccccccc}
1 & \longrightarrow & K^\times & \longrightarrow & \mathrm{GL}_d(K) & \longrightarrow & \mathrm{PGL}_d(K) & \longrightarrow & 1 \\
& & \downarrow{\scriptstyle p^r} & & \downarrow{\scriptstyle F_r} & & \downarrow{\scriptstyle F_r} & & \\
1 & \longrightarrow & K^\times & \longrightarrow & \mathrm{GL}_d(K) & \longrightarrow & \mathrm{PGL}_d(K) & \longrightarrow & 1
\end{array}
$$

where the first vertical map is given by $a \mapsto a^{p^r}$ and the third one is induced by the first two. As the G-action on K commutes with the map $a \mapsto a^{p^r}$, this is in fact a diagram of groups with G-action, and therefore it gives rise to the commutative diagram of boundary maps

$$
\begin{array}{ccc}
H^1(G, \mathrm{PGL}_d(K)) & \longrightarrow & H^2(G, K^\times) \\
\downarrow{\scriptstyle F_{r*}} & & \downarrow{\scriptstyle p^r} \\
H^1(G, \mathrm{PGL}_d(K)) & \longrightarrow & H^2(G, K^\times).
\end{array}
$$

Since A is of period p^r by assumption and the horizontal maps have trivial kernel, the diagram shows that the class $[A] \in H^1(G, \mathrm{PGL}_d(K))$ is annihilated by the map F_{r*}. Thus if $z : G \to \mathrm{PGL}_d(K)$ is a 1-cocycle representing $[A]$, the 1-cocycle $z_r := F_r \circ z$ is a coboundary.

Now we consider the morphism of projective spaces $f_r : \mathbf{P}_K^{d-1} \to \mathbf{P}_K^{d-1}$ defined by $(x_0, x_1, \ldots, x_d) \mapsto (x_0^{p^r}, x_1^{p^r}, \ldots, x_d^{p^r})$. It commutes with the action of G on both sides. It also commutes with the action of $\mathrm{PGL}_d(K)$ if we consider the action on the target space as the one induced by composing the usual action of $\mathrm{PGL}_d(K)$ by F_r. Twisting the G-action by the above 1-cocycle z and considering G-invariants as in the proof of Theorem 5.2.1, we obtain a morphism of Severi–Brauer varieties $f : X \to X_r$ over k. Here by construction X is the Severi–Brauer variety associated with A. On the other hand, the twisted G-action on the target space amounts to considering the usual $\mathrm{PGL}_d(K)$-action on $\mathbf{P}_K^{d-1}$ and twisting the Galois action by the cocycle z_r. Since this cocycle is a coboundary, the Severi–Brauer variety X_r is split over k and therefore has a k-rational point P_r.

After base change to $\bar{k}$, the morphism $f_{\bar{k}}$ identifies with the surjective morphism $f_r : \mathbf{P}_{\bar{k}}^{d-1} \to \mathbf{P}_{\bar{k}}^{d-1}$, and hence there exists a $\bar{k}$-point $\overline{P}$ with $f_{\bar{k}}(\overline{P}) = P_r$. The homogeneous coordinates of $\overline{P}$ lie in a purely inseparable extension of k of height $\leq r$. Thus if P is the closed point of X below $\overline{P}$, its residue field $k(P)$ is a purely inseparable extension of k of height $\leq r$ that splits X and hence also A. To show that $k(P)$ has a system of generators as in the first paragraph, it will suffice to show that it can be generated by at most $d - 1$ elements over k. Viewing K as a separable subextension of $\bar{k}|k$, we see that $Kk(P) \cong K \otimes_k k(P)$, and this field is the residue field of a unique closed point of X_K lying above P. Since $X_K \cong \mathbf{P}_K^{d-1}$, this point may be described by a d-tuple $(a_0, \ldots, a_{d-1})$ of homogeneous coordinates, where we may assume $a_0 = 1$. Thus $Kk(P) = K(a_1, \ldots, a_{d-1})$, or in other words the purely inseparable extension $Kk(P)|K$ can be generated by $d - 1$ elements. Since K is separable over k, the same holds for the extension $k(P)|k$ by Propositions A.8.9 and A.8.10 of the Appendix, which concludes the proof. $\square$

9.2 Differential forms and p-torsion in the Brauer group

We now discuss another method for describing the p-torsion part of the Brauer group of k. The basic idea is the following. For m prime to p, a fundamental tool for studying $_m\mathrm{Br}(k)$ was furnished by the exact sequence coming from multiplication by m on $k_s^\times$, which is surjective with kernel μ_m. In contrast to this, for $m = p$ the multiplication by p map is injective, and it has a nontrivial cokernel which can be described using differential forms, via the dlog map.

To define this map, consider for an arbitrary extension $K|k$ the module $\Omega^1_K = \Omega^1_{K/\mathbf{Z}}$ of absolute differentials over K. The map dlog $: K^\times \to \Omega^1_K$ is then defined by sending $y \in K^\times$ to the logarithmic differential form dy/y. This is a homomorphism of abelian groups whose kernel is $K^{\times p}$; denote by $\nu(1)_K$ its image. For $K = k_s$ we therefore have an exact sequence

$$0 \to k_s^\times \xrightarrow{p} k_s^\times \xrightarrow{\text{dlog}} \nu(1)_{k_s} \to 0. \tag{9.4}$$

This is in fact an exact sequence of $\mathrm{Gal}(k_s|k)$-modules, so taking the associated long exact sequence yields an isomorphism

$$H^1(k, \nu(1)) \xrightarrow{\sim} {}_p\mathrm{Br}(k). \tag{9.5}$$

To proceed further, we would like to have a more explicit presentation of $H^1(k, \nu(1))$. Assume we had a surjective map on $\Omega^1_{k_s}$ whose *kernel* is precisely the image $\nu(1)_{k_s}$ of the dlog map. Then we would have another short exact sequence of the form

$$0 \to \nu(1)_{k_s} \to \Omega^1_{k_s} \to \Omega^1_{k_s} \to 0,$$

and by the associated long exact sequence $H^1(k, \nu(1))$ would arise as a quotient of $\left(\Omega^1_{k_s}\right)^G$ which in fact equals Ω^1_k, as one sees from Proposition A.8.9 of the Appendix. Thus all in all we would get a presentation of ${}_p\mathrm{Br}(k)$ by differential forms.

The required map comes from the theory of the (inverse) Cartier operator. To define it, recall first a construction from linear algebra in characteristic $p > 0$. Given a K-vector space V, we may equip the underlying abelian group of V with another K-vector space structure ${}^pV$ in which $a \in K$ acts via $a \cdot w := a^pw$. A K-linear map $V \to {}^pW$ is sometimes called a p-*linear* map from V to W. Recall also (from the Appendix) that the subgroup $B^1_K \subset \Omega^1_K$ is defined as the image of the universal derivation $d : K \to \Omega^1_K$. Though d is only a k-linear map, the induced map $d : {}^pK \to {}^p\Omega^1_K$ is already K-linear, in view of the formula $a^pdb = d(a^pb)$, where the right-hand side is d applied to the product of $b \in {}^pK$ with a. Thus ${}^pZ^1_K$ and ${}^pB^1_K$ are K-subspaces of ${}^p\Omega^1_K$.

Lemma 9.2.1 *There exists a unique morphism of K-vector spaces*

$$\gamma : \Omega^1_K \to {}^p\Omega^1_K/{}^pB^1_K$$

satisfying $\gamma(da) = a^{p-1}da \bmod B^1_K$ *for all* $a \in K$. *Moreover, we have* $d \circ \gamma = 0$, *where* $d : \Omega^1_K \to \Omega^2_K$ *is the differential of the de Rham complex.*

Proof Recall from the Appendix that the K-vector space Ω^1_K has a presentation by symbols of the form da for $a \in K$ subject to the relations

$d(a + b) = da + db$ and $d(ab) = adb + bda$. Define γ on the elements da by the formula above and extend by linearity. To see that γ is well defined, we have to show that it annihilates all elements of the form $d(a + b) - da - db$ or $d(ab) - adb - dba$. For elements of the second type, we compute

$$\gamma(adb + bda) = a^p b^{p-1} db + b^p a^{p-1} da = (ab)^{p-1}(adb + bda) = \gamma(d(ab)).$$

For elements of the first type, we have to see that $(a + b)^{p-1}(da + db) - a^{p-1} da - b^{p-1} db$ belongs to B_K^1. Notice first that the relation

$$d((x+y)^p) = p(x+y)^{p-1} d(x+y) = p(x^{p-1} dx + y^{p-1} dy) + \sum_{i=1}^{p-1} \binom{p}{i} d(x^i y^{p-i})$$

holds in the space of absolute differentials of the polynomial ring $\mathbf{Z}[x, y]$, which is the free $\mathbf{Z}[x, y]$-module generated by dx and dy according to Appendix, Example A.8.2. Since all binomial coefficients in the sum are divisible by p, after dividing by p it follows that

$$(x + y)^{p-1}(dx + dy) - x^{p-1} dx - y^{p-1} dy \in B_{\mathbf{Z}[x,y]}^1.$$

We obtain the required identity in Ω_K^1 by specialization via the homomorphism $\mathbf{Z}[x, y] \to K$ defined by $x \mapsto a, y \mapsto b$. The last statement follows from the equality $da \wedge da = 0$. □

For historical reasons, the resulting map is called the *inverse Cartier operator*. We now have the following theorem due to Jacobson and Cartier.

Theorem 9.2.2 *For every field K of characteristic $p > 0$ the sequence of maps*

$$1 \to K^\times \xrightarrow{p} K^\times \xrightarrow{\text{dlog}} \Omega_K^1 \xrightarrow{\gamma - 1} {}^p\Omega_K^1/{}^pB_K^1$$

is exact.

We postpone the proof of the theorem to the next section, and now consider its application to our problem of presenting elements in $H^1(k, \nu(1))$ by differential forms. The solution is based on the following corollary.

Lemma 9.2.3 *Let k be a field of characteristic $p > 0$ with separable closure k_s. The sequence*

$$1 \to \nu(1)_{k_s} \longrightarrow \Omega_{k_s}^1 \xrightarrow{\gamma - 1} \Omega_{k_s}^1/B_{k_s}^1 \to 0$$

is an exact sequence of Gal $(k_s|k)$*-modules.*

The superscripts p disappeared from the last term because we are not interested here in its k_s-vector space structure.

Proof That the maps dlog and $\gamma - 1$ are Galois equivariant follows from their construction. So in view of Theorem 9.2.2, it remains to prove surjectivity of $\gamma - 1$. Recall first that the Artin–Schreier map $\wp : k_s \to k_s$ defined by $x \mapsto x^p - x$ is surjective. Hence given a 1-form $adb \in \Omega^1_{k_s}$, we may find $x \in k_s$ with $x^p - x = ab$. Then

$$(\gamma - 1)(xb^{-1}db) = x^p b^{-p} b^{p-1} db - xb^{-1} db = (x^p - x)b^{-1}db = adb$$

according to the defining properties of the operator γ, whence the required surjectivity. □

We can now prove the following theorem which seems to have been first noticed by Kato.

Theorem 9.2.4 *There exists a canonical isomorphism*

$$\Omega^1_k / (B^1_k + (\gamma - 1)\Omega^1_k) \xrightarrow{\sim} {}_p\mathrm{Br}\,(k).$$

Proof Denote by G the Galois group $\mathrm{Gal}\,(k_s|k)$. The exact sequence of Lemma 9.2.3 gives rise to the long exact sequence

$$(\Omega^1_{k_s})^G \xrightarrow{\gamma - 1} (\Omega^1_{k_s}/B^1_{k_s})^G \to H^1(k, \nu(1)) \to H^1(k, \Omega^1_{k_s}).$$

The G-module $\Omega^1_{k_s}$ is a k_s-vector space, so $H^1(k, \Omega^1_{k_s}) = 0$ by the additive form of Hilbert's Theorem 90 (Lemma 4.3.11). On the other hand, Proposition A.8.9 of the Appendix implies $(\Omega^1_{k_s})^G = \Omega^1_k$, and hence also $(B^1_{k_s})^G = B^1_k$, since the differential of the de Rham complex is G-equivariant by construction. As $H^1(k, B^1_{k_s}) = 0$ (again by Lemma 4.3.11), it follows that $(\Omega^1_{k_s}/B^1_{k_s})^G = \Omega^1_k/B^1_k$, and finally we get an isomorphism

$$\Omega^1_k / (B^1_k + (\gamma - 1)\Omega^1_k) \xrightarrow{\sim} H^1(k, \nu(1)),$$

whose composition with the isomorphism (9.5) yields the isomorphism of the theorem. □

One can make the isomorphism of the above theorem quite explicit.

Proposition 9.2.5 *The isomorphism of Theorem 9.2.4 sends the class of a 1-form $adb \in \Omega^1_k$ to the class of the cyclic algebra $[ab, b)$ in ${}_p\mathrm{Br}\,(k)$.*

Proof As in the proof of Lemma 9.2.3 we find $x \in k_s$ with $x^p - x = ab$ and a 1-form $xb^{-1}db$ satisfying $(\gamma - 1)(xb^{-1}db) = adb$. The image of adb

by the coboundary map $\left(\Omega^1_{k_s}/B^1_{k_s}\right)^G \to H^1(k, \nu(1))$ is represented by the 1-cocycle $c_\sigma : \sigma \mapsto \sigma(xb^{-1}db) - xb^{-1}db = (\sigma(x)-x)\mathrm{dlog}(b)$. The character $\tilde{\chi} : \sigma \mapsto \sigma(x) - x \in \mathbf{Z}/p\mathbf{Z}$ is precisely the one defining the extension of k given by the Artin–Schreier polynomial $x^p - x - ab$, and the cocycle c_σ represents the image of the pair $(\tilde{\chi}, b)$ by the cup-product

$$H^1(k, \mathbf{Z}/p\mathbf{Z}) \times H^0(k, k_s^\times) \to H^1(k, k_s^\times \otimes \mathbf{Z}/p\mathbf{Z})$$

followed by the isomorphism

$$H^1(k, k_s^\times \otimes \mathbf{Z}/p\mathbf{Z}) \xrightarrow{\sim} H^1(k, \nu(1)) \tag{9.6}$$

coming from exact sequence (9.4). Writing δ for the coboundary $H^1(k, \mathbf{Z}/p\mathbf{Z}) \to H^2(k, \mathbf{Z})$ and δ' for the coboundary $H^1(k, k_s^\times \otimes \mathbf{Z}/p\mathbf{Z}) \to H^2(k, k_s^\times)$ (which identifies to a coboundary coming from (9.4) via the isomorphism (9.6)), we have $\delta'(\tilde{\chi} \cup b) = \delta(\tilde{\chi}) \cup b$ by Proposition 3.4.8. But the latter class in $\mathrm{Br}\,(k)$ is represented by the cyclic algebra $[ab, b)$, as recalled at the beginning of the previous chapter. □

Remarks 9.2.6

1. The 1-forms adb generate Ω^1_k as an abelian group, so one may try to define a map $\Omega^1_k \to {}_p\mathrm{Br}\,(k)$ by sending a finite sum $\sum a_i db_i$ to $\sum [\,[a_i b_i, b_i)\,]$. An easy computation shows that this map annihilates all elements of the form $d(a + b) - da - db$ and $d(ab) - adb - bda$, so it indeed induces a well-defined map $\Omega^1_k \to {}_p\mathrm{Br}\,(k)$. This gives an elementary construction of the map inducing the isomorphism of Theorem 9.2.4.

2. Theorem 9.2.4 and Proposition 9.2.5 together give another proof of Teichmüller's theorem in the case $r = 1$. But we have seen in the previous section that the general case follows from this by an easy induction argument. So we obtain a proof of Teichmüller's theorem which does not use the theory of central simple algebras – but relies, of course, on the nontrivial theorem of Jacobson and Cartier.

9.3 Logarithmic differentials and flat p-connections

In this section we prove Theorem 9.2.2. Following Katz, our main tool in the argument will be the study, for a differential form $\omega \in \Omega^1_K$, of the map $\nabla_\omega : K \to \Omega^1_K$ defined by

$$\nabla_\omega(a) = da + a\omega. \tag{9.7}$$

It follows from this definition that the 1-form ω is logarithmic if and only if $\nabla_\omega(a) = 0$ for some $a \in K^\times$; indeed, the latter condition is equivalent to $\omega = -a^{-1}da = \mathrm{dlog}(a^{-1})$.

The map ∇_ω is a basic example for a connection on K. For later purposes, we introduce this notion in a more general context. We shall work in the following setup: K will be a field of characteristic $p > 0$, and k a subfield of K containing K^p. The extension $K|k$ is then a purely inseparable extension of height 1. The most important case will be when $k = K^p$, for then we have $\Omega^1_{K|k} = \Omega^1_K$.

Now define a *connection* on a finite dimensional K-vector space V to be a homomorphism $\nabla : V \to \Omega^1_{K|k} \otimes_K V$ of abelian groups satisfying

$$\nabla(av) = a\nabla(v) + da \otimes v$$

for all $a \in K$ and $v \in V$.

A connection ∇ gives rise to a K-linear map $\nabla_* : \mathrm{Der}_k(K) \to \mathrm{End}_k(V)$ sending a derivation D to the map $\nabla_*(D)$ obtained as the composite

$$\nabla_*(D) : V \xrightarrow{\nabla} \Omega^1_{K|k} \otimes_K V \xrightarrow{D \otimes id} K \otimes_K V \cong V,$$

where D is regarded as a K-linear map $\Omega^1_{K|k} \to K$ via the isomorphism $\mathrm{Der}_k(K) \cong \mathrm{Hom}_K(\Omega^1_{K|k}, K)$. Note that though ∇_* is K-linear, the element $\nabla_*(D)$ is only a k-endomorphism in general (by the defining property of connections), but not a K-endomorphism. A straightforward computation yields the formula

$$\nabla_*(D)(av) = D(a)v + a\nabla_*(D)(v) \qquad (9.8)$$

for all $v \in V$ and $a \in K$.

Example 9.3.1   Given $\omega \in \Omega^1_{K|k}$, the map $\nabla_\omega : K \to \Omega^1_{K|k}$ defined by (9.7) is a connection on the 1-dimensional K-vector space K. To see this, we compute

$$\nabla_\omega(ab) = adb + bda + ab\omega = a(db + b\omega) + bda = a\nabla_\omega(b) + da \otimes b,$$

as required. The induced map $\nabla_{\omega*} : \mathrm{Der}_k(K) \to \mathrm{End}_k(K)$ sends D to the map $a \mapsto D(a) + aD(\omega)$.

Now recall the following facts from the Appendix. The K-vector space $\mathrm{End}_k(V)$ carries a Lie algebra structure over k, with Lie bracket defined by $[\phi, \psi] = \phi \circ \psi - \psi \circ \phi$. This Lie bracket and the p-operation sending an endomorphism ϕ to its p-th iterate $\phi^{[p]}$ equip $\mathrm{End}_k(V)$ with the structure of a p-Lie algebra over k (see the Appendix for the precise definition). The K-vector space $\mathrm{Der}_k(K)$ is a k-subspace of $\mathrm{End}_k(K)$ preserved by the Lie bracket and the p-operation of $\mathrm{End}_k(K)$, therefore it is a p-Lie subalgebra. It is then a natural condition for a connection to require that the map ∇_* respects the p-Lie algebra structures on $\mathrm{Der}_k(K)$ and $\mathrm{End}_k(V)$. Accordingly, we say that the connection ∇ is *flat* or *integrable* if ∇_* is a homomorphism of Lie algebras over k, i.e. if $\nabla_*([D_1, D_2]) = [\nabla_*(D_1), \nabla_*(D_2)]$ for all $D_1, D_2 \in \mathrm{Der}_k(K)$, and that ∇ is a *p-connection* if $\nabla_*(D^{[p]}) = (\nabla_*(D))^{[p]}$ for all $D \in \mathrm{Der}_k(K)$.

Remark 9.3.2 One may introduce important invariants which measure the defect for a connection ∇ from being flat or a p-connection. The first of these is its *curvature* $K_*(\nabla)$, defined as the map $\mathrm{Der}_k(K) \times \mathrm{Der}_k(K) \to \mathrm{End}_k(V)$ sending the pair (D_1, D_2) to $[\nabla_*(D_1), \nabla_*(D_2)] - \nabla_*([D_1, D_2])$. One may check that $K_*(\nabla)(D_1, D_2)$ equals the composite

$$V \xrightarrow{\nabla} \Omega^1_{K|k} \otimes_K V \xrightarrow{\nabla_1} \Omega^2_{K|k} \otimes_K V \xrightarrow{(D_1 \wedge D_2) \otimes id} K \otimes_K V \cong V,$$

where $\nabla_1(\omega \otimes v) := d\omega \otimes v - \omega \wedge \nabla(v)$. Therefore $K_*(\nabla)$ is the map induced on derivations by $\nabla_1 \circ \nabla : V \to \Omega^2_{K|k} \otimes_K V$; one often defines the curvature as being the latter map. The second invariant is the *p-curvature* $\psi_*(\nabla)$, defined as the map $\mathrm{Der}_k(K) \to \mathrm{End}_k(V)$ sending D to $\nabla_*(D^{[p]}) - (\nabla_*(D))^{[p]}$. We shall not investigate any case where one of these invariants is nonzero.

It turns out that when the differential form ω is logarithmic, the connection ∇_ω on K is a flat p-connection. We shall prove this as part of the following theorem, which is the main result of this section.

Theorem 9.3.3 *Given an extension $K|k$ of fields of characteristic $p > 0$ with $K^p \subset k$, the following are equivalent for a differential form $\omega \in \Omega^1_{K|k}$:*

1. *The 1-form ω is logarithmic, i.e. $\omega = \mathrm{dlog}(a)$ for some $a \in K^\times$.*
2. *We have $\gamma(\omega) = \omega \mod B^1_{K|k}$.*
3. *The connection ∇_ω is a flat p-connection.*

In order to give a sense to statement (2), we have to extend the definition of the inverse Cartier operator to relative differentials for the extension $K|k$. To do so, apply Corollary A.8.13 of the Appendix with $k_0 = K^p$ to obtain a split exact sequence of K-vector spaces

$$0 \to K \otimes_k \Omega^1_k \to \Omega^1_K \to \Omega^1_{K|k} \to 0. \tag{9.9}$$

The composite map $\Omega^1_K \xrightarrow{\gamma} \Omega^1_K/B^1_K \to \Omega^1_{K|k}/B^1_{K|k}$ vanishes on $K \otimes_k \Omega^1_k$, and hence the operator γ induces a relative operator $\Omega^1_{K|k} \to \Omega^1_{K|k}/B^1_{K|k}$ which we again denote by γ.

Of course, for $k = K^p$ we get back the γ of the previous section. In this case the equivalence (1) $\Leftrightarrow$ (2) is a restatement of Theorem 9.2.2. Indeed, the implication (1) $\Rightarrow$ (2) yields that the sequence of Theorem 9.2.2 is a complex, and it is obviously exact at the first term. Exactness at the second term follows from implication (2) $\Rightarrow$ (1).

For the proof we first investigate the implication (2) $\Rightarrow$ (3). It will result from the following slightly more general proposition:

Proposition 9.3.4 *Let* $\omega \in \Omega^1_{K|k}$ *be a differential form.*

1. *If* $d\omega = 0$, *then the connection* ∇_ω *is flat.*
2. *If* $\gamma(\omega) = \omega \bmod B^1_{K|k}$, *then* ∇_ω *is a p-connection.*

Note that the condition in part (2) implies $d\omega = 0$ as well, in view of the last statement of Proposition 9.2.1. For the proof we need the following lemma on derivations.

Lemma 9.3.5 *Let* $\omega \in \Omega^1_{K|k}$ *be a differential form.*

1. *For all derivations* $D_1, D_2 \in \mathrm{Der}_k(K)$ *we have*

$$(D_1 \wedge D_2)(d\omega) = D_1(D_2(\omega)) - D_2(D_1(\omega)) - [D_1, D_2](\omega).$$

2. *If* $\omega \in B^1_{K|k}$, *we have*

$$D^{[p]}(\omega) - D^{[p-1]}(D(\omega)) = 0.$$

3. *For general* ω *and all derivations* $D \in \mathrm{Der}_k(K)$ *we have*

$$D(\omega)^p = D^{[p]}(\gamma(\omega)) - D^{[p-1]}(D(\gamma(\omega))).$$

Note that the right-hand side of the formula in part (3) is well defined in view of part (2).

Proof It is enough to check (1) on generators of $\Omega^1_{K|k}$, so we may assume $\omega = a\,db$, so that $d\omega = da \wedge db$. Therefore on the one hand we have

$$(D_1 \wedge D_2)(d\omega) = D_1(da)D_2(db) - D_2(da)D_1(db)$$
$$= D_1(a)D_2(b) - D_2(a)D_1(b),$$

where we first regard the D_i as linear maps $\Omega^1_{K|k} \to K$, and then as derivations $K \to K$. On the other hand, we compute

$$D_1(D_2(adb)) - D_2(D_1(adb)) - [D_1, D_2](adb)$$
$$= D_1(aD_2(b)) - D_2(aD_1(b)) - a[D_1, D_2](b),$$

which may be rewritten as

$$D_1(a)D_2(b) + aD_1(D_2(b)) - D_2(a)D_1(b) - aD_2(D_1(b))$$
$$- a\big(D_1(D_2(b)) - D_2(D_1(b))\big),$$

so that after cancelling terms we again get $D_1(a)D_2(b) - D_2(b)D_1(a)$, as desired.

To check (2), one simply remarks that for $\omega = da$ one has

$$D^{[p]}(\omega) - D^{[p-1]}(D(\omega)) = D^{[p]}(a) - D^{[p-1]}(D(a)) = 0.$$

For (3), we may assume by p-linearity of γ that $\omega = da$ for some $a \in K$, and so we have to check

$$(D(a))^p = D^{[p]}(a^{p-1}da) - D^{[p-1]}(D(a^{p-1}da)).$$

As already remarked, $D^{[p]}$ is a derivation, so that $D^{[p]}(a^{p-1}da) = a^{p-1}D^{[p]}(a)$. After this substitution, the formula reduces to Hochschild's formula (Proposition A.7.1 of the Appendix). □

For the proof of the proposition it is convenient to introduce for $a \in K$ the notation L_a for the element in $\mathrm{End}_k(K)$ given by multiplication by a, as in the Appendix. Recall also that for $D \in \mathrm{Der}_k(K)$ we have the equality

$$[D, L_a] = L_{D(a)} \tag{9.10}$$

in $\mathrm{End}_k(K)$ because of the computation $[D, L_a](x) = D(ax) - aD(x) = D(a)x$.

Proof of Proposition 9.3.4 In the notation above, the formula for $\nabla_{\omega*}$ in Example 9.3.1 reads $\nabla_{\omega*}(D) = D + L_{D(\omega)}$. Hence to prove (1) we may write

$$[\nabla_{\omega*}(D_1), \nabla_{\omega*}(D_2)] = [D_1, D_2] + [D_1, L_{D_2(\omega)}] - [D_2, L_{D_1(\omega)}]$$
$$- [L_{D_1(\omega)}, L_{D_2(\omega)}].$$

Here the last term vanishes as the elements $D_1(\omega), D_2(\omega) \in K$ commute, so using equality (9.10) we may write

$$[\nabla_{\omega*}(D_1), \nabla_{\omega*}(D_2)] = [D_1, D_2] + L_{D_1(D_2(\omega))} - L_{D_2(D_1(\omega))},$$

or else

$$[\nabla_{\omega*}(D_1), \nabla_{\omega*}(D_2)] = \nabla_{\omega*}([D_1, D_2]) - L_{[D_1, D_2](\omega)}$$
$$+ L_{D_1(D_2(\omega))} - L_{D_2(D_1(\omega))}.$$

But according to Lemma 9.3.5 (1) we have

$$L_{D_1(D_2(\omega))} - L_{D_2(D_1(\omega))} - L_{[D_1, D_2](\omega)} = L_{(D_1 \wedge D_2)(d\omega)},$$

which is zero by our assumption $d\omega = 0$.

To handle part (2), we compute $(\nabla_{\omega*}(D))^{[p]} - \nabla_{\omega*}(D^{[p]})$ as

$$(D + L_{D(\omega)})^{[p]} - (D^{[p]} + L_{D^{[p]}(\omega)})$$
$$= D^{[p]} + L_{D(\omega)^p} + L_{D^{[p-1]}(D(\omega))} - (D^{[p]} + L_{D^{[p]}(\omega)})$$

using Proposition A.7.2 from the Appendix. But by Lemma 9.3.5 (2) and our assumption we may write

$$L_{D(\omega)^p - D^{[p]}(\omega) + D^{[p-1]}(D(\omega))} = L_{D(\omega)^p - D^{[p]}(\gamma(\omega)) + D^{[p-1]}(D(\gamma(\omega)))},$$

which vanishes by Lemma 9.3.5 (3). □

We now turn to implication (3) $\Rightarrow$ (1) in Theorem 9.3.3. Recall that we are working over a field extension $K|k$ with $K^p \subset k$. Given a K-vector space V equipped with a connection $\nabla : V \to \Omega^1_{K|k} \otimes_K V$, set

$$V^\nabla := \{v \in V : \nabla(v) = 0\}.$$

The defining property of connections implies that V^∇ is a k-subspace of V. In geometric language, it is the space of 'horizontal sections' of the connection ∇. In accordance with the remarks at the beginning of this section, our goal is to prove that for a flat p-connection we have $V^\nabla \neq 0$.

In the case when the extension $K|k$ is *finite*, this is assured by the following descent statement which can be regarded as an analogue of Speiser's lemma (Lemma 2.3.8) for finite purely inseparable extensions of height 1.

Theorem 9.3.6 *Let $K|k$ be a finite extension with $K^p \subset k$, and let V be a K-vector space equipped with a flat p-connection ∇. Then the natural map*

$$K \otimes_k V^\nabla \to V$$

is an isomorphism.

The following proof is taken from the book of Springer [1].

Proof Take a p-basis $a_1, \ldots, a_m$ of the extension $K|k$. Then the da_i form a basis of the K-vector space $\Omega^1_{K|k}$ (by Proposition A.8.11 of the Appendix). Let ∂_i be the derivation defined by sending da_i to 1 and da_j to 0 for $i \neq j$, and set $D_i := a_i \partial_i$. By construction, the D_i satisfy

$$[D_i, D_j] = 0 \quad \text{for all} \quad i \neq j, \quad \text{and} \quad D_i^{[p]} = D_i \quad \text{for all} \quad i.$$

Now consider the elements $\nabla_*(D_i) \in \text{End}_K(V)$ for $1 \leq i \leq m$. Since ∇ is a flat connection, we have $[\nabla_*(D_i), \nabla_*(D_j)] = \nabla_*([D_i, D_j]) = 0$ for $i \neq j$, since $[D_i, D_j] = 0$. This means that the endomorphisms $\nabla_*(D_i)$ pairwise commute, and hence they are simultaneously diagonalizable by a well-known theorem of linear algebra. Moreover, since ∇ is a p-connection, we have $\nabla_*(D_i)^{[p]} = \nabla_*(D_i^{[p]}) = \nabla_*(D_i)$, which implies that the eigenvalues of the $\nabla_*(D_i)$ all lie in $\mathbf{F}_p$. Represent elements of $\mathbf{F}_p^m$ by vectors $(\lambda_1, \ldots, \lambda_m)$, and for each $(\lambda_1, \ldots, \lambda_m)$ put

$$V_{\lambda_1, \ldots, \lambda_m} = \{v \in V : \nabla_*(D_i)v = \lambda_i v \text{ for } 1 \leq i \leq m\}.$$

By our remark on simultaneous diagonalization, we may write V as the direct sum of the $V_{\lambda_1, \ldots, \lambda_m}$, and moreover $V^\nabla = V_{0, \ldots, 0}$ by definition. On the other hand, as the a_i form a p-basis of $K|k$, we have a direct sum decomposition $K \otimes_k V^\nabla \cong \bigoplus a_1^{\lambda_1} \cdots a_m^{\lambda_m} V^\nabla$. But $a_1^{\lambda_1} \cdots a_m^{\lambda_m} V^\nabla$ is none but $V_{\lambda_1, \ldots, \lambda_m}$, and the theorem follows. $\qquad \square$

Remarks 9.3.7

1. Given a k-vector space W with $V \cong K \otimes_k W$, one may define a connection ∇_W on V by setting $\nabla_W(a \otimes w) := da \otimes w$ and extending k-linearly. Then for all $D \in \text{Der}_k(V)$ the K-endomorphism $\nabla_{W*}(D)$ sends $a \otimes w$ to $D(a) \otimes w$. Using this formula one immediately checks that ∇_W is a flat p-connection. Moreover, in the case $W = V^\nabla$ one checks easily that $\nabla_{V^\nabla} = \nabla$. So we may rephrase the theorem by saying that the functor $W \mapsto (W, \nabla_W)$ induces an equivalence of categories between the category of k-vector spaces and that of K-vector spaces equipped with a flat p-connection, the inverse being given by the functor $V \mapsto V^\nabla$.

2. A direct ancestor of the above descent statement is the following analogue of the Galois correspondence for finite purely inseparable extensions $K|k$ of height 1 due to Jacobson. For a Lie subalgebra $\mathfrak{g} \subset \text{Der}_k(K)$ let $K^{\mathfrak{g}} \subset K$ be the intersection of the kernels of the derivations in $\mathfrak{g}$. Then the map $\mathfrak{g} \mapsto K^{\mathfrak{g}}$ induces a bijection between the Lie subalgebras of $\text{Der}_k(K)$ stable under $D \mapsto D^{[p]}$ and the subextensions of $K|k$, and moreover $[K : K^{\mathfrak{g}}] = p^{\dim K^{\mathfrak{g}}}$. The proof is similar to that of

336 Symbols in positive characteristic

Theorem 9.3.6; see Jacobson [2], vol. II, Theorem 8.43 or Springer [1], Theorem 11.1.15. Gerstenhaber [1] has extended this correspondence to infinite purely inseparable extensions of height 1, by defining an analogue of the Krull topology.

We now come to:

Proof of Theorem 9.3.3 The implication (1) ⇒ (2) follows from the easy calculation $\gamma(a^{-1}da) = a^{-p}a^{p-1}da = a^{-1}da$ for all $a \in K^\times$. As already remarked, the implication (2) ⇒ (3) follows from Proposition 9.3.4, so it remains to see (3) ⇒ (1). This we first prove in the case when $K|k$ is a finite extension. Indeed, applying Theorem 9.3.6 we conclude that K^{∇_ω} is a k-vector space of dimension 1; in particular, it is nonzero. Therefore we find a nonzero y in K with $\nabla_\omega(y) = 0$. But as already remarked at the beginning of the section, $\nabla_\omega(y) = 0$ is equivalent to $\omega = \mathrm{dlog}(y^{-1})$. To treat the general case, write K as a direct limit of subfields K_λ finitely generated over $\mathbf{F}_p$ and set $k_\lambda := k \cap K_\lambda$. Given $\omega \in \Omega^1_{K|k}$, we find some K_λ as above so that ω comes from an element $\omega_\lambda \in \Omega^1_{K_\lambda|k_\lambda}$. If ∇_ω is a flat p-connection on K, then so is ∇_{ω_λ} on K_λ. As K_λ is finitely generated and $K_\lambda^p \subset k_\lambda$, the extension $K_\lambda|k_\lambda$ is finite, and therefore the previous case yields $y \in K_\lambda \subset K$ with $\omega_\lambda = \mathrm{dlog}(y)$ in $\Omega^1_{K_\lambda|k_\lambda}$. But then $\omega = \mathrm{dlog}(y)$ in $\Omega^1_{K|k}$. □

9.4 Decomposition of the de Rham complex

As a preparation for our study of the higher dimensional differential symbol, we now discuss properties of the de Rham complex over fields of characteristic $p > 0$. Let $K|k$ be again a field extension with $K^p \subset k$. Recall from the Appendix that the de Rham complex is a complex of the shape

$$\Omega^\bullet_{K|k} = (K \xrightarrow{d} \Omega^1_{K|k} \xrightarrow{d} \Omega^2_{K|k} \xrightarrow{d} \Omega^3_{K|k} \xrightarrow{d} \dots).$$

Assume moreover that $K|k$ is finite of degree p^r, and choose a p-basis $b_1, \dots, b_r$. According to Proposition A.8.11 of the Appendix, the elements $db_1, \dots db_r$ form a K-basis of the vector space $\Omega^1_{K|k}$, and hence the i-fold exterior products $db_{\lambda_1} \wedge \dots \wedge db_{\lambda_i}$ form a K-basis of $\Omega^i_{K|k}$. In particular, this implies that $\Omega^i_{K|k} = 0$ for $i > r$. According to the Appendix, the differential $d : \Omega^i_{K|k} \to \Omega^{i+1}_{K|k}$ coincides with the universal derivation d for $i = 0$, and satisfies $d(\omega_1 \wedge \omega_2) = d\omega_1 \wedge \omega_2 + (-1)^i \omega_1 \wedge d\omega_2$ for $\omega_1 \in \Omega^i_{K|k}$ and $\omega_2 \in \Omega^j_{K|k}$.

The products $b_1^{\alpha_1} \dots b_r^{\alpha_r}$ with $0 \le \alpha_i \le p - 1$ form a basis of the k-vector space K. This implies $K \cong k(b_1) \otimes_k \dots \otimes_k k(b_r)$, as the extensions $k(b_i)|k$ are linearly disjoint. On the other hand, the elements $b_1^{\alpha_1} \dots b_r^{\alpha_r} db_{\lambda_1} \wedge \dots \wedge db_{\lambda_i}$

form a k-basis of $\Omega^i_{K|k}$, so the above description of differentials and the definition of tensor products of complexes (Chapter 3, Section 3.4) imply:

Proposition 9.4.1 *For $K|k$ and $b_1, \ldots, b_r$ as above, the de Rham complex $\Omega^\bullet_{K|k}$ considered as a complex of k-vector spaces decomposes as a tensor product*

$$\Omega^\bullet_{K|k} \cong \Omega^\bullet_{k(b_1)|k} \otimes_k \cdots \otimes_k \Omega^\bullet_{k(b_r)|k}.$$

This decomposition will be one of our main tools in the study of the de Rham complex. The other one is the following general lemma on tensor products of complexes which is a special case of the Künneth formula.

Lemma 9.4.2 *Let $A^\bullet$ and $B^\bullet$ be complexes of vector spaces over the same field k, concentrated in nonnegative degrees. Then for all $i \geq 0$ the natural maps*

$$\bigoplus_{p+q=i} H^p(A^\bullet) \otimes_k H^q(B^\bullet) \to H^i(A^\bullet \otimes_k B^\bullet)$$

are isomorphisms.

Proof Denote by $Z^\bullet(A)$ (resp. $B^\bullet(A)$) the subcomplexes of $A^\bullet$ obtained by restricting to the subspaces $Z^i(A)$ (resp. $B^i(A)$) of A^i in degree i; note that all differentials in these complexes are zero. Denoting by $B^\bullet(A)[1]$ the shifted complex with $B^i(A)[1] = B^{i+1}(A)$, one has an exact sequence of complexes

$$0 \to Z^\bullet(A) \to A^\bullet \to B^\bullet(A)[1] \to 0.$$

Tensoring with $B^\bullet$ yields

$$0 \to Z^\bullet(A) \otimes_k B^\bullet \to A^\bullet \otimes_k B^\bullet \to B^\bullet(A)[1] \otimes_k B^\bullet \to 0. \qquad (9.11)$$

This sequence is again exact, by the following argument. For each $n \geq 0$ we define the truncated complex $B^\bullet_{\leq n}$ by setting all terms of degree $> n$ in $B^\bullet$ to 0. A straightforward induction on n using the exact sequences $0 \to B^\bullet_{\leq n-1} \to B^\bullet_{\leq n} \to B^n[-n] \to 0$ (where $B^n[-n]$ has a single nonzero term in degree n) then implies that the sequences

$$0 \to Z^\bullet(A) \otimes_k B^\bullet_{\leq n} \to A^\bullet \otimes_k B^\bullet_{\leq n} \to B^\bullet(A)[1] \otimes_k B^\bullet_{\leq n} \to 0$$

are exact for all n, whence the exactness of (9.11). Now part of the long exact sequence associated with (9.11) reads

$$H^i(B^\bullet(A) \otimes_k B^\bullet) \to H^i(Z^\bullet(A) \otimes_k B^\bullet) \to H^i(A^\bullet \otimes_k B^\bullet) \to$$
$$\to H^{i+1}(B^\bullet(A) \otimes_k B^\bullet) \to H^{i+1}(Z^\bullet(A) \otimes_k B^\bullet). \qquad (9.12)$$

As $B^\bullet(A) \to Z^\bullet(A)$ is an injective map of complexes with trivial differentials, so is the map $B^\bullet(A) \otimes_k B^\bullet \to Z^\bullet(A) \otimes_k B^\bullet$. Thus the last map in (9.12) is injective, whence the exactness of the sequence

$$H^i(B^\bullet(A) \otimes_k B^\bullet) \to H^i(Z^\bullet(A) \otimes_k B^\bullet) \to H^i(A^\bullet \otimes_k B^\bullet) \to 0.$$

Again using the triviality of differentials in the complexes $Z^\bullet(A)$ and $B^\bullet(A)$, we may identify the first map here with the map

$$\bigoplus_{p+q=i} B^p(A^\bullet) \otimes_k H^q(B^\bullet) \to \bigoplus_{p+q=i} Z^p(A^\bullet) \otimes_k H^q(B^\bullet),$$

whence the lemma.   □

We now come to applications of the above observations. The first one is another basic result of Cartier concerning the operator γ, usually called the *Cartier isomorphism* in the literature. To be able to state it, we first extend the definition of γ to higher differential forms by setting

$$\gamma(\omega_1 \wedge \cdots \wedge \omega_i) := \gamma(\omega_1) \wedge \cdots \wedge \gamma(\omega_i),$$

where on the right-hand side the maps γ are defined as in Proposition 9.2.1 (and extended to the relative case as in the previous section). For $i = 0$ we put $\gamma(a) := a^p$. In this way we obtain K-linear maps

$$\gamma : \Omega^i_{K|k} \to {}^pZ^i_{K|k}/{}^pB^i_{K|k}$$

for all $i \geq 0$. Note that the differentials of the complex ${}^p\Omega^\bullet_K$ are K-linear by a similar argument as in degree 0, so ${}^pB^i_{K|k} \subset {}^pZ^i_{K|k}$ is a K-subspace.

Theorem 9.4.3 (Cartier) *For all extensions $K|k$ as above and all $i \geq 0$ the map γ is an isomorphism.*

Proof Assume first $K|k$ is a finite extension of degree p^r, and fix a p-basis $b_1, \ldots b_r$ of $K|k$ as above. Using the above explicit description of $\Omega^\bullet_{K|k}$ and the construction of γ we see that it is enough to check the following:

- $H^0(\Omega^\bullet_{K|k})$ is the 1-dimensional k-vector space generated by 1;
- $H^1(\Omega^\bullet_{K|k})$ is the r-dimensional k-vector space generated by the $b_i^{p-1}db_i$;
- $H^i(\Omega^\bullet_{K|k}) \cong \Lambda^i H^1(\Omega^\bullet_{K|k})$ for $i > 1$.

Now using Proposition 9.4.1, Lemma 9.4.2 and induction on r we see that it is enough to check these statements for $r = 1$, i.e. $K = k(b)$ with $b^p \in k$. In this case, K and $\Omega^1_{K|k}$ are p-dimensional k-vector spaces with bases $\{b^i : 0 \leq i \leq p-1\}$ and $\{b^i db : 0 \leq i \leq p-1\}$, respectively. Moreover, one has $\Omega^i_{K|k} = 0$ for $i > 1$, and the differential $d : K \to \Omega^1_{K|k}$ sends b^i to $ib^{i-1}db$. It follows that $H^0(\Omega^\bullet_{K|k})$ and $H^1(\Omega^\bullet_{K|k})$ are 1-dimensional over k, generated by 1 and $b^{p-1}db$, respectively, and $\Lambda^i H^1(\Omega^\bullet_{K|k}) = 0$ for $i > 0$, which shows that the three required properties are satisfied.

For the general case, write K as a direct limit of subfields K_λ finitely generated over $\mathbf{F}_p$. Each K_λ is a finite extension of K^p_λ and therefore also of $k \cap K_\lambda$, so that the case just discussed applies, and the theorem follows by passing to the limit. □

The theorem enables us to define for all $i \geq 0$ the *Cartier operator* $C : {}^pZ^i_{K|k}/{}^pB^i_{K|k} \to \Omega^i_{K|k}$ as the inverse of γ. It is also customary to regard it as a linear map $C : {}^pZ^i_{K|k} \to \Omega^i_{K|k}$ defined on closed differential forms. With these notations, Theorem 9.4.3 and Theorem 9.2.2 respectively yield the following characterization of exact and logarithmic differential forms, which is the form of Cartier's results often found in the literature.

Corollary 9.4.4

1. *An i-form* $\omega \in \Omega^i_{K|k}$ *is exact, i.e. lies in* $B^i_{K|k}$, *if and only if* $d\omega = C(\omega) = 0$.
2. *A 1-form* $\omega \in \Omega^1_{K|k}$ *is logarithmic, i.e. lies in the image* $\nu(1)_{K|k}$ *of the dlog map, if and only if* $d\omega = 0$ *and* $C(\omega) = \omega$.

Remarks 9.4.5

1. The second statement of the corollary has an analogue for higher differentials as follows. Assume $k = K^p$ and $\omega \in \Omega^i_K$ is an i-form satisfying $d\omega = 0$ and $C(\omega) = \omega$. Then there is a finite separable extension $L|K$ such that the image of ω in Ω^i_L is a sum of elements of the form $dx_1/x_1 \wedge \cdots \wedge dx_i/x_i$ with $x_1, \ldots, x_i \in L^\times$. This is an unpublished result of Bloch; see Illusie [1], Part I, Theorem 2.4.2 for an exposition. In Theorem 9.6.1 below we shall prove the difficult fact due to Kato that it is in fact possible to take $L = K$ in the above statement.
2. The definition of the inverse Cartier operator γ works over an arbitrary integral domain A of characteristic $p > 0$ in exactly the same way. One can then prove that if A is a finitely generated *smooth* algebra over a perfect field k of characteristic $p > 0$, then the map γ induces an isomorphism of A-modules $\Omega^i_A \xrightarrow{\sim} {}^pH^i(\Omega^\bullet_A)$ for all $i \geq 0$. The idea (due to Grothendieck) is to treat first the case of the polynomial ring $k[x_1, \ldots, x_r]$ which is similar to the case of fields treated above, the module $\Omega^1_{k[x_1,\ldots,x_r]|k}$ being a free $k[x_1, \ldots, x_r]$-module over the dx_i and the x_i forming a 'p-basis' of

$k[x_1, \ldots x_r]$ over $k[x_1^p, \ldots x_r^p]$. The general case then follows from the well-known fact of algebraic geometry (Mumford [1], III.6, Theorem 1) according to which on a smooth k-variety each point has an open neighbourhood equipped with an étale morphism to some affine space over k, together with the fact an étale morphism induces an isomorphism on modules of differentials. See Katz [1] or Illusie ([1],[2]) for details and a globalization for smooth varieties.

Another consequence of the tensor product decomposition of the de Rham complex is a direct sum decomposition we shall use later. We consider again an extension $K|k$ and a p-basis $b_1, \ldots, b_r$ as at the beginning of this section, and fix a multiindex $\alpha := (\alpha_1, \ldots \alpha_r)$ with $0 \le \alpha_i \le p - 1$. For each $i \ge 0$ consider the k-subspace $\Omega^i_{K|k}(\alpha) \subset \Omega^i_{K|k}$ generated by the elements $b_1^{\alpha_1} \ldots b_r^{\alpha_r}(db_{\lambda_1}/b_{\lambda_1}) \wedge \cdots \wedge (db_{\lambda_i}/b_{\lambda_i})$ for $1 \le \lambda_1 \le \cdots \le \lambda_i \le r$; for $i = 0$ this just means the 1-dimensional subspace generated by $b_1^{\alpha_1} \ldots b_r^{\alpha_r}$. Since the $(db_{\lambda_1}/b_{\lambda_1}) \wedge \cdots \wedge (db_{\lambda_i}/b_{\lambda_i})$ form a K-basis of $\Omega^i_{K|k}$ just like the $db_{\lambda_1} \wedge \cdots \wedge db_{\lambda_i}$, the k-vector space $\Omega^i_{K|k}$ decomposes as a direct sum of the subspaces $\Omega^i_{K|k}(\alpha)$ for all possible α. The case $\alpha = (0, \ldots, 0)$ is particularly important; we shall abbreviate it by $\alpha = 0$.

Proposition 9.4.6   *Let $K|k$ and α be as above.*

1. *The differentials $d : \Omega^i_{K|k} \to \Omega^{i+1}_{K|k}$ map each $\Omega^i_{K|k}(\alpha)$ to $\Omega^{i+1}_{K|k}(\alpha)$, giving rise to subcomplexes $\Omega^\bullet_{K|k}(\alpha)$ of $\Omega^\bullet_{K|k}$.*
2. *One has a direct sum decomposition $\Omega^\bullet_{K|k} \cong \bigoplus_\alpha \Omega^\bullet_{K|k}(\alpha)$.*
3. *The complex $\Omega^\bullet_{K|k}(0)$ has zero differentials, and the complexes $\Omega^\bullet_{K|k}(\alpha)$ are acyclic for $\alpha \ne 0$.*

Proof   Using Proposition 9.4.1 (and Lemma 9.4.2 for the third statement) we reduce all three statements to the case $r = 1$. In this case the first two statements are immediate, and the third one follows by the same calculation as at the end of the proof of Theorem 9.4.3.   □

9.5 The Bloch–Gabber–Kato theorem: statement and reductions

As a field K of characteristic p has p-cohomological dimension ≤ 1, the Galois symbol $h^n_{K,p}$ is a trivial invariant for $n > 1$. However, the discussion of Section 9.2 suggests the investigation of another invariant, the *differential symbol*. To define it, introduce for all $n \ge 0$ the notation

$$\nu(n)_K := \ker(\gamma - \mathrm{id} : \Omega_K^n \to \Omega_K^n/B_K^n).$$

We shall omit the subscript K if clear from the context. Note that $\nu(0) \cong \mathbf{F}_p$ as we defined γ to be p-th power map for 0-forms, and for $n = 1$ we get back the $\nu(1)_K$ of Section 9.2. Since γ takes its values in Z_K^n/B_K^n, it follows that $\nu(n)_K \subset Z_K^n$. Moreover, since γ is defined for $n > 1$ as the n-th exterior power map of the operator γ on Ω_k^1, one has a natural map $\nu(1)^{\otimes n} \to \nu(n)$. This allows one to define

$$\mathrm{dlog} : (K^\times)^{\otimes n} \to \nu(n)$$

for $n > 1$ by taking the n-th tensor power of the map $\mathrm{dlog} : K^\times \to \nu(1)_K$ and then composing with the map above.

Lemma 9.5.1 *The map* dlog *factors through the quotient* $K_n^M(K)/pK_n^M(K)$, *and thus defines a map*

$$\psi_K^n : K_n^M(K)/pK_n^M(K) \to \nu(n)_K$$

sending $\{y_1, y_2, \ldots, y_n\}$ *to* $(dy_1/y_1) \wedge \cdots \wedge (dy_n/y_n)$ *for all* $y_1, \ldots, y_n \in K^\times$.

Proof As $p\nu(n) = 0$, the only point to be checked is the Steinberg relation. It holds because for $y \neq 0, 1$

$$\mathrm{dlog}(y) \wedge \mathrm{dlog}(1 - y) = \frac{1}{y(1-y)} dy \wedge d(1-y) = -\frac{1}{y(1-y)} dy \wedge dy = 0.$$

$\square$

We can now state the following basic theorem, whose surjectivity statement is due to Kato, and whose injectivity statement was proven independently by Bloch–Kato and Gabber (unpublished).

Theorem 9.5.2 (Bloch–Gabber–Kato) *Let* K *be a field of characteristic* $p > 0$. *For all integers* $n \geq 0$, *the differential symbol*

$$\psi_K^n : K_n^M(K)/pK_n^M(K) \to \nu(n)_K$$

is an isomorphism.

Remarks 9.5.3

1. The surjectivity statement of the theorem generalizes Theorem 9.2.2 to higher differential forms.

2. Using the theorem above, Bloch and Kato proved the bijectivity of the Galois symbol $h_{k,p}^n$ for n arbitrary and k a complete discrete valuation field of characteristic 0 with residue field of characteristic p (Bloch–Kato [1], §5; see also Colliot-Thélène [4] for a detailed survey). This statement is now a special case of the main result of Voevodsky [2].

The proof of the theorem will be given in the next two sections. We now discuss an extension that describes the groups $K_n^M(K)/p^r K_n^M(K)$ for $r > 1$ by means of a generalized differential symbol. The target of this symbol is constructed via Illusie's theory of the de Rham–Witt complex which is briefly summarized at the end of the Appendix. We thus take up the notation introduced there.

Assume K is a field of characteristic $p > 0$ as before. For each $r > 0$ there are maps

$$\mathrm{dlog}:\ K^\times \to W_r\Omega_K^1 \tag{9.13}$$

sending $x \in K^\times$ to $[x]^{-1}d[x]$, where $[x]$ is the class $(x, 0, \ldots, 0) \in W_r(K)$. Using the product structure of the complex $W_r\Omega_K^\bullet$, we may consider for each $n > 0$ the submodule $W_r\Omega_{K,\log}^n \subset W_r\Omega_K^n$ formed by elements whose image in $W_r\Omega_L^n$ for some finite separable extension $L|K$ is a sum of elements of the shape $\mathrm{dlog}(x_1) \cdots \mathrm{dlog}(x_n)$ with $x_1, \ldots, x_n \in L^\times$. By Remark 9.4.5 (1) we have $W_1\Omega_{K,\log}^n = \nu(n)_K$.

The multiplicative structure of the complexes $W_i\Omega_K^\bullet$ implies that for each $n, i > 0$ we have a map

$$\mathrm{dlog}: (K^\times)^{\otimes n} \to W_i\Omega_{K,\log}^n$$

given by the n-th tensor power of the map (9.13) with $i = r$. Since the product structure on $W_i\Omega_K^\bullet$ is anticommutative, the same argument as in the proof of Lemma 9.5.1 shows that there exist induced maps

$$\psi_K^{n,i}:\ K_n^M(K)/p^i K_n^M(K) \to W_i\Omega_{K,\log}^n$$

for all $n, i > 0$, the case $i = 1$ being the map ψ_K^n in Theorem 9.5.2. The said theorem now implies:

Corollary 9.5.4 *The maps $\psi_K^{n,i}$ are isomorphisms for all $n, i > 0$.*

To derive the corollary we need the following facts on logarithmic de Rham–Witt groups.

Facts 9.5.5 According to Illusie [1], Part I, Proposition 3.4, multiplication by p^i on $W_{r+i}\Omega_K^n$ factors through the quotient $W_r\Omega_K^n$ for all $i > 0$, inducing an injective map $W_r\Omega_K^n \to W_{r+i}\Omega_K^n$. Its restriction to logarithmic differentials sits in an exact sequence

$$0 \to W_r\Omega_{K,\log}^n \to W_{r+i}\Omega_{K,\log}^n \xrightarrow{R^r} W_i\Omega_{K,\log}^n \to 0. \qquad (9.14)$$

Here surjectivity of the map R^r follows from the definitions, and the sequence is a complex as $W_i\Omega_K^n$ is annihilated by p^i. For a proof of exactness at the middle term, see Colliot-Thélène–Sansuc–Soulé [1], Lemma 3 on page 779, itself based on Illusie [1], formula 5.7.4. Note that in these references the above facts are proven for smooth $\mathbf{F}_p$-algebras. However, we may deduce the corresponding statements for K by writing it as a directed union of smooth $\mathbf{F}_p$-subalgebras A using Proposition A.8.8 of the Appendix, and then passing to the direct limit over the $W_r\Omega_{A,\log}^n$.

Proof of Corollary 9.5.4 We use induction on i, the case $i = 1$ being Theorem 9.5.2. By exact sequence (9.14) for $r = 1$ and the construction of the maps $\psi_K^{n,i}$ we have a commutative diagram with exact rows

$$
\begin{array}{ccccccc}
K_n^M(K)/p & \longrightarrow & K_n^M(K)/p^{i+1} & \longrightarrow & K_n^M(K)/p^i & \longrightarrow & 0 \\
\downarrow{\psi_K^n} & & \downarrow{\psi_K^{n,i+1}} & & \downarrow{\psi_K^{n,i}} & & \\
0 \longrightarrow & \Omega_{K,\log}^n & \longrightarrow & W_{i+1}\Omega_{K,\log}^n & \longrightarrow & W_i\Omega_{K,\log}^n & \longrightarrow 0
\end{array}
$$
$$(9.15)$$

recalling that $W_1\Omega_{K,\log}^n \cong \Omega_{K,\log}^n$. According to the snake lemma, if ψ_K^n and $\psi_K^{n,i}$ are both isomorphisms, then so is $\psi_K^{n,i+1}$. This completes the inductive step. □

The following corollary will serve in Section 9.8 below, where a much stronger result will be established.

Corollary 9.5.6 *For each $n > 0$ the p-primary torsion subgroup of $K_n^M(K)$ is divisible.*

Proof Since the maps $\psi_K^{n,i}$ are isomorphisms for all i, diagram (9.15) in the above proof shows that the maps $K_n^M(K)/p \to K_n^M(K)/p^{i+1}$ are injective for all i. Continuing the Tor exact sequence of the upper row of (9.15) on the left, we see that consequently the maps $\mathrm{Tor}_1(\mathbf{Z}/p^{i+1}, K_n^M(K)) \to \mathrm{Tor}_1(\mathbf{Z}/p^i, K_n^M(K))$ are surjective (see Remark 3.1.10 (2)). But this is equivalent to saying that the multiplication-by-p maps $_{p^{i+1}}K_n^M(K) \to {}_{p^i}K_n^M(K)$ are surjective. □

To conclude this section, we perform some preliminary constructions and reductions to be used in the proof of Theorem 9.5.2.

First we define trace maps $\operatorname{tr}_{K'|K} : \Omega_{K'}^n \to \Omega_K^n$ for finite separable extensions $K'|K$ (they also exist for inseparable extensions, but the construction is more complicated and will not be used). The construction is based on the fact that for $K'|K$ separable one has $\Omega_{K'}^n \cong K' \otimes_K \Omega_K^n$ by Proposition A.8.9 of the Appendix. Hence the field trace $\operatorname{tr} : K' \to K$ induces a trace map $\operatorname{tr} = \operatorname{tr} \otimes \operatorname{id} : K' \otimes_K \Omega_K^n \to \Omega_K^n$ which is the map we were looking for. The composite $\operatorname{tr} \circ \operatorname{id} : \Omega_K^n \to \Omega_{K'}^n \to \Omega_K^n$ is multiplication by $[K' : K]$, and one has the projection formula $\operatorname{tr}(\omega_1 \wedge (\omega_2)_{K'}) = \operatorname{tr}(\omega_1) \wedge \omega_2$ for $\omega_1 \in \Omega_{K'}^n$, $\omega_2 \in \Omega_K^m$. Moreover, the trace map commutes with differentials and the Cartier operator by construction, so it restricts to a trace map $\operatorname{tr} : \nu(n)_{K'} \to \nu(n)_K$.

Lemma 9.5.7 *Given a finite separable extension $K'|K$, the diagram*

$$
\begin{array}{ccc}
K_n^M(K')/pK_n^M(K') & \xrightarrow{\ \psi_{K'}^n\ } & \nu(n)_{K'} \\[2mm]
{\scriptstyle N_{K'|K}} \downarrow & & \downarrow {\scriptstyle \operatorname{tr}} \\[2mm]
K_n^M(K)/pK_n^m(K) & \xrightarrow{\ \psi_K^n\ } & \nu(n)_K
\end{array}
$$

commutes.

Proof Let us consider first the case $n = 1$. Let $y \in K'^\times$, and let K_s be a separable closure of K. As the maps $K^\times \to K_s^\times$ and $\Omega_K^1 \to \Omega_{K_s}^1$ are injective, we may reason in K_s. There the element $N_{K'|K}(y)$ decomposes as a product $N_{K'|K}(y) = \prod \sigma(y)$, where the product is taken over the K-embeddings of K' into K_s. As these embeddings commute with the dlog map, it follows that inside $\Omega_{K_s}^1$ one has $\operatorname{dlog}(N_{K'|K}(y)) = \sum \sigma(\operatorname{dlog}(y))$, which is none but $\operatorname{tr}(\operatorname{dlog}(y))$. In the case $n > 1$ we consider a maximal prime to p extension $K^{(p)}|K$ (which is of course separable). The tensor product $K^{(p)} \otimes_K K'$ is a finite direct product of extensions $K_i|K^{(p)}$, and the induced norm map on K_n^M is the sum of the norm maps for these extensions by Lemma 7.3.6. Similarly, the trace map on differentials becomes the sum of the traces for the $K_i|K^{(p)}$. Since moreover the map $\Omega_K^n \to \Omega_{K^{(p)}}^n$ is injective, we are reduced to the case $K = K^{(p)}$. But then according to the Bass–Tate lemma (Corollary 7.2.10) it is enough to treat symbols of the form $\{a_1, \ldots, a_n\}$ with $a_1 \in K'$ and $a_i \in K$ for $i > 1$, and the statement reduces to the case $n = 1$ by the projection formula. $\qquad \square$

The trace construction implies the following reduction statement.

Proposition 9.5.8 *Let $K'|K$ be a finite extension of degree prime to p. Then the natural map $\ker(\psi_K^n) \to \ker(\psi_{K'}^n)$ is injective, and the trace map induces a surjection* $\operatorname{coker}(\psi_{K'}^n) \to \operatorname{coker}(\psi_K^n)$.
In particular, if $\psi_{K'}^n$ is injective (resp. surjective), then so is ψ_K^n.

Proof The statement about the kernel follows from the injectivity of the map $K_n^M(K)/pK_n^M(K) \to K_n^M(K')/pK_n^M(K')$ already noted during the proof of Proposition 7.5.8. That about the cokernel comes from the previous lemma, noting that the trace map is surjective, as so is the composite $[K' : K] = \mathrm{tr} \circ \mathrm{id}$, the degree $[K' : K]$ being prime to p and $\nu(n)_K$ being p-torsion.

$\square$

On the other hand, the following reduction statement is immediate by writing K as a union of its finitely generated subfields.

Lemma 9.5.9 *If the differential symbol ψ_F^n is injective (resp. surjective) for all subfields $F \subset K$ which are finitely generated over $\mathbf{F}_p$, then so is ψ_K^n.*

9.6 Surjectivity of the differential symbol

In this section we prove the surjectivity of the differential symbol. Recall the statement:

Theorem 9.6.1 (Kato) *For all $n > 0$, the map* $\mathrm{dlog} : (K^\times)^{\otimes n} \to \nu(n)$ *is surjective, i.e. the group $\nu(n)$ is additively generated by the elements of the shape $dx_1/x_1 \wedge \cdots \wedge dx_n/x_n$.*

Before embarking on the proof, we establish an innocent-looking result of linear algebra that will be needed later.

Proposition 9.6.2 *Let E be a field of characteristic p, and let $F = E(b)$ be a purely inseparable extension of degree p. Consider an E-linear map $g : F \to E$. Then up to replacing E by a finite extension $E'|E$ of degree prime to p, F by FE' and g by the induced map, there exists $c \in F^\times$ satisfying $g(c^i) = 0$ for $1 \leq i \leq p - 1$.*

We begin the proof of the proposition with:

Lemma 9.6.3 *Let $F|E$ be a field extension as in Lemma 9.6.2. Then up to replacing E by E' and F by FE' as above we may write each element of $\Omega^1_{F|E} \setminus dF$ in the form $u\,dy/y$ with suitable $u \in E^\times$ and $y \in F^\times$.*

Proof One may write each element $\omega \in \Omega^1_{F|E}$ as $\omega = a(db/b)$ for suitable $a, b \in E^\times$. As the E-vector space $\Omega^1_{F|E}/dF$ is 1-dimensional (see the proof of Theorem 9.4.3), the condition $\omega \notin dF$ implies that its image spans $\Omega^1_{F|E}/dF$. In particular, there exists $\rho \in E^\times$ with $a^p(db/b) = \rho a(db/b)$ in $\Omega^1_{F|E}/dF$. Up to replacing E by the prime-to-p extension $E(u)$ for some u with $\rho = u^{p-1}$, we may assume that such a u exists in $E^\times$, so division by u^p yields $(u^{-1}a)^p(db/b) = u^{-1}a(db/b)$ in $\Omega^1_{F|E}/dF$. In other words, $u^{-1}\omega = u^{-1}a(db/b)$ lies in the kernel of $\gamma - 1$, and therefore $u^{-1}\omega = dy/y$ according to Theorem 9.3.3. The lemma follows. $\quad\square$

Proof of Proposition 9.6.2 Take an isomorphism ϕ between the 1-dimensional F-vector spaces F and $\Omega^1_{F|E}$. The E-subspaces $\ker(g) \subset F$ and $dF \subset \Omega^1_{F|E}$ are both of codimension 1, hence up to modifying ϕ by multiplication with an element in $F^\times$ we may assume that $\phi(\ker(g)) = dF$ (this is immediately seen by looking at the dual spaces). Let a be an element of $F \setminus \ker(g)$. Then $\phi(a)$ generates $\Omega^1_{F|E}$ as an F-vector space, and its mod dF class generates the E-vector space $\Omega^1_{F|E}/dF$. By the lemma above we have $\phi(a) = u\,dy/y$ for suitable $y \in F^\times$ and $u \in E^\times$; up to modifying a we may assume $u = 1$. Now by the above choice of ϕ we have the equivalences $g(x) = 0 \Leftrightarrow xa \in \ker(g) \Leftrightarrow x\,dy/y \in dF$ for all $x \in F$. On the other hand, we have seen during the proof of Theorem 9.4.3 that the elements $y^i dy$ for $0 \le i \le p-2$ span dF. Thus $c = y$ is a good choice. $\quad\square$

Now return to the field K of Theorem 9.6.1, and let $k \subset K$ be a subfield containing K^p. Assume moreover that $K|k$ is a finite extension of degree p^r, and take a p-basis $\{b_1, \ldots, b_r\}$ of $K|k$. Then for $1 \le i \le r$ the db_i/b_i form a K-basis of $\Omega^1_{K|k}$. For $n > 1$ we shall use the explicit basis of $\Omega^n_{K|k}$ given as follows. Denote by S_n the set of strictly increasing functions from $\{1, \ldots, n\}$ to $\{1, \ldots, r\}$, and for all $s \in S_n$ set

$$\omega_s = db_{s(1)}/b_{s(1)} \wedge \cdots \wedge db_{s(n)}/b_{s(n)}.$$

Then the ω_s for $s \in S_n$ form a K-basis of $\Omega^n_{K|k}$. Equip the set S_n with the lexicographic ordering, i.e. for $s, t \in S_n$ set $t < s$ if there exists m in $\{1, \ldots, n\}$ so that $t(i) = s(i)$ for all $i < m$ and $t(m) < s(m)$. Denote by $\Omega^n_{K|k,<s}$ the K-subspace of $\Omega^n_{K|k}$ generated by the ω_t with $t < s$, and put

$B^n_{K|k,<s} := d(\Omega^{n-1}_{K|k,<s})$. We shall adopt analogous notations for subfields of K.

We may now state the key proposition which will be also useful for proving the injectivity of the differential symbol.

Proposition 9.6.4 *Let $K|k$ be a finite extension of degree p^r as above. Fix $s \in S_n$ and (with the notations above) assume $a \in K$ satisfies*

$$(a^p - a)\omega_s \in \Omega^n_{K|k,<s} + B^n_{K|k}. \tag{9.16}$$

Then up to replacing K by a finite extension of degree prime to p we have

$$a\omega_s \in \Omega^n_{K|k,<s} + \mathrm{Im}\,(\mathrm{dlog}).$$

Before proving the proposition, we derive Theorem 9.6.1.

Proof of Theorem 9.6.1 By Lemma 9.5.9 we may assume K is finitely generated over $\mathbf{F}_p$, so that setting $k = K^p$ the extension $K|k$ is finite. Assume $\omega \in \nu(n)(K)$ is not in the image of the dlog map. Since Ω^n_K is the direct sum of the $\Omega^n_{K,s}$, we may then find a smallest s (with respect to the lexicographic order) so that $\omega = \omega' + \eta$, with $\omega' \in \Omega^n_{K,<s+1}$ and $\eta \in \mathrm{Im}\,(\mathrm{dlog})$. Write $\omega' = a\omega_s + \omega''$ with suitable $a \in K$ and $\omega'' \in \Omega^n_{K,<s}$. As by construction γ maps $\Omega^n_{K,<s}$ to its image in Ω^n_K/B^n_K, applying $\gamma - 1$ yields $(a^p - a)\omega_s \in \Omega^n_{K,<s} + B^n_K$, noting that $(\gamma-1)\,\eta$ and $(\gamma-1)\,\omega$ both lie in B^n_K. By Proposition 9.6.4, after passing to a finite prime to p extension $K'|K$ we have $a\omega_s \in \Omega^n_{K',<s} + \mathrm{Im}\,(\mathrm{dlog})$, so that $\omega \in \Omega^n_{K',<s} + \mathrm{Im}\,(\mathrm{dlog})$ as well. Since a p-basis of $K|k$ provides a p-basis of $K'|K'^p$ as well, by taking traces and using the formula $\mathrm{tr} \circ \mathrm{id} = [K' : K]$ in a by now familiar way we see that the minimal s for K' cannot be greater than for K, and the above argument shows that it is actually strictly smaller. Thus after taking finitely many finite prime to p extensions we arrive at an extension where the dlog map is surjective, and then the theorem follows from Proposition 9.5.8. □

We now come to the proof of the proposition, which is a fairly long calculation. The exposition is largely based on the notes of Colliot-Thélène [4].

Proof of Proposition 9.6.4 With the notations adopted before the proposition define subfields k_0, k_1 and k_2 of K by

$$k_0 := k(b_1, b_2, ..., b_{s(1)-1}), \quad k_1 := k_0(b_{s(1)}), \text{ and}$$
$$k_2 := k_0(b_{s(1)}, b_{s(1)+1}, \ldots, b_{s(n)}).$$

We have $[k_2 : k_0] = p^N$, where N stands for the number of integers j such that $s(1) \le j \le s(n)$.

We first show that under the assumption of the proposition the element a belongs to k_2. To see this, denote by $m(1) < \cdots < m(N-n)$ those integers in the interval $[s(1), s(n)]$ which are *not* of the shape $s(j)$, and introduce the elements

$$\omega_m := db_{m(1)}/b_{m(1)} \wedge \cdots \wedge db_{m(N-n)}/b_{m(N-n)} \in \Omega_K^{N-n} \text{ and}$$

$$\omega_{\max} := \omega_m \wedge \omega_s \in \Omega_K^N$$

(for $n = 1$ write simply $\omega_m = 1$ and $\omega_{\max} = \omega_s$). Applying d to the equation (9.16) yields $da \wedge \omega_s \in B_{K,<s}^{n+1}$. But the set $\omega_m \wedge B_{K,<s}^{n+1} \subset \Omega_K^{N+1}$ maps to 0 in $\Omega_{K|k_0}^N$, because the components of an element in $B_{K,<s}^{n+1}$ either are of the form $db_j/b_j \wedge w'$ with $j < s(1)$ and hence have trivial image in $\Omega_{K|k_0}^{n+1}$, or contain some $db_{m(i)}/b_{m(i)}$. In particular, $da \wedge \omega_{\max} = 0$ in $\Omega_{K|k_0}^{N+1}$. Now if a were not in k_2, then $b_{s(1)}, \ldots, b_{s(n)}$ and a would form part of a p-basis of $K|k_0$, which would contradict $da \wedge \omega_{\max} = 0$.

So we have $a \in k_2$, and hence by the choice of k_2 we may consider $a\omega_s$ as an element of $\Omega_{k_2|k_0}^n$. We may thus rewrite our assumption (9.16) as

$$(a^p - a)\omega_s = \omega + d\omega_1 \quad \text{with} \quad \omega \in \Omega_{k_2|k_0,<s}^n, \ \omega_1 \in \Omega_{k_2|k_0}^{n-1}. \tag{9.17}$$

The next step is to replace $b_{s(1)}$ by another generator c of the extension $k_1|k_0$ which has a useful additional technical property. For this we first take the wedge product of both sides by ω_m, whence using $\omega \wedge \omega_m = 0$ in $\Omega_{k_2|k_0}^N$ (same argument as in the previous paragraph) we obtain

$$(a^p - a)\omega_{\max} = d\omega_1 \wedge \omega_m = d(\omega_1 \wedge \omega_m) \tag{9.18}$$

(recall that $d\omega_m = 0$ because ω_m is logarithmic), so $(a^p - a)\omega_{\max} \in B_{k_2|k_0}^N$.

Consider now the k_0-linear map $k_1 \to \Omega_{k_2|k_0}^N / B_{k_2|k_0}^N$ sending $x \in k_1$ to the class of $xa\omega_{\max}$. As $[k_2 : k_0] = p^N$, the space $\Omega_{k_2|k_0}^N / B_{k_2|k_0}^N$ is 1-dimensional over k_0 and moreover generated by the class of $\omega_{\max}$ (recall again the proof of Theorem 9.4.3). Thus by applying Proposition 9.6.2 to $E = k_0$, $F = k_1$ and the above k_0-linear map we find $c \in k_1^\times$ with

$$c^i a\omega_{\max} \in B_{k_2|k_0}^N \quad \text{for all} \quad 1 \le i \le p-1 \tag{9.19}$$

up to replacing k_0 (and hence ultimately K) by a finite prime to p extension. Moreover, the c we found does not lie in k_0. Indeed, if it did, $c\,a\,\omega_{\max} \in B_{k_2|k_0}^N$ would imply $a\omega_{\max} \in B_{k_2|k_0}^N$. Taking (9.18) into account, we would then get $a^p\omega_{\max} \in B_{k_2|k_0}^N$ and finally $\omega_{\max} \in B_{k_2|k_0}^N$, which is impossible.

Thus the above c generates the extension $k_1|k_0$. Henceforth we use c together with the set $\{b_i : i > s(1)\}$ as a p-basis of $k_2|k_0$. Using this p-basis we may rewrite $a\omega_s$ in the following way. For $n = 1$ we have $a\omega_s = a'dc/c$

for suitable $a' \in k_1$. For $n > 1$ we define $s' \in S_{n-1}$ by setting $s'(i) = s(i+1)$ for $1 \le i \le n - 1$. Then for suitable $a' \in k_2$ we have

$$a\omega_s = a'(dc/c) \wedge \omega_{s'} \in \Omega^n_{k_2|k_0}. \tag{9.20}$$

In the case $n = 1$ we shall show that $a' \in \mathbf{F}_p$, which will conclude the proof. For $n > 1$ our goal is to show that $\omega_{s'}$ satisfies

$$(a'^p - a')\omega_{s'} \in \Omega^{n-1}_{K|k, <s'} + B^{n-1}_{K|k}. \tag{9.21}$$

Once this is established, we can apply induction on n to find

$$a'\omega_{s'} = v' + x' \tag{9.22}$$

for suitable $v' \in \Omega^{n-1}_{K|k, <s'}$ and $x' \in \mathrm{Im}\,(\mathrm{dlog})$. Note that $dc/c \wedge v' \in \Omega^n_{K|k, <s}$ because c and $b_{s(1)}$ differ only by a constant in $k_1 \subset K$. On the other hand, the kernel of the projection $\Omega^n_{K|k} \to \Omega^n_{K|k_0}$ lies in $\Omega^n_{K|k, <s}$ by definition of k_0, so that (9.20) may be rewritten as

$$a\omega_s = a'(dc/c) \wedge \omega_{s'} + \omega_1 \in \Omega^n_{K|k} \quad \text{with} \quad \omega_1 \in \Omega^n_{K|k, <s}.$$

In view of this formula the proposition will follow from (9.22) after wedge product with dc/c.

In the direction of (9.21) we first investigate what additional property the special choice of c implies for a'. For this, write k_2 as a direct sum $k_2 = k_1 \oplus V$, where $V \subset k_2$ is the natural complement of k_1 generated by nontrivial products of the basis elements in $\{b_i : i > s(1)\}$. Write $a' = \sum_{j=0}^{p-1} \alpha_j c^j + a'_1$ with $a'_1 \in V$, $\alpha_i \in k_0$. We contend that here $\alpha_j = 0$ for $j > 0$. Indeed, if $\alpha_j \neq 0$ for some $j > 0$, consider the element

$$c^{p-j} a'(dc/c) \wedge \omega_{s'} \wedge \omega_m \in \Omega^N_{k_2|k_0} \tag{9.23}$$

and apply the decomposition of Proposition 9.4.6 (2) to the extension $k_2|k_0$ and the p-basis $\{c, b_i : i > s(1)\}$. By our assumption on α_j the element (9.23) has a nontrivial component in $\Omega^N_{k_2|k_0}(0)$. On the other hand, (9.19) with $i = p - j$ together with (9.20) imply that the element (9.23) lies in $B^N_{k_2|k_0}$. This however contradicts the first part of Proposition 9.4.6 (3) according to which $\Omega^N_{k_2|k_0}(0)$ has trivial differentials. We have thus proven that

$$a' = \alpha_0 + a'_1 \quad \text{with} \quad \alpha_0 \in k_0, a'_1 \in V. \tag{9.24}$$

This being said, we apply $\gamma - 1 : \Omega^n_{k_2|k_0} \to \Omega^n_{k_2|k_0}/B^n_{k_2|k_0}$ to the equation (9.20) and obtain

$$(a^p - a)\omega_s = (a'^p - a')(dc/c) \wedge \omega_{s'} \mod B^n_{k_2|k_0},$$

whence by comparison with (9.17)

$$(a'^p - a')(dc/c) \wedge \omega_{s'} \in \Omega^n_{k_2|k_0, <s} \text{ mod } B^n_{k_2|k_0}. \qquad (9.25)$$

Wedge product with ω_m therefore shows that the element

$$(a'^p - a')(dc/c) \wedge \omega_{s'} \wedge \omega_m \qquad (9.26)$$

lies in $B^n_{k_2|k_0}$, since $\omega_m \wedge \Omega^n_{k_2|k_0, <s} = 0$ as already noted. In the decomposition of Proposition 9.4.6, the component of the element (9.26) lying in $\Omega^N_{k_2|k_0}(0)$ is

$$(a'^p - \alpha_0)(dc/c) \wedge \omega_{s'} \wedge \omega_m,$$

with α_0 as in (9.24). On the other hand, by Proposition 9.4.6 (3) we have $\Omega^N_{k_2|k_0}(0) \cap B^N_{k_2|k_0} = 0$. Since $(dc/c) \wedge \omega_{s'} \wedge \omega_m$ generates the 1-dimensional k_2-vector space $\Omega^N_{k_2|k_0}$, we conclude that $a'^p - \alpha_0$ must be 0. Comparison with (9.24) yields

$$a'^p - a' = -a'_1. \qquad (9.27)$$

All the above holds in the case $n = 1$ as well, with the modification that one should omit the component $\omega_{s'}$ everywhere. Since for $n = 1$ we have $k_2 = k_1$ and hence $V = 0$, equation (9.27) then reads $a'^p - a' = 0$, i.e. $a' \in \mathbf{F}_p$, as required.

We return to the case $n > 1$. Applying the differential d to (9.25) and taking (9.27) into account we obtain

$$d(a'_1 \omega_{s'}) \wedge (dc/c) \in B^{n+1}_{k_2|k_0, <s}.$$

Since dc/c differs from $db_{s(1)}/b_{s(1)}$ by a constant in $k_1 \subset k_2$, we have

$$\Omega^n_{k_2|k_0, <s} \subset \Omega^{n-1}_{k_2|k_0, <s'} \wedge (dc/c).$$

Thus we find $\omega \in \Omega^{n-1}_{k_2|k_0, <s'}$ such that $d(a'_1 \omega_{s'}) \wedge (dc/c) = d\omega \wedge (dc/c)$, or in other words

$$d(a'_1 \omega_{s'} - \omega) \wedge (dc/c) = 0.$$

As dc/c is part of a basis of $\Omega^1_{k_2|k_0}$, this is only possible if

$$d(a'_1 \omega_{s'} - \omega) = (dc/c) \wedge \omega_1 \in \Omega^n_{k_2|k_0}$$

for a suitable $\omega_1 \in \Omega^{n-1}_{k_2|k_0}$ (see the end of the proof of Proposition 9.7.2 below for details on this type of argument). So since $c \in k_1$, the element $d(a'_1 \omega_{s'} - \omega)$ vanishes in $\Omega^n_{k_2|k_1}$.

Consider the decomposition of $\Omega^{n-1}_{k_2|k_1}$ coming from the p-basis $\{b_i : i > s(1)\}$ as in Proposition 9.4.6. Since $a'_1 \notin k_1$, we see that $a'_1 \omega_{s'}$ has trivial projection to $\Omega^{n-1}_{k_2|k_1}(0)$. This may not be the case for ω, but since

$d(\Omega^{n-1}_{k_2|k_1}(0)) = 0$ by Proposition 9.4.6 (3), we may modify ω by an element of $\Omega^{n-1}_{k_2|k_1}(0)$ without affecting the condition $d(a'_1\omega_{s'} - \omega) = 0$. Thus we may assume that $a'_1\omega_{s'} - \omega$ avoids $\Omega^{n-1}_{k_2|k_1}(0)$, and therefore $a'_1\omega_{s'} - \omega \in B^{n-1}_{k_2|k_1}$ by the statement of Proposition 9.4.6 (3) about the other components. Since $\omega \in \Omega^{n-1}_{k_2|k_0,<s'}$, we obtain

$$a'_1\omega_{s'} \in \Omega^{n-1}_{k_2|k_1,<s'} + B^{n-1}_{k_2|k_1},$$

so by (9.27)

$$(a'^p - a')\omega_{s'} \in \Omega^{n-1}_{k_2|k_1,<s'} + B^{n-1}_{k_2|k_1}.$$

As the kernel of the projection $\Omega^{n-1}_K \to \Omega^{n-1}_{K|k_1}$ is contained in $\Omega^{n-1}_{K,<s'}$, this implies (9.21) and concludes the proof.   □

Remark 9.6.5 We record for later use the following by-product of the above proof. Under the assumption of Proposition 9.6.4 there exist elements $a' \in K$, $\tau \in \Omega^n_{K,<s}$ and $c \in k_1$ such that up to replacing K by a finite extension K' of degree prime to p we have

$$a\omega_s = a'\omega_{s'} \wedge (dc/c) + \tau \quad \text{and} \quad (a'^p - a')\omega_{s'} \in \Omega^{n-1}_{K,<s'} + B^{n-1}_K$$

for $n > 1$, and for $n = 1$ we have $a\omega_s = a'(dc/c) + \tau$ with $a' \in \mathbf{F}_p$. This comes from (9.20) and (9.21), the element τ being an element in the kernel of $\Omega^n_K \to \Omega^n_{K|k_0}$ which is a subset of $\Omega^n_{K,<s}$ by definition of k_0.

Note moreover that the finite extension $K'|K$ we allow may be chosen to be a tower of Galois extensions. This is seen as follows: $K'|K$ arises by adjoining the $(p-1)$-st root u of an element in K in the proof of Proposition 9.6.2. This may not be Galois over K, but embeds in $K(\zeta)(u)|K$, where ζ is a primitive $(p-1)$-st root of unity. The latter extension still has prime to p degree and is a tower of Galois extensions.

9.7 Injectivity of the differential symbol

Now that the surjectivity statement of Theorem 9.5.2 is established, we turn to injectivity. For brevity's sake introduce the notation

$$k_n(K) := K^M_n(K)/pK^M_n(K).$$

We then have to prove:

Theorem 9.7.1 *The differential symbol $\psi^n_K : k_n(K) \to \nu(n)_K$ is injective.*

The first step in the proof is the following analogue of Proposition 7.5.5 (1).

Proposition 9.7.2   *Assume that the differential symbols $\psi_K^n : k_n(K) \to \nu(n)_K$ and ψ_L^{n-1} are injective, where $L|K$ is an arbitrary finite extension. Then so is $\psi_{K(t)}^n : k_n(K(t)) \to \nu(n)_{K(t)}$ for a purely transcendental extension $K(t)$.*

Proof   We first construct a commutative diagram

$$
\begin{array}{ccccccccc}
0 & \longrightarrow & k_n(K) & \longrightarrow & k_n(K(t)) & \xrightarrow{\oplus \partial_P} & \displaystyle\bigoplus_{P \in (\mathbf{A}_K^1)_0} k_{n-1}(\kappa(P)) & \longrightarrow & 0 \\
& & \psi_K^n \downarrow & & \psi_{K(T)}^n \downarrow & & \oplus i_P \downarrow & & \\
0 & \longrightarrow & \Omega_{K[t]}^n & \longrightarrow & \Omega_{K(t)}^n & \longrightarrow & \displaystyle\bigoplus_{P \in (\mathbf{A}_K^1)_0} \Omega_{K(t)}^n / \Omega_{K[t]P}^n. & &
\end{array}
$$

Here the upper row is the Milnor exact sequence modulo p (Theorem 7.2.1), and the lower row comes from the localization property of differentials (Appendix, Proposition A.8.3 (2)), noting that $K[t]$ is the intersection of the $K[t]_P$. The maps i_P are defined as the composite of $\psi_{\kappa(P)}^{n-1}$ with the map $j_P : \Omega_{\kappa(P)}^{n-1} \to \Omega_{K(t)}^n / \Omega_{K[t]P}^n$ given by

$$
j_P(x_0 \, dx_1 \wedge \cdots \wedge dx_{n-1}) = \widetilde{x}_0 \, d\widetilde{x}_1 \wedge \cdots \wedge d\widetilde{x}_{n-1} \wedge \pi_P^{-1} d\pi_P,
$$

where π_P is a local parameter at P and the $\widetilde{x}_i$ are arbitrary liftings of the x_i to $K[t]_P$. The map does not depend on the choice of the liftings $\widetilde{x}_i$, since an easy calculation shows that changing $\widetilde{x}_i$ to $\widetilde{x}_i + u\pi_P$ with some unit in $K[t]_P$ changes the right-hand side by an element in $\Omega_{K[t]P}^n$ (the factor π_P^{-1} gets cancelled). Another easy calculation shows that the right-hand square commutes up to a factor $(-1)^{n-1}$; commutativity of the left square is straightforward.

Now the left vertical map is injective by assumption; if we show injectivity of the maps i_P, that of $\psi_{K(T)}^n$ will follow. As $\psi_{\kappa(P)}^{n-1}$ is injective by assumption, it remains to establish the injectivity of j_P. By definition, this is equivalent to the injectivity of the map $x_0 \, dx_1 \wedge \cdots \wedge dx_{n-1} \mapsto \widetilde{x}_0 \, d\widetilde{x}_1 \wedge \cdots \wedge d\widetilde{x}_{n-1} \wedge d\pi_P$. By Proposition A.8.14 of the Appendix $\Omega_{K[t]P}^1$ is a free $K[t]_P$-module on a basis consisting of $d\pi_P$ and some other elements da_i; a basis of $\Omega_{K[t]P}^n$ is then given by n-fold exterior products of these forms. Hence for an element $\omega \in \Omega_{K[t]P}^{n-1}$ the relation $\omega \wedge d\pi_P = 0$ can only hold if when writing ω as a linear combination of basis elements only those involving $d\pi_P$ have nonzero coefficient. But then the image of ω in $\Omega_{\kappa(P)}^{n-1}$ is 0, as required. $\square$

The idea of the proof of Theorem 9.7.1 in the general case is now the following. By Lemma 9.5.9 it is enough to consider the case of a field F finitely generated over $\mathbf{F}_p$. We apply induction on n, the case $n = 0$ being obvious. If d denotes the transcendence degree of F, then by Corollary A.3.5 of the Appendix there exists a scheme-theoretic point of codimension 1 on the affine space $\mathbf{A}_{\mathbf{F}_p}^{d+1}$ whose local ring R has residue field isomorphic to F. Note that R is a discrete valuation ring (being integrally closed of Krull dimension 1) whose fraction field is purely transcendental over $\mathbf{F}_p$. To proceed further, we need:

Construction 9.7.3 Let K be the fraction field of the above R, and M its maximal ideal. Define $k_n(R)$ to be the kernel of the residue map $\partial^M : k_n(K) \to k_{n-1}(F)$; note that it is generated by symbols whose entries are units in R by Proposition 7.1.7. It follows that the differential symbol ψ_K^n restricts to a map $\psi_R^n : k_n(R) \to \nu(n)_R$, where $\nu(n)_R$ is the kernel of the operator $\gamma - 1 : \Omega_R^n \to \Omega_R^n/B_R^n$ (see Remark 9.4.5). Denote by $k_n(R, M)$ the kernel of the specialization map $s_R^M : k_n(R) \to k_n(F)$ (which does not depend on generators of M by Remark 7.1.9), and by $\nu(n)_{R,M}$ that of the reduction map $\rho_R : \nu(n)_R \to \nu(n)_F$. The easy compatibility $\psi_R^n \otimes_R F = \psi_F^n \circ s_R^M$ implies that ψ_R^n restricts to a map $\psi_{R,M}^n : k_n(R, M) \to \nu(n)_{R,M}$.

Lemma 9.7.4 *With notations as above, assume that the differential symbol* $\psi_{R,M}^n : k_n(R, M) \to \nu(n)_{R,M}$ *is surjective. Then the symbol* h_F^n *is injective.*

Proof We have the commutative diagram with exact rows

$$
\begin{array}{ccccccccc}
0 & \longrightarrow & k_n(R,M) & \longrightarrow & k_n(R) & \longrightarrow & k_n(F) & \longrightarrow & 0 \\
& & \downarrow{\scriptstyle \psi_{R,M}^n} & & \downarrow{\scriptstyle \psi_R^n} & & \downarrow{\scriptstyle \psi_F^n} & & \\
0 & \longrightarrow & \nu(n)_{R,M} & \longrightarrow & \nu(n)_R & \longrightarrow & \nu(n)_F & &
\end{array}
$$

By Proposition 9.7.2 and our inductive assumption on n the middle vertical map is injective, so the lemma follows by diagram chase. $\quad\square$

Thus in order to prove Theorem 9.7.1 it suffices to prove the surjectivity of $k_n(R, M) \to \nu(n)_{R,M}$. In the course of the proof we shall be forced to make finite extensions of the fraction field of R, and the integral closure of R in these extensions will not be local any more, but in general only a semi-local Dedekind ring, i.e. a Dedekind ring with finitely many maximal ideals. We therefore have to extend the statement to these.

Construction 9.7.5 If R is a semi-local Dedekind ring with maximal ideals $M_1, \ldots, M_r$, denote by $I = \cap M_i$ its Jacobson radical. By the Chinese Remainder Theorem $R/I \cong \bigoplus R/M_i$; in particular, it is a direct sum of fields. Therefore we may define $K_n^M(R/I)$ as the direct sum of the $K_n^M(R/M_i)$ and denote its mod p quotient by $k_n(R/I)$. The group $k_n(R) \subset k_n(K)$ is defined as the intersection of the kernels of the residue maps $k_n(K) \to k_{n-1}(R/M_i)$

associated with the localizations $R_i := R_{M_i}$ of R; it is generated by symbols coming from the units of R. The group $k_n(R, I)$ is defined as the kernel of the direct sum of specialization maps $\oplus s^M_{R_i} : k_n(R) \to \bigoplus k_n(R/M_i)$. As in the case of discrete valuation rings, the symbol ψ^n_K restricts to a symbol $k_n(R) \to \nu(n)_R$; it is compatible with the direct sum of the symbols ψ^n_{R/M_i} via the specialization maps. Hence it restricts to a symbol $\psi^n_{R,I} : k_n(R, I) \to \nu(n)_{R,I}$, where the latter group is the kernel of the map $\nu(n)_R \to \oplus \nu(n)_{R/M_i}$.

Therefore the statement to be proven is:

Proposition 9.7.6 *Let k be a perfect field of characteristic $p > 0$, and R a semi-local Dedekind domain which is obtained as a localization of a finitely generated k-algebra. Then the differential symbol*

$$\psi^n_{R,I} : k_n(R, I) \to \nu(n)_{R,I}$$

is surjective.

We first prove some preliminary lemmas.

Lemma 9.7.7 *For R as in the proposition the R-module $\Omega^1_R = \Omega^1_{R|k}$ is free of finite rank.*

Proof For each maximal ideal M of R the localization R_M is a discrete valuation ring, hence $\Omega^1_{R_M|k}$ is free of finite rank by Proposition A.8.6 of the Appendix. It follows that $\Omega^1_{R|k}$ is a finitely generated projective R-module, by Matsumura [1], 7.12 together with Proposition A.8.3 (2) of the Appendix. But then it must be free, since R is semi-local (see Matsumura [1], Theorem 2.5 for the local case; the same proof works in general).   □

Lemma 9.7.8 *Let K be the fraction field of R, and let $K'|K$ be a finite extension of degree prime to p which is a tower of Galois extensions. Let R' be the normalization of R in K, and denote by I' the Jacobson radical of R'. If the differential symbol $\psi^n_{R',I'}$ is surjective, then so is $\psi^n_{R,I}$.*

Proof This will follow by the same norm argument as in Proposition 9.5.8, once we show that the norm and trace maps of Lemma 9.5.7 restrict to $k_n(R', I')$ and $\nu(n)_{R',I'}$, giving rise to a commutative diagram

$$
\begin{array}{ccc}
k_n(R', I') & \xrightarrow{\psi^n_{R',I'}} & \nu(n)_{R',I'} \\
{\scriptstyle N_{R'|R}} \downarrow & & \downarrow {\scriptstyle \mathrm{tr}} \\
k_n(R, I) & \xrightarrow{\psi^n_{R,I}} & \nu(n)_{R,I}.
\end{array}
\qquad (9.28)
$$

To check this, note first that the norm map $N_{K'|K} : k_n(K') \to k_n(K)$ restricts to a norm map $k_n(R') \to k_n(R)$ by Proposition 7.4.1 applied to the finitely many residue maps coming from the maximal ideals of R'. It further restricts to a map $k_n(R', I') \to k_n(R, I)$ using Corollary 7.4.2 (note that it applies since the ramification is tame at each maximal ideal of R by our assumption on $K'|K$ and Corollary A.6.6 of the Appendix). On the other hand, consider the diagram

$$
\begin{array}{ccccc}
k_n(R') & \xrightarrow{\psi_{R'}^n} & \Omega_{R'}^n & \longrightarrow & \Omega_{K'}^n \\
{\scriptstyle N_{R'|R}}\downarrow & & & & \downarrow{\scriptstyle \text{tr}} \\
k_n(R) & \xrightarrow{\psi_R^n} & \Omega_R^n & \longrightarrow & \Omega_K^n
\end{array}
$$

where the outer rectangle commutes by virtue of Lemma 9.5.7. It follows from Lemma 9.7.7 that $\Omega_{R'}^n$ and Ω_R^n are free of finite rank, and therefore the horizontal arrows in the right half of the diagram are injective. From this we obtain that the trace map on $\Omega_{K'}^n$ restricts to a map $\mathrm{Im}(\psi_{R'}^n) \to \mathrm{Im}(\psi_R^n)$. Using compatibilities with specialization maps one checks that this map sends $\mathrm{Im}(\psi_{R',I'}^n)$ into $\mathrm{Im}(\psi_{R,I}^n)$. By our assumption we have $\mathrm{Im}(\psi_{R',I'}^n) = \nu_{R',I'}^n$, whence the existence of the commutative diagram (9.28). $\qquad\square$

Another important ingredient in the proof will be the following integral version of Theorem 9.2.2.

Lemma 9.7.9 *Let $R \supset T$ be an extension of semi-local Dedekind rings which arise as localizations of finitely generated algebras over a perfect field k of characteristic $p > 0$. Assume that the arising extension $K|K_0$ of fraction fields is finite, and moreover $R^p \subset T$. Then the sequence*

$$
0 \to R^\times / T^\times \xrightarrow{\mathrm{dlog}} \Omega_{R|T}^1 \xrightarrow{\gamma_R - 1} \Omega_{R|T}^1 / B_{R|T}^1
$$

is exact, where $R^\times$ (resp. $T^\times$) denotes the units in R (resp. T).

Proof Given $\omega \in \ker(\gamma_R - 1)$, we have $\omega = \mathrm{dlog}(f)$ in $\Omega_{K|K_0}^1$ for some f in $K^\times$ by Theorem 9.3.3. Note that R is a principal ideal domain, with prime elements the generators t_i of the finitely many maximal ideals $M_1, \ldots, M_r$ (see Matsumura [1], Ex. 11.7). Up to multiplying f with a sufficiently high power of $(t_1 \ldots t_r)^p$ we may assume $f \in R$. We now show that $f \in R^\times$, which is equivalent to showing that $f \notin M_i$ for all i. Assume $f \in M_i$ for some i. Then $f = u t_i^m$ for some $m > 0$ with some $u \in R \setminus M_i$; up to dividing f by a power of t_i^p we may assume $(m, p) = 1$. But then $\mathrm{dlog}(f) = (du/u) + m(dt_i/t_i)$ in $\Omega_{K|K_0}^1$. Now notice that dt_i/t_i does not lie in $\Omega_{R_{M_i}|(R_{M_i} \cap K_0)}^1$.

Indeed, applying Proposition A.8.3 (4) of the Appendix with $A = k$, $B = R_{M_i}$ and $I = M_i R_{M_i}$ yields that $\Omega^1_{R_{M_i}|k} \otimes_{R_{M_i}} R/M_i$ has a basis consisting of dt_i and a basis of $\Omega^1_{(R/M_i)|k}$. Hence by Nakayama's lemma (Lang [3], Chapter X, Lemma 4.3) these elements generate $\Omega^1_{R_{M_i}|k}$ and a fortiori $\Omega^1_{R_{M_i}|(R_{M_i} \cap K_0)}$. If dt_i/t_i were an element of this module, then writing it in terms of the above basis we would get $dt_i \in M_i \Omega^1_{R_{M_i}|(R_{M_i} \cap K_0)}$, a contradiction. All in all, we obtain that ω and du/u lie in $\Omega^1_{R_{M_i}|(R_{M_i} \cap K_0)}$ but $m(dt_i/t_i)$ doesn't, which contradicts $\omega = \mathrm{dlog}(f)$ and thus proves the lemma.   □

Finally an easy lemma from Milnor K-theory:

Lemma 9.7.10   *Let M be a maximal ideal of R and R_M the associated localization. Then for all integers $m, m' > 0$ the product operation*

$$k_m(K) \otimes k_{m'}(K) \to k_{m+m'}(K)$$

sends $k_m(R_M, MR_M) \otimes k_{m'}(K)$ into $k_{m+m'}(R_M, MR_M)$.

Proof   Note that R_M is a discrete valuation ring. By induction we may assume that $m' = 1$. The group $k_1(K)$ is generated by $k_1(R_M)$ and a local parameter π for M. Since $k_m(R_M, MR_M) \otimes k_1(R_M) \subset k_{m+1}(R_M, MR_M)$, it remains to show that $k_m(R_M, MR_M) \otimes \{\pi\} \subset k_{m+1}(R_M, MR_M)$. Take α in $k_n(R_M, MR_M)$. Then $\partial_M(\{\pi, \alpha\}) = s^M_{-\pi}(\alpha) = 0$ in $k_n(R/M)$, and therefore $\{\pi, \alpha\} \in k_n(R_M)$. But $s^M_\pi(\{\pi, \alpha\}) = 0$ as well, so $\{\alpha, \pi\} = (-1)^{m-1}\{\pi, \alpha\}$ lies in $k_{n+1}(R_M, MR_M)$.   □

We now begin the proof of Proposition 9.7.6. The first step is again to choose a suitable p-basis of K.

Lemma 9.7.11   *There exist units $b_1, \ldots, b_{r-1}$ in R and a generator b_r of I so that $db_1, \ldots, db_r$ form a basis of the free R-module Ω^1_R, and the mod I images of $db_1, \ldots db_{r-1}$ form a basis of $\Omega^1_{R/I}$.*

Here recall that R/I is a direct sum of fields.

Proof   Take a p-basis $\bar{b}_1, \ldots, \bar{b}_{r-1}$ of R/I. By multiplying with a suitable nonsingular matrix with entries in $\mathbf{F}_p^\times$ we may ensure that each $\bar{b}_i$ has nonzero components in the direct sum decomposition $R/I \cong \oplus R/M_i$. This implies that if we take arbitrary liftings $b_1, \ldots, b_{r-1}$ of the $\bar{b}_i$, then the b_i are units in R. Moreover, an argument using the exact sequence

$$I/I^2 \xrightarrow{\delta} \Omega^1_{R|k} \otimes_R R/I \to \Omega^1_{(R/I)|k} \to 0$$

as in the second half of the proof of Lemma 9.7.9 shows that for any generator b_r of I the elements $db_1, \ldots, db_r$ generate Ω_R^1. Note that the map δ in the above sequence must be injective, for if it were 0, then so would be the map $M_i/M_i^2 \rightarrow \Omega_{R_{M_i}|k}^1 \otimes_{R_{M_i}} R/M_i$ for any maximal ideal M_i of R, which would contradict Corollary A.8.7 of the Appendix. Thus $db_1, \ldots, db_r$ is a minimal generating system, and hence a basis of the free R-module Ω_R^1. □

Fix $b_1, \ldots, b_r$ as in the lemma above. By Propositions A.8.3 (2) and A.8.11 of the Appendix they form a p-basis of the extension $K|k$. From now on we take up the notations of the previous section. In particular, we put $\omega_s = (db_{s(1)}/b_{s(1)}) \wedge \cdots \wedge (db_{s(n)}/b_{s(n)})$ for a strictly increasing function $s : \{1, \ldots, n\} \rightarrow \{1, \ldots, r\}$. Recall that the ω_s form a k-basis of Ω_K^n. An element $\sum a_s \omega_s \in \Omega_K$ lies in $\nu(n)(K)$ if and only if $\sum (a_s^p - a_s)\omega_s \in B_K^n$. Moreover, by our choice of the p-basis $b_1, \ldots b_r$ the element $\sum a_s \omega_s$ lies in $\nu(n)(R, I)$ if and only if $a_s \in I$ for all s.

Our goal is then to prove:

Proposition 9.7.12 *Fix $a \in I$ and $s \in S_n$, and assume that*

$$(a^p - a)\omega_s \in \Omega_{K|k,<s}^n + B_{K|k}^n.$$

Then up to replacing K by some finite prime to p extension $K'|K$ which is a tower of Galois extensions, and R by its normalization in K', we have

$$a\omega_s \in \Omega_{K|k,<s}^n + \mathrm{Im}\,(\psi_{R,I}^n).$$

This proposition implies Proposition 9.7.6 by exactly the same argument as Proposition 9.6.4 implies Theorem 9.6.1. So all that remains is to give its proof.

Proof As in the proof of Proposition 9.6.4, write $k_0 := k(b_1, \ldots, b_{s(1)-1})$ and $k_1 := k_0(b_{s(1)})$. According to Remark 9.6.5, there exist elements $a' \in K$, $\tau \in \Omega_{K,<s}^n$ and $c \in k_1$ such that up to replacing K by a finite extension of degree prime to p we have

$$a\omega_s = a'\omega_{s'} \wedge (dc/c) + \tau \qquad (9.29)$$

and

$$(a'^p - a')\omega_{s'} \in \Omega_{K,<s'}^{n-1} + B_K^{n-1} \qquad (9.30)$$

for $n > 1$, and for $n = 1$ we have $a\omega_s = a'(dc/c) + \tau$ with $a' \in \mathbf{F}_p$. We have seen during the proof of Proposition 9.6.4 that this statement implies $a\omega_s \in \mathrm{Im}\,(\psi_K^n) \bmod \Omega_{K|k,<s}^n$ by induction, but in the present case we have

to ensure that $a\omega_s \in \mathrm{Im}\,(\psi_{R,I}^n) \bmod \Omega_{K|k,<s}^n$. This will be done by suitably modifying the element c.

Write

$$dc/c = \sum_{i=1}^{s(1)} \gamma_i(db_i/b_i)$$

in $\Omega_{k_1}^1$. It follows from (9.29) that $a = (-1)^{n-1}a'\gamma_{s(1)}$. In particular, since $a \in I$ and $a' \in \mathbf{F}_p$ for $n = 1$, we have $\gamma_{s(1)} \in I$ for $n = 1$. In the general case define ideals

$$J = \bigcap_{\gamma_{s(1)} \in M} M, \quad L = \bigcap_{J \not\subset M} M$$

in R, where M runs over the maximal ideals of R. Let R_J, R_L be the respective localizations, so that JR_J and LR_L are the Jacobson radicals. For $n = 1$ we have $I = J$ and $L = R$. In the general case we have $a' \in LR_L$ by construction.

Write $T := R_J \cap k_0$, and consider the commutative diagram

$$
\begin{array}{ccc}
\Omega_{R_J}^1 & \longrightarrow & \Omega_K^1 \\
\downarrow & & \downarrow \\
\Omega_{R_J|T}^1 & \longrightarrow & \Omega_{K|k_0}^1
\end{array}
$$

where the vertical maps are the natural projections and the horizontal maps are injective by the localization property of differentials. The forms dc/c and $\gamma_{s(1)}(db_{s(1)}/b_{s(1)})$ have the same image in $\Omega_{K|k_0}^1$; denote it by θ. Since dc/c is in the kernel of $\gamma_K - \mathrm{id}$, this implies that θ lies in the kernel of the operator $\gamma_{R_J|T} - \mathrm{id}$, the restriction of $\gamma_{K|k_0} - \mathrm{id}$ to $\Omega_{R_J|T}^1$. Now put $H = R_J/JR_J$ and $P = T/J \cap T$. As $\gamma_{s(1)} \in J$ by definition of J, the image of θ in $\Omega_{H|P}^1$ is trivial. Consider the commutative diagram with exact rows

$$
\begin{array}{c}
0 \\
\downarrow \\
(1 + JR_J)/(T^\times \cap (1 + JR_J)^\times) \\
\downarrow
\end{array}
$$

$$
\begin{array}{ccccccc}
0 & \longrightarrow & R_J^\times/T^\times & \xrightarrow{\mathrm{dlog}} & \Omega_{R_J|T}^1 & \xrightarrow{\gamma-\mathrm{id}} & \Omega_{R_J|T}^1/B_{R_J|T}^1 \\
& & \downarrow & & \downarrow & & \\
0 & \longrightarrow & H^\times/P^\times & \xrightarrow{\mathrm{dlog}} & \Omega_{H|P}^1 & &
\end{array}
$$

where exactness of the middle row follows from Lemma 9.7.9. As θ is annihilated by both maps starting from $\Omega^1_{R_J|T}$, the diagram shows that there exists $\delta \in 1 + JR_J$ such that $\theta = \mathrm{dlog}(\delta)$ in $\Omega^1_{R_J|T}$, whence

$$dc/c = d\delta/\delta + \eta \qquad (9.31)$$

in Ω^1_K with a suitable $\eta \in R_J\mathrm{Im}\,(\Omega^1_T \to \Omega^1_{R_J})$. In particular, η is in the K-span of the image of $\Omega^1_{k_0}$ and thus lies in $\Omega^1_{K,<s(1)}$. In the case $n = 1$ the proof ends here, because the equality $J = I$ implies that δ mod p lies in $k_1(R, I)$ and so $d\delta/\delta \in \mathrm{Im}\,(\psi^1_{R,I})$. In the case $n > 1$ we only have that δ mod p lies in $k_1(R_J, JR_J)$. On the other hand, since $a' \in LR_L$ as noted before, we may use (9.30) to apply induction on n to $a'\omega_{s'}$ and obtain

$$a'\omega_{s'} \in \mathrm{Im}\,(\psi^{n-1}_{R_L,LR_L}) + \Omega^{n-1}_{K,<s'}. \qquad (9.32)$$

Since $\eta \in \Omega^1_{K,<s(1)}$, this implies $a'\omega_{s'} \wedge \eta \in \Omega^n_{K,<s}$, so that from (9.29), (9.31) and (9.32) we obtain

$$a\omega_s = \beta \wedge (d\delta/\delta) \quad \mathrm{mod} \quad \Omega^n_{K,<s}$$

for some $\beta \in \mathrm{Im}\,(\psi^{n-1}_{R_L,LR_L})$. Since δ mod p lies in $k_1(R_J, J)$, the proposition follows if we check that the natural product map $k_{n-1}(K) \otimes k_1(K) \to k_n(K)$ sends $k_{n-1}(R_L, LR_L) \otimes k_1(R_J, JR_J)$ to $k_n(R, I)$. Writing $k_n(R, I)$ as the intersection of the groups $k_n(R_M, MR_M)$ where M runs over the maximal ideals of R, it suffices to show that the image is contained in $k_n(R_M, MR_M)$ for each M. But by construction each R_M contains either R_J or R_L, so the claim follows from Lemma 9.7.10. $\square$

In case the reader felt a bit lost among the numerous reduction steps of the above proof of Theorem 9.7.1, we recapitulate the logical structure of the argument: Proposition 9.7.12 just proven implies Proposition 9.7.6, which together with Proposition 9.7.2 and induction on n implies Theorem 9.7.1 via Lemma 9.7.4. Thus Theorem 9.7.1 is proven, and with it the theorem of Bloch, Kato and Gabber.

9.8 Application to p-torsion in Milnor K-theory

The Bloch–Gabber–Kato theorem makes it possible to generalize Suslin's theorem on the absence of p-torsion in K_2 of fields of characteristic p (Theorem 8.4.8) to Milnor K-groups of arbitrary degree.

Theorem 9.8.1 (Izhboldin)   *The groups $K_n^M(k)$ have no p-torsion for a field k of characteristic $p > 0$.*

The following consequence of the Bloch–Gabber–Kato theorem constitutes the first step towards the proof of Theorem 9.8.1.

Lemma 9.8.2   *Let k be a field of characteristic $p > 0$, and let $K|k$ be a degree p cyclic Galois extension. The sequence*

$$0 \to K_n^M(k)/pK_n^M(k) \xrightarrow{\iota_{K|k}} K_n^M(K)/pK_n^M(K) \xrightarrow{\sigma-1} K_n^M(K)/pK_n^M(K)$$

is exact, where σ is a generator of the Galois group $\mathrm{Gal}\,(K|k)$.

Proof   Consider the commutative diagram

$$
\begin{array}{ccccccc}
0 & \longrightarrow & \Omega_k^n & \longrightarrow & \Omega_K^n & \xrightarrow{\sigma-1} & \Omega_K^n \\
& & \Big\downarrow{\gamma-\mathrm{id}} & & \Big\downarrow{\gamma-\mathrm{id}} & & \Big\downarrow{\gamma-\mathrm{id}} \\
0 & \longrightarrow & \Omega_k^n/B_k^n & \longrightarrow & \Omega_K^n/B_K^n & \xrightarrow{\sigma-1} & \Omega_K^n/B_K^n
\end{array}
$$

where the rows are exact since $\Omega_K^n \cong \Omega_k^n \otimes_k K$ by virtue of Proposition A.8.9 of the Appendix. The diagram commutes because the differentials of the de Rham complex and the inverse Cartier operator are Galois-equivariant. So by taking the kernels of the vertical maps we obtain an exact sequence

$$0 \to \nu(n)_k \to \nu(n)_K \xrightarrow{\sigma-1} \nu(n)_K. \tag{9.33}$$

The Bloch–Gabber–Kato theorem identifies this exact sequence with that of the lemma.   □

The proof of Theorem 9.8.1 will proceed by induction on n. The key ingredient is a partial case of Hilbert's Theorem 90 which we shall prove in its full form later in this section. To state the partial result, let $K|k$ be a cyclic Galois extension of degree p, and σ a generator in the Galois group. Like in the proof of Theorem 8.4.1, for a field extension $F|k$ consider the complex

$$K_n^M(K \otimes_k F) \xrightarrow{\sigma-1} K_n^M(K \otimes_k F) \xrightarrow{N_{K \otimes_k F|F}} K_n^M(F).$$

Now introduce the following notation:

$$P_{p,n}(F) := K_n^M(K \otimes_k F)\{p\} \cap \ker(N_{K \otimes_k F|F}),$$
$$I_{p,n}(F) := K_n^M(K \otimes_k F)\{p\} \cap \mathrm{Im}(\sigma-1),$$
$$V_{p,n}(F) := P_{p,n}(F)/I_{p,n}(F),$$

where $K_n^M(K \otimes_k F)\{p\}$ denotes the p-primary torsion subgroup of $K_n^M(K \otimes_k F)$. With the conventions above we have:

Proposition 9.8.3 *Assume that the p-torsion subgroup* $_p K_{n-1}^M(F)$ *in* $K_{n-1}^M(F)$ *is trivial for all extensions* $F|k$. *Then the group* $V_{p,n}(k)$ *is trivial.*

The strategy of the proof is analogous to that of Theorem 8.4.1, so we structure the presentation in the same way. The first step is:

Lemma 9.8.4 *Assume that* k *has no nontrivial finite extension of degree prime to* p, *and the norm map* $N_{K|k} : K^\times \to k^\times$ *is surjective. Then* $V_{p,n}(k) = 0$.

Proof In fact, in this case we can prove directly that the sequence

$$K_n^M(K) \xrightarrow{\sigma-1} K_n^M(K) \xrightarrow{N_{K|k}} K_n^M(k)$$

is exact. This is done exactly as in the proof of Proposition 8.4.3: we define a retraction

$$K_n^M(k) \to K_n^M(K)/(\sigma - 1)K_n^M(K)$$

of the norm map $N_{K|k}$ by sending $\{b_1, b_2 \dots, b_n\} \in K_n^M(k)$ to $\{a_1, b_2, \dots, b_n\} \in K_n^M(k)$, where $a_1 \in K^\times$ satisfies $N_{K|k}(a_1) = b_1$. The same calculation as in the proof of Proposition 8.4.3 shows that this map is well defined and respects the Steinberg relation; its surjectivity follows from the Bass–Tate lemma. □

Next we prove the analogue of Proposition 8.4.4:

Lemma 9.8.5 *If* $k'|k$ *is an algebraic extension of degree prime to* p, *the map* $V_{p,n}(k) \to V_{p,n}(k')$ *is injective.*

Proof Arguing as in the proof of Proposition 8.4.4, we reduce to proving $pV_{p,n}(k) = 0$. This holds because for $\alpha \in P_{p,n}(k)$ we have

$$p\alpha = p\alpha - N_{K|k}(\alpha) = p\alpha - \sum_{i=0}^{p-1}\sigma^i\alpha = \sum_{i=0}^{p-1}(1-\sigma^i)\alpha = (\sigma-1)\sum_{i=0}^{p-1}\sum_{j=0}^{i-1}(-\sigma^j\alpha)$$

which is an element of $I_{p,n}(k)$. □

The next step is analogous to Proposition 8.4.5. The proof, however, does not use the K-theory of Severi–Brauer varieties. Instead, information coming from the Bloch–Gabber–Kato theorem is applied at this point.

Lemma 9.8.6   *Assume* $_pK_{n-1}^M(F) = 0$ *for all extensions* $F|k$. *If* X *is a Severi–Brauer variety over* k *split by the extension* $K|k$, *the map* $V_{p,n}(k) \to V_{p,n}(k(X))$ *is injective.*

Proof   We first treat the case where X is already split over k. Then $k(X)$ is a finitely generated purely transcendental extension of k. By induction on the transcendence degree we are reduced to proving the injectivity of $V_{p,n}(k) \to V_{p,n}(k(t))$ for a simple transcendental extension $k(t)|k$. Consider the base change extension $K(t)|K$ equipped with the Galois action induced from that of $K|k$. By Theorem 7.2.1 we have a split exact sequence

$$0 \to K_n^M(K) \to K_n^M(K(t)) \xrightarrow{\partial^M} \bigoplus_{P \in \mathbf{P}_0^1 \setminus \{\infty\}} K_{n-1}^M(\kappa(P)) \to 0$$

where the groups $K_{n-1}^M(\kappa(P))$ have no p-torsion by the inductive assumption. It follows that the injection $K_n^M(K) \to K_n^M(K(t))$ induces an isomorphism $K_n^M(K)\{p\} \xrightarrow{\sim} K_n^M(K(t))\{p\}$ on p-primary torsion subgroups. Since this isomorphism is compatible with the Galois actions on K and $K(t)$, it induces an isomorphism $I_{p,n}(k) \to I_{p,n}(k(t))$. The injectivity of $V_{p,n}(k) \to V_{p,n}(k(t))$ follows by passing to the quotient.

In the general case, the Severi–Brauer variety X_K is split, and therefore $K_n^M(K) \to K_n^M(K(X))$ is injective by the argument of the previous paragraph. By definition, the group $V_{p,n}(k)$ is a subgroup of $K_n^M(K)/I_{p,n}(k)$ and similarly, the group $V_{p,n}(k(X))$ is a subgroup of $K_n^M(K(X))/I_{p,n}(k(X))$. It therefore suffices to show the surjectivity of the map $I_{p,n}(k) \to I_{p,n}(k(X))$.

We introduce the intermediate field $L := k(X)^p k$ between k and $k(X)$ and prove the surjectivity of the natural maps $I_{p,n}(k) \to I_{p,n}(L)$ and $I_{p,n}(L) \to I_{p,n}(k(X))$ separately. Consider the maximal height 1 purely inseparable extension $\tilde{k}|k$ encountered in Fact 9.1.6. We have $\tilde{k}(X)^p = L$, and moreover X splits over $\tilde{k}$ by Lemma 9.1.7. It follows that $\tilde{k}(X) \cong \tilde{k}(t_1, \ldots, t_m)$ with some indeterminates t_i, and hence $L = \tilde{k}(X)^p \cong k(t_1^p, \ldots, t_m^p)$ is purely transcendental over k. By the first part of the proof this implies the surjectivity of the map $I_{p,n}(k) \to I_{p,n}(L)$.

Finally, the extension $k(X)|L$ satisfies $k(X)^p \subset L$. This implies the surjectivity of the map $I_{p,n}(L) \to I_{p,n}(k(X))$ as follows. Pick an element α in $I_{p,n}(k(X))$. By definition, there exists $\beta \in K_n^M(K(X))$ with $\alpha = \sigma(\beta) - \beta$. Since $\alpha \in K_n^M(K(X))\{p\}$, it maps to 0 in $K_n^M(K(X))/pK_n^M(K(X))$ by Corollary 9.5.6. Hence by Lemma 9.8.2 there exist $\gamma \in K_n^M(k(X))$ and $\delta \in K_n^M(K(X))$ with $\beta = \iota_{K(X)|k(X)}(\gamma) + p\delta$. But since $\iota_{K(X)|k(X)}(\gamma)$ is fixed by σ, we have $p(\sigma(\delta) - \delta) = \sigma(\beta) - \beta = \alpha$, which means that α is divisible by p in $I_{p,n}(k(X))$. Thus by a straightforward induction it is also divisible by p^n. But the multiplication-by-p map on $k(X)^\times$ induces the multiplication-by-p^n map on $K_n^M(k(X))$ by n-linearity of Milnor K-theory, from which we conclude that α is in the image of the natural map $I_{p,n}(k(X)^p) \to I_{p,n}(k(X))$. Therefore it also comes from $I_{p,n}(L)$. □

Proof of Proposition 9.8.3 We consider the same tower of fields

$$k = F_0 \subset F_1 \subset F_2 \subset F_3 \subset \cdots \subset F_\infty = \bigcup_n F_n$$

as in the proof of Theorem 8.4.1: for n odd the field F_n is the maximal prime-to-p extension of F_{n-1}, and for n even F_n is the composite of function fields of all Severi–Brauer varieties associated with cyclic algebras split by $KF_{n-1}|F_{n-1}$. By applying Lemmas 9.8.5 and 9.8.6 successively, we conclude that the natural map $V_{p,n}(k) \to V_{p,n}(F_\infty)$ is injective. As shown during the proof of Theorem 8.4.1, the field norm $N_{KF_\infty|F_\infty}$ is surjective, and F_∞ has no nontrivial extension of degree prime to p. It remains to apply Lemma 9.8.4. □

Proof of Theorem 9.8.1 We use induction on n for all fields of characteristic p, the case $n = 1$ being obvious. Consider a purely transcendental extension $k(t)$ of k. Let κ be the fixed field of the automorphism $\sigma : k(t) \to k(t)$ that is the identity on k and sends t to $t + 1$. The extension $k(t)|\kappa$ is a degree p cyclic Galois extension, and κ contains k.

Given $\alpha \in {}_pK_n^M(k)$, its image α_κ in $K_n^M(\kappa)$ is still a p-torsion element, and moreover $\alpha_{k(t)} := \iota_{k(t)|k}(\alpha)$ satisfies

$$N_{k(t)|\kappa}(\alpha_{k(t)}) = N_{k(t)|\kappa}(\iota_{k(t)|\kappa}(\alpha_\kappa)) = p\alpha_\kappa = 0.$$

Thanks to the inductive assumption, we may apply Proposition 9.8.3 to the extension $k(t)|\kappa$, and find $\beta \in K_n^M(k(t))$ satisfying $(\sigma - 1)(\beta) = \alpha_{k(t)}$. Now consider the specialization homomorphism $s_{t-1}^M : K_n^M(k(t)) \to K_n^M(k)$ splitting the exact sequence of Theorem 7.2.1. By its construction, we have $s_{t-1}^M \circ \sigma = s_{t-1}^M$ for the automorphism σ considered above, and therefore

$$s_{t-1}^M(\alpha_{k(t)}) = s_{t-1}^M((\sigma - 1)(\beta)) = 0.$$

But $s_{t-1}^M(\alpha_{k(t)}) = \alpha$, which proves the theorem. □

Remark 9.8.7 In Theorem 8.1 of Geisser–Levine [1] there is another proof of Theorem 9.8.1 based on a vanishing theorem for motivic cohomology groups. It uses only the modulo p differential symbol, but not its de Rham–Witt extension.

As a consequence of the results of this section, we obtain the exact sequence of Hilbert's Theorem 90 for the Milnor K-theory of the extension $K|k$, which in characteristic $p > 0$ can be extended on the left (unlike in characteristic 0).

Theorem 9.8.8   *Let $K|k$ be a degree p cyclic Galois extension of fields of characteristic $p > 0$. The sequence*

$$0 \to K_n^M(k) \xrightarrow{\iota_{K|k}} K_n^M(K) \xrightarrow{\sigma-1} K_n^M(K) \xrightarrow{N_{K|k}} K_n^M(k) \qquad (9.34)$$

of Milnor K-groups is exact, where σ is a generator of the Galois group $G = \mathrm{Gal}\,(K|k)$.

Proof   Since $N_{K|k} \circ \iota_{K|k}$ is multiplication by p, the kernel of $\iota_{K|k}$ is p-torsion, hence trivial by Theorem 9.8.1. To prove exactness at the second term, pick $\alpha \in \ker(\sigma - 1)$. By Lemma 9.8.2 exactness holds modulo p, so we find $\beta \in K_n^M(k)$ and $\gamma \in K_n^M(K)$ with $\alpha = i_{K|k}(\beta) + p\gamma$. But then $\sigma(p\gamma) = p\gamma$, or in other words $p(\sigma(\gamma) - \gamma) = 0$. By Theorem 9.8.1 this implies $\sigma(\gamma) = \gamma$. But then

$$p\gamma = N\gamma := \sum_{i=0}^{p-1} \sigma^i(\gamma).$$

Since $N = \iota_{K|k} \circ N_{K|k}$, we conclude that $p\gamma \in \mathrm{Im}(\iota_{K|k})$ and thus also $\alpha \in \mathrm{Im}(\iota_{K|k})$, as was to be shown.

To prove exactness at the third term, consider the exact sequence of G-modules

$$0 \to K_n^M(K) \xrightarrow{p} K_n^M(K) \to \nu(n)_K \to 0 \qquad (9.35)$$

coming from Theorem 9.5.2 and Theorem 9.8.1.

Taking G-invariants, we obtain an exact sequence

$$0 \to K_n^M(k) \xrightarrow{p} K_n^M(k) \to \nu(n)_k \to 0$$

since $K_n^M(K)^G \cong K_n^M(k)$ by the exactness of sequence (9.34) at the first two terms, and moreover $\nu(n)_K^G = \nu(n)_k$ by exact sequence (9.33). Since Theorem 9.5.2 also holds over k, we have surjectivity on the right.

But part of the long exact G-cohomology sequence coming from (9.35) reads

$$K_n^M(K)^G \to \nu(n)_K^G \to H^1(G, K_n^M(K)) \xrightarrow{p} H^1(G, K_n^M(K)).$$

Since G has order p, we have $pH^1(G, K_n^M(K)) = 0$, and hence the exact sequence implies $H^1(G, K_n^M(K)) = 0$ by surjectivity of $K_n^M(k) \to \nu(n)_k$. Finally, we have $H^1(G, K_n^M(K)) \cong \ker(N)/\mathrm{Im}(\sigma - 1)$ by the cohomology of cyclic groups. Applying the formula $N = \iota_{K|k} \circ N_{K|k}$ once more, we obtain the required exactness. $\qquad\square$

EXERCISES

1. Show that Proposition 9.1.2 is true for an arbitrary finite extension $K|k$, i.e. that the boundary map $\delta_n : H^1(k, \mathrm{PGL}_n) \to \mathrm{Br}\,(k)$ always induces a bijection

$$\ker(H^1(k, \mathrm{PGL}_n) \to H^1(K, \mathrm{PGL}_n)) \cong \mathrm{Br}\,(K|k),$$

where $n = [K : k]$.

2. Assume that $\mathrm{char}(k) = 2$. Let $A_1 = [a_1, b_1)$ and $A_2 = [a_2, b_2)$ be two central simple algebras of degree 2. Construct an explicit simple purely inseparable field extension $K|k$ which splits both A_1 and A_2.

3. Let K be a field of characteristic zero complete with respect to a discrete valuation, with residue field κ of characteristic $p > 0$. Denote by A the valuation ring of K, by $\widetilde{A}$ the maximal unramified extension of A, and by G the absolute Galois group $\mathrm{Gal}\,(\kappa_s|\kappa)$.

   (a) Construct a natural isomorphism $H^2(G, \widetilde{A}^\times) \xrightarrow{\sim} H^2(G, \kappa_s^\times) \cong \mathrm{Br}\,(\kappa)$ and a canonical embedding $\rho : \mathrm{Br}\,(\kappa) \hookrightarrow \mathrm{Br}\,(K)$. [*Hint:* Argue as in the proof of Proposition 6.3.1.]
   (b) Show that ρ maps Brauer classes of cyclic κ-algebras to Brauer classes of cyclic K-algebras.
   (c) Assume moreover that K contains the p-th roots of unity. Show without using the Merkurjev–Suslin theorem that the subgroup $\rho({}_p\mathrm{Br}\,(\kappa))$ lies in the image of the Galois symbol $h_{K,p}^2 : K_2^M(K)/pK_2^M(K) \to {}_p\mathrm{Br}\,(K)$.

4. Keeping the assumptions and the notations of the previous exercise, let A_1 be a central simple k-algebra, and let A_2 be a central simple K-algebra such that $[A_2] = \rho([A_1])$.

   (a) Show that $\mathrm{ind}_K(A_2)$ divides $\mathrm{ind}_k(A_1)$. [*Hint:* Use Hensel's lemma and Proposition 4.5.1.]
   (b) Let $K'|K$ be a finite Galois extension which splits A_2, and let κ' be its residue field. Show that κ' is a splitting field of A_1.
   (c) Conclude that if A_2 is isomorphic to a cyclic algebra, then so is A_1. [In particular, if $p = 2$ or 3 and A_1 has degree p, then A_1 is cyclic by Proposition 1.2.3 and Chapter 7, Exercise 9.]

5. (Katz) Let assumptions and notations be as in Theorem 9.3.6. This exercise gives an explicit formula for the projection $P : V \to V^\nabla$ whose existence is implied by the direct sum decomposition of V constructed in the proof of that theorem. Let $a_1, \ldots, a_m$ be a p-basis of $K|k$ and ∂_i the derivation sending da_i to 1 and da_j to 0 for $i \neq j$. Define the map $P : V \to V$ by

$$P = \sum_\lambda \prod_{i=1}^m \left(\frac{(-a_i)^{\lambda_i}}{\lambda_i!} \right) \prod_{i=1}^m \nabla_*(\partial_i)^{\lambda_i},$$

where the sum is taken over all m-tuples $\lambda = (\lambda_1, .., \lambda_m) \in \mathbf{F}_p^m$. Show that $\mathrm{Im}\,(P) \subset V^\nabla$ and P induces the identity map on V^∇.

6. Let K be a field of characteristic $p > 0$ satisfying $[K : K^p] = p$. Let b be a generator of the extension $K|K^p$; then db generates the 1-dimensional K-vector space Ω^1_K. Define a map $C_b : \Omega^1_K \to \Omega^1_K$ as follows. Write $\omega \in \Omega^1_K$ uniquely as $\omega = f\,db$ with $f \in K$, and let $c_0, \ldots, c_{p-1} \in K$ be the unique elements with $f = \sum_{i=0}^{p-1} c_i^p b^i$. Then put $C_b(\omega) := c_{p-1}\,db$.

   (a) Show that C_b equals the Cartier operator $C : \Omega^1_K \to \Omega^1_K$. [Note that $\Omega^1_K = Z^1_K$ in this case.]
   (b) (Tate) Verify directly that C_b does not depend on the choice of b.

   [*Remark:* The above construction of Tate [1] was the first explicit appearance of the Cartier operator. Serre [1] showed using this description that in the case when K is the function field of a curve X over an algebraically closed field, the restriction of the map C to global 1-forms identifies to the dual of the Frobenius map on $H^1(X, \mathcal{O}_X)$ via the duality that bears his name.]

7. (suggested by Bouw and Wewers) This exercise gives a simple proof of a special case of Theorem 9.2.2. Let k be an algebraically closed field of characteristic $p > 0$, and consider the rational function field $k(t)$ as a subfield of $k((t))$. Let $f \in k(t)$ be a rational function with Laurent series expansion $f = \sum c_i t^i$, and put $\omega = f\,dt$.

   (a) Show that the following are equivalent:

      1. $\omega \in \ker(\gamma - \mathrm{id})$;
      2. $f^p = -\partial_t^{p-1}(f)$, where ∂_t is derivation with respect to t;
      3. $c_i^p = c_{(i+1)p-1}$ for all $i \in \mathbf{Z}$.

   (b) Deduce that if $\omega \in \ker(\gamma - \mathrm{id})$, then $c_i = 0$ for $i < -1$ and $c_{-1} \in \mathbf{F}_p$.
   (c) Let $a_1, \ldots, a_m \in k$ be the finite poles of f and $c_{-1,1}, \ldots, c_{-1,m}$ the corresponding residues. Deduce from (2) that if $\omega \in \ker(\gamma - \mathrm{id})$, then $\omega = \sum c_{-1,j}(t - a_j)^{-1}\,dt$ and moreover $c_{-1,j} \in \mathbf{F}_p$. [*Hint:* Use the fact from algebraic geometry that there are no differential forms which are everywhere regular on the projective line.]
   (d) Conclude that $\omega \in \ker(\gamma - \mathrm{id})$ if and only if $\omega = \mathrm{dlog}(P)$ for a suitable polynomial $P \in k[t]$.

8. Let k be algebraically closed of characteristic $p > 0$, and let $k(y)|k(x)$ be the extension of rational function fields given by the Artin–Schreier equation $y^p - y = x$. Consider the differential 1-form $\omega = yx^{-1}dx$ in $\Omega^1_{k(y)}$.

   (a) Show that ω is a logarithmic 1-form.
   (b) Find an explicit polynomial $P \in k[y]$ such that $\omega = \mathrm{dlog}(P)$. [*Hint:* Use the previous exercise and partial fraction decomposition.]

Appendix: a breviary of algebraic geometry

This Appendix is strictly utilitarian: we have assembled here some basic notions from algebraic geometry and related algebra needed in the main text (except for the first three sections of Chapter 8, which are more advanced), mostly with references to standard textbooks. Accordingly, the treatment here is far from being the most general or elegant one; its sole purpose is to present the needed facts as quickly as possible. Readers should consult it at their peril.

A.1 Affine and projective varieties

In the present-day literature, by an *algebraic variety* one usually means a separated scheme of finite type over a field, together with a possible integrality condition. In most of this book we only need the notion of affine and projective varieties, which may be defined in a more elementary way. As in the standard texts they are usually discussed only over an algebraically closed base field, we briefly recall the basics.

In what follows k will be a field, and $\bar{k}$ a fixed algebraic closure of k. Points of *affine n-space* $\mathbf{A}^n_{\bar{k}}$ over $\bar{k}$ may be identified with $\bar{k}^n$. An *affine closed subset* of $\mathbf{A}^n_{\bar{k}}$ is defined as the locus of common zeroes of a finite set of polynomials $f_1, \ldots, f_m \in \bar{k}[x_1, \ldots, x_n]$; we denote it by $X = V(f_1, \ldots, f_m)$. The f_i are of course not uniquely determined by X. We say that X is *defined over* k if there is a representation $X = V(f_1, \ldots, f_m)$ with $f_i \in k[x_1, \ldots x_n]$ for all i. In this case the quotient ring $\mathcal{O}(X) := k[x_1, \ldots, x_n]/(f_1, \ldots, f_m)$ is called the *coordinate ring* of X. Moreover, if the ideal $I(X) := (f_1, \ldots, f_m)$ is a prime ideal in $k[x_1, \ldots x_n]$ and the ring $\mathcal{O}(X) \otimes_k \bar{k}$ has no nilpotent elements, we say that X is an *affine variety* defined over k.

Remark A.1.1 The condition that $\mathcal{O}(X) \otimes_k \bar{k}$ has no nilpotent elements is not always part of the definition of a variety in the literature. It is, however, imposed in basic texts on algebraic groups such as Springer [1]. Note that if $\mathcal{O}(X)$ itself has no nilpotents and k is perfect, then neither has $\mathcal{O}(X) \otimes_k \bar{k}$ (see e.g. Matsumura [1], Theorem 26.3). This property does not hold over non-perfect fields.

Note that if X is an affine variety over k, it is not necessarily an affine variety over $\bar{k}$ (think of the ideal $(x^2 + 1)$ in $\mathbf{R}[x]$). Therefore from now on we employ the notation $X_{\bar{k}}$ for X considered as an affine closed set over $\bar{k}$. More generally, for an algebraic extension $k \subset L \subset \bar{k}$ we define the *base change* X_L of X to L as the closed set defined over L by $(f_1, \ldots, f_m)$, with the f_i regarded as polynomials with coefficients in L. This looks bizarre at first sight, but notice the difference in the coordinate rings: $\mathcal{O}(X_L) \cong \mathcal{O}(X) \otimes_k L$.

In general, the partially ordered set of prime ideals in $k[x_1, \ldots, x_n]$ containing $I(X)$ has finitely many minimal elements; the varieties determined by these are called the *irreducible components* of X over k. Over $\bar{k}$ the irreducible components are irreducible closed subsets in $\mathbf{A}_{\bar{k}}^n$, i.e. they cannot be written as a union of two proper closed subsets. The system of affine closed sets defined over k is closed under finite unions and arbitrary intersections, and hence it defines a topology on $\mathbf{A}_k^n$ called the *Zariski topology*. Subsets of $\mathbf{A}_k^n$ are always equipped with the induced topology. If $Y \subset X$ are affine varieties and Y is closed in the Zariski topology of X, we say that Y is a *closed subvariety* of X.

Over $\bar{k}$, a point $P = (a_1, \ldots, a_n)$ of $\mathbf{A}_{\bar{k}}^n$ corresponds to the maximal ideal $(x_1 - a_1, \ldots x_n - a_n)$ in $\bar{k}[x_1, \ldots, x_n]$, so it is an affine variety. By Hilbert's Nullstellensatz (Matsumura [1], Theorem 5.3) all maximal ideals in $\bar{k}[x_1, \ldots x_n]$ are of this form. Now let X be an affine closed set over k. In the theory of schemes one considers all prime ideals containing $I(X)$ as points of X; we refer to these as *scheme-theoretic points*. A scheme-theoretic point P such that $(k[x_1, \ldots, x_n]/P) \otimes_k \bar{k}$ has no nilpotents corresponds to a closed subvariety Y of X. In this situation we say that P is the *generic point* of Y. Scheme-theoretic points correspond bijectively to prime ideals of the coordinate ring $\mathcal{O}(X)$ of X. Of particular interest are the scheme-theoretic points associated with maximal ideals; these are called *closed points*. By a more general form of the Nullstellensatz (same reference as above), their residue fields are finite extensions L of k; in this situation we speak of *points defined over L*. In the case $L = k$ they are called *k-rational points* or *k-points* for short. By the above, over $\bar{k}$ all closed points are $\bar{k}$-rational.

We now move on to projective varieties. Points of *projective n-space* $\mathbf{P}_{\bar{k}}^n$ over $\bar{k}$ may be identified with the elements in $\bar{k}^{n+1} \setminus \{(0, \ldots, 0)\}$ modulo the equivalence relation $(a_0, \ldots, a_n) \sim (\lambda a_1, \ldots \lambda a_{n+1})$ for $\lambda \neq 0$. A *projective closed subset* $X = V(F_1, \ldots, F_m)$ in $\mathbf{P}^n_k$ is defined as the locus of common zeroes of a finite set of *homogeneous* polynomials $F_1, \ldots, F_m \in \bar{k}[x_0, \ldots, x_n]$. We say that X is *defined over* k if there is a representation with the F_i lying in $k[x_0, \ldots x_n]$ for all i; the quotient $\mathcal{O}(X) := k[x_0, \ldots, x_n]/(F_1, \ldots, F_m)$ is its *homogeneous coordinate ring*. Projective closed subsets define the Zariski topology of $\mathbf{P}^n$. If the ideal $I(X) := (F_1, \ldots, F_m)$ is a prime ideal in $k[x_1, \ldots x_{n+1}]$ and $\mathcal{O}(X) \otimes_k \bar{k}$ has no nilpotent elements, we say that X is a *projective variety* over k. Henceforth by a 'variety' we shall mean an affine or a projective variety, but in fact all results stated for varieties will hold with the more general definition alluded to at the beginning of this section. Also, notice that the spaces $\mathbf{A}^n_k$ and $\mathbf{P}^n_k$ are both k-varieties (they are even defined over the prime field of k), so we drop the subscripts k and $\bar{k}$ in what follows if clear from the context.

Projective n-space $\mathbf{P}^n$ has a standard open covering by $n + 1$ copies of $\mathbf{A}^n$, defined by $D_+(x_i) := \mathbf{P}^n \setminus V(x_i)$. The inclusions $D_+(x_i) \hookrightarrow \mathbf{P}^n$ are homeomorphisms in the Zariski topology. Therefore each projective closed set X has a standard open covering by affine closed sets $X^{(i)} := X \cap D_+(x_i)$. We define its scheme-theoretic points (resp. closed points, k-rational points) as the union of those of the $X^{(i)}$; one checks that this notion does not depend on the choice of i. Scheme-theoretic points correspond to closed subvarieties in the projective case as well (this follows from the easy topological fact that $Y \subset X$ is closed if and only if the $Y \cap D_+(x_i)$ are closed in the $X \cap D_+(x_i)$), so the notion of generic point extends to the projective case.

Proposition A.1.2 *Assume that k is separably closed. Then each variety defined over k has a k-rational point. Moreover, k-rational points are dense in the Zariski topology of X.*

Proof See e.g. Springer [1], Theorem 11.2.7. □

The *local ring* $\mathcal{O}_{X,P}$ of a scheme-theoretic point P on an affine variety X is defined as the localization of $\mathcal{O}(X)$ by the prime ideal P. If $Y \subset X$ is the closed subvariety coming from P, one also speaks of the *local ring of the subvariety* Y and uses the notation $\mathcal{O}_{X,Y}$ for $\mathcal{O}_{X,P}$. The residue field $\kappa(P)$ of this local ring is the *residue field* of the point P; for closed points it is a finite extension of k. In the case of a closed point P we say that P is a *smooth point* if the Jacobian matrix of a minimal system of generators of P has a maximal subdeterminant whose image in $\kappa(P)$ is nonzero; one checks that this does not depend on the choice of the generators. The nonsmooth points are called *singular points*. They form a Zariski closed subset in X (meaning that the singular points of $X_{\bar{k}}$ form a closed subset defined over k) defined by the vanishing of the maximal subdeterminants. This subset is the *singular locus* of X. Under our assumption that $\mathcal{O}(X) \otimes_k \bar{k}$ has no nilpotent elements it is always a proper closed subset of X (see e.g. Springer [1], Theorem 4.3.3). If it is empty, we say that X is *smooth over* k.

For a projective variety X one defines the local ring of a scheme-theoretic point P as $\mathcal{O}_{X^{(i)},P}$ for an $X^{(i)}$ containing P; one checks that it does not depend on i. Likewise, the notion of smooth closed points extends to the projective case. The *function field* $k(X)$ of a variety X defined over k is defined as the common fraction field of its local rings.

A.2 Maps between varieties

A *rational function* f on a variety X is an element of its function field $k(X)$. If P is a closed point and $f \in \mathcal{O}_{X,P}$, we may define its *evaluation* $f(P)$ at P as the image of f in the residue field $\kappa(P)$. If P is a k-rational point, then $\kappa(P) = k$ and this is an honest evaluation map. Indeed, the local ring $\mathcal{O}_{X,P}$ being a localization of $\mathcal{O}(X)$, we may represent f by a quotient of polynomials whose denominator does not vanish at P, and the maximal ideal of $\mathcal{O}_{X,P}$ consists of functions vanishing at P.

Let X be a variety (affine or projective) over k. A *rational map* $X \to \mathbf{P}^n$ is given by an $(n+1)$-tuple $\phi = (f_0, \ldots, f_n) \in k(X)^{n+1}$ of rational functions, not all identically 0. Two $(n+1)$-tuples $(f_0, \ldots, f_n)$ and $(g_0, \ldots, g_n)$ define the same rational map if there exists a rational function $g \in k(X)$ with $f_i = gg_i$ for all i. We say that ϕ is *regular* at a closed point P of X if it may be represented by an $(n+1)$-tuple $(f_0, \ldots, f_n)$ with $f_i \in \mathcal{O}_{X,P}$ for all i and $f_i(P) \neq 0$ for some i. When ϕ is regular at all closed points of X, we say that ϕ is a *morphism*. If $Y \subset \mathbf{P}^n$ is another variety (in the affine case embed it in $\mathbf{P}^n$ by identifying $\mathbf{A}^n$ with $D_+(x_0)$), by a rational map (resp. morphism) $X \to Y$ we mean a rational map (resp. morphism) $X \to \mathbf{P}^n$ as above, so that moreover after passing to $\bar{k}$ we have $(f_0(P), \ldots, f_n(P)) \in Y_{\bar{k}}$ for all closed points $P \in X_{\bar{k}}$ where ϕ is regular.

In the case when X is projective, we may represent rational functions on X by quotients of homogeneous polynomials of the same degree (see Shafarevich [2], Chapter I, Section 4.3). Hence by multiplying with a common denominator, we may represent rational maps $X \to \mathbf{P}^n$ by an $(n+1)$-tuple of homogeneous polynomials of the same degree d.

Example A.2.1 Fix positive integers n and d, and put $N = \binom{n+d}{d} - 1$. We define the *d-uple* (or *Veronese*) *embedding* $\phi_d : \mathbf{P}^n \to \mathbf{P}^N$ by setting

$\phi_d = (x_0^d, \ldots, x_0^{i_0} \cdots x_n^{i_n}, \ldots, x_n^d)$, where we have listed in lexicographic order all monomials of degree d in $x_0, \ldots, x_n$. One checks that ϕ_d is a morphism which embeds $\mathbf{P}^n$ as a closed subvariety into $\mathbf{P}^N$ (see Shafarevich [2], Chapter I, Section 4.4). The homogeneous coordinate ring of $\phi_d(X)$ may be identified with the free k-algebra generated by monomials of degree d.

In the case when $\phi : X \to Y$ is a morphism so that there is a morphism $\psi : Y \to X$ with $\phi \circ \psi$ and $\psi \circ \phi$ identity maps, we say that ϕ is an *isomorphism* between X and Y. Not surprisingly, isomorphisms $X \to X$ are called *automorphisms*.

Example A.2.2 All automorphisms of $\mathbf{P}^n$ are linear, i.e. defined by linear polynomials. In other words, we may identify the automorphism group of *the projective variety* $\mathbf{P}^n$ with the group $\mathrm{PGL}_n(k)$. See Hartshorne [1], Example II.7.1.1 for a proof.

A rational map $\phi : X \to Y$ is said to be *birational* if its image in Y is dense in the Zariski topology (after passing to $\bar{k}$) and there is a rational map $\psi : Y \to X$ with $\phi \circ \psi$ and $\psi \circ \phi$ identity maps at all points where they are regular. Note that these points form open subsets in X, resp. Y: they are the complements of the zero loci of the possible denominators of the rational functions involved. Thus a birational map is actually bijective on points in a Zariski open subset. Another criterion for birationality is that the map $\phi^* : k(Y) \to k(X)$ induced by composing rational functions with ϕ is an isomorphism (see Shafarevich [2], Chapter II, Section 4.3). We need the simplest nontrivial example of a birational map, that of blowing up a point in projective space.

Example A.2.3 Assume that k is algebraically closed. Given two integers $m, n > 0$, set $N := nm + n + m$. One defines the *product* $\mathbf{P}^n \times \mathbf{P}^m$ of projective spaces as the closed subvariety in $\mathbf{P}^N$ obtained as the image of the *Segre embedding* $S_{n,m} : \mathbf{P}^n \times \mathbf{P}^m \to \mathbf{P}^N$. By definition, $S_{n,m}$ sends the pair $((x_0, \ldots, x_n), (y_0, \ldots, y_m))$ of points given in homogeneous coordinates to $(x_0 y_0, \ldots, x_i y_j, \ldots, x_n y_m)$ (with the lexicographic order). For the fact that this is (set-theoretically) an embedding with Zariski closed image we refer to Shafarevich [2], Chapter I, Section 5.1. We keep the coordinates $(x_0, \ldots, x_n, y_0, \ldots, y_m)$ for points in the product.

Now set $m = n - 1$ and $P_0 = (0, \ldots, 0, 1) \in \mathbf{P}^n$. The *blowup* $B_{P_0}(\mathbf{P}^n)$ of $\mathbf{P}^n$ at P_0 is the closed subvariety of $\mathbf{P}^n \times \mathbf{P}^{n-1}$ defined by the polynomials $x_i y_j - x_j y_i$ for all possible $0 \leq i, j \leq n - 1$. The first projection $\pi : \mathbf{P}^n \times \mathbf{P}^{n-1} \to \mathbf{P}^n$ restricted to $B_{P_0}(\mathbf{P}^n)$ is a surjective morphism. It is also birational, for outside P_0 an inverse is given by mapping $(x_0, \ldots, x_n)$ to

$(x_0, \ldots, x_n, x_0, \ldots, x_{n-1})$. However, the inverse image $\pi^{-1}(P_0) \subset B_{P_0}(\mathbf{P}^n)$ is the hyperplane $\{P_0\} \times \mathbf{P}^{n-1}$.

Of course, by composition with an automorphism of $\mathbf{P}^n$ one may define the blowup $B_P(\mathbf{P}^n)$ at any point P. Given a subvariety $X \subset \mathbf{P}^n$ of dimension d containing P as a smooth point, one proves (Shafarevich [2], Chapter II, Section 4.3) that the inverse image of X in $B_P(\mathbf{P}^n)$ consists of two components: the hyperplane $\{P\} \times \mathbf{P}^{n-1}$ and a projective variety which one defines to be the blowup $B_P(X)$ of X at P. Moreover, the restriction of π to $B_P(X)$ is a birational morphism onto X which is an isomorphism outside P, and the preimage of P is a subvariety isomorphic to $\mathbf{P}^{d-1}$. It is called the *exceptional divisor*.

For a discussion of blowups valid over a more general base, see Hartshorne [1], Section II.7.

A.3 Function fields and dimension

Let k be a field and $K|k$ a finitely generated extension. Recall that the *transcendence degree* tr.deg.$(K|k)$ of $K|k$ is defined as the cardinality of a maximal subset of elements of K algebraically independent over k; such a maximal subset is called a *transcendence basis* of $K|k$. If K is the function field of a variety X over k, we define the *dimension* of X to be tr.deg.$(K|k)$. Varieties of dimension 0 are points defined over a finite extension of k; those of dimension 1 are called *curves* and those of dimension 2 *surfaces*. For a closed subvariety $Y \subset X$ we define the *codimension* of Y in X to be $\dim(X) - \dim(Y)$.

On the other hand, for a ring A we have the notion of *Krull dimension* for A: it is the maximum of lengths d of strictly decreasing chains $P_0 \supset P_1 \supset \cdots \supset P_d$ of prime ideals in A. Similarly, one defines the *height* of a prime ideal P in A as the maximum of lengths d of strictly decreasing chains $P \supset P_1 \supset \cdots \supset P_d$ of prime ideals in A. Note that the height of P equals the Krull dimension of the localization A_P.

Proposition A.3.1 *Let X be an affine variety over a field k, and let $\mathcal{O}(X)$ be its coordinate ring. The dimension of X equals the Krull dimension of $\mathcal{O}(X)$, and the codimension of each closed subvariety $Y \subset X$ equals the height of the corresponding prime ideal in $\mathcal{O}(X)$.*

Proof See Matsumura [1], Theorem 5.6. □

The notion of Krull dimension extends to a variety X over k: it is defined as the maximum of lengths d of strictly decreasing chains of $Z_1 \supset \cdots \supset Z_d$

of proper nonempty subvarieties of X. For affine varieties we get back the previous notion via the dictionary between subvarieties and prime ideals. One may then derive from the previous proposition:

Corollary A.3.2 *The dimension of a k-variety X equals its Krull dimension.*

Proof See Mumford [1], I.7, Corollary 2. □

We define the dimension of a closed subset in $\mathbf{A}^n$ or $\mathbf{P}^n$ to be the maximal dimension of its irreducible components.

Corollary A.3.3 *Let $f_1, \ldots f_r$ be homogeneous polynomials in $k[t_0, \ldots, t_n]$. If $r \leq n$, the closed subset $V(f_1, \ldots f_r) \subset \mathbf{P}^n$ is nonempty.*

Proof The previous corollary implies that in the chain

$$V(f_1) \supset V(f_1, f_2) \supset \cdots \supset V(f_1, \ldots, f_r)$$

of closed subsets of $\mathbf{P}^n$ the dimension drops by at most 1 in each step. Since $\mathbf{P}^n$ is easily seen to have dimension n, the set $V(f_1, \ldots f_r)$ must be nonempty. □

Recall that the extension $K|k$ is said to be *separably generated* if there exists a transcendence basis $\{t_1, \ldots t_d\}$ of $K|k$ so that the finite extension $K|k(t_1, \ldots t_d)$ is separable. The basic theorem on separably generated extensions is the following.

Proposition A.3.4 *If k is perfect, then every finitely generated extension $K|k$ is separably generated.*

Proof See Matsumura [1], Theorems 26.2 and 26.3 or van der Waerden [1], §155. □

We need the following consequence of this in the main text.

Corollary A.3.5 *Let k be perfect, and let $K|k$ be a finitely generated extension of transcendence degree d. Then K arises as the residue field of a scheme-theoretic point of codimension 1 on $\mathbf{A}_k^{d+1}$.*

Proof By the proposition and the theorem of the primitive element we may write $K = k(t_1, \ldots t_d, a)$ with a separable over $k(t_1, \ldots, t_d)$. Clearing denominators in the minimal polynomial of a we obtain an irreducible polynomial $f \in k[t_1, \ldots t_d, t]$ such that K is the fraction field of $k[t_1, \ldots t_d, t]/(f)$. The prime ideal $(f) \subset k[t_1, \ldots, t_d, t]$ defines the required point. □

A.4 Divisors

Throughout the whole section, X will denote a variety over a field k *all of whose local rings are unique factorization domains.* For example, this is the case when X is smooth over a perfect field, the only one needed in this book.

Denote by $\mathrm{Div}(X)$ the free abelian group with basis the irreducible closed subvarieties in X that are of codimension 1. The elements of $\mathrm{Div}(X)$ are called *(Weil) divisors,* and $\mathrm{Div}(X)$ itself is the *group of divisors* on X. An element $D \in \mathrm{Div}(X)$ is thus of the form $D = \Sigma m_Y Y$ with some $m_Y \in \mathbf{Z}$ and Y irreducible of codimension 1; the union of the Y appearing with nonzero coefficients is called the *support* of D.

Given a codimension 1 irreducible subvariety $Y \subset X$, the local ring $\mathcal{O}_{X,Y}$ is of dimension 1 and it is a unique factorization domain by assumption, so it is a discrete valuation ring (Matsumura [1], Theorem 11.2). Denote by $v_Y : k(X) \to \mathbf{Z} \cup \{\infty\}$ the associated discrete valuation.

Lemma A.4.1 *For a given nonzero rational function $f \in k(X)^{\times}$ there are only finitely many $Y \subset X$ as above with $v_Y(f) \neq 0$.*

Proof   See Hartshorne [1], Lemma II.6.1.   □

In view of the lemma, we may define the *divisor of the rational function* $f \in k(X)^{\times}$ by

$$\mathrm{div}(f) := \sum v_Y(f) \, Y \in \mathrm{Div}(X),$$

where the sum is over all codimension 1 irreducible subvarieties in X. The additivity of discrete valuations implies that in this way we get a group homomorphism $\mathrm{div}: k(X)^{\times} \to \mathrm{Div}(X)$. We denote the cokernel of this morphism by $\mathrm{Pic}\,(X)$ and call it the *Picard group* of X. Two divisors D_1 and D_2 are said to be *linearly equivalent* if they have the same class in the Picard group.

Remark A.4.2   The term 'Picard group' is nowadays mostly used for the group of isomorphism classes of line bundles on a variety, and the group defined above is sometimes called the *divisor class group* (e.g. in Hartshorne [1]). However, under our assumption on X the two notions coincide (see Hartshorne [1], Prop. 6.11 and 6.13).

Proposition A.4.3   *Assume moreover that X is projective. Then the sequence*

$$1 \to k^{\times} \to k(X)^{\times} \xrightarrow{\mathrm{div}} \mathrm{Div}(X) \to \mathrm{Pic}\,(X) \to 0$$

is exact.

Proof Everything results from the definitions, except exactness at $k(X)^\times$, which is a consequence of the fact that on a projective variety all regular functions are constant (see Hartshorne [1], Theorem I.3.4 for a proof over an algebraically closed base field; the general case follows immediately). □

The next proposition gives the Picard groups of affine and projective space. Note that the Picard group of a point is by definition 0.

Proposition A.4.4

1. *For any X as above there is an isomorphism* $\text{Pic}(X) \xrightarrow{\sim} \text{Pic}(X \times \mathbf{A}^1)$. *Hence* $\text{Pic}(\mathbf{A}^d) = 0$ *for all $d > 0$.*
2. *For all $d > 0$ we have* $\text{Pic}(\mathbf{P}^d) \cong \mathbf{Z}$, *a positive generator being given by the class of a hyperplane.*

Proof See Hartshorne [1], Propositions II.6.6 and II.6.4, or apply Propositions 8.2.5 and 8.2.6 of this book (with $i = d - 1$ and $n = 1 - d$). □

For a divisor class $[D]$ on $\mathbf{P}^d$ one defines its *degree* as its image in $\mathbf{Z}$ by the isomorphism of statement (2) above.

Remark A.4.5 Assume that k is perfect, $\bar{k}$ is an algebraic closure of k, and $\overline{X} := X \times_k \bar{k}$ is irreducible. The Galois group $G := \text{Gal}(\bar{k}|k)$ permutes the codimension 1 subvarieties of $\overline{X}$, and thus it acts on the group $\text{Div}(\overline{X})$. Moreover, the map div is compatible with the Galois actions on $\bar{k}(\overline{X})^\times$ and $\text{Div}(\overline{X})$, so there is an induced action on $\text{Pic}(\overline{X})$. Needless to say, all the above holds at the level of finite Galois extensions of the base field as well.

For X projective there is a relation between divisors and rational maps via the notion of a *linear system*. First some terminology: we say that a divisor is *positive* and write $D \geq 0$ if $D = \Sigma m_Y Y$ with all $m_Y \geq 0$. The *complete linear system* $|D|$ associated with an arbitrary divisor D on X is then defined as the set of positive divisors linearly equivalent to D. This set carries the structure of a projective space over k. Indeed, write $L(D)$ for the k-vector space of functions $f \in k(X)^\times$ satisfying $\text{div}(f) + D \geq 0$; it has finite dimension over k by Shafarevich [2], Chapter VI, Corollary 1 of Section 3.4. For $f \in L(D)$ the element $D_f = \text{div}(f) + D$ lies in $|D|$, and conversely, each element of $|D|$ is of the above form. Since by Proposition A.4.3 the divisor $\text{div}(f)$ determines f up to a constant in $k^\times$, elements in $|D|$ are in bijection with the projectivization of the k-vector space $L(D)$.

A *linear system* $\mathcal{D}$ on X is then by definition a projective linear subspace of some $|D|$ as above; it comes from a subspace $M_\mathcal{D}$ of $L(D)$. Choosing a k-basis $(f_0, \ldots, f_m)$ of $M_\mathcal{D}$ defines a rational map $\phi_\mathcal{D} : X \to \mathbf{P}^m$; the choice of a different basis yields a map which differs by an automorphism of $\mathbf{P}^m$. If the linear system is *base point free*, i.e. if there is no closed point of X contained in the support of all divisors in $\mathcal{D}$, then the map is actually a morphism. The divisors in $\mathcal{D}$ are then the pullbacks to X of the divisors on $\phi_{\mathcal{D}}(X)$ obtained by intersecting $\phi_{\mathcal{D}}(X)$ with hyperplanes in $\mathbf{P}^m$ (see Shafarevich [2], Chapter III, Section 1.5 or Hartshorne, Remark II.7.8.1 for more details). For instance, given a hyperplane H on $\mathbf{P}^n$ and an integer $d > 0$, the complete linear system $|dH|$ on $\mathbf{P}^n$ yields the d-uple embedding of Example A.2.1. A divisor D is called *ample* if the linear system $|mD|$ induces a closed embedding into some projective space for a suitable $m > 0$.

In the case when X is a *curve*, divisors on X are finite $\mathbf{Z}$-linear combinations of closed points on X. Thus it makes sense to define the *degree* of a divisor $D = \Sigma n_P P$ on X as $\deg(D) := \Sigma n_P [\kappa(P) : k]$. In this way we obtain a homomorphism deg: $\mathrm{Div}(X) \to \mathbf{Z}$ of abelian groups called the *degree map*, not to be confused with the one defined above for projective spaces. The fundamental fact about it is:

Proposition A.4.6   *Let X be a smooth projective curve. For each $f \in k(X)^\times$ one has* $\deg(\mathrm{div}(f)) = 0$.

Proof   See Hartshorne [1], Corollary II.6.10; the proposition is stated there under the extra assumption that k is algebraically closed, but the proof works in general. $\square$

As a consequence we obtain that the degree map factors through the Picard group and induces a map $\mathrm{Pic}(X) \to \mathbf{Z}$.

A.5 Complete local rings

Let A be a ring and $I \subset A$ an ideal. The *I-adic completion* $\widehat{A}$ of A is defined as the inverse limit of the natural inverse system of quotients A/I^n for all $n \geq 0$ (see Chapter 4 for basics about inverse limits). More generally, the I-adic completion of an A-module N is defined as the inverse limit of the A-modules $N/I^n N$. We say that A is *I-adically complete* if the natural map $A \to \widehat{A}$ is an isomorphism. In the case when A is a local ring with maximal ideal M we simply say completion instead of M-adic completion. Similarly, for semi-local rings (i.e. rings with finitely many maximal ideals) by completion we mean completion with respect to the intersection of maximal ideals. In fact, if A is a semi-local ring with maximal ideals $M_1, \ldots, M_r$, then the completion of A is isomorphic to the direct product of the completions of the localizations A_{M_i} (Matsumura [1], Theorem 8.15).

For a Noetherian local ring the map $A \rightarrow \widehat{A}$ is always injective by the Krull intersection theorem (Matsumura [1], Theorem 8.10). Another useful fact about Noetherian local rings is:

Lemma A.5.1 *If A is a Noetherian local ring and $J \subset A$ is an ideal, then the completion of the A-module J is isomorphic to $J\widehat{A}$, and moreover $\widehat{A/J} \cong \widehat{A}/J\widehat{A}$.*

Proof See Matsumura [1], Theorem 8.11. □

The following is a delicate relation between integral closure and normalization due to Zariski. Following Grothendieck [4], we formulate it using the general concept of excellent rings defined in §7.8 of *loc. cit.*, but for our applications it suffices to know that local rings of varieties over some field are excellent (Grothendieck [4], (7.8.3) (ii), (iii)).

Theorem A.5.2 *If A is an excellent Noetherian local domain, then the completion $\widehat{A}$ has no nilpotent elements, and the integral closure $\bar{A}$ of A is a finitely generated A-module. Moreover, the completion of the semi-local ring $\bar{A}$ is isomorphic to the integral closure of $\widehat{A}$ in its total ring of fractions.*

Proof In the case when A is a local ring of a variety over a *perfect* field, a proof is given in the last section of Zariski–Samuel [1]. The general case follows from Grothendieck [4], (7.6.1) and (7.8.3) (vi). □

There is a beautiful structure theory for complete local rings. We first state the case of equal characteristic discrete valuation rings, which is the one we use most often.

Proposition A.5.3 *Let A be a complete discrete valuation ring with fraction field K and residue field κ. If $\mathrm{char}(K) = \mathrm{char}(\kappa)$, then A is isomorphic to the formal power series ring $\kappa[[t]]$.*

Proof See Serre [2], Section II.4 for a direct proof, or the references for Theorem A.5.4 below. □

The general structure theorem is the following.

Theorem A.5.4 (Cohen Structure Theorem) *Let A be a complete Noetherian local domain with residue field κ and fraction field K.*

1. *There exists a complete discrete valuation ring B with residue field κ and a power series ring $C = B[[t_1, \ldots, t_d]]$ such that $C \subset A$ and A is a finitely generated C-module.*
2. *If A is regular of dimension $d + 1$ and $\text{char}(K) = \text{char}(\kappa)$, then we may find C as above with $C \cong A$. In particular, A is a formal power series ring in $d + 1$ variables over κ.*

Proof See Matsumura [1], Theorems 29.4 and 29.7; this reference also contains a complete discussion of the mixed characteristic case. $\square$

We finally discuss Hensel's lemma. In the literature various related results go under this name, of which the most common one is perhaps the following.

Proposition A.5.5 *Let A be a complete discrete valuation ring with maximal ideal M, and let f be a polynomial with coefficients in A. If $a \in A$ satisfies $f(a) = 0 \bmod M$ and $f'(a) \neq 0 \bmod M$, then there exists $b \in A$ with $f(b) = 0$ and $a - b \in M$.*

Proof The proof is by Newton's approximation method; see e.g. Lang [3], Chap. XII, Prop. 7.6, Serre [2], Chap. II. Prop. 7, or the proof of Proposition A.5.6 below. $\square$

In fact, the proposition holds more generally for an arbitrary complete local ring. However, in the following more refined version for a system of polynomials in several variables we need to restrict to discrete valuation rings.

Proposition A.5.6 *Let A be a complete discrete valuation ring with maximal ideal M, and let $f_1, \ldots, f_n \in A[x_1, \ldots, x_n]$ be a system of polynomials. Assume that $P = (a_1, \ldots, a_n) \in A^n$ satisfies*

$$f_i(a_1, \ldots a_n) \in M^N \quad \text{for} \quad 1 \leq i \leq n$$

with $N > 2\nu$, where $\nu \geq 0$ is the valuation of the Jacobian matrix $J(P)$ of the f_i at P. Then one may find $(b_1, \ldots b_n) \in A^n$ with

$$f_i(b_1, \ldots b_n) = 0 \text{ for all } 1 \leq i \leq n, \text{ and } a_j - b_j \in M^{N-\nu} \text{ for } 1 \leq j \leq n.$$

Proof It will be enough to construct vectors $P^{(q)} = (b_1^{(q)}, \ldots, b_n^{(q)})$ for all $q \geq 0$ such that $b_j^{(0)} = a_j$ for $1 \leq j \leq n$, and moreover

1. $f_i(b_1^{(q)}, \ldots b_n^{(q)}) \in M^{N+q}$ for all $1 \leq i \leq n$,
2. $b_j^{(q)} - b_j^{(q+1)} \in M^{N-\nu+q}$ for all $1 \leq j \leq n$, and
3. $J(P^{(q)})$ has valuation ν.

These then converge in A^n to a vector $(b_1, \ldots b_n)$ with the required properties. Using induction on q and reindexing we reduce to the case $q = 1$. Write F for the column vector formed by the f_i and $F(P)$ for the column vector of their evaluations at $P = (a_1, \ldots, a_n)$. Look for $P^{(1)}$ in the form $P^{(1)} = P + t^{N-\nu}Q$, where t is a generator of M and $Q \in A^n$ is a vector to be determined. Taylor's formula in several variables yields an equation

$$F(P^{(1)}) = F(P) + t^{N-\nu}J(P)Q + t^{2N-2\nu}R \qquad (A.1)$$

with a suitable column vector $R \in A^n$. By assumption $F(P) = t^N S$ and $J(P) = t^\nu u$ for some column vector $S \in A^n$ and unit $u \in A \setminus M$. As u is a unit, we may choose Q in such a way that all entries of the vector $S + uQ$ lie in M. With this choice the right-hand side of (A.1) lies in M^{N+1}, so that condition 1 holds for $q = 1$. Condition 2 being automatic, it remains to check condition 3 for $q = 1$. For this we take the Jacobian matrix of both sides in equation (A.1), and obtain $J(P^{(1)}) = J(P) \bmod M^{N-\nu}$. As $J(P)$ has valuation ν and $\nu < N - \nu$ by assumption, condition 3 follows. $\qquad \square$

A.6 Discrete valuations

Recall that a *discrete valuation* on a field K is a map $K \to \mathbf{Z} \cup \{\infty\}$ satisfying the conditions (1) $v(x) = \infty \Leftrightarrow x = 0$, (2) $v(xy) = v(x) + v(y)$, and (3) $v(x + y) \geq \min(v(x), v(y))$. The elements $x \in K$ with $v(x) \geq 0$ form a subring $A_v \subset K$ called the valuation ring of v. This is a local ring whose maximal ideal M_v consists of the elements of positive valuation. The ideal M_v is principal, each element of minimal valuation being a generator. We refer to generators of M_v as *local parameters* for v. A local ring arising as the valuation ring of some discrete valuation of its fraction field is called a *discrete valuation ring*. Other characterizations of discrete valuation rings are:

Lemma A.6.1 *The following are equivalent for an integral domain A.*

1. *A is a discrete valuation ring.*
2. *A is a local principal ideal domain which is not a field.*
3. *A is an integrally closed local domain of Krull dimension 1.*

Proof See Matsumura [1], Theorem 11.2. $\qquad \square$

We collect here some basic results concerning extensions of discrete valuations to finite extensions of the field K. The most important situation where these facts are applied in this book is the following.

Example A.6.2   Let $\phi : X \to Y$ be a finite morphism of smooth curves over a perfect field k. Recall that this means that each closed point Q of Y has a Zariski open neighbourhood $U \subset Y$ so that both U and $V := \phi^{-1}(U)$ are isomorphic to affine varieties, and the coordinate ring $\mathcal{O}(V)$ becomes a finitely generated $\mathcal{O}(U)$-module via the map $\mathcal{O}(U) \to \mathcal{O}(V)$ induced by pulling back functions via ϕ. A nonconstant morphism of smooth projective curves is always finite (see Shafarevich [2], Section II.5, Theorem 8).

By the finiteness condition on ϕ the induced extension $L|K$ is finite, and so is the fibre $\phi^{-1}(Q)$ for the point Q above. The local ring $\mathcal{O}_{Y,Q}$ of Q is a discrete valuation ring, hence it induces a discrete valuation v_Q of the function field K of Y, and similarly the closed points in the fibre $\phi^{-1}(Q)$ induce discrete valuations on the function field L of X. They are exactly the finitely many possible extensions of v_P to L. This holds because X is the *normalization* of Y in L (see Shafarevich [2], Section II.5, Ex. 1), and on the other hand each discrete valuation ring with fraction field L containing $\mathcal{O}_{Y,Q}$ must contain the integral closure of $\mathcal{O}_{Y,Q}$ in L, being an integrally closed domain.

In general, given a finite extension $L|K$, a discrete valuation v on K may always be extended to a discrete valuation w of L (Matsumura [1], Theorem 9.3 (i)), and there are only finitely many such extensions (*loc. cit.*, Corollary to Theorem 11.7). The maximal ideals of their valuation rings satisfy $M_w \cap K = M_v$, hence there is an induced extension $\kappa(w)|\kappa(v)$ of residue fields. When the extension $L|K$ is Galois, the Galois group acts on the extensions w of v to L via $(\sigma, w) \mapsto w \circ \sigma$.

Proposition A.6.3   *Let $L|K$ be a finite Galois extension with group G, and v a discrete valuation on K.*

1. *The Galois group G acts transitively on the set of discrete valuations w extending v to L.*
2. *For each w the induced extension of residue fields $\kappa(w)|\kappa(v)$ is normal. In particular, if it is separable, then it is a Galois extension.*

Proof   See e.g. Serre [2], Chapter I, Propositions 19 and 20. $\square$

Given a discrete valuation v on K, the completion $\widehat{A}_v$ of its valuation ring A_v is again a discrete valuation ring by Lemma A.5.1; denote its fraction field by $\widehat{K}_v$ and keep the notation v for the canonical extension of v to $\widehat{K}_v$. In the case when $\widehat{K}_v = K$ we say that K is *complete with respect to v*.

Proposition A.6.4 *In the above situation let $L|K$ be a finite extension, and assume that the integral closure of A_v in L is a finitely generated A_v-module.*

1. There is a decomposition

$$L \otimes_K \widehat{K}_v \cong \bigoplus_{w|v} \widehat{L}_w,$$

where w runs over the extensions of v to L.

2. If moreover $L|K$ is Galois with group G, then each $\widehat{L}_w|\widehat{K}_v$ is Galois as well, with Galois group isomorphic to the stabilizer of w under the action of G on the set of discrete valuations extending v.

Proof See Serre [2], Chapter II, Theorem 1 and its Corollary 4. □

Remark A.6.5 The assumption of the proposition is satisfied if $L|K$ is separable or if A_v is a localization of a finitely generated algebra over a field (Matsumura [1], Lemmas 1 and 2 of §33 together with Shafarevich [2], Appendix, §8).

Part (2) of the proposition implies:

Corollary A.6.6 *If $L|K$ is a finite extension that can be written as a tower of Galois extensions and w is an extension of a discrete valuation v of K to L, then the degree of the field extension $\widehat{L}_w|\widehat{K}_v$ divides $[L:K]$.*

The statement of the corollary is not true for an arbitrary finite extension (the reader may construct a counterexample).

Returning to an arbitrary finite extension $L|K$ and w extending v, the value group $v(K^\times)$ is a subgroup of finite index in $w(L^\times)$. This index is called the *ramification index* $e(w|v)$ of w above v. The extension is *unramified* if $e(w|v) = 1$ and the residue field extension $\kappa(w)|\kappa(v)$ is separable. More generally, the ramification is called *tame* if the extension $\kappa(w)|\kappa(v)$ is separable and the characteristic of $\kappa(v)$ does not divide $e(w|v)$; otherwise it is *wild*.

Proposition A.6.7 *Let K be a field equipped with a discrete valuation v, and let $L|K$ be a finite extension. Assume that the integral closure of the valuation ring A_v of v in L is a finitely generated A_v-module. Then*

$$\sum_{w|v} e(w|v)[\kappa(w):\kappa(v)] = [L:K],$$

where w runs over the extensions of v to L.

Proof See Serre [2], Section I.4, Proposition 10. □

Proposition A.6.8   *Let K be complete with respect to a discrete valuation v, and $L|K$ a finite extension.*

1. *There is a unique discrete valuation w of L extending v, L is complete with respect to w, and its valuation ring A_w is the integral closure of A_v in L.*
2. *With the notation $f := [\kappa(w) : \kappa(v)]$ one has $w = (1/f)(v \circ N_{L|K})$.*
3. *If the extension $\kappa(w)|\kappa(v)$ is separable, there is a unique unramified extension $N|K$ contained in L satisfying $[N : K] = f$.*
4. *If moreover the ramification is tame, then L is a radical extension of N (i.e. it may be obtained by adjoining a root of a polynomial of the form $x^m - a$).*

Proof   For (1), see Serre [2], Section II.2, Proposition 3; for (2), Corollary 4 of that proposition; for (3), *loc. cit.*, Section III.5, Corollary 2 to Theorem 2, and for (4), Neukirch [1], Chapter II, Theorem 7.7 (and its proof).   □

Remark A.6.9   Under the assumptions of the proposition, the finiteness condition in Proposition A.6.7 is satisfied (see Serre [2], Section II.2, Proposition 3), hence the formula $e(w|v)[\kappa(w) : \kappa(v)] = [L : K]$ holds. It follows that given a tower $M|L|K$ of finite extensions such that $M|L$ and $L|K$ are unramified, then so is $M|K$. Moreover, if $L_1|K$ and $L_2|K$ are unramified extensions within a given separable closure K_s of K, then so is their composite $L_1 L_2$ (Neukirch [1], Chapter II, Corollary 7.3). The composite of all finite unramified extensions of K within K_s is the *maximal unramified extension K_{nr}* of K.

If in the situation of the proposition we assume moreover that the extensions $L|K$ and $\kappa(w)|\kappa(v)$ are Galois, there is a natural group homomorphism $\mathrm{Gal}\,(L|K) \to \mathrm{Gal}\,(\kappa(w)|\kappa(v))$. This map is surjective (Serre [2], Chapter I, Proposition 20); its kernel I is called the *inertia group* of w.

Proposition A.6.10   *Assume moreover that the extension $L|K$ is a tamely ramified Galois extension. Then I is a cyclic subgroup isomorphic to the group $\mu(\kappa(w))$ of roots of unity contained in $\kappa(w)$. Explicitly, this isomorphism maps σ to the image of $\sigma(\pi)\pi^{-1}$ in $\kappa(w)$, where π is an arbitrary local parameter for w.*

Proof   See Serre [2], Chapter IV, Proposition 7 and its corollaries.   □

The proposition implies that for a tamely ramified Galois extension the conjugation action of $\mathrm{Gal}\,(L|K)$ on I induces an action of $\Gamma := \mathrm{Gal}\,(\kappa(w)|\kappa(v))$ on I. The explicit description of the isomorphism $I \cong \mu(\kappa(w))$ implies:

Corollary A.6.11 *The action of* Γ *on* I *corresponds via the isomorphism* $I \cong \mu(\kappa(w))$ *to the natural* Γ-*action on* $\mu(\kappa(w))$. *In particular, if* $\mu(\kappa(w))$ *is contained in* $\kappa(v)$, *then* I *is central in* $\mathrm{Gal}\,(L|K)$.

We conclude with an extension statement to transcendental extensions.

Proposition A.6.12 *Let* A *be a discrete valuation ring with maximal ideal* M *and fraction field* K. *There is a discrete valuation ring* B *whose fraction field is the rational function field* $K(t)$, *whose maximal ideal is* MB, *and moreover* $B \cap K = A$.

Proof Let M be the maximal ideal of A, and take B to be the localization of $A[t]$ by the prime ideal $M[t]$. It is a discrete valuation ring, as it satisfies the criterion of Lemma A.6.1 (3). The other requirements are immediate. $\qquad\square$

A.7 Derivations

Let $A \subset B$ be an extension of rings, and M a B-module. A *derivation* of B into M is a homomorphism of abelian groups $D : B \to M$ satisfying the *Leibniz rule*: $D(b_1 b_2) = b_1 D(b_2) + b_1 D(b_2)$ for all $b_1, b_2 \in B$. The derivation D is an *A-derivation* if $D(a) = 0$ for all $a \in A$; note that this implies $D(ab) = aD(b)$ for all $a \in A$ and $b \in B$ by the Leibniz rule. We denote the set of A-derivations $B \to M$ by $\mathrm{Der}_A(B, M)$; this carries a natural B-module structure given by $(bD)(x) = b \cdot D(x)$ for all $b \in B$. In the case $B = M$ we shall write $\mathrm{Der}_A(B)$ instead of $\mathrm{Der}_A(B, B)$.

Given a derivation $D \in \mathrm{Der}_A(B)$ and an integer $n > 0$, we denote by $D^{[n]}$ the n-th iterate $D \circ \cdots \circ D$. Iterating the Leibniz rule then yields

$$D^{[n]}(ab) = \sum_{i=0}^{n} \binom{n}{i} D^{[i]}(a) D^{[n-i]}(b).$$

Assuming $pB = 0$ for a prime p, the above formula with $n = p$ implies

$$D^{[p]}(ab) = D^{[p]}(a)b + aD^{[p]}(b), \qquad (A.2)$$

i.e. that $D^{[p]}$ is again a derivation.

We also need two other important formulae in characteristic $p > 0$. The best reference for these are the notes of Seshadri [1]; since they are not easily accessible, we give details for the reader's convenience.

Proposition A.7.1 (Hochschild) *Let A be an integral domain of characteristic $p > 0$. For all derivations $D \in \mathrm{Der}_{\mathbf{F}_p}(A)$ and elements $a \in A$ we have*

$$D(a)^p = a^{p-1} D^{[p]}(a) - D^{[p-1]}(a^{p-1} D(a)).$$

Proof   The idea is to reduce to the universal case. Let $A' = \mathbf{F}_p[x_0, x_1, x_2, \dots]$ be the ring of polynomials in infinitely many variables $x_0, x_1, \dots$ Define a derivation D' on A' by $D'(x_n) = x_{n+1}$ for all $n \geq 0$. There exists a unique morphism $\phi : A' \to A$ such that $\phi(x_0) = a$ and $\phi(x_n) = D^{[n]}(a)$ for all $n \geq 1$. We have $D \circ \phi = \phi \circ D'$, so it is enough to prove the formula with $A = A'$, $D = D'$ and $a = x_0$. We shall first prove that the element

$$Q := x_0^{p-1} D^{[p]}(x_0) - D^{[p-1]}(x_0^{p-1} D(x_0))$$

lies in A^p. For this, notice first that $D(Q) = 0$, since

$$D(x_0^{p-1} D^{[p]}(x_0)) = (p-1)x_0^{p-2} D(x_0) D^{[p]}(x_0) + x_0^{p-1} D^{[p+1]}(x_0) = D^{[p]}(x_0^{p-1} D(x_0)),$$

using the fact that $D^{[p]}$ is a derivation. Now assume Q is not in A^p, and denote by i the smallest integer such that $Q \in \mathbf{F}_p[x_0, \dots, x_i, x_{i+1}^p, x_{i+2}^p, \dots]$. Denoting by ∂_j the partial derivation with respect to x_j, we have $0 = D(Q) = \sum_j (\partial_j Q) D(x_j) = \sum_{j=0}^{i} (\partial_j Q) x_{j+1}$. By minimality of i, we have here $\partial_i Q \neq 0$, whence $x_{i+1} \in \mathbf{F}_p(x_0, x_1, \dots, x_i, x_{i+1}^p, \dots)$, which is impossible, and therefore $Q \in A^p$. Now by definition Q is a homogeneous polynomial of degree p in the x_i. On the other hand, consider the grading on $A = \mathbf{F}_p[x_0, x_1, \dots]$ in which x_i has degree i. Then D transforms elements of degree i in elements of degree $i + 1$, from which it follows that Q has degree p in this grading as well. Denoting by $\widetilde{Q}$ the unique element of A with $\widetilde{Q}^p = Q$, we get that $\widetilde{Q}$ has degree 1 for both the traditional and the new grading. This is only possible if $\widetilde{Q} = m x_1$ for some $m \in \mathbf{F}_p$. On the other hand, the leading term of Q as a polynomial in $x_1 = D(x_0)$ is $-(p-1)! \, x_1^p$, which equals x_1^p by Wilson's theorem. So $m = 1$ and $Q = x_1^p = D(x_0)^p$, as desired. $\square$

For the next formula we consider elements of $\mathrm{Der}_{\mathbf{F}_p}(A)$ as $\mathbf{F}_p$-linear maps on A, and for $a \in A$ we denote by $L_a : A \to A$ the $\mathbf{F}_p$-linear endomorphism $A \to A$ given by $L_a(x) = ax$.

Proposition A.7.2   *Given an integral domain of characteristic $p > 0$, an element $a \in A$ and a derivation $D \in \mathrm{Der}_{\mathbf{F}_p}(A)$, the identity*

$$(D + L_a)^{[p]} = D^{[p]} + L_a^{[p]} + L_{D^{[p-1]}(a)}$$

holds in $\mathrm{End}_{\mathbf{F}_p}(A)$.

The proof uses the notion of *Jacobson polynomials*, which are defined as follows. Given a not necessarily commutative ring R with unit, for each element $w \in R$ we may define a map $\mathrm{ad}_w : R \to R$ by $\mathrm{ad}_w(x) = wx - xw$. Applying the above to the polynomial ring $R[t]$ in place of R (where t is to commute with all elements of R) and taking two elements $u, v \in R$, we find noncommutative polynomials in two variables $s_i(U, V)$ with $\mathbf{Z}$-coefficients satisfying

$$(\mathrm{ad}_{tu+v})^{[p-1]}(u) = \sum_{i=1}^{p-1} i s_i(u, v) t^{i-1}, \tag{A.3}$$

where $\mathrm{ad}_{tu+v}^{[p-1]}$ stands for the $(p-1)$-st iterate of ad_{tu+v}. Working in the free noncommutative $\mathbf{Z}$-algebra generated by two elements U, V, one sees that the polynomials $s_i(U, V)$ do not depend on the choice of u, v.

Lemma A.7.3 (Jacobson) *Assume moreover that $pR = 0$. Then the identity*

$$(u + v)^p = u^p + v^p + \sum_{i=1}^{p-1} s_i(u, v)$$

holds for all $u, v \in R$.

Proof For all $w \in R[t]$ introduce the endomorphisms $L_w, R_w : R[t] \to R[t]$ defined by $L_w(x) = wx$ and $R_w(x) = xw$, respectively. We have $R_w \circ L_w = L_w \circ R_w$ and $\mathrm{ad}_w = L_w - R_w$. Raising to the $(p-1)$-st power, we get from the binomial formula

$$\mathrm{ad}_w^{[p-1]} = \sum_{i=0}^{p-1} (-1)^{p-1-i} \binom{p-1}{i} L_w^{[i]} R_w^{[p-1-i]} = \sum_{i=0}^{p-1} L_w^{[i]} R_w^{[p-1-i]}$$

in End_R, as the binomial coefficient is congruent to $(-1)^i$ in characteristic p. In particular, we have

$$\mathrm{ad}_w^{[p-1]}(u) = \sum_{i=0}^{p-1} w^i u w^{p-1-i}. \tag{A.4}$$

On the other hand, expanding $(tu + v)^p$ with respect to t we find noncommutative polynomials $s_i'(U, V) \in R[U, V]$ satisfying the identity

$$(tu + v)^p = t^p u^p + v^p + \sum_{i=1}^{p-1} s_i'(u, v) t^i. \tag{A.5}$$

Differentiating with respect to t yields

$$\sum_{i=0}^{p-1}(tu+v)^i u(tu+v)^{p-1-i} = \sum_{i=1}^{p-1} i s_i'(u,v)t^{i-1}.$$

(Note that the multiplication in $R[t]$ is noncommutative!) Comparing with formula (A.4) for $w = tu+v$ and applying the defining identity (A.3) of Jacobson polynomials we get $s_i'(u,v) = s_i(u,v)$, so the lemma follows from (A.5) by setting $t = 1$. □

Proof of Proposition A.7.2 We shall apply the lemma with $R = \mathrm{End}_{\mathbf{F}_p}(A)$, $u = D$ and $v = L_a$. The calculation

$$\mathrm{ad}_D(L_a)(x) = DL_a(x) - L_a D(x) = D(ax) - aD(x) = D(a)x = L_{D(a)}(x)$$

shows that $\mathrm{ad}_D(L_a) = L_{D(a)}$, and therefore

$$\mathrm{ad}_{tD+L_a}(D) = (tD+L_a)D - D(tD+L_a) = -\mathrm{ad}_D(L_a) = -L_{D(a)}.$$

Using the fact that L_a and $L_{D(a)}$ commute, we obtain from the above

$$\mathrm{ad}_{tD+L_a}^{[2]}(D) = -\mathrm{ad}_{tD+L_a}(L_{D(a)}) = -\mathrm{ad}_{tD}(L_{D(a)}) = -t L_{D^{[2]}(a)}.$$

Iterating the argument yields

$$\mathrm{ad}_{tD+L_a}^{[p-1]}(D) = -t^{p-2} L_{D^{[p-1]}(a)} = (p-1)t^{p-2} L_{D^{[p-1]}(a)}.$$

A comparison with formula (A.3) shows that $s_i(D, L_a) = 0$ for $i < p-1$ and $s_{p-1}(D, L_a) = L_{D^{[p-1]}(a)}$, so the proposition follows from the lemma. □

Finally we say a few words about the Lie algebra structure on derivations. Assume k is a field, and V is a k-vector space. Then $\mathrm{End}_k(V)$ carries a *Lie bracket* defined by $[\phi, \psi] = \phi \circ \psi - \psi \circ \phi$. This Lie bracket is k-bilinear and satisfies $[\phi, \phi] = 0$ as well as the Jacobi identity $[[\phi, \psi], \rho]] + [[\psi, \rho], \phi]] + [[\rho, \phi], \psi] = 0$, so it gives $\mathrm{End}_k(V)$ the structure of a Lie algebra. Given $\phi \in \mathrm{End}_k(V)$, the map $\mathrm{ad}(\phi) : \psi \mapsto [\phi, \psi]$ is a Lie algebra endomorphism. If k has characteristic p, we also have the p-operation $\phi \mapsto \phi^{[p]}$ sending an endomorphism to its p-th iterate. Obviously $(a\phi)^{[p]} = a^p \phi^{[p]}$ for all $a \in k$, $\mathrm{ad}(\phi^{[p]}) = \mathrm{ad}(\phi)^{[p]}$, and moreover $(\phi+\psi)^{[p]} = \phi^{[p]} + \psi^{[p]} + \sum s_i(\phi, \psi)$ by Lemma A.7.3.

In fact, one may check using the definition of Jacobson polynomials that the $s_i(\phi, \psi)$ lie in the Lie subalgebra of $\mathrm{End}_k(V)$ generated by ϕ and ψ, i.e. may be obtained from ϕ and ψ by means of a formula involving additions and Lie brackets. For a (rather complicated) explicit formula, see

Demazure–Gabriel [1], II, §7, No. 3. Via this observation one may extend the definition of Jacobson polynomials to arbitrary Lie algebras of characteristic p. In the literature a Lie algebra over a field of characteristic p equipped with a p-operation satisfying the above three properties is called a *restricted p-Lie algebra* or a *p-Lie algebra* for short.

Now consider the case when $V = K$ is a field extension of k. Then every k-derivation $D : K \to K$ is also an element of $\mathrm{End}_k(K)$. The Lie bracket on $\mathrm{End}_k(K)$ preserves $\mathrm{Der}_k(K)$, in view of the computation

$$
\begin{aligned}
[D_1, D_2](ab) &= D_1(D_2(ab)) - D_2(D_1(ab)) \\
&= D_1(D_2(a)b) + D_1(aD_2(b)) - D_2(D_1(a)b) - D_2(aD_1(b)) \\
&= (D_1 D_2(a))b + a(D_1 D_2(b)) - (D_2 D_1(a))b - a(D_2 D_1(b)) \\
&= ([D_1, D_2](a))b + a([D_1, D_2](b)).
\end{aligned}
$$

Moreover, the p-th iterate of a derivation is again a k-derivation according to formula (A.2), so the p-operation of $\mathrm{End}_k(K)$ also preserves $\mathrm{Der}_k(K)$. All in all, $\mathrm{Der}_k(K)$ is a p-Lie subalgebra of $\mathrm{End}_k(K)$.

A.8 Differential forms

One defines differential forms by the following universal property.

Proposition A.8.1 *Let $A \subset B$ be an extension of commutative rings. There exist a B-module $\Omega^1_{B|A}$ and an A-derivation $d : B \to \Omega^1_{B|A}$ so that for all B-modules M the map $\phi \to \phi \circ d$ induces an isomorphism*

$$
\mathrm{Hom}_B(\Omega^1_{B|A}, M) \xrightarrow{\sim} \mathrm{Der}_A(B, M),
$$

functorial in M.

Proof Let $F(B)$ be the free B-module generated by symbols db for all $b \in B$. Define $\Omega^1_{B|A}$ as the quotient of $F(B)$ by the submodule generated by elements of the form da, $d(b_1 + b_2) - db_1 - db_2$ or $d(b_1 b_2) - b_1 db_2 - b_2 db_1$ for some $a \in A$ or $b_1, b_2 \in B$, and define d by sending b to db. Verification of the required properties is straightforward. $\qquad\square$

The B-module $\Omega^1_{B|A}$ is called the module of *differential forms* of B relative to A. In the case $A = \mathbf{Z}$ we set $\Omega^1_{B|\mathbf{Z}} =: \Omega^1_B$, and call it the module of *absolute differential forms*. As $\Omega^1_{B|A}$ is defined by a universal property, it is unique up to unique isomorphism. In Matsumura [1] another construction for $\Omega^1_{B|A}$ is given, which yields the same module by this remark.

Example A.8.2   Assume that B arises as the quotient of the polynomial ring $A[x_1, \ldots, x_n]$ by an ideal $(f_1, \ldots, f_m)$. Then $\Omega^1_{B|A}$ is the quotient of the free B-module generated by the dx_i modulo the submodule generated by the elements $\sum_i \partial_i f_j dx_i$ $(1 \le j \le m)$, where ∂_i denotes the partial derivative with respect to x_i. This follows immediately from the above construction.

Proposition A.8.3   *The module of differentials* $\Omega^1_{B|A}$ *enjoys the following basic properties.*

1. *(Base change) For an A-algebra A' one has $\Omega^1_{B \otimes_A A'|A'} \cong \Omega^1_{B|A} \otimes_A A'$.*
2. *(Localization) Given a multiplicative subset S of B, one has*

$$\Omega^1_{B_S|A} \cong \Omega^1_{B|A} \otimes_B B_S.$$

3. *(First exact sequence) A tower of ring extensions $A \subset B \subset C$ gives rise to an exact sequence of C-modules*

$$\Omega^1_{B|A} \otimes_B C \to \Omega^1_{C|A} \to \Omega^1_{C|B} \to 0.$$

4. *(Second exact sequence) A short exact sequence $0 \to I \to B \to C \to 0$ of A-algebras gives rise to an exact sequence*

$$I/I^2 \xrightarrow{\delta} \Omega^1_{B|A} \otimes_B C \to \Omega^1_{C|A} \to 0$$

of C-modules, where the map δ sends a class $x \bmod I^2$ to $dx \otimes 1$. (Note that the B-module structure on I/I^2 induces a C-module structure.)

Proof   For (1) and (2), see Matsumura [1], Ex. 25.4; for (3) and (4), see Theorems 25.1 and 25.2 of *loc. cit.*   □

Corollary A.8.4   *If $K|k$ is a finitely and separably generated field extension of transcendence degree r, then $\Omega^1_{K|k}$ has dimension r over K.*

Proof   Write K as the fraction field of a quotient $k[t_1, \ldots, t_{d+1}]/(f)$ with f separable as in the proof of Corollary A.3.5 and apply part (2) of the proposition together with Example A.8.2.   □

Corollary A.8.5   *A finite extension $K|k$ is separable if and only if $\Omega^1_{K|k} = 0$.*

Proof   The 'only if' part is contained in the previous corollary. To treat the 'if' part, denote by $k_0|k$ the maximal separable subextension of $K|k$. If $K \ne k_0$, then K is generated as a k_0-algebra by finitely many elements of the form $\sqrt[r]{a}$ with some $a \in k_0$ and $r > 0$, where $p = \mathrm{char}(k)$. The minimal polynomial of these elements has trivial differential in characteristic p, and so

$\Omega^1_{K|k_0} \neq 0$ by Example A.8.2. But $\Omega^1_{K|k_0}$ is a quotient of $\Omega^1_{K|k}$, whence the corollary. □

The following statement gives a relation between smoothness, differentials and regularity.

Proposition A.8.6 *Let k be a perfect field, A an integral domain which is a finitely generated k-algebra, and P a prime ideal of A. Denote by d the Krull dimension of A and by m the height of P. Then the following are equivalent:*

1. *The module of differentials $\Omega^1_{A_P|k}$ is free of rank d over A_P;*
2. *A_P is a regular local ring (i.e. $\dim_{\kappa(P)} P/P^2 = m$).*

For $d = m$ these are both equivalent to the condition that the closed point P is smooth.

Proof For the first statement, see Matsumura [1], Lemma 1, p. 216 and Theorem 30.3. The case $d = m$ follows easily from the first part using Nakayama's lemma (see Mumford [1], Section III.4). □

Corollary A.8.7 *Under the equivalent conditions of the proposition the sequence*

$$0 \to P/P^2 \to \Omega^1_{A_P|k} \otimes_{A_P} \kappa(P) \to \Omega^1_{\kappa(P)|k} \to 0$$

is exact, where $\kappa(P)$ is the residue field of P.

Proof In view of Proposition A.8.3 (4) only injectivity at the left should be checked, and this follows from Corollary A.8.4 and a dimension count using the proposition above. □

Call an integral domain A that is an algebra over a perfect field k *smooth* if A is finitely generated and every prime ideal $P \subset A$ satisfies the equivalent conditions of Proposition A.8.6. This is equivalent to A being the coordinate ring of a smooth affine k-variety.

Proposition A.8.8 *Let k be a perfect field, and $K|k$ an arbitrary field extension. The smooth k-subalgebras of K form a direct system with respect to the natural inclusion maps whose direct limit is K.*

Proof Every element of K is contained in a finitely generated k-subalgebra A. By Proposition A.3.4 the fraction field of A is separably generated over k, so the smoothness condition of Proposition A.8.6 is satisfied at the ideal (0) by Matsumura [1], Theorem 26.9. Then the corollary to Matsumura [1],

Theorem 30.5 implies that there is a localization A_f of A by powers of an element f which is a smooth k-algebra. This shows that K is the union of its smooth k-subalgebras. To see that the latter form a direct system, observe that any two smooth k-subalgebras $A_1, A_2 \subset K$ generate a finitely generated k-subalgebra $A_3 \subset K$. By the previous argument a suitable localization of A_3 will be smooth. $\square$

We now turn to differential forms over fields. The first result is:

Proposition A.8.9   *If $K|k$ is an extension of fields and $L|K$ is a finite separable extension, then $\Omega^1_{L|k} \cong \Omega^1_{K|k} \otimes_K L$.*

Proof   See Matsumura [1], Theorem 25.3. $\square$

Next a result on purely inseparable extensions.

Proposition A.8.10   *For a finite purely inseparable field extension $K|k$ the dimension of the K-vector space $\Omega^1_{K|k}$ equals the minimal number of k-algebra generators of K.*

Proof   If $a_1, \ldots a_r$ generate the k-algebra K, then $\dim_K \Omega^1_{K|k} \leq r$ by Example A.8.2. Conversely, if $a_1, \ldots, a_r$ are such that $da_1, \ldots, da_r$ generate $\Omega^1_{K|k}$, set $K_0 := k(a_1, \ldots a_r)$. Proposition A.8.3 (3) gives an exact sequence

$$\Omega^1_{K_0|k} \otimes_{K_0} K \to \Omega^1_{K|k} \to \Omega^1_{K|K_0} \to 0,$$

showing $\Omega^1_{K|K_0} = 0$. By Corollary A.8.5 this means that $K|K_0$ is separable, and therefore $K_0 = K$. $\square$

Now consider a not necessarily finite extension $K|k$ of fields of characteristic $p > 0$ satisfying $K^p \subset k$. In this case one calls a system of elements $\{b_\lambda : \lambda \in \Lambda\}$ a *p-basis* for the extension $K|k$ if the products $b_{\lambda_1}^{\alpha_1} \ldots b_{\lambda_m}^{\alpha_m}$ for all finite subsets $\{\lambda_1, \ldots, \lambda_m\} \subset \Lambda$ and exponents $0 \leq \alpha_i \leq p - 1$ yield a basis of the k-vector space K, i.e. they form a k-linearly independent generating system.

Proposition A.8.11   *Let $K|k$ be an extension of fields of characteristic $p > 0$ satisfying $K^p \subset k$. A system of elements $\{b_\lambda : \lambda \in \Lambda\}$ is a p-basis of $K|k$ if and only if the system $\{db_\lambda : \lambda \in \Lambda\}$ is a basis of the K-vector space $\Omega^1_{K|k}$.*

Proof   See Matsumura [1], Theorem 26.5. $\square$

Corollary A.8.12 *Every finite extension $K|k$ of fields of characteristic $p > 0$ satisfying $K^p \subset k$ has a p-basis.*

Proof Choose a K-basis for $\Omega^1_{K|k}$ (using Zorn's lemma). □

Corollary A.8.13 *Let $k_0 \subset k$ be a subfield with $K^p \subset k_0$. Then the sequence of K-vector spaces*

$$0 \to \Omega^1_{k|k_0} \otimes_k K \to \Omega^1_{K|k_0} \to \Omega^1_{K|k} \to 0$$

is exact. Moreover, the choice of a p-basis of $K|k$ induces a splitting of the exact sequence.

Proof The only nonobvious points are exactness at $\Omega^1_{K|k_0}$ and the existence of the splitting. For these choose a p-basis $\{x_\lambda : \lambda \in \Lambda_{k|k_0}\}$ of $k|k_0$ and a p-basis $\{x_\lambda : \lambda \in \Lambda_{K|k}\}$ of $K|k$. Together they form a p-basis of $K|k_0$, and the corollary follows from the proposition. □

As an application of the above, we compute the module of absolute differentials for local rings of the affine line over a field K of characteristic $p > 0$. Note that the case when K is perfect already follows from Example A.8.2 and the localization property of differentials, so the interesting case is when $K \neq K^p$.

Proposition A.8.14 *For each closed point P of the affine line $\mathbf{A}^1_K$ the $K[t]_P$-module $\Omega^1_{K[t]_P}$ is free on a basis containing $d\pi_P$, where π_P is a local parameter at P contained in $K[t]$.*

Proof We first show the freeness of the $K[t]$-module $\Omega^1_{K[t]}$, which will imply the corresponding property over the localization $K[t]_P$ by Proposition A.8.3 (2). Applying Proposition A.8.3 (3) with $A = \mathbf{Z}$, $B = K$ and $C = K[t]$ we get an exact sequence

$$\Omega^1_K \otimes_K K[t] \to \Omega^1_{K[t]} \to \Omega^1_{K[t]|K} \to 0.$$

By Example A.8.2 the $K[t]$-module $\Omega^1_{K[t]|K}$ is free of rank one generated by dt, and therefore it identifies to a direct summand of $\Omega^1_{K[t]}$. It remains therefore to show the injectivity of the map $\Omega^1_K \otimes_K K[t] \to \Omega^1_{K[t]}$. As the kernel of this map is in any case a torsion free module over the principal ideal ring $K[t]$, this is equivalent to showing the injectivity of the map $\Omega^1_K \otimes_K K(t) \to \Omega^1_{K[t]} \otimes_{K[t]} K(t)$, which is the same as the natural map $\Omega^1_K \otimes_K K(t) \to \Omega^1_{K(t)}$ by Proposition A.8.3 (2). The claim then follows from Corollary A.8.13.

Concerning the statement about $d\pi_P$ we distinguish two cases. If π_P is a separable polynomial in $K[t]$, then the derivative $\partial_t \pi_P$ is prime to π_P in $K[t]$, and therefore a unit in $K[t]_P$. The formula $d\pi_P = (\partial_t \pi_P)dt$ then shows that $d\pi_P$ is a generator of $\Omega^1_{K[t]_P|K}$, so by the above argument it yields a basis of $\Omega^1_{K[t]_P}$ together with a basis of Ω^1_K. In the case when π_P is an inseparable polynomial we may write $\pi_P = f(t^{p^r})$ for a suitable $r > 0$ and separable polynomial f. Applying Proposition A.8.3 (3) with $A = \mathbf{Z}$, $B = K[t^{p^r}]$ and $C = K[t]$ and noting the isomorphism $\Omega^1_{K[t]|K[t^{p^r}]} \cong \Omega^1_{K[t]|k}$ we see as above that $\Omega^1_{K[t]}$ is free on a basis consisting of dt and a basis of $\Omega^1_{K[t^{p^r}]}$. But by the separable case $d\pi_P$ may be taken as a basis element in the latter, and the proof is complete. □

We now discuss higher differential forms. For an integer $i > 0$ the module of differential i-forms $\Omega^i_{B|A}$ is defined as the i-th exterior power $\Lambda^i \Omega^1_{B|A}$; for $i = 0$ we put $\Omega^0_{B|A} := B$ by convention. They form the terms of a complex of A-modules, the *de Rham complex*:

$$\Omega^\bullet_{B|A} = (B \xrightarrow{d} \Omega^1_{B|A} \xrightarrow{d} \Omega^2_{B|A} \xrightarrow{d} \Omega^3_{B|A} \xrightarrow{d} \dots).$$

For simplicity, we define the differentials $d : \Omega^i_{B|A} \to \Omega^{i+1}_{B|A}$ only in the case when $\Omega^1_{B|A}$ is freely generated as a B-module by elements db_λ, the only one we need. In this case, d sends an i-form $b\,db_{\lambda_1} \wedge \dots \wedge db_{\lambda_i}$ to $db \wedge db_{\lambda_1} \wedge \dots \wedge db_{\lambda_i}$. Note that for $i = 0$ this gives back the universal derivation $d : B \to \Omega^1_{B|A}$; the fact that we have a complex is obvious. An easy calculation shows that the differential d satisfies the identity

$$d(\omega_1 \wedge \omega_2) = d\omega_1 \wedge \omega_2 + (-1)^i \omega_1 \wedge d\omega_2 \qquad (A.6)$$

for $\omega_1 \in \Omega^i_{B|A}$ and $\omega_2 \in \Omega^j_{B|A}$; in fact, this identity and the fact that it yields the universal derivation for $i = 0$ characterize d. The submodules $\ker(d)$ (resp. $\mathrm{Im}\,(d)$) of $\Omega^1_{B|A}$ are denoted by $Z^i_{B|A}$ and $B^i_{B|A}$, respectively, and are classically called the module of *closed* (resp. *exact*) i-forms. The quotient $H^i(\Omega^\bullet_{B|A}) := Z^i_{B|A}/B^i_{B|A}$ is the *i-th de Rham cohomology group* of B over A.

Finally, we describe a generalization of the de Rham complex due to Illusie [1]. To do so, we first need to summarize some basic facts about Witt vectors. Fix a prime p, and recall (e.g. from §II.6 of Serre [2]) that given a commutative ring A, the set $W(A)$ of its p-Witt vectors is formed by infinite sequences $a = (a_0, a_1, a_2, \dots)$ of elements of A. This set carries a ring structure defined by certain explicit polynomials $S_r, P_r \in \mathbf{Z}[X_0, \dots, X_r, Y_0, \dots, Y_r]$ so that

$$a + b = (a_0 + b_0, S_1(a, b), S_2(a, b), \dots), \quad a \cdot b = (a_0 b_0, P_1(a, b), P_2(a, b), \dots).$$

There are operators F (Frobenius) and V (Verschiebung) given by

$$Fa = (a_0^p, a_1^p, a_2^p, \dots), \quad Va = (0, a_0, a_1, \dots)$$

and one can show that $FV = VF$ equals multiplication by p. For $r > 0$ we denote by $W_r(A)$ the quotient $W(A)/V^r W(A)$. For $r = 1$ this gives $W_1(A) \cong A$, and the ring $W(A)$ is the inverse limit of the quotients $W_r(A)$ with respect to the natural projections $R : W_{r+1}(A) \to W_r(A)$. The operators F and V induce operators $F : W_{r+1}(A) \to W_r(A)$ and $V : W_r(A) \to W_{r+1}(A)$ satisfying $FV = VF = p$. Finally, given a unit $a \in A$, we denote by $[a]$ the Witt vector $(a, 0, 0, \dots)$ and also its images in the truncations $W_r(A)$. The map $a \mapsto [a]$ satisfies $[ab] = [a] \cdot [b]$. In the important special case $A = \mathbf{F}_p$ one has $W(\mathbf{F}_p) \cong \mathbf{Z}_p$ and $W_r(\mathbf{F}_p) \cong \mathbf{Z}/p^r\mathbf{Z}$.

Now assume A is an $\mathbf{F}_p$-algebra and fix an integer $r > 0$. In Illusie [1] there is a construction for a complex

$$W_r\Omega_A^\bullet := (W_r\Omega_A^0 \xrightarrow{d} W_r\Omega_A^1 \xrightarrow{d} W_r\Omega_A^2 \xrightarrow{d} \cdots)$$

of abelian groups, functorial in A, called the *r-truncated de Rham–Witt complex* of A. This complex is moreover equipped with a product structure

$$W_r\Omega_A^i \times W_r\Omega_A^j \to W_r\Omega_A^{i+j}$$

satisfying $x \cdot y = (-1)^{ij} y \cdot x$ for $x \in W_r\Omega_A^i, y \in W_r\Omega_A^j$ (and $x \cdot x = 0$ for i odd, $p = 2$) as well as the compatibility

$$d(x \cdot y) = dx \cdot y + (-1)^i x \cdot dy$$

analogous to (A.6). Moreover, one has $W_r\Omega_A^0 = W_r(A)$, which implies thanks to the product structure that each term $W_r\Omega_A^i$ has a $W_r(A)$-module structure, and consequently it is annihilated by p^r. In the case $r = 1$ one recovers the usual de Rham complex $\Omega_{A/\mathbf{F}_p}^\bullet$. There are surjective restriction maps $R : W_{r+1}\Omega_A^\bullet \to W_r\Omega_A^\bullet$ extending the projections $R : W_{r+1}(A) \to W_r(A)$ that turn the $W_r\Omega_A^\bullet$ into an inverse system. Furthermore, there exist additive operators $F : W_{r+1}\Omega_A^\bullet \to W_r\Omega_A^\bullet$ and $V : W_r\Omega_A^\bullet \to W_{r+1}\Omega_A^\bullet$ extending the Frobenius and Verschiebung operators on truncated Witt vectors that are compatible with the restriction maps R and satisfy $FV = VF = p$. The inverse system of the $W_r\Omega_A^\bullet$ with its additional structure has many other important properties, but we do not need them in this text.

In the case when A is a smooth $\mathbf{F}_p$-algebra (or a direct limit of such), there is the following quick construction for $W_r\Omega_A^\bullet$ (see Illusie [1], Theorem 4.2):

$$W_r\Omega_A^\bullet \cong \Omega_{W(A)|\mathbf{Z}_p}^\bullet / (T + \mathrm{Fil}^r \Omega_{W(A)|\mathbf{Z}_p}^\bullet),$$

where T is the subcomplex of $\Omega_{W(A)|\mathbf{Z}_p}^\bullet$ formed by p-primary torsion elements and $\mathrm{Fil}^r \Omega_{W(A)|\mathbf{Z}_p}^\bullet$ is the kernel of the natural map $\Omega_{W(A)|\mathbf{Z}_p}^\bullet \to \Omega_{W_r(A)|(\mathbf{Z}/p^r\mathbf{Z})}^\bullet$.

Bibliography

Albert, Adrian A.

[1] On the Wedderburn norm condition for cyclic algebras, *Bull. Amer. Math. Soc.* **37** (1931), 301–312.

[2] Normal division algebras of degree four over an algebraic field, *Trans. Amer. Math. Soc.* **34** (1932), 363–372.

[3] Simple algebras of degree p^e over a centrum of characteristic p, *Trans. Amer. Math. Soc.* **40** (1936), 112–126.

[4] *Structure of Algebras*, American Mathematical Society Colloquium Publications, vol. XXIV, 1939.

[5] Tensor products of quaternion algebras, *Proc. Amer. Math. Soc.* **35** (1972), 65–66.

Amitsur, Shimshon A.

[1] Generic splitting fields of central simple algebras, *Ann. of Math.* *(2)* **62** (1955), 8–43.

[2] On central division algebras, *Israel J. Math.* **12** (1972), 408–420.

Amitsur, Shimshon, Rowen, Louis H. and Tignol, Jean-Pierre

[1] Division algebras of degree 4 and 8 with involution, *Israel J. Math.* **33** (1979), 133–148.

Amitsur, Shimshon and Saltman, David

[1] Generic Abelian crossed products and p-algebras, *J. Algebra* **51** (1978), 76–87.

Antieau, Benjamin and Williams, Ben

[1] Unramified division algebras do not always contain Azumaya maximal orders, *Invent. Math.* **197** (2014), 47–56.

[2] The prime divisors of the period and index of a Brauer class, *J. Pure Appl. Algebra* **219** (2015), 2218–2224.

[3] Prime decomposition for the index of a Brauer class, *Ann. Sc. Norm. Super. Pisa Cl. Sci.*, XVII(2017), 277–285.

Arason, Jón Kristinn
[1] Cohomologische Invarianten quadratischer Formen, *J. Algebra* **36** (1975), 448–491.
[2] A proof of Merkurjev's theorem, in *Quadratic and Hermitian forms (Hamilton, Ont., 1983)*, CMS Conf. Proc., 4, Amer. Math. Soc., Providence, 1984, 121–130.

Artin, Emil
[1] Kennzeichnung des Körpers der reellen algebraischen Zahlen, *Abh. Math. Sem. Hamburg* **3** (1924), 319–323.

Artin, Emil and Schreier, Otto
[1] Eine Kennzeichnung der reell abgeschlossenen Körper, *Abh. Math. Sem. Hamburg* **5** (1927), 225–231.

Artin, Emil and Tate, John
[1] *Class Field Theory*, 2nd edition, Addison-Wesley, Redwood, 1990.

Artin, Michael
[1] Brauer-Severi varieties, in *Brauer groups in ring theory and algebraic geometry (Wilrijk, 1981)*, Lecture Notes in Math. **917**, Springer-Verlag, Berlin-New York, 1982, 194–210.

Artin, Michael and Mumford, David
[1] Some elementary examples of unirational varieties which are not rational, *Proc. London Math. Soc. (3)* **25** (1972), 75–95.

Asok, Aravind
[1] Rationality problems and conjectures of Milnor and Bloch-Kato, *Compositio Math.* **149** (2013), 1312–1326.

Atiyah, Michael Francis and Macdonald, Ian G.
[1] *Introduction to Commutative Algebra*, Addison-Wesley, Reading, 1969.

Atiyah, Michael Francis and Wall, Charles Terence Clegg
[1] Cohomology of groups, in *Algebraic Number Theory (J. W. S. Cassels and A. Fröhlich, eds.)*, Academic Press, London, 1967, 94–115.

Auslander, Maurice and Brumer, Armand
[1] Brauer groups of discrete valuation rings, *Indag. Math.* **30** (1968), 286–296.

Ax, James
[1] A field of cohomological dimension 1 which is not C_1, *Bull. Amer. Math. Soc.* **71** (1965), 717.

[2] Proof of some conjectures on cohomological dimension, *Proc. Amer. Math. Soc.* **16** (1965), 1214–1221.

Bass, Hyman, Milnor, John and Serre, Jean-Pierre
[1] Solution of the congruence subgroup problem for SL_n ($n \geq 3$) and Sp_{2n} ($n \geq 2$), *Inst. Hautes Études Sci. Publ. Math.* **33** (1967), 59–137.

Bass, Hyman and Tate, John
[1] The Milnor ring of a global field, in *Algebraic K-theory II*, Lecture Notes in Math. **342**, Springer-Verlag, Berlin, 1973, 349–446.

Beauville, Arnaud
[1] *Surfaces algébriques complexes*, Astérisque No. 54, Société Mathématique de France, Paris, 1978; English translation: *Complex Algebraic Surfaces*, London Mathematical Society Lecture Note Series, vol. 68, Cambridge University Press, 1983.

Berhuy, Grégory and Frings, Christoph
[1] On the second trace form of central simple algebras in characteristic two, *Manuscripta Math.* **106** (2001), 1–12.

Bloch, Spencer
[1] K_2 and algebraic cycles, *Ann. of Math. (2)* **99** (1974), 349–379.
[2] *Lectures on Algebraic Cycles*, Duke University Mathematics Series IV, Duke University, Durham, 1980. Second edition: New Mathematical Monographs, vol. 16, Cambridge University Press, 2010.
[3] Torsion algebraic cycles, K_2, and Brauer groups of function fields, in *The Brauer Group (Les Plans-sur-Bex, 1980)*, Lecture Notes in Math. **844**, Springer-Verlag, Berlin, 1981, 75–102.

Bloch, Spencer and Kato, Kazuya
[1] p-adic étale cohomology, *Inst. Hautes Études Sci. Publ. Math.* **63** (1986), 107–152.

Bogomolov, Fedor A.
[1] The Brauer group of quotient spaces of linear representations (Russian), *Izv. Akad. Nauk SSSR Ser. Mat.* **51** (1987), 485–516, 688; English translation in *Math. USSR-Izv.* **30** (1988), 455–485.
[2] Brauer groups of the fields of invariants of algebraic groups (Russian), *Mat. Sb.* **180** (1989), 279–293; English translation in *Math. USSR-Sb.* **66** (1990), 285–299.

Brauer, Richard
[1] Untersuchungen über die arithmetischen Eigenschaften von Gruppen linearer Substitutionen I, *Math. Zeit.* **28** (1928), 677–696; II, *ibid.* **31** (1930), 733–747.

[2] Über die algebraische Struktur von Schiefkörpern, *J. reine angew. Math.* **166** (1932), 241–252.

Brauer, Richard, Hasse, Helmut and Noether, Emmy,
[1] Beweis eines Hauptsatzes in der Theorie der Algebren, *J. reine angew. Math.* **167** (1932), 399–404.

Brussel, Eric
[1] Noncrossed products and nonabelian crossed products over $Q(t)$ and $Q((t))$, *Amer. J. Math.* **117** (1995), 377–393.

Cartan, Henri and Eilenberg, Samuel
[1] *Homological Algebra*, Princeton University Press, Princeton, 1956.

Cartier, Pierre
[1] Questions de rationalité des diviseurs en géométrie algébrique, *Bull. Soc. Math. France* **86** (1958), 177–251.

Cassels, John William Scott and Fröhlich, Albrecht (eds.)
[1] *Algebraic Number Theory*, Academic Press, London, 1967.

Clemens, Herbert and Griffiths, Phillip
[1] The intermediate Jacobian of the cubic threefold, *Ann. of Math. (2)* **95** (1972), 281–356.

Châtelet, François
[1] Variations sur un thème de H. Poincaré, *Ann. Sci. Éc. Norm. Sup. III. Sér.* **61** (1944), 249–300.

Chevalley, Claude
[1] Démonstration d'une hypothèse de M. Artin, *Abh. Math. Semin. Hamb. Univ.* **11** (1935), 73–75.

Chu, Huah and Kang, Ming-chang
[1] Rationality of p-group actions, *J. Algebra* **237** (2001), 673–690.

Chu, Huah, Hu, Shou-Yen, Kang, Ming-chang and Prokhorov, Yuri
[1] Noether's problem for groups of order 32, *J. Algebra* **320** (2008), 3022–3035.

Colliot-Thélène, Jean-Louis
[1] Hilbert's Theorem 90 for K_2, with application to the Chow groups of rational surfaces, *Invent. Math.* **71** (1983), 1–20.
[2] Les grands thèmes de François Châtelet, *Enseign. Math. (2)* **34** (1988), 387–405.
[3] Cycles algébriques de torsion et K-théorie algébrique, in *Arithmetic algebraic geometry (Trento, 1991)*, Lecture Notes in Math. **1553**, Springer-Verlag, Berlin, 1993, 1–49.

[4] Cohomologie galoisienne des corps valués discrets henséliens, d'après K. Kato et S. Bloch, in *Algebraic K-Theory and Its Applications (Trieste, 1997)* (H. Bass, A. Kuku, C. Pedrini, eds.), World Scientific, River Edge, 1999, 120–163.

[5] Fields of cohomological dimension one versus C_1-fields, in *Algebra and Number Theory* (R. Tandon, ed.), Hindustan Book Agency, New Delhi, 2005, 1–6.

Colliot-Thélène, Jean-Louis, Hoobler, Raymond and Kahn, Bruno

[1] The Bloch-Ogus–Gabber theorem, in *Algebraic K-theory (Toronto, 1996)*, Fields Inst. Commun., vol. 16, Amer. Math. Soc., Providence, 1997, 31–94.

Colliot-Thélène, Jean-Louis, and Madore, David A.

[1] Surfaces de del Pezzo sans point rationnel sur un corps de dimension cohomologique un, *J. Inst. Math. Jussieu* **3** (2004), 1–16.

Colliot-Thélène, Jean-Louis and Ojanguren, Manuel

[1] Variétés unirationnelles non rationnelles: au-delà de l'exemple d'Artin et Mumford, *Invent. Math.* **97** (1989), 141–158.

Colliot-Thélène, Jean-Louis, Ojanguren, Manuel and Parimala, Raman

[1] Quadratic forms over fraction fields of two-dimensional Henselian rings and Brauer groups of related schemes, in *Algebra, Arithmetic and Geometry*, Tata Inst. Fund. Res. Stud. Math., vol. 16, Bombay, 2002, 185–217.

Colliot-Thélène, Jean-Louis and Sansuc, Jean-Jacques

[1] The rationality problem for fields of invariants under linear algebraic groups (with special regards to the Brauer group), in *Proceedings of the International Colloquium on Algebraic Groups and Homogeneous Spaces (Mumbai, 2004)* (V. Mehta, ed.), TIFR Mumbai, Narosa Publishing House, 2007, 113–186.

Colliot-Thélène, Jean-Louis, Sansuc, Jean-Jacques and Soulé, Christophe

[1] Torsion dans le groupe de Chow de codimension deux, *Duke Math. J.* **50** (1983), 763–801.

Demazure, Michel and Gabriel, Pierre

[1] *Groupes algébriques I*, Masson, Paris and North-Holland, Amsterdam, 1970.

Dennis, R. Keith and Stein, Michael R.

[1] K_2 of discrete valuation rings, *Adv. in Math.* **18** (1975), 182–238.

Dickson, Lawrence J.

[1] Linear algebras, *Trans. Amer. Math. Soc.* **13** (1912), 59–73.

Dieudonné, Jean
[1] Les déterminants sur un corps non commutatif, *Bull. Soc. Math. France* **71** (1943), 27–45.
[2] Sur une généralisation du groupe orthogonal à quatre variables, *Arch. Math.* **1** (1949), 282–287.

Draxl, Peter
[1] *Skew Fields*, London Mathematical Society Lecture Note Series, vol. 81, Cambridge University Press, 1983.

Elman, Richard, Karpenko, Nikita and Merkurjev, Alexander
[1] *The Algebraic and Geometric Theory of Quadratic Forms*, Colloquium Publications, vol. 56, American Mathematical Society, Providence, 2008.

Elman, Richard and Lam, Tsit Yuen
[1] Pfister forms and K-theory of fields, *J. Algebra* **23** (1972), 181–213.

Faddeev, Dmitri K.
[1] Simple algebras over a field of algebraic functions of one variable (Russian), *Trudy Mat. Inst. Steklova* **38** (1951), 321–344; English translation in *Amer. Math. Soc. Transl.* (2) **3** (1956), 15–38.
[2] On the theory of homology in groups (Russian), *Izv. Akad. Nauk SSSR. Ser. Mat.* **16** (1952), 17–22.
[3] On the theory of algebras over the field of algebraic functions of one variable (Russian), *Vestnik Leningrad. Univ. Ser. Mat. Meh. Astr.* **12** (1957), 45–51.

Fesenko, Ivan B. and Vostokov, Sergey V.
[1] *Local Fields and Their Extensions*, 2nd edition, Translations of Mathematical Monographs, vol. 121, American Mathematical Society, Providence, 2002.

Fischer, Ernst
[1] Die Isomorphie der Invariantenkörper der endlicher Abelschen Gruppen linearer Transformationen, *Nachr. Akad. Wiss. Göttingen Math.-Phys.* 1915, 77–80.
[2] Zur Theorie der endlichen Abelschen Gruppen, *Math. Ann.* **77** (1915), 81–88.

Florence, Mathieu
[1] On the symbol length of p-algebras, *Compositio Math.* **149** (2013), 1353–1363.

Fulton, William
[1] *Intersection Theory*, 2nd edition, Springer-Verlag, Berlin, 1998.

Garibaldi, Skip, Merkurjev, Alexander and Serre, Jean-Pierre
[1] *Cohomological invariants in Galois cohomology*, University Lecture Series **28**, American Mathematical Society, Providence, RI, 2003.

Geisser, Thomas and Levine, Marc
[1] The K-theory of fields in characteristic p, *Invent. Math.* **139** (2000), 459–493.

Gerstenhaber, Murray
[1] On infinite inseparable extensions of exponent one, *Bull. Amer. Math. Soc.* **71** (1965), 878–881.

Graber, Tom, Harris, Joe and Starr, Jason
[1] Families of rationally connected varieties, *J. Amer. Math. Soc.* **16** (2003), 57–67.

Greenberg, Marvin J.
[1] Rational points in Henselian discrete valuation rings, *Inst. Hautes Études Sci. Publ. Math.* **31** (1966), 59–64.
[2] *Lectures On Forms In Many Variables*, W. A. Benjamin, New York-Amsterdam, 1969.

Grothendieck, Alexander
[1] Technique de descente et théorèmes d'existence en géométrie algébrique I: Généralités. Descente par morphismes fidèlement plats, *Sém. Bourbaki*, exp.190 (1960); reprinted by Société Mathématique de France, Paris, 1995.
[2] *Revêtements étales et groupe fondamental (SGA 1)*, Lecture Notes in Mathematics, vol. 224, Springer-Verlag, Berlin, 1971; reprinted as vol. 3 of Documents Mathématiques, Société Mathématique de France, Paris, 2003.
[3] Le groupe de Brauer I, II, III, in *Dix exposés sur la cohomologie des schémas*, North-Holland, Amsterdam/Masson, Paris, 1968, 46–188.
[4] *Éléments de géométrie algébrique IV: Étude locale des schémas et des morphismes de schémas*, 2e partie, *Inst. Hautes Études Sci. Publ. Math.* **24**, (1965).

Hartshorne, Robin
[1] *Algebraic Geometry*, Graduate Texts in Mathematics, vol. 52, Springer-Verlag, New York-Heidelberg, 1977.

Heuser, Ansgar
[1] Über den Funktionenkörper der Normfläche einer zentral einfachen Algebra, *J. reine angew. Math.* **301** (1978), 105–113.

Hochschild, Gerhard
[1] Cohomology of restricted Lie algebras, *Amer. J. Math.* **76** (1954), 555–580.
[2] Simple algebras with purely inseparable splitting fields of exponent 1, *Trans. Amer. Math. Soc.* **79** (1955), 477–489.

Hochschild, Gerhard and Nakayama, Tadasi
[1] Cohomology in class field theory, *Ann. of Math.* **55** (1952), 348–366.

Hoshi, Akinari, Kang, Ming-chang and Kunyavskiĭ, Boris
[1] Noether's problem and unramified Brauer groups, *Asian J. Math.* **17** (2013), 689–713.

Iskovskih, Vasiliy A. and Manin, Yuri I.
[1] Three-dimensional quartics and counterexamples to the Lüroth problem (Russian), *Mat. Sb.* **86 (128)** (1971), 140–166.

Illusie, Luc
[1] Complexe de de Rham-Witt et cohomologie cristalline, *Ann. Sci. École Norm. Sup.* (4) **12** (1979), 501–661.
[2] Frobenius et dégénérescence de Hodge, in J.-P. Demailly et al., *Introduction à la théorie de Hodge*, Soc. Math. France, Paris, 1996, 113–168.

Izhboldin, Oleg T.
[1] On p-torsion in K_*^M for fields of characteristic p, *Adv. Soviet Math.* **4** (1991), 129–144.

Jacob, Bill
[1] Indecomposable division algebras of prime exponent, *J. reine angew. Math.* **413** (1991), 181–197.

Jacobson, Nathan
[1] Abstract derivations and Lie algebras, *Trans. Amer. Math. Soc.* **42** (1937), 206–224.
[2] *Basic Algebra I-II*, 2nd edition, W. H. Freeman & Co., New York, 1989.
[3] *Finite-dimensional Division Algebras over Fields*, Springer-Verlag, Berlin, 1996.

Jahnel, Jörg
[1] The Brauer-Severi variety associated with a central simple algebra: a survey, available from the server https://www.math.uni-bielefeld.de/lag/, preprint No. 52, 2000.

de Jong, Aise Johan
[1] The period-index problem for the Brauer group of an algebraic surface, *Duke Math. J.* **123** (2004), 71–94.

de Jong, Aise Johan and Starr, Jason
[1] Every rationally connected variety over the function field of a curve has a rational point, *Amer. J. Math.* **125** (2003), 567–580.

Kahn, Bruno
[1] La conjecture de Milnor (d'après V. Voevodsky), Séminaire Bourbaki, exp. 834, *Astérisque* **245** (1997), 379–418.
[2] Motivic cohomology of smooth geometrically cellular varieties, in *Algebraic K-theory* (W. Raskind and C. Weibel, eds.), Proc. Sympos. Pure Math. **67**, Amer. Math. Soc., Providence, 1999, 149–174.
[3] Quelques remarques sur le u-invariant, *Sém. Théorie des Nombres de Bordeaux* **2** (1990), 155–161.

Kang, Ming-Chang
[1] Constructions of Brauer-Severi varieties and norm hypersurfaces, *Canad. J. Math.* **42** (1990), 230–238.

Kato, Kazuya
[1] A generalization of local class field theory by using K-groups I, *J. Fac. Sci. Univ. Tokyo* **26** (1979), 303–376; II, *J. Fac. Sci. Univ. Tokyo* **27** (1980), 603–683.
[2] Galois cohomology of complete discrete valuation fields, in *Algebraic K-theory II (Oberwolfach, 1980)*, Lecture Notes in Math., **967**, Springer-Verlag, Berlin-New York, 1982, 215–238.
[3] Residue homomorphisms in Milnor K-theory, in *Galois groups and their representations (Nagoya, 1981)*, Adv. Stud. Pure Math., **2**, North-Holland, Amsterdam, 1983, 153–172.
[4] Milnor K-theory and the Chow group of zero cycles, in *Applications of Algebraic K-Theory to Algebraic Geometry and Number Theory* (S. Bloch et al., eds.), Contemp. Math., **55/2**, Amer. Math. Soc., Providence, 1986, 241–253.

Kato, Kazuya and Kuzumaki, Takako
[1] The dimension of fields and algebraic K-theory, *J. Number Theory* **24** (1986), 229–244.

Katz, Nicholas M.
[1] Nilpotent connections and the monodromy theorem: Applications of a result of Turrittin, *Inst. Hautes Études Sci. Publ. Math.* **39** (1970), 175–232.
[2] Algebraic solutions of differential equations (p-curvature and the Hodge filtration), *Invent. Math.* **18** (1972), 1–118.

Kersten, Ina
[1] *Brauergruppen von Körpern*, Vieweg, Braunschweig, 1990.

[2] Noether's problem and normalization, *Jahresber. Deutsch. Math.-Verein.* **100** (1998), 3–22.

Kerz, Moritz
[1] The Gersten conjecture for Milnor K-theory, *Invent. Math.* **175** (2009), 1–33.

Klingen, Norbert
[1] A short remark on the Merkurjev-Suslin theorem, *Arch. Math.* **48** (1987), 126–129.

Kneser, Martin
[1] Konstruktive Lösung p-adischer Gleichungssysteme, *Nachr. Akad. Wiss. Göttingen Math.-Phys.* (1978), 67–69.

Knus, Max-Albert
[1] Sur la forme d'Albert et le produit tensoriel de deux algèbres de quaternions, *Bull. Soc. Math. Belg. Sér. B* **45** (1993), 333–337.

Knus, Max-Albert, Merkurjev, Alexander, Rost, Markus and Tignol, Jean-Pierre
[1] *The Book of Involutions*, American Mathematical Society Colloquium Publications, vol. 44, American Mathematical Society, Providence, 1998.

Kollár János
[1] Severi-Brauer varieties; a geometric treatment, preprint `arXiv:1606. 04368`, 2016.

Krull, Wolfgang
[1] Galoissche Theorie der unendlichen algebraischen Erweiterungen, *Math. Ann.* **100** (1928), 687–698.

Lam, Tsit-Yuen
[1] *Introduction to Quadratic Forms over Fields*, Graduate Studies in Mathematics, vol. 67, American Mathematical Society, Providence, 2005.

Lang, Serge
[1] On quasi-algebraic closure, *Ann. of Math. (2)* **55** (1952), 373–390.
[2] Algebraic groups over finite fields, *Amer. J. Math.* **78** (1956), 555–563.
[3] *Algebra*, 3rd edition, Addison-Wesley, Redwood, 1993.

Lenstra, Hendrik W.
[1] Rational functions invariant under a finite abelian group, *Invent. Math.* **25** (1974), 299–325.
[2] K_2 of a global field consists of symbols, in *Algebraic K-theory*, Lecture Notes in Math., vol. 551, Springer-Verlag, Berlin, 1976, pp. 69–73.

Lichtenbaum, Stephen

[1] The period-index problem for elliptic curves, *Amer. J. Math.* **90** (1968), 1209–1223.

Lieblich, Max

[1] Twisted sheaves and the period-index problem, *Compositio Math.* **144** (2008), 1–31.

[2] Period and index in the Brauer group of an arithmetic surface (with an appendix by Daniel Krashen), *J. reine angew. Math.* **659** (2011), 1–41.

[3] The period-index problem for fields of transcendence degree 2, *Ann. of Math.* **182** (2015), 391–427.

Mammone, Pascal and Merkurjev, Alexander S.

[1] On the corestriction of p^n-symbol, *Israel J. Math.* **76** (1991), 73–79.

Mammone, Pascal and Tignol, Jean-Pierre

[1] Dihedral algebras are cyclic, *Proc. Amer. Math. Soc.* **101** (1987), 217–218.

Matsumura, Hideyuki

[1] *Commutative Ring Theory*, Cambridge Studies in Advanced Mathematics, vol. 8, Cambridge University Press, 1989.

Matzri, Eliyahu

[1] All dihedral division algebras of degree five are cyclic, *Proc. Amer. Math. Soc.* **136** (2008), 1925–1931.

McKinnie, Kelly

[1] Noncyclic and indecomposable p-algebras, PhD thesis, University of Texas at Austin, 2006.

Merkurjev, Alexander S.

[1] On the norm residue symbol of degree 2 (Russian), *Dokl. Akad. Nauk SSSR* **261** (1981), 542–547.

[2] K_2 of fields and the Brauer group, in *Applications of Algebraic K-Theory to Algebraic Geometry and Number Theory* (S. Bloch *et al.*, eds.), Contemp. Math., vol. 55/1, Amer. Math. Soc., Providence, 1986, 529–546.

[3] Kaplansky's conjecture in the theory of quadratic forms (Russian), *Zap. Nauchn. Sem. Leningrad. Otdel. Mat. Inst. Steklov.* **175** (1989), 75–89, 163–164; English translation in *J. Soviet Math.* **57** (1991), 3489–3497.

[4] K-theory of simple algebras, in *K-theory and Algebraic Geometry: Connections with Quadratic Forms and Division Algebras* (B. Jacob and J. Rosenberg, eds.), Proc. Sympos. Pure Math., vol. 58, Part 1, Amer. Math. Soc., Providence, 1995, 65–83.

[5] On the norm residue homomorphism of degree two, *Proceedings of the St. Petersburg Mathematical Society*, vol. XII, Amer. Math. Soc. Transl. Ser. 2, vol. 219, American Mathematical Society, Providence, 2006, 103–124.

[6] Brauer groups of fields, *Comm. Algebra* **11** (1983), 2611–2624.

Merkurjev, Alexander S. and Suslin, Andrei A.

[1] K-cohomology of Severi-Brauer varieties and the norm residue homomorphism (Russian), *Izv. Akad. Nauk SSSR Ser. Mat.* **46** (1982), 1011–1046, 1135–1136.

[2] Norm residue homomorphism of degree three (Russian), *Izv. Akad. Nauk SSSR Ser. Mat.* **54** (1990), 339–356; English translation in *Math. USSR Izv.* **36** (1991), 349–367.

[3] The group K_3 for a field (Russian), *Izv. Akad. Nauk SSSR Ser. Mat.* **54** (1990), 522–545; English translation in *Math. USSR Izv.* **36** (1991), 541–565.

Milne, James S.

[1] Duality in the flat cohomology of a surface, *Ann. Sci. École Norm. Sup.* **9** (1976), 171–201.

[2] *Étale Cohomology*, Princeton Mathematical Series, vol. 33, Princeton University Press, 1980.

[3] *Arithmetic Duality Theorems*, Academic Press, Boston, 1986.

[4] Jacobian varieties, in *Arithmetic Geometry* (G. Cornell and J. H. Silverman, eds.), Springer-Verlag, New York, 1986, 167–212.

Milnor, John W.

[1] Algebraic K-theory and quadratic forms, *Invent. Math.* **9** (1969/1970), 318–344.

[2] *Introduction to algebraic K-theory*, Annals of Mathematics Studies No. 72, Princeton University Press, 1971.

Mumford, David

[1] *The Red Book of Varieties and Schemes*, Lecture Notes in Mathematics **1358**, Springer-Verlag, Berlin, 1988.

Murre, Jacob P.

[1] Applications of algebraic K-theory to the theory of algebraic cycles, in *Algebraic geometry, Sitges (Barcelona), 1983*, Lecture Notes in Math. **1124**, Springer-Verlag, Berlin, 1985, 216–261.

Neukirch, Jürgen

[1] *Algebraic Number Theory*, Grundlehren der Mathematischen Wissenschaften, vol. 322, Springer-Verlag, Berlin, 1999.

Neukirch, Jürgen, Schmidt, Alexander and Wingberg, Kay
[1] *Cohomology of Number Fields*, Grundlehren der Mathematischen Wissenschaften, vol. 323, Springer-Verlag, Berlin, 2000.

Peyre, Emmanuel
[1] Unramified cohomology of degree 3 and Noether's problem, *Invent. Math.* **171** (2008), 191–225.

Pfister, Albrecht
[1] *Quadratic Forms with Applications to Algebraic Geometry and Topology*, London Mathematical Society Lecture Note Series, vol. 217, Cambridge University Press, 1995.

Pierce, Richard
[1] *Associative Algebras*, Graduate Texts in Mathematics, vol. 88, Springer-Verlag, New York-Berlin, 1982.

Platonov, Vladimir P.
[1] On the Tannaka–Artin problem, *Dokl. Akad. Nauk SSSR* **221** (1975), 1038–1041.

Quillen, Daniel
[1] Higher algebraic K-theory I, in *Algebraic K-theory I: Higher K-theories*, Lecture Notes in Math. **341**, Springer-Verlag, Berlin 1973, 85–147.

Rehmann, Ulf, Tikhonov, Sergey V. and Yanchevskiĭ, Vyacheslav I.
[1] Symbol algebras and the cyclicity of algebras after a scalar extension (Russian), *Fundam. Prikl. Mat.* **14** (2008), 193–209; English translation in *J. Math. Sci. (N. Y.)* **164** (2010), 131–142.

Ribes, Luis and Zalesskii, Pavel
[1] *Profinite groups*, Second edition, Ergebnisse der Mathematik und ihrer Grenzgebiete, vol. 40, Springer-Verlag, Heidelberg, 2010.

Rieffel, Marc A.
[1] A general Wedderburn theorem, *Proc. Nat. Acad. Sci. USA* **54** (1965), 1513.

Roquette, Peter
[1] On the Galois cohomology of the projective linear group and its applications to the construction of generic splitting fields of algebras, *Math. Ann.* **150** (1963), 411–439.
[2] Splitting of algebras by function fields of one variable, *Nagoya Math. J.* **27** (1966), 625–642.
[3] Class Field Theory in characteristic p: Its origin and development, in *Class Field Theory - Its Centenary and Prospect* (K. Miyake, ed.), Advanced Studies in Pure Mathematics **30**, Tokyo, 2000, 549–631.

[4] *The Brauer-Hasse-Noether Theorem in Historical Perspective,* Schriften der Mathematisch-Naturwissenschaftlichen Klasse der Heidelberger Akademie der Wissenschaften, vol. 15, Springer-Verlag, Berlin, 2005.

Rosset, Shmuel and Tate, John
[1] A reciprocity law for K_2-traces, *Comment. Math. Helv.* **58** (1983), 38–47.

Rost, Markus
[1] Chow groups with coefficients, *Doc. Math.* **1** (1996), 319–393.
[2] The chain lemma for Kummer elements of degree 3, *C. R. Acad. Sci. Paris Sér. I Math.* **328** (1999), 185–190.
[3] Hilbert's Satz 90 for K_3 for quadratic extensions, preprint, 1988, available from the author's homepage.
[4] Chain lemma for splitting fields of symbols, preprint, 1998, available from the author's homepage.

Rowen, Louis Halle
[1] Cyclic division algebras, *Israel J. Math.* **41** (1982), 213–234; Correction, *Israel J. Math.* **43** (1982), 277–280.
[2] *Ring Theory II*, Academic Press, Boston, 1988.
[3] Are p-algebras having cyclic quadratic extensions necessarily cyclic? *J. Algebra* **215** (1999), 205–228.

Rowen, Louis Halle and Saltman, David J.
[1] Dihedral algebras are cyclic, *Proc. Amer. Math. Soc.* **84** (1982), 162–164.

Saltman, David J.
[1] Generic Galois extensions and problems in field theory, *Adv. in Math.* **43** (1982), 250–283.
[2] Noether's problem over an algebraically closed field, *Invent. Math.* **77** (1984), 71–84.
[3] *Lectures on Division Algebras,* American Mathematical Society, Providence, 1999.
[4] Division algebras over p-adic curves, *J. Ramanujan Math. Soc.* **12** (1997), 25–47; correction, *ibid.* **13** (1998), 125–129.

Scharlau, Winfried
[1] Über die Brauer-Gruppe eines algebraischen Funktionenkörpers in einer Variablen, *J. reine angew. Math.* **239/240** (1969), 1–6.
[2] *Quadratic and Hermitian Forms,* Grundlehren der Mathematischen Wissenschaften, vol. 270, Springer-Verlag, Berlin, 1985.

Segre, Beniamino
[1] Questions arithmétiques sur les variétés algébriques, *Colloques Internat. CNRS,* vol. 24 (1950), 83–91.

Serre, Jean-Pierre
[1] Sur la topologie des variétés algébriques en caractéristique p, in *Symposium internacional de topologia algebraica*, Mexico City, 1958, 24–53.
[2] *Corps locaux*, Hermann, Paris, 1962; English translation: *Local Fields*, Springer-Verlag, 1979.
[3] *Lectures on the Mordell-Weil Theorem*, Vieweg, Braunschweig, 1989.
[4] *Cohomologie Galoisienne*, 5e éd., révisée et complétée. Lecture Notes in Mathematics **5**, Springer-Verlag, Berlin, 1994; English translation: *Galois Cohomology*, Springer-Verlag, Berlin, 2002.

Seshadri, Conjeerveram Srirangachari
[1] L'opérateur de Cartier. Applications, *Séminaire Chevalley*, année 1958/59, exposé 6.

Severi, Francesco
[1] Un nuovo campo di ricerche nella geometria sopra una superficie e sopra una varietà algebrica, *Mem. Accad. Ital., Mat.* **3** (1932), 1–52.

Shafarevich, Igor R.
[1] On the Lüroth problem (Russian), *Trudy Mat. Inst. Steklov* **183** (1990), 199–204; English translation in *Proc. Steklov Inst. Math.* (1991), No. 4, 241–246.
[2] *Basic Algebraic Geometry I-II*, Springer-Verlag, Berlin, 1994.

Shatz, Stephen S.
[1] *Profinite Groups, Arithmetic, and Geometry*, Annals of Mathematics Studies, No. 67, Princeton University Press, 1972.

Soulé, Christophe
[1] K_2 et le groupe de Brauer (d'après A. S. Merkurjev et A. A. Suslin), *Séminaire Bourbaki*, exp. 601, Astérisque **105–106** (1983), Soc. Math. France, Paris, 79–93.

Speiser, Andreas
[1] Zahlentheoretische Sätze aus der Gruppentheorie, *Math. Zeit.* **5** (1919), 1–6.

Springer, Tonny A.
[1] *Linear Algebraic Groups*, 2nd edition, Progress in Mathematics, vol. 9, Birkhäuser, Boston, MA, 1998.

Sridharan, Ramaiyengar
[1] (in collaboration with Raman Parimala) *2-Torsion in Brauer Groups: A Theorem of Merkurjev*, notes from a course held at ETH Zürich in 1984/85, available from the homepage of M.-A. Knus.

Srinivas, Vasudevan
[1] *Algebraic K-Theory*, 2nd edition, Progress in Mathematics, vol. 90, Birkhäuser, Boston, 1996.

Suslin, Andrei A.
[1] Algebraic K-theory and the norm residue homomorphism (Russian), in *Current problems in mathematics* **25** (1984), 115–207.
[2] Torsion in K_2 of fields, *K-Theory* **1** (1987), 5–29.
[3] SK_1 of division algebras and Galois cohomology, *Adv. Soviet Math.* **4**, Amer. Math. Soc., Providence, 1991, 75–99.

Suslin, Andrei A. and Joukhovitski, Seva
[1] Norm varieties, *J. Pure Appl. Algebra* **206** (2006), 245–276.

Suslin, Andrei A. and Voevodsky, Vladimir
[1] Bloch-Kato conjecture and motivic cohomology with finite coefficients, in *The Arithmetic and Geometry of Algebraic Cycles* (B. Brent Gordon *et al.*, eds.), Kluwer, Dordrecht, 2000, 117–189.

Suzuki, Michio
[1] *Group Theory I*, Grundlehren der Mathematischen Wissenschaften, vol. 247, Springer-Verlag, New York, 1982.

Swan, Richard G.
[1] Invariant rational functions and a problem of Steenrod, *Invent. Math.* **7** (1969), 148–158.
[2] Noether's problem in Galois theory, in *Emmy Noether in Bryn Mawr*, Springer-Verlag, New York-Berlin, 1983, 21–40.
[3] Higher algebraic K-theory, in *K-Theory and Algebraic Geometry: Connections with Quadratic Forms and Division Algebras* (B. Jacob and J. Rosenberg, eds.), Proc. Sympos. Pure Math., vol. 58, Part I, Amer. Math. Soc., Providence, 1995, 247–293.

Tabuada, Gonçalo and van den Bergh, Michel
[1] Noncommutative motives of Azumaya algebras, *J. Inst. Math. Jussieu* **14** (2015), 379–403.

Tate, John
[1] Genus change in inseparable extensions of function fields, *Proc. Amer. Math. Soc.* **3** (1952), 400–406.
[2] Global class field theory, in *Algebraic Number Theory* (J. W. S. Cassels and A. Fröhlich, eds.), Academic Press, London, 1967, 162–203.
[3] Symbols in arithmetic, in *Actes du Congrès International des Mathématiciens (Nice, 1970)*, Tome 1, Gauthier-Villars, Paris, 1971, 201–211.

[4] Relations between K_2 and Galois cohomology, *Invent. Math.* **36** (1976), 257–274.

[5] On the torsion in K_2 of fields, in *Algebraic number theory (Kyoto, 1976)*, Japan Soc. Promotion Sci., Tokyo, 1977, 243–261.

Teichmüller, Oswald
[1] p-Algebren, *Deutsche Math.* **1** (1936), 362–388.

Tignol, Jean-Pierre
[1] Algèbres indécomposables d'exposant premier, *Adv. in Math.* **65** (1987), 205–228.

[2] On the corestriction of central simple algebras, *Math. Z.* **194** (1987), 267–274.

Tignol, Jean-Pierre and Wadsworth, Adrian
[1] *Value Functions on Simple Algebras, and Associated Graded Rings*, Springer-Verlag, Heidelberg, New York, 2015.

Tits, Jacques
[1] Sur les produits tensoriels de deux algèbres de quaternions, *Bull. Soc. Math. Belg. Sér. B* **45** (1993), 329–331.

Tregub, Semion L.
[1] Birational equivalence of Brauer-Severi manifolds, *Uspekhi Mat. Nauk* **46** (1991), 217–218; English translation in *Russian Math. Surveys* **46** (1992), 229.

Tsen, Chiung-Tse
[1] Divisionsalgebren über Funktionenkörpern, *Nachr. Akad. Wiss. Göttingen Math.-Phys.* 1933, 335–339.

Voevodsky, Vladimir
[1] Motivic cohomology with $\mathbf{Z}/2$-coefficients, *Publ. Math. Inst. Hautes Études Sci.* **98** (2003), 59–104.

[2] On motivic cohomology with $\mathbf{Z}/l$-coefficients, *Ann. of Math.* **174** (2011), 401–438.

Voevodsky, Vladimir, Suslin, Andrei A. and Friedlander, Eric M.
[1] *Cycles, Transfers, and Motivic Homology Theories*, Ann. of Math. Studies, vol. 143, Princeton Univ. Press, 2000.

Voskresenskiĭ, Valentin E.
[1] On the question of the structure of the subfield of invariants of a cyclic group of automorphisms of the field $\mathbf{Q}(x_1, \cdots, x_n)$ (Russian), *Izv. Akad. Nauk SSSR Ser. Mat.* **34** (1970), 366–375; English translation in *Math. USSR Izv.* **4** (1971), 371–380.

[2] *Algebraic Groups and Their Birational Invariants,* Translations of Mathematical Monographs, vol. 179, American Mathematical Society, Providence, 1998.

Wadsworth, Adrian

[1] Merkurjev's elementary proof of Merkurjev's theorem, in *Applications of Algebraic K-Theory to Algebraic Geometry and Number Theory* (S. Bloch *et al.*, eds.), Contemp. Math., vol. 55/2, Amer. Math. Soc., Providence, 1986, 741–776.

van der Waerden, Bartel Leendert

[1] *Algebra* I (7te Auflage) II (5te Auflage), Springer-Verlag, Berlin, 1966–67. English translation: Springer-Verlag, New York, 1991.

Wang, Shianghaw

[1] On the commutator group of a simple algebra, *Amer. J. Math.* **72** (1950), 323–334.

Warning, Ewald

[1] Bemerkung zur vorstehenden Arbeit von Herrn Chevalley, *Abh. Math. Semin. Hamb. Univ.* **11** (1935), 76–83.

Wedderburn, Joseph Henry Maclagan

[1] A theorem on finite algebras, *Trans. Amer. Math. Soc.* **6** (1905), 349–352.
[2] On hypercomplex numbers, *Proc. London Math. Soc.* **6** (1908), 77–118.
[3] On division algebras, *Trans. Amer. Math. Soc.* **22** (1921), 129–135.

Weibel, Charles

[1] *An Introduction to Homological Algebra,* Cambridge Studies in Advanced Mathematics, vol. 38, Cambridge University Press, 1994.
[2] *The K-book: Introduction to Algebraic K-Theory,* Graduate Studies in Mathematics, vol. 145, American Mathematical Society, Providence, 2013.

Weil, André

[1] Sur la théorie du corps de classes, *J. Math. Soc. Japan* **3** (1951), 1–35.
[2] The field of definition of a variety, *Amer. J. Math.* **78** (1956), 509–524.
[3] *Basic Number Theory,* Basic number theory, 3rd edition, Grundlehren der Mathematischen Wissenschaften, vol. 144, Springer-Verlag, New York-Berlin, 1974.

Witt, Ernst

[1] Über ein Gegenbeispiel zum Normensatz, *Math. Zeit.* **39** (1934), 462–467.

[2] Schiefkörper über diskret bewerteten Körpern, *J. reine angew. Math.* **176** (1936), 153–156.

Zariski, Oscar and Samuel, Pierre

[1] *Commutative Algebra*, vol. II, Van Nostrand, Princeton, 1960.

Index

413

Printed in the United States
by Baker & Taylor Publisher Services